Eric van Damme

Stability and Perfection of Nash Equilibria

Second, Revised and Enlarged Edition

With 105 Figures

Springer-Verlag Berlin Heidelberg GmbH

Prof. Dr. Eric van Damme

CentER for Economic Research
Tilburg University, PO Box 90153
5000 LE Tilburg, The Netherlands

ISBN 978-3-540-53800-4

Library of Congress Cataloging-in-Publication Data
Damme, Eric van. Stability and perfection of Nash equilibria/Eric van Damme. – 2nd, rev. and enl. ed.
Includes bibliographical references and index.
ISBN 978-3-540-53800-4 ISBN 978-3-642-58242-4 (eBook)
DOI 10.1007/978-3-642-58242-4
1. Game theory. 2. Equilibrium (Economics) I. Title. HB144.D36 1991 91-10656 339.5–dc20 CIP

2nd printing 1996
Originally published by Springer-Verlag Berlin Heidelberg New York in 1991

Typesetting: With a system of the Springer Produktions-Gesellschaft, Berlin.
Dataconversion: Brühlsche Universitätsdruckerei, Giessen.

42/3020-543210

To Jeroen, Jessica and Jean-Paul

The rules of rational behavior must provide definitely for the possibility of irrational conduct on the part of others. In other words: Imagine that we have discovered a set of rules for all participants – to be termed as "optimal" or "rational" – each of which is indeed optimal provided that the other participants conform. Then the question remains as to what will happen if some of the participants do not conform. If that should turn out to be advantageous for them – and, quite particularly, disadvantageous to the conformists – then the above "solution" would seem very questionable. We are in no position to give a positive discussion of these things as yet – but we want to make it clear that under such conditions the "solution," or at least its motivation, must be considered as imperfect and incomplete. In whatever way we formulate the guiding principles and the objective justification of "rational behavior," provisos will have to be made for every possible conduct of "the others." Only in this way can a satisfactory and exhaustive theory be developed. But if the superiority of "rational behavior" over any other kind is to be established, then its description must include rules of conduct for all conceivable situations – including those where "the others" behaved irrationally, in the sense of the standards which the theory will set for them.

John von Neumann and Oskar Morgenstern
Theory of Games and Economic Behavior,
Princeton University Press, Princeton, N.J.
(2nd ed. 1947, p. 32)

Preface to the Second Edition

I have been pleased with the favourable reception of the first edition of this book and I am grateful to have the opportunity to prepare this second edition. In this revised and enlarged edition I corrected some misprints and errors that occurred in the first edition (fortunately I didn't find too many) and I added a large number of notes that give the reader an impression of what kind of results have been obtained since the first edition was printed and that give an indication of the direction the subject is taking. Many of the notes discuss (or refer to papers discussing) applications of the refinements that are considered. Of course, it is the quantity and the quality of the insights and the applications that lend the refinements their validity. Although the guide to the applications is far from complete, the notes certainly allow the reader to form a good judgement of which refinements have really yielded new insights. Hence, as in the first edition, I will refrain from speculating on which refinements of Nash equilibria will survive in the long run. To defend this position let me also cite Binmore [1990] who compares writing about refinements to the Herculean task of defeating the nine-headed Hydra which grew too heads for each that was struck off.

It is a pleasure to have the opportunity to thank my secretary, Marjoleine de Wit, who skilfully and, as always, cheerfully typed the manuscript and did the proofreading.

Tilburg, March 1991

Preface to the First Edition

The last decade has seen a steady increase in the application of concepts from noncooperative game theory to such diverse fields as economics, political science, law, operations research, biology and social psychology. As a byproduct of this increased activity, there has been a growing awareness of the fact that the basic noncooperative solution concept, that of Nash equilibrium, suffers from severe drawbacks. The two main shortcomings of this concept are the following:

(i) In extensive form games, off the equilibrium path a Nash strategy may prescribe behavior that is manifestly irrational. (Specifically, Nash equilibria may involve incredible threats),
(ii) Nash equilibria need not be robust with respect to small perturbations in the data of the game.

Confronted with the growing evidence to the detriment of the Nash concept, game theorists were prompted to search for more refined equilibrium notions with better properties and they have come up with a wide array of alternative solution concepts. This book surveys the most important refinements that have been introduced. Its objectives are fourfold

(i) to illustrate desirable properties as well as drawbacks of the various equilibrium notions by means of simple specific examples,
(ii) to study the relationships between the various refinements,
(iii) to derive simplifying characterizations, and
(iv) to discuss the plausibility of the assumptions underlying the concepts.

The book is addressed primarily to researchers who want to apply game theory, but who do not know their way through the myriad of noncooperative solution concepts. It can also be used as the basis for an advanced course in game theory at the graduate level. It will be successful if it enables the reader to sift the grain from the corn and if it can direct him to the concepts that are really innovative.

Acknowledgements. Many colleagues and friends helped to shape my thinking on these topics in the last several years. I am sure that, had I listened to their invaluable advice more carefully I would have understood the subject better and I would have written a better book. Especially I want to express my thanks to Stef Tijs, Jaap Wessels, Jan van der Wal, Reinhard Selten, Werner Güth, Roger Myerson and Ehud Kalai for their help at crucial stages of the development. Of course, the views expressed are not necessarily theirs, and I alone am responsible for errors and mistakes.

Sincere appreciation is also extended to my wife Suzan, Ellen Jansen, Lieneke Lekx, Netty Zuidervaart and Rolf-Peter Schneider who shared the effort in typing the several versions of the manuscript. A special thanks also goes to the editorial staff of Springer-Verlag, especially to Werner Müller for exerting just sufficient pressure to get this book finished. Finally, I thank Jeroen and Suzan for their patience, understanding and encouragement while the book was being written.

Bonn, July 1987

Eric E. C. van Damme

Organization

The Chaps. 1–6 of this book are virtually identical to the monograph "Refinements of the Nash equilibrium concept" that I published with Springer-Verlag in 1983. Werner Müller, the economics editor of Springer kindly asked me to extend that monograph with some chapters illustrating the various equilibrium concepts in specific examples. I was pleased to honor that request and I gladly took this opportunity to clear up some inaccuracies and to include some recent developments.

This book consists of 4 parts:

Part 1 (Chap. 1) provides a general introduction. It is argued that the solution of a noncooperative game should be a Nash equilibrium but that not every Nash equilibrium is eligible for the solution. Various refinements of the Nash concept are introduced informally and simple examples are used to illustrate these concepts.

Part 2 (Chaps. 2–5) deals with normal form games. A great variety of refined equilibrium concepts is introduced and relationships between these refinements are derived, as well as characterizations of several of them. For a quick overview, the reader should consult the Survey Diagrams 1 and 2 at the end of the book. A main result, however, is that for normal form games there is actually little need to refine Nash's concept since generically all Nash equilibria satisfy all properties one could hope for.

Part 3 (Chap. 6) provides an introduction to extensive form games. Formal definitions are given and elementary properties of several concepts (such as (subgame) perfect equilibria and sequential equilibria) are derived. The main result is that a proper equilibrium of the normal form induces a sequential equilibrium in the extensive form. However, normal form properness does not eliminate all "unreasonable" equilibria.

Part 4 (Chaps. 7–10) is devoted to specific applications, illustrating the strength (resp. weakness) of the various concepts. These chapters are independent of each other and familiarity with the basic notions from Chaps. 2 and 6 suffices to follow the discussion.

The main theme in Chap. 7 is "how to implement concepts from cooperative game theory by noncooperative methods?" The power of the subgame perfectness concept is illustrated by means of a simple fair division problem, by means of the Rubinstein bargaining model (which implements the Nash solution) and by means of the Moulin model (that implements the Kalai/Smorodinsky solution). Furthermore, Nash's bargaining model is used to illustrate the essential equilibrium concept.

In Chap. 8 we prove the Folk Theorem, which states that the set of cooperative outcomes of the one-shot game coincides with the set of noncooperative outcomes of the repeated game. This chapter shows that the subgame perfectness concept has certain drawbacks and that it is not as restrictive as one might initially think.

In Chap. 9 we turn to the biological branch of game theory. Here, the main solution concept is that of evolutionarily stable strategies (ESS), i.e. of symmetric Nash equilibrium strategies that satisfy a certain condition of neighborhood stability. The relationships between (refinements of) the ESS concept and concepts of strategic stability are studied. (An overview is given in Diagram 3.) The ESS concept is very restrictive, however, (especially in extensive games) and it is shown that to coarsen this concept one can use the same methods as those that are used to refine the Nash concept.

In Chap. 10 we return to the question of whether a game is adequately represented by its normal form, i.e. whether knowledge of the normal form is sufficient to define rational behavior. In particular, it is investigated what kind of restrictions the Kohlberg/Mertens concept of stability imposes on the beliefs in a sequential equilibrium. Special attention is paid to signalling games, for which several "intuitive criteria" for eliminating "unintuitive equilibria" are discussed (see Diagram 4 for an overview). Many examples are given to illustrate the various concepts.

Notational Conventions

We have tried to use standard notations as much as possible. The notation that is specific to normal form games (resp. extensive form games) is introduced in Sect. 2.1 (resp. 6.1). Chapter 9 uses the notation that is conventional in the biological branch of game theory and this differs from our general notation for normal form games. At this point, we confine ourselves to introducing some terminology that is used throughout the book and that may not be completely standard.

$\mathbb{N}$ denotes the set of the positive integers $\{1, 2, \ldots\}$ (positive will always mean strictly greater than 0). When dealing with an n-person game we will frequently write N for $\{1, \ldots, n\}$.

$\mathbb{R}$ denotes the set of real numbers and $\mathbb{R}^m$ is the m dimensional Euclidean space. For $x, y \in \mathbb{R}^m$, we write $x \leqq y$ if $x_i \leqq y_i$ for all i. Furthermore, $x < y$ means $x_i < y_i$ for all i. We write $\mathbb{R}^m_+$ for the set of all $x \in \mathbb{R}^m$ which satisfy $0 \leqq x$ and $\mathbb{R}^m_{++}$ for the set of all $x \in \mathbb{R}^m$ for which $0 < x$. Euclidean distance on $\mathbb{R}^m$ is denoted by ϱ and λ denotes Lebesgue measure on $\mathbb{R}^m$.

The set of all mappings from A to B is denoted by $\mathscr{F}(A, B)$. If $f \in \mathscr{F}(\mathbb{R}^m_{++}, \mathbb{R}^l)$, then "$y$ is a limit point of a sequence $\{f(x)\}_{x \downarrow 0}$" is used as an abbreviation for "there exists a sequence $\{x(t)\}_{t \in \mathbb{N}}$ such that $x(t)$ converges to 0 and $f(x_t)$ converges to y as t tends to infinity".

If A is a subset of some Euclidean space, then conv A denotes its convex hull and 2^A denotes the powerset of A.

Let A and B be subsets of Euclidean spaces. A correspondence from A to B is an element of $\mathscr{F}(A, 2^B)$. The correspondence F from A to B is said to be upper semi-continuous if it has a closed graph, i.e. if $\{(x, F(x));\ x \in A\}$ is closed.

The number of elements of a finite set A is denoted by $|A|$. If A is finite, and $f \in \mathscr{F}(A, \mathbb{R})$ then $f(A) := \sum_{a \in A} f(a)$.

Indices can occur as subscripts or superscripts. Lower indices usually refer to players. Upper indices usually stem from a certain numbering. For instance, when dealing with an n-person normal form game, we write s_i^k for the probability which the mixed strategy s_i of player i assigns to the kth pure strategy of this player. To avoid misunderstandings between exponents and indices, we will sometimes write the basis of a power between brackets. Hence, $(s_i^k)^2$ denotes the square of s_i^k.

Definitions are indicated by using italics. The symbol := is used to define quantities. The symbol □ denotes the end of a proof.

Contents

1 Introduction

In this chapter, we illustrate (by means of a series of examples) why the Nash equilibrium concept has to be refined and several refinements which have been proposed in the literature are introduced in an informal way. (No formal definitions are given in this chapter). First, in Sect. 1.1, it is motivated why the solution of a noncooperative game has to be a Nash equilibrium. In Sects. 1.2 – 1.4, we consider games in extensive form and discuss the following refinements of the Nash equilibrium concept: subgame perfect equilibria, sequential equilibria, perfect equilibria and stable equilibria. In Sects. 1.5 and 1.6, we consider refinements of the Nash equilibrium concept for normal form games, such as perfect equilibria, proper equilibria, persistent equilibria, essential equilibria and regular equilibria.

1.1 Informal Description of Games and Game Theory

In this section, an informal description of a (strategic) game and of Game Theory is given. For a thorough introduction to Game Theory, the reader is referred to Luce and Raiffa [1957], Owen [1968], Harsanyi [1977], Rosenmüller [1981], Shubik [1983], Moulin [1985] or Friedman [1986].[1]

Game Theory is a mathematical theory which deals with conflict situations. A conflict situation (game) is a situation in which two or more individuals (players) interact and thereby jointly determine the outcome. Each participating player can partially control the situation, but no player has full control. Each player has certain personal preferences over the set of possible outcomes and strives to obtain that outcome which is most profitable to him. It is assumed that these preferences can be described by a von Neumann-Morgenstern utility function, hence, that each player is characterized by a numerical function whose expected value he tries to maximize.

Game Theory is a *normative theory*: it aims to prescribe what each player in a game should do in order to promote his interests optimally, i.e. which strategy each player should play such that his partial influence on the situation benefits him most. Hence, the aim of Game Theory is to provide a solution (i.e. to give a characterization (definition) of "rational behavior") for every game.

The foundation of Game Theory was laid in an article by John von Neumann in 1928 (von Neumann [1928]), but the theory received widespread attention only after the publication of the fundamental book von Neumann and Morgenstern [1947]. In this book, the aim of Game Theory is described as follows:

(...) We wish to find the mathematically complete principles which define "rational behavior" for the participants in a social economy, and to derive from them the general characteristics of that behavior. And while the principles ought to be perfectly general – i.e., valid in all situations – we may be satisfied if we can find solutions, for the moment, only in some characteristic special cases.

von Neumann and Morgenstern
[1947], p. 31

Traditionally, games have been divided into two classes: cooperative games and noncooperative games. In this book, we restrict ourselves to noncooperative games. A *noncooperative game* is a game in which there are no possibilities for communication, correlation or (pre)commitment, except for those that are explicitly allowed by the rules. (Hence, all relevant aspects should be captured by the rules of the game). A solution of such a game is a set of recommendations, which tell each player how to behave in every situation that may arise. This solution should be consistent, i.e. no player should have an incentive to deviate from his recommendation. Hence, a solution must be *self-enforcing*: As long as the others obey their recommendations, it should not be in my interest to deviate. In game theoretic terminology this means that the solution should be a *Nash equilibrium* (Nash [1950], [1951]), i.e. a strategy combination with the property that no layer can gain by unilaterally deviating from it. Let us illustrate this by means of the game of Fig. 1.1.1, which is the so called prisoners' dilemma game, probably the most discussed game of the literature.

The rows of the table represent the possible choices T and B for player 1, the columns represent the choices L and R of player 2. In each cell the upper left entry is the payoff to player 1, while the lower right entry is the payoff to player 2. The rules of the game are as follows: It is a one-shot game (each player has to make a choice just once), the players have to make their choices simultaneously and independently of each other, communication nor binding agreements are possible. (Hence, the picture tells the story.)

The most attractive strategy combination of the game of Fig. 1.1.1 is (T, L). However, a sensible theory cannot prescribe this strategy pair as the solution. Namely, suppose it is suggested to play (T, L). Then each player has an incentive to disobey his recommendation as long as he expects his opponent to obey it: By unilaterally deviating, player 1 obtains 11, which is more than the 10 that he gets if he obeys his recommendation. Hence, the solution (T, L) is inconsistent with the assumption that players try to maximize expected utility. The strategy pair (T, L)

	L	R
T	10, 10	0, 11
B	11, 0	3, 3

Fig. 1.1.1. Prisoners' dilemma

is not self-enforcing but self-destabilizing: both players have an incentive to deviate from it. In this game, only (B, R) has the property that no player can gain by unilaterally deviating, hence, only (B, R) is a Nash equilibrium. Therefore, Game Theory has to prescribe (B, R) as the solution in a noncooperative context.

Of course, if binding agreements would be possible, then (T, L) could be the solution: the players could just sign a binding contract that prohibits them from deviating. In this case, however, noncooperative modelling would require that one gives a complete specification of what is physically possible, hence, we would have an entirely different game and probably a different noncooperative solution (also see Harsanyi and Selten [1988], Chap. 1).

Note that in this specific example, communication does not bring anything: As long as commitment is impossible, the solution is (B, R). For general games, however, the solution might be different when communication is allowed. Namely, in the latter case the relevant solution concept is correlated equilibrium (Aumann [1974], for refinements see Myerson [1986a, 1986b]) and not Nash equilibrium. (Actually a correlated equilibrium is nothing but a Nash equilibrium of an extended game in which communication has been modelled in detail). In this book, we will stick to the purely noncooperative context, hence, we will not consider correlated equilibria.[2]

The discussion above clearly shows that the solution of a noncooperative game has to be a Nash equilibrium since every other strategy combination is self-destabilizing if binding agreements are not possible.[3] In general, however, a game may possess more than one Nash equilibrium and, therefore, the *core problem of noncooperative game theory* can be formulated as: given a game with more than one Nash equilibrium, which one of these should be chosen as the solution of the game? This core problem will not be solved in this monograph, but we will show that some Nash equilibria are better qualified to be chosen as the solution than others. Namely, we will show that not every Nash equilibrium has the property of being self-enforcing. The next 5 sections illustrate how such equilibria can arise and how game theoretists have tried to eliminate them. For convenience, Nash equilibria which fail to be self-enforcing will be called 'unreasonable' or 'nonsensible'.

1.2 Dynamic Programming

There are several ways in which a game can be described. One way is to summarize the rules of the game by indicating the choices available to each player, the information a player has when it is his turn to move, and the payoffs each player receives at the end of the game. A game described in this way is referred to as a *game in extensive form* (see Sect. 6.1). Usually, such a game is represented by a tree, following Kuhn [1953]. Another way of representing a game is by listing all the strategies (complete plans of action) each player has available together with the payoffs associated with the various strategy combinations. A game described in this way is called a *game in normal form* (see Sect. 2.1). In Sects. 1.2–1.4, we confine ourselves to games in extensive form. Normal form games will be considered in Sects. 1.5 and 1.6.

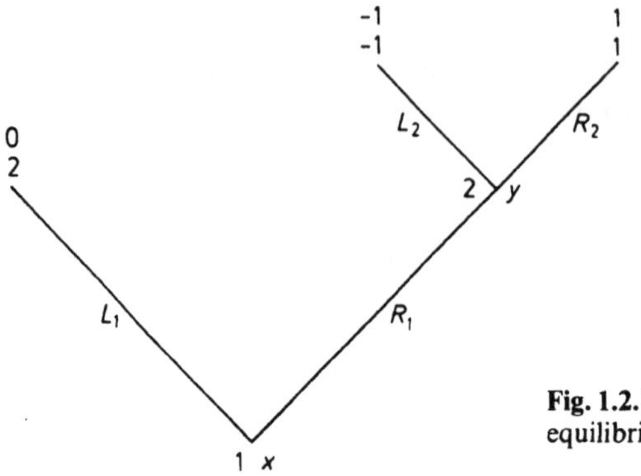

Fig. 1.2.1. An extensive form game with a Nash equilibrium which is not self-enforcing

As an example of a game in extensive form, consider the game of Fig. 1.2.1. The rules of this game are as follows. The game starts at the root x of the tree, where player 1 has to move. He can choose between L_1 and R_1. Player 2, who can choose between L_2 and R_2, has to move only after player 1 has chosen R_1. The payoffs to the players are represented at the endpoints of the tree, the upper number being the payoff to player 1. So, for example, if player 1 chooses L_1, then player 1 receives 0 and player 2 receives 2. The game is played just once.

The game of Fig. 1.2.1 possesses two (pure) Nash equilibria (or shortly equilibria), viz. (L_1, L_2) and (R_1, R_2). The equilibrium (L_1, L_2), however, is not self-enforcing. Namely, suppose the players have agreed to play* (L_1, L_2). If player 1 expects that player 2 will keep to the agreement, then indeed it is optimal for him to play L_1. But should player 1 expect that player 2 will keep to the agreement? The answer is no: since R_2 yields player 2 a higher payoff than L_2 does *if y is reached*, player 2 will play R_2 *if he actually has to make a choice*. Therefore, it is better for player 1 to play R_1 and so he will also violate the agreement and play R_1. So, although (L_1, L_2) is an equilibrium, it is not self-enforcing and, therefore, it is not qualified to be chosen as the solution of the game of Fig. 1.2.1.

Hence, the solution of this game is the equilibrium (R_1, R_2), which is indeed self-enforcing.

The equilibrium (L_1, L_2) of the game of Fig. 1.2.1 can be interpreted as a *threat equilibrium*: player 2 threatens player 1 that he will punish him by playing L_2 if he does not play L_1. Above we argued that this threat is not credible since player 2 will not execute it in the event: Facing the *fait accompli* that player 1 has chosen R_1 it is better for player 2 to play R_2. Note that here we use the basic feature of noncooperative games: no commitments are possible, except from those explicitly allowed by the rules of the game. Notice that the situation changes drastically if player 2 has the possibility to commit himself before the beginning of the game. In this case it is optimal for player 2 to commit himself to L_2, thereby forcing player 1 to play L_1.

To avoid misunderstandings, let us stress again that we do not think that commitments are not possible in conflict situations. We merely hold the view that,

* We will frequently use the phrase "suppose it has been agreed to play..." rather than the clumsier, but more accurate phrase "suppose the game theoretic recommendation is...".

if such commitments are possible, they should explicitly be incorporated in the model (cf. the above discussion of prisoners' dilemma). The great strategic importance of the possibility of committing oneself in games was first pointed out in Schelling [1960].

The game of Fig. 1.2.1 is an example of what is called an *extensive form game with perfect information.* A game is said to have perfect information if the following two conditions are satisfied:

there are no simultaneous moves, and (1.2.1)

at each decision point it is known which choices have previously been made. (1.2.2)

The argument used to exclude the equilibrium (L_1, L_2) in the game of Fig. 1.2.1 generalizes to all games with perfect information: Since in a noncooperative game there are no possibilities for commitment, once the decision point x is reached, the part of the game tree which does not come after x has become strategically irrelevant and, therefore, the decision at x should be based only on that part of the tree which comes after x. This implies that for games with perfect information only those equilibria which can be found by *dynamic programming* (Bellman [1957]), i.e. by inductively working backwards in the game tree, are sensible (i.e. self-enforcing) (cf. Kuhn [1953], Corollary 1).[4]

The game of Fig. 1.2.2 shows that this has the consequence that a sensible equilibrium may be *payoff dominated* by a non-sensible one. The unique equilibrium found by dynamic programming is (L_1r_1, R_2), i.e. player 1 plays L_1 at his first decision point, r_1 at his second, and player 2 plays R_2. Note that we require a strategy of player 1 to prescribe a choice at his second decision point also in the case in which this player chooses L_1 at his first decision point. The significance of this requirement will become clear in Sect. 1.4.[5] The equilibrium (L_1r_1, R_2) yields both players a payoff 1. Another equilibrium is (R_1l_1, L_2). This equilibrium yields both players a payoff 2. This one, however, is not sensible since player 1 cannot commit himself to playing l_1 at his second decision point: both players know that player 1 will play r_1 if this point is actually reached. Therefore, it is illusory of the players to think that they can get a payoff 2. If player 1 chooses R_1, he will end up with a payoff 0.

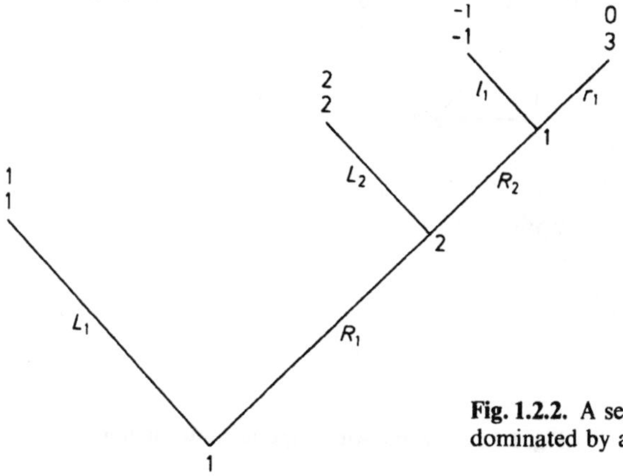

Fig. 1.2.2. A sensible equilibrium may be payoff dominated by a non-sensible one

1.3 Subgame Perfect Equilibria

For games without perfect information one cannot employ the straightforward dynamic programming approach which works so well for games with perfect information. In this section, we will illustrate a slightly more sophisticated dynamic programming approach to exclude equilibria that are not self-enforcing in games without perfect information.

As an example of a game without perfect information, consider the game of Fig. 1.3.1. In this game player 1 cannot discriminate between z and z' (i.e. he does not get to hear whether player 2 has played L_2 or R_2). This is denoted by a dotted line connecting z and z'. The set $\{z, z'\}$ is called an *information set* of player 1.

The straightforward dynamic programming approach fails in this example: in z player 1 should play l_1 and in z' he should play r_1. Hence, he faces a dilemma, since he does not know whether he is in z or in z'. For this game, the more sophisticated approach amounts to nothing else than going one step further backwards in the game tree. Namely, notice that the subtree starting at y constitutes a game of its own, called the *subgame* starting at y. An equilibrium of the game is sensible only if it prescribes an equilibrium also in this subgame. Namely, otherwise at least one player would have an incentive to deviate once the subgame is actually reached. It is easily seen that the subgame has only one equilibrium, viz. (r_1, R_2). Hence, player 1 should play r_1 at his information set $\{z, z'\}$ and player 2 should play R_2. Once this is established, it follows that player 1 should play R_1 at x. Hence, $(R_1 r_1, R_2)$ is the only sensible equilibrium of the game of Fig. 1.3.1. Notice that this is not the only equilibrium: $(L_1 l_1, L_2)$ is also an equilibrium of this game. This equilibrium, however, is not sensible since it involves the incredible threat of player 2 to play L_2.

It was first pointed out explicitly in Selten [1965] that the above argument is valid for every noncooperative game: Since commitments are not possible, behavior in a subgame can depend only on the subgame itself and, therefore, for an equilibrium to be sensible, it is necessary that this equilibrium induces an

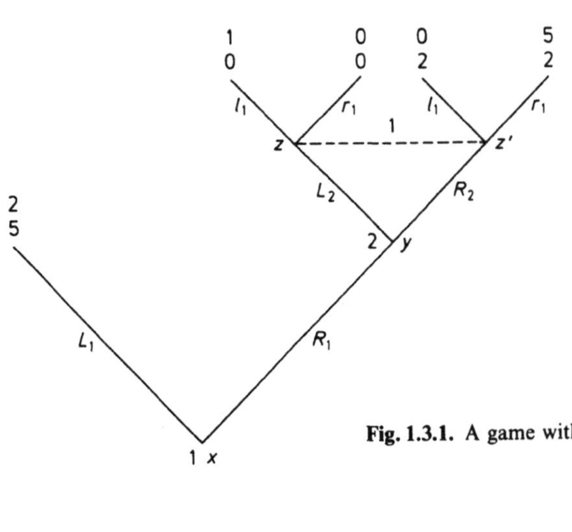

Fig. 1.3.1. A game with imperfect information

	L_2	M_2	R_2
L_1	10, 10	0, 11	0, 0
M_1	11, 0	3, 3	1, 0
R_1	0, 0	0, 1	0, 0

Fig. 1.3.2. A normal form game Γ, which is a slight modification of the game of Fig. 1.1.1

equilibrium in every subgame. Equilibria which possess this property are called *subgame perfect equilibria*, following Selten [1975].

As a second illustration of the subgame perfectness concept, we will now consider a finite repetition of the modified prisoners' dilemma game of Fig. 1.3.2. For general results concerning subgame perfectness in repeated games, the reader is referred to Chap. 8.

Notice that the game Γ results from the game of Fig. 1.1.1 by adding for each player a dominated strategy. Also in Γ the strategy pair (L_1, L_2) is the most attractive one, but this pair is not an equilibrium. The unique equilibrium of Γ is (M_1, M_2).

Now consider the game $\Gamma(2)$, which consists of playing Γ twice in succession. In $\Gamma(2)$ each player tries to maximize the sum of the payoffs he receives at stage 1 and stage 2 and at the second stage each player gets to hear which choices have been made at the first stage.

At the second stage of $\Gamma(2)$ everything which has happened at the first stage had become strategically irrelevant and, therefore, the behavior at stage 2 can depend only on Γ. Hence, at stage 2 the players should play (M_1, M_2), the unique equilibrium of Γ. But, once this has been established, it follows that the players also should play (M_1, M_2) at the first stage. Hence, there is only one subgame perfect equilibrium of $\Gamma(2)$: (M_1, M_2) should be played at both stages.

However, $\Gamma(2)$ has a phletora of equilibria which are not subgame perfect. An example of such an equilibrium is the strategy combination (φ_1, φ_2), where φ_i $(i \in \{1, 2\})$ is given by (1.3.1):

$$\varphi_i \begin{cases} \text{at stage 1:} & \text{play } L_i \\ \text{at stage 2:} & \begin{cases} \text{play } M_i, \text{ if } (L_1, L_2) \text{ has been played at stage 1,} \\ \text{play } R_i, \text{ otherwise.} \end{cases} \end{cases} \tag{1.3.1}$$

In this equilibrium, each player threatens the other that he will be punished at the second stage if he does not cooperate at the first stage. If both players believe the threats, the "cooperative outcome" (L_1, L_2) will result at the first stage. This equilibrium, however, is not sensible, since a player should not believe the other player's threat. If player 2 plays the strategy φ_2 of (1.3.1), then player 1, knowing that it is not optimal for player 2 to execute the threat, should play M_1 at the first stage.

In the last couple of years, many papers have been published in which the subgame perfectness concept is applied (some earlier applications are Selten [1965, 1973, 1977, 1978] and Ståhl [1972, 1977]). We will return to this concept in Chaps. 7 and 8. In Chap. 7, we will consider a class of games for which the subgame perfectness concept is very effective in excluding unreasonable equilibria. In Chap. 8, however, it will be shown that, for repeated games, the concept is not as restrictive as one might initially think.

1.4 Sequential Equilibria and Perfect Equilibria

It was first pointed out in Selten [1975] that a subgame perfect equilibrium may also prescribe irrational (non-maximizing) behavior at information sets which are not reached when the equilibrium is played. Consequently, a subgame perfect equilibrium need not be sensible. The 3-person game of Fig. 1.4.1, which is taken from Selten [1975], Sect. 6, can illustrate this fact.

Since there are no subgames in the game of Fig. 1.4.1, every equilibrium is subgame perfect (for the formal definition of a subgame see (6.1.16)). One equilibrium of this game is (L_1, R_2, R_3). However, this equilibrium is not sensible since player 2 will violate an agreement to play (L_1, R_2, R_3) in case his information set is actually reached. Namely, if player 2 plays L_2, then player 3 will not find out that the agreement is violated (he cannot discriminate between z and

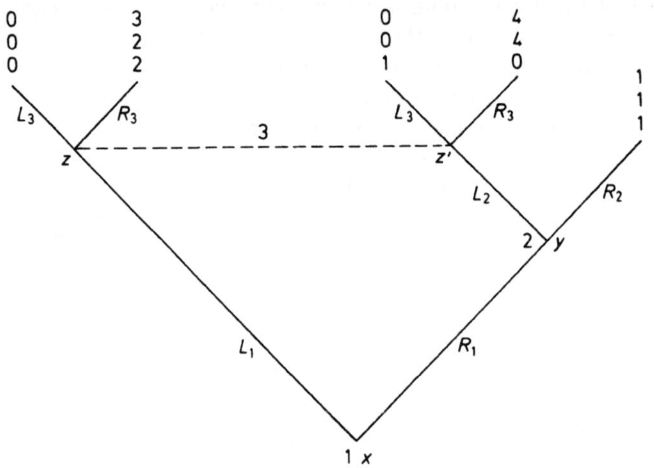

Fig. 1.4.1. A subgame perfect equilibrium need not to be sensible. The numbers at the endpoints of the tree represent the payoffs to the players; the upper number is the payoff to player 1, the second one is the payoff to player 2, etc.

z') and, therefore, this player will still play R_3. Hence, playing L_2 yields player 2 a payoff 4, which is more than R_2 yields and, therefore, this player will play L_2 if his information set is actually reached. Realizing this, player 1 will play R_1 (which yields him a payoff 4), rather than L_1 (which yields only 3). Hence, an agreement to play (L_1, R_2, R_3) is not self-enforcing and, therefore, the equilibrium (L_1, R_2, R_3) is not sensible. (It can be shown that any sensible equilibrium has player 1 playing R_1, player 2 playing R_2, and player 3 playing L_3 with a probability at least 3/4, see Selten [1975]).

The Nash equilibrium concept requires that each player chooses a strategy which maximizes his expected payoff, assuming that the other players will play in accordance with the equilibrium. The reason that the equilibrium (L_1, R_2, R_3) in the game of Fig. 1.4.1 is not sensible is the following: If (L_1, R_2, R_3) is played, the information set of player 2 is not reached and, therefore, the expected payoff of this player does not depend on his own strategy. This obviously implies that every strategy maximizes his expected payoff. However, since player 2 has to move only if the point y is actually reached, he should not let himself be guided by his *a priori* expected payoff, but by his *expected payoff after* y. The a priori expected payoff is based on the assumption that player 1 plays L_1, but, if y is reached this has shown to be wrong and player 2 should incorporate this in computing his expected payoff.

The discussion above shows that, for a subgame perfect equilibrium to be sensible, it is necessary that this equilibrium prescribes at each information set which is a singleton a choice which maximizes the expected payoff after that information set. Note that the restriction to singleton information sets is necessary to ensure that the expected payoff after the information set is well-defined. This restriction, however, has the consequence that not all subgame perfect equilibria which satisfy this additional condition are sensible. This is illustrated by means of the game of Fig. 1.4.2.

A subgame perfect equilibrium of this game which, moreover, satisfies the above condition is (A, R_2). This equilibrium is not sensible since it is always better for player 2 to play L_2 if his information set is reached. (Note that we can draw this conclusion without being able to compute the expected payoff of player 2 after

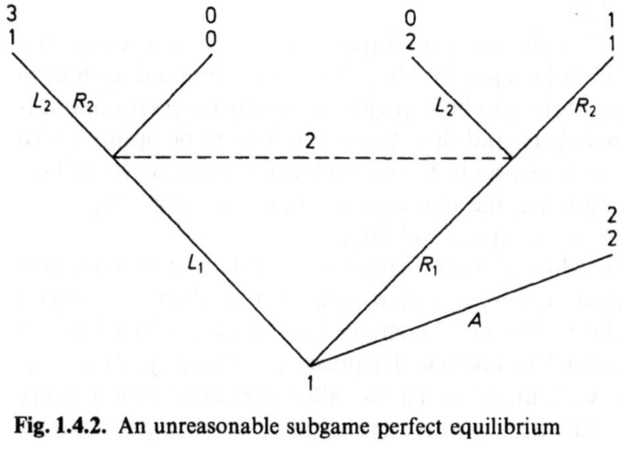

Fig. 1.4.2. An unreasonable subgame perfect equilibrium

his information set.) Realizing this, player 1 should play L_1 and therefore (L_1, L_2) is the only sensible equilibrium of the game of Fig. 1.4.2.

The examples in this section illustrate that a sensible (self-enforcing) equilibrium has to prescribe rational (maximizing) behavior at all information sets, including the ones which can be reached only after a deviation from the equilibrium. The problem, however, is: how should rational behavior at an information set with prior probability zero be defined. (Note that at such a set, if it is not a singleton, the conditional expected payoff is not well-defined.) In the literature two related solutions to this problem have been proposed, one in Selten [1975] (the concept of *perfect equilibria*) and one in Kreps and Wilson [1982a] (the concept of *sequential equilibria*). Let us first explain the concept of sequential equilibria.

The basic assumption underlying the sequential equilibrium concept is that the players are rational in the sense of Savage [1954], i.e. that a player who has to make a choice in the face of uncertainty will construct a personal probability for every event about which he is uncertain and that he will maximize expected utility with respect to these probabilities. To be more precise, suppose the players in an extensive form game have agreed to play an equilibrium φ and assume that a player nevertheless finds himself in an information set which could not be reached when φ is actually played. In this case, the player will try to reconstruct what has gone wrong, i.e. where a deviation from the equilibrium has occurred. In general, this player will not be able to reconstruct completely what has gone wrong and, therefore, he will not be able to tell in which point of his information set he actually is. However, he will represent his uncertainty by a posterior probability distribution on the nodes in this information set (his so called *beliefs* at the information set) and having constructed his beliefs, he will take a choice which maximizes his expected utility with respect to these beliefs, assuming that in the remainder of the game the players will play according to φ. A *sequential equilibrium* is then defined as an equilibrium φ which has the property that, if the players behave as indicated above, no player has an incentive to deviate from φ at any of his information sets. To be more precise: a strategy combination is a sequential equilibrium if there exist (consistent) beliefs such that each player's strategy prescribes at every information set a choice which is optimal with respect to these beliefs (see Def. 6.3.1).

In the game of Fig. 1.4.2 only the equilibrium (L_1, L_2) is a sequential equilibrium. No matter which beliefs player 2 has, it is always optimal for him to play L_2. Note that for an equilibrium to be sequential it is only necessary that it is optimal with respect to *some* beliefs, and that it does not have to be optimal with respect to *all* beliefs or even with respect to the most plausible ones. As we will see at the end of this section, this has the consequence that not every sequential equilibrium is sensible (also see Chaps. 6 and 10).

In Selten [1975] a somewhat different approach is followed to eliminate unreasonable subgame perfect equilibria. Selten assumes that there is always a small probability that a player will take a choice by *mistake*, which has the consequence that every choice will be taken with a positive probability. Therefore, in an extensive form game with mistakes (a so called *perturbed game*) every information set will be reached with a positive probability, which implies that an equilibrium of such a game will prescribe rational behavior at every information set. The assumption that mistakes occur only with a very small probability leads Selten to define a *perfect equilibrium* as an equilibrium which can be obtained as a limit point of a sequence of equilibria of disturbed games in which the mistake probabilities go to zero. Hence, an equilibrium is perfect if each player's equilibrium strategy is not only optimal against the equilibrium strategies of his opponents, but if it is also optimal against *some* slight perturbations of these strategies (see Def. 6.4.2).

In the game of Fig. 1.4.2 only the equilibrium (L_1, L_2) is perfect. Namely, in a perturbed game associated with this game, player 1 will take the choices L_1 and R_1 with a positive probability (if only by mistake) and, therefore, the information set of player 2 will actually be reached, which forces player 2 to play L_2.

It can be proved that every game possesses at least one perfect equilibrium (Theorem 6.4.4) and that every perfect equilibrium is a sequential equilibrium (see Theorem 6.4.3). However, not every sequential equilibrium is perfect. To illustrate the difference between the two concepts, consider the following slight modification of the game of Fig. 1.4.2: Player 1 receives 3 if he plays A, all other payoffs remain as in Fig. 1.4.2. As before, one can see that player 2 has to play L_2. For player 1, both L_1 and A are best replies against L_2 and, therefore, in a sequential equilibrium player 1 can play any combination of L_1 and A. The only perfect equilibrium, however, is (A, L_2). The reason is that if player 1 plays A he is sure of getting 3, whereas if he plays L_1 he can expect only slightly less than 3 since player 2 with a small probability will make a mistake and play R_2.

In Kreps and Wilson [1982a] it is shown that there is not much difference between the solutions generated by the sequential equilibrium concept and the solutions generated by the perfectness concept. They proved that almost all sequential equilibria are perfect (Kreps and Wilson [1982a] Theorem 3; for a more exact formulation of this result, see Theorem 6.4.3). However, verifying whether a given equilibrium is sequential is easier than to verify whether it is perfect.

Two questions concerning the concepts of sequential and perfect equilibria remain to be answered:

(i) Don't we exclude any sensible equilibria by restricting ourselves to sequential (resp. perfect) equilibria?

(ii) Is every sequential (resp. perfect) equilibrium sensible?

In our view, the first question certainly has to be answered affirmatively for sequential equilibria: if an equilibrium is not sequential, then at least one player has an incentive to deviate from the equilibrium at some of his information sets and, therefore, this equilibrium is not self-enforcing.[6] Whether this question should be answered affirmatively for perfect equilibria depends on one's personal viewpoint of how seriously the possibility of mistakes should be taken.

The second question, however, has to be answered negatively: many perfect (and, hence, sequential) equilibria are not sensible. Loosely speaking this is caused by the fact that some sequential (resp. perfect) equilibria are sustained only by implausible beliefs (resp. implausible mistake probabilities). This may be illustrated by means of the game of Fig. 1.4.3.

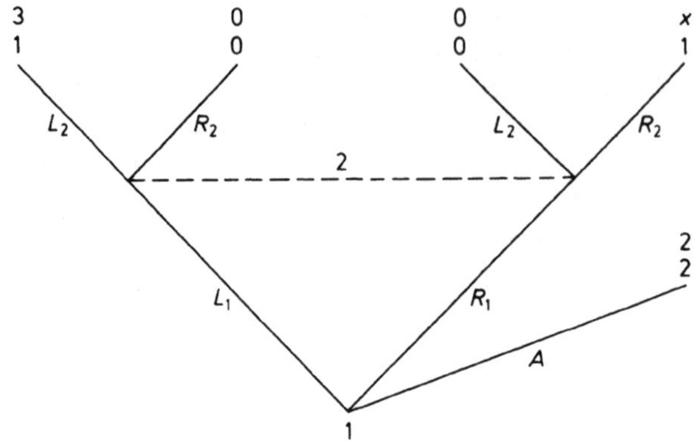

Fig. 1.4.3. Not every perfect (sequential) equilibrium is sensible

We claim that (A, R_2) is a perfect (hence, sequential) equilibrium of the game of Fig. 1.4.3 if $x \leqq 2$. Clearly, this is a Nash equilibrium: No player can gain by deviating unilaterally. Note that player 2 is not reached when A is played. Whenever 2 is reached, he could think that 1 has chosen R_1 and in this case playing R_2 is justified. Since such beliefs of player 2 are not excluded by the sequential equilibrium concept, we have that (A, R_2) is indeed sequential. This also is a perfect equilibrium: If player 2 expects the mistake R_1 to occur with a larger probability than the mistake L_1, then he should indeed play R_2. However, in our view, this equilibrium is not sensible. To support the equilibrium, player 2 should believe that player 1 has chosen R_1, but, if $x \leqq 0$, then R_1 is dominated by both L_1 and A. Hence, player 2 should believe that 1 has chosen his worst strategy and this does not seem reasonable. (If $x \leqq 0$, then (A, R_2) can be eliminated by invoking properness (see Sect. 1.5): R_1 is a more costly mistake than L_1 is, hence, according to properness the mistake L_1 should be considered much more likely.) If $0 < x \leqq 2$, then R_1 is no longer dominated by L_1, but it is still dominated by A. Note that L_1 is not dominated. Hence, if one accepts that a rational player will not choose a dominated strategy, then it follows that player 2 (whenever he is reached) should conclude that player 1 has chosen L_1. Consequently, 2 should choose L_2 and the only sensible equilibrium is (L_1, L_2).

The above example shows that there exist sequential equilibria that can be supported only by implausible beliefs and this raises the question whether one can devise general criteria to eliminate such equilibria. An affirmative answer to this question will be given in Chap. 10: Such equilibria can be eliminated by requiring that the equilibrium path should be stable against all small perturbations in the equilibrium strategies. (This stability criterion has first been proposed in Kohlberg and Mertens [1986]; it is easily seen that only (L_1, L_2) is stable in the game of Fig. 1.4.3.)

1.5 Perfect, Proper and Persistent Equilibria

If we have a game in which each player has to make a choice just once and if, moreover, the players make their choices simultaneously and independently of each other, then we speak of a *normal form game*. An example of such a game is the prisoners' dilemma game of Fig. 1.1.1. A normal form game can be considered as a special kind of extensive form game, but, on the other hand, with each extensive form game, one can associate a game in normal form (von Neumann and Morgenstern [1944], Kuhn [1953]). In the next two sections, it will be shown that also for normal form games it is necessary to refine the Nash equilibrium concept in order to obtain sensible solutions and in several examples the refinements which have been proposed for this class of games will be illustrated. These refinements will be of a slightly different kind than the ones we considered for games in extensive form. Namely, for extensive form games, the basic reason why one has to refine the equilibrium concept is that a Nash equilibrium may prescribe irrational behavior at unreached parts of the game tree. In a normal form game, however, every player has to make a choice, so that there are no unreached information sets. Yet, we will see that it is necessary to refine the equilibrium concept for normal form games, due to the fact that *an equilibrium* of such a game *need not be robust*. As an example of an equilibrium which is not robust consider the game of Fig. 1.5.1.

This game has two equilibria: the strategy combinations (L_1, L_2) and (R_1, R_2). In our view, the latter equilibrium is not a sensible one. This strategy combination satisfies Nash's equilibrium condition only since this condition presumes that each player will completely ignore all parts of the payoff matrix to which his opponent's strategy assigns zero probability. We feel, however, that a player should not ignore this information and that he, therefore, should play L. To be sure, if player 2 plays R_2, then player 1 cannot gain by playing L_1. However, by doing so, he cannot loose either and, as a matter of fact, if player 2 by mistake would play L_2, then player 1 is actually better off by playing L_1. Similarly, we have that player 2 can only gain by playing L_2. Therefore, if it is proposed to play (R_1, R_2), both players have an incentive to deviate. So, an agreement to play (R_1, R_2) is self-destabilizing and, therefore, this equilibrium is not sensible. The only sensible equilibrium of the game of Fig. 1.5.1 is the perfect equilibrium (L_1, L_2). If it is suggested to play this equilibrium, no player has an incentive whatever to disobey the recommendation.

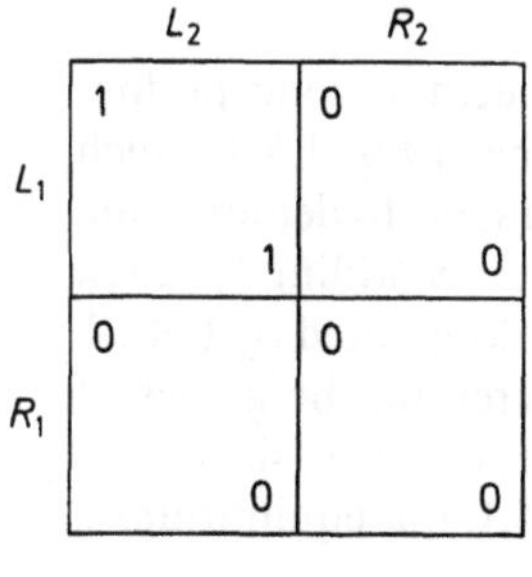

Fig. 1.5.1. The equilibrium (R_1, R_2) is not robust

	L_2	R_2
L_1	1, 1	10, 0
R_1	0, 10	10, 10

Fig. 1.5.2. A perfect equilibrium may be payoff dominated by a non-perfect one

	L_2	R_2	A_2
L_1	1, 1	0, 0	-1, -2
R_1	0, 0	0, 0	0, -2
A_1	-2, -1	-2, 0	-2, -2

Fig. 1.5.3. Adding dominated strategies may enlarge the set of perfect equilibria

If one takes the possibility of the players making mistakes seriously, then, for normal form games, one can only consider perfect equilibria as being reasonable. If an equilibrium fails to be perfect it is unstable with respect to small perturbations of the equilibrium and, therefore, at least one player will have an incentive to deviate from it. By restricting oneself to perfect equilibria, however, one may eliminate equilibria with attractive payoffs, as is shown by the game of Fig. 1.5.2.

This game has two equilibria, viz. (L_1, L_2) and (R_1, R_2). The equilibrium (R_1, R_2) yields both players the highest payoff. The game of Fig. 1.5.2 has exactly the same structure as the game of Fig. 1.5.1 (each player can only gain by playing L) and, therefore, in this game, the equilibrium (R_1, R_2) is as unstable as it is in the game of Fig. 1.5.1. If the players expect mistakes to occur with a small probability, then no player can really expect a payoff 10: the only perfect equilibrium is (L_1, L_2).

It was first pointed out in Myerson [1978] that the perfectness concept does not eliminate all intuitively unreasonable equilibria. The game of Fig. 1.5.3, which is a slight modification of the example given by Myerson, can serve to demonstrate this. Notice that this game results from the game of Fig. 1.5.1 by adding for each player a strategy A. One might argue that, since A is strictly dominated by both L and R, this strategy is strategically irrelevant and that, therefore, the games of Fig. 1.5.1. and 1.5.3 have the same sets of reasonable equilibria. Hence, since (L_1, L_2) is the only reasonable equilibrium of the game of Fig. 1.5.1, this equilibrium is also the unique reasonable equilibrium of the game of Fig. 1.5.3. However, the sets of perfect equilibria do not coincide for these games: in the game of Fig. 1.5.3 also the equilibrium (R_1, R_2) is perfect. Namely, if the players have agreed to play (R_1, R_2) and if each player expects that the mistake A will occur with a larger probability than the mistake L, then it is indeed optimal for each player to play R. Hence, adding strictly dominated strategies may change the set of perfect equilibria.

Myerson considers it to be an undesirable property of the perfectness concept that adding strictly dominated strategies may change the set of perfect equilibria and, therefore, he introduced a further refinement of the perfectness concept: the *proper equilibrium* (Myerson [1978], see Def. 2.3.1). The basic idea underlying the properness concept is that a player will make his mistakes in a more or less rational way, i.e. that he will make a more costly mistake with a much smaller probability than a less costly one, as a consequence of the fact that he will try much harder to prevent a more costly mistake.

According to the philosophy of the properness concept, in the game of Fig. 1.5.3, the players should not expect the mistake A to occur with a larger probability than the mistake L: since A is strictly dominated by L, each player will try harder to prevent the mistake A than he will try to prevent the mistake L and as a result A will occur with a smaller probability than L (in Myerson's view, the probability of A will even be of smaller order than the probability of L (cf. Def. 2.3.1)). If indeed the mistake L occurs with a larger probability than the mistake A, then each player will prefer to play L, hence, the equilibrium (R_1, R_2) is not proper. The only proper equilibrium of the game of Fig. 1.5.3 is (L_1, L_2): Once the players have agreed to play (L_1, L_2), no player has an incentive whatever to deviate from the equilibrium.

Myerson has shown that every normal form game possesses at least one proper equilibrium and that every proper equilibrium is perfect (Myerson [1978], see Theorem 2.3.3). A problem concerning this concept is that it is not clear whether the basic assumption underlying it (a more costly mistake is chosen with a probability which is of smaller order than the probability of a less costly one) can be justified. Myerson himself did not give a justification for this assumption. In Chaps. 4 and 5, we will investigate whether this assumption can be justified.

The game of Fig. 1.5.4 shows that not all proper equilibria possess the same degree of robustness. This game has several equilibria. Three of these are (L_1, L_2), (M_1, M_2) and (R_1, R_2). It is easily seen that the equilibrium (R_1, R_2) is not perfect: if mistakes might occur, each player will prefer M to R. The equilibria (L_1, L_2) and (M_1, M_2) are both perfect and even proper, but, the equilibrium (L_1, L_2) is much more robust than the equilibrium (M_1, M_2). Namely, if the players have agreed to play (L_1, L_2), then as long as mistakes occur with a probability smaller than $\frac{1}{2}$, each player is still willing to choose L. However, if the players have agreed to play (M_1, M_2), then each player is willing to keep to the agreement only if he expects that the mistake R will occur with a probability at least as big as the probability of the mistake L.

In Okada [1981a] a refinement of the perfectness concept, the *strictly perfect equilibrium*, is introduced which is based on the idea that a sensible equilibrium should be stable against *arbitrary slight perturbations* (see Def. 2.2.7). Obviously,

	L_2	M_2	R_2
L_1	2, 2	1, 1	0, 0
M_1	1, 1	1, 1	1, 1
R_1	0, 0	1, 1	1, 1

Fig. 1.5.4. Not all proper equilibria are equally robust

	L_2	M_2	R_2
L_1	1, 1	1, 0	0, 0
R_1	1, 1	0, 0	1, 0

Fig. 1.5.5. A game without strictly perfect equilibria

for the game of Fig. 1.5.4, the equilibrium (L_1, L_2) is strictly perfect whereas (M_1, M_2) is not. At first sight, it does not seem to be unreasonable to require that the solution of a game should be a strictly perfect equilibrium. The game of Fig. 1.5.5 shows that this cannot always be required, since there exist games without strictly perfect equilibria.

In the game of Fig. 1.5.5, every strategy pair in which player 2 plays L_2 is an equilibrium. None of these equilibria is strictly perfect: if player 1 expects that the mistake M_2 will occur more often than the mistake R_2, he should play L_1; if he expects this mistake to occur with a smaller probability, he should play R_1.

One might take the point of view that one really cannot hope for a general solution concept to be single-valued and that, in the game of Fig. 1.5.5, one should be willing to accept the set of all equilibria as the solution. This point of view is taken in Kohlberg and Mertens [1986]. In that paper, a set-valued analogon of the strict perfectness concept is proposed of which it is shown that it produces satisfactory answers in extensive form games as well as in games in normal form.[7] This stability concept of Kohlberg and Mertens will be considered in Chap. 10.

In the concluding example, we illustrate the concept of *persistent equilibria* that has been introduced in Kalai and Samet [1984]. This concept does not play a prominent role in the monograph, but it certainly deserves further study.[8]

The strategy pair in which both players choose $(\frac{1}{2}, \frac{1}{2})$ is an equilibrium of both games of Fig. 1.5.6. In fact, this is a strictly perfect equilibrium of both games. (It is easily seen that a completely mixed equilibrium is always strictly perfect, hence

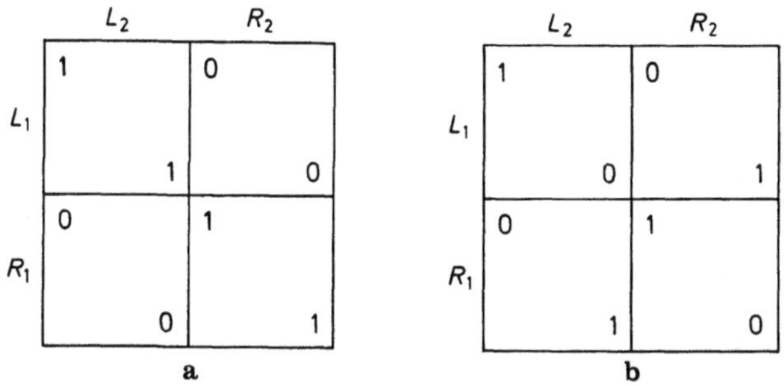

Fig. 1.5.6. The mixed equilibrium is not persistent in **a**, but it is persistent in **b**

proper and perfect.) Yet, intuitively one would say that the mixed equilibrium of game b is more stable than the mixed equilibrium of game a. If there are small trembles in the equilibrium strategies, then it seems that any dynamic adjustment process[9] will lead to either (L_1, L_2) or (R_2, R_2) in game a. In game b, however, there are no equilibria with smaller support that threaten the interior equilibrium. (In fact, in this game the Brown/Robinson process converges to the interior equilibrium.) The point is that the interior equilibrium is "minimal" in b, but not in a and intuitively one feels that "nonminimal" equilibria are not that stable. It is this idea of minimality that is formalised by the concept of persistent equilibria (see Sect. 2.3): In game a only (L_1, L_2) and (R_1, R_2) are persistent, in b the completely mixed equilibrium is persistent.

1.6 Essential Equilibria and Regular Equilibria

In the previous section, we considered refinements of the Nash equilibrium concept which are based on the idea that a sensible equilibrium should be stable against slight perturbations of the equilibrium strategies. One could also argue that a sensible equilibrium should be stable against *slight perturbations in the payoffs* of the game. Namely, one can maintain that these payoffs can be determined only somewhat inaccurately. A refinement of the equilibrium concept, based on this idea is the *essential equilibrium* concept, introduced in Wu Wen-Tsün and Jiang Jia-He [1962]. An equilibrium φ of a game Γ is said to be essential if every game with payoffs near to Γ has an equilibrium near to φ. Intuitively, it will be clear that an essential equilibrium is very stable. This indeed will be proved in Chap 2, where we will, for instance, show that every essential equilibrium is strictly perfect. As a consequence, not every game possesses an essential equilibrium. Indeed the payoffs in the game of Fig. 1.5.5 can be perturbed in such a way that either L_1 or R_1 is the unique best reply against L_2 and, therefore, this game does not have an essential equilibrium. Hence, we cannot require that a sensible equilibrium should be essential. Moreover, even in games which have essential equilibria, it is not always true that an essential equilibrium should be preferred to a non-essential one. This is illustrated by the game of Fig. 1.6.1.

	L_2	M_2	R_2
L_1	1, 1	0, 0	0, 0
M_1	0, 0	2, 2	2, 2
R_1	0, 0	2, 2	2, 2

Fig. 1.6.1. An essential equilibrium is not necessarily preferable to a non-essential equilibrium

	L_2	R_2
L_1	2, 2	4, 1
R_1	4, 1	3, 3

Fig. 1.6.2. Instability of equilibria in mixed strategies

The unique essential equilibrium of this game is (L_1, L_2). However, M and R are just duplicates of each other and, once it has been agreed to play some combination of M and R, no player has an incentive to deviate. Hence, the essential equilibrium concept is too restrictive.

It should be noted that the set of all equilibria with payoff $(2, 2)$ is an essential set of equilibria and the reader might be inclined to think that the set-valued analogue of essentiality (this is called hyper-stability in Kohlberg and Mertens [1986]) will always lead to reasonable equilibria. This is not the case, however, since hyper-stable sets may include dominated strategies. (If one deletes R_2 in the game of Fig. 1.5.5 then only R_1 is dominated, but any hyper-stable set includes R_1.)

In this chapter, it was forcibly argued that the solution of a noncooperative game has to be self-enforcing and, therefore, a Nash equilibrium. In many examples we have seen that not all Nash equilibria are self-enforcing: there exist equilibria at which at least one player has an incentive to deviate. Now suppose we have an equilibrium at which no player can gain by deviating. Is this equilibrium necessarily self-enforcing? The answer seems to be no: although no player may have an incentive to deviate, it may be the case that no player has an incentive to play his equilibrium strategy either. This situation occurs for equilibria in mixed strategies as is illustrated by means of the game of Fig. 1.6.2.

This game has a unique equilibrium, which is in mixed strategies. The equilibrium strategy of player 1 is $(2/3 L_1, 1/3 R_1)$ and the equilibrium strategy of player 2 is $(1/3L_2, 2/3R_2)$. The equilibrium yields player 1 a payoff 10/3 and player 2 a payoff 5/3. This equilibrium seems unstable, since, if player 2 plays $(1/3L_2, 2/3R_2)$, then player 1 receives a payoff of 10/3, no matter what he does and, therefore, he can shift to any other strategy without penalty. So, what is his incentive to play his equilibrium strategy? The same remark applies to player 2: if player 1 plays his equilibrium strategy, player 2 receives 5/3 no matter what he does and, therefore, he can also shift to any strategy without penalty.

One could even argue (as is done in Aumann and Maschler [1972, Sect. 2]) that, in the game of Fig. 1.6.2, the players have an incentive to deviate from their equilibrium strategies. Namely, if the equilibrium is played, player 1 receives 10/3, which is just the *maximin value* of this game for player 1, i.e. the payoff which player 1 can *guarantee* himself. However, the equilibrium strategy of player 1 does not guarantee 10/3, it only yields 10/3 if player 2 plays his equilibrium strategy. In order to guarantee 10/3 player 1 should play his *maximin strategy*, which is $(1/3L_1, 2/3R_1)$. So, if player 1 knows that he cannot obtain more that 10/3, why should not he play his maximin strategy which guarantees 10/3? The same remark applies to player 2 and so he also could have an incentive to play his maximin strategy, rather than his equilibrium strategy.

Aumann and Maschler write that they do not know what to recommend in this situation (since the maximin strategies are not in equilibrium), but they prefer the maximin strategies (Aumann and Maschler [1972, Sect. 2]). In Harsanyi [1977] (especially in Sect. 7.7.) it is argued that the players should indeed play their maximin strategies (also see van Damme [1980a]), but Harsanyi has changed his position in favour of the equilibrium strategies (Harsanyi and Selten [1988], Chap. 1).

From the discussion above it will be clear that the instability of equilibria in mixed strategies poses a serious problem. This problem is serious indeed, since many games possess only equilibria in mixed strategies. In Harsanyi [1973a] it is shown that this instability is only a seeming instability. Harsanyi argues that a player can never know the payoffs (utilities) of another player exactly since these are subject to random disturbances, due to stochastic fluctuations in this player's mood or taste. Therefore, a conflict situation, rather than by an ordinary game, is more adequately modelled by a so called *disturbed game*, i.e.. a game in which each player, although knowing his own payoffs exactly, knows the payoffs of the other players only somewhat inexactly. Harsanyi shows that for such a disturbed game every equilibrium is essentially in pure strategies and is, therefore, stable (also see Theorem 5.4.2). Harsanyi also shows that almost every equilibrium of an ordinary normal form game (whether in pure or in mixed strategies) can be obtained as the limit of equilibria of disturbed games, in which the disturbances go to zero, i.e. in which each player's information about the other players' payoffs becomes better and better. Hence, for almost all equilibria in mixed strategies the instability disappears if we take account of the actual uncertainty each player has about the other players' payoffs. Upon a closer investigation (see Theorem 5.6.2) it turns out that the equilibria which are stable in this sense are the *regular equilibria*, which have been introduced in Harsanyi [1973b]. A regular equilibrium is defined as an equilibrium which has the property that the Jacobian of a certain mapping associated with the game evaluated at this equilibrium is nonsingular (see

Def. 2.5.1). These regular equilibria will play a prominent role in the monograph. It will be shown that regular equilibria possess all robustness properties one reasonably can expect equilibria to possess: they are perfect, proper and even strictly perfect (i.e. stable) and essential. (However, a regular equilibrium need not be persistent: All equilibria in the game of Fig. 1.5.5 are regular.)

Unfortunately not all normal form games possess regular equilibria, but it can be shown that for almost all normal form games all equilibria are indeed regular (Theorem 2.6.2). These results indicate that for generic normal form games there is actually little need to refine the Nash equilibrium concept.[10] For extensive form games, however, the situation is quite different as we will see in Chaps. 6 and 10. Almost any nontrivial extensive form game has equilibria that are 'unreasonable'.

Notes

1. Since the publication of the first edition some new good textbooks in game theory have appeared and a couple of others are about to appear. The following cover the whole spectrum from the elementary to the advanced level: Dixit and Nalebuff [1990], Fudenberg and Tirole [1991], Kreps [1990a, b], Myerson [1990] and Rasmusen [1989]. The important role of game theory in industrial organization is illustrated by Tirole [1988], while McMillan [1989] focuses on the use of game theory in management. An overview of some applications of game theory in macroeconomics is given in Persson and Tabelini [1990], and the strategic theory of international trade is surveyed in McMillan [1986]. Good introductory surveys of noncooperative game theory are Tirole [1988, Chapter 11] and Fudenberg and Tirole [1989]. Kohlberg [1989] provides an excellent overview of refinements of Nash equilibrium, i. e. on the topic of this book. A very nice overview of the history and development of game theory is given in Aumann [1987a]. Surveys of all important areas of game theory will appear in Aumann and Hart [1991]. The reader is also urged to consult Aumann [1987b], a survey that focuses on methodology, on the relation between game theory and the real world, and on the question 'what is game theory trying to accomplish?'.
2. Some recent papers dealing with communication in games of complete information are Farrell [1988], Myerson [1989] and Rabin [1990]. These papers incorporate the idea that language has a focal meaning. Farrell [1987], Farrell and Saloner [1988], Matthews [1989] and Matthews and Postlewaite [1989] illustrate the importance of preplay communication in specific settings. Cooper et al. [1989] report experimental results on the role of cheap talk in the battle of the sexes game. For papers dealing with communication in incomplete information games see note 18 to Chap. 10.
3. Bernheim [1984] and Pearce [1984] have argued that for a strategy combination to be 'reasonable' it is not necessary that it is a Nash equilibrium. They proposed the weaker notion of 'rationalizability' as a necessary condition for 'reasonableness'. Also see Aumann [1987c], Battigalli [1990], Bernheim [1986], Gul [1989] and Rubinstein and Wolinsky [1990].

4. The technique of dynamic programming (hence the concept of subgame perfect equilibrium) is based on the assumption of persistent rationality. Backward induction requires that, when a player has evidence that an opponent does not play according to the backward induction solution, he disregards this information and believes that from now on the opponent will always play as backward induction dictates. This assumption has been criticized in Basu [1988, 1990], Bicchieri [1989], Binmore [1987, 1988], Reny [1988a, 1988b] and Rosenthal [1981]. In these papers it is argued that in the game of Fig. 1.2.2 player 2 may justify playing L_2 with positive probability by the argument that since player 1 has played 'irrationally' at his first move he may also play 'irrationally' at his second move.
5. A strategy of player i is usually interpreted as a complete plan of action for this player. However, it also serves as a description of the opponents' expectations of player i's behavior. The strategy concept is scrutinized in Rubinstein [1988].
6. This statement needs qualification. As Kreps and Wilson already pointed out themselves, it is not completely clear that the consistency notion used in the definition of sequential equilibrium is the appropriate one. See the notes to Chap. 6 for further details.
7. The stability concept from Kohlberg and Mertens [1986] was not entirely satisfactory. A modification of the concept was proposed in Mertens [1980a].
8. A related concept of cyclically stable sets has been proposed in Gilboa and Matsui [1989].
9. Recently the questions of whether 'learning' will lead to a Nash equilibrium and of what kind of learning processes will lead to which type of equilibrium have received considerable attention. A sample of the papers in this area is Canning [1989, 1990], Fudenberg and Levine [1990], Fudenberg and Kreps [1988], Friedmann and Rosenthal [1986], Kalai and Lehrer [1990], Milgrom and Roberts [1989a, b] and Selten [1988a].
10. The most refined equilibrium notion that will be discussed in this book is the strict equilibrium, i.e. an equilibrium in pure strategies in which each player actually looses if he deviates unilaterally. Experiments reported in Van Huyck et al. [1988] suggest that not even all strict equilibria are self-enforcing. Consider the 2×2 bimatrix game Γ in which each player chooses between α and β: If both choose α both receive the payoff 1, if both choose β each receives the payoff 2, if choices don't match each receives 0. The results of Van Huyck et al. [1988] suggest that if it is recommended to the players to play (α, α), then even if one gives them all the arguments for why (α, α) is a good equilibrium, the players will (without having to communicate) succeed to jointly deviate to (β, β). Carlsson and Van Damme [1990] present a noncooperative theory that only allows (β, β) as the solution of Γ. In general this theory selects the risk dominant equilibrium (Harsanyi and Selten [1988]) as the unique solution of a 2×2 bimatrix game. Note that there may be a conflict between risk dominance and payoff dominance, see, for example, Aumann [1989] which also argues that not every strict Nash equilibrium is self-enforcing.

2 Games in Normal Form

For normal form games the Nash equilibrium concept has to be refined since a Nash equilibrium of such a game need not be robust, i.e. it may be unstable against small perturbations in the data of the game. In this chapter, we will consider various refinements of the Nash concept for this class of games, all of which require an equilibrium to satisfy some particular robustness condition.

In Sect. 2.2, perfect equilibria (Selten [1975]) are considered. These are equilibria which are stable against some slight perturbations in the equilibrium strategies. It is shown that every normal form game possesses at least one perfect equilibrium and several properties of such equilibria are derived. In this section also strictly perfect equilibria (Okada [1981a]), i.e. equilibria which are stable against arbitrary slight perturbations in the equilibrium strategies, are considered.

In Sect. 2.3, the concepts of proper equilibria (Myerson [1978]) and weakly proper equilibria are studied. These concepts require an equilibrium to be stable against perturbations of the equilibrium strategies which are more or less rational, i.e. which assign the preponderance of weight to the better strategies. Both concepts are refinements of the perfectness concept. Furthermore, the concepts of strictly proper equilibria (a refinement of strict perfectness) and of persistent equilibria (Kalai and Samet [1984]) are introduced.

In Sect. 2.4, we consider essential equilibria (Wu Wen-Tsün and Jiang Jia-He [1962]), i.e. equilibria which are stable against arbitrary slight perturbations in the payoffs of the game, and we show that every essential equilibrium is strictly perfect. Furthermore, the concept of strongly stable equilibria (Kojima, Okada and Shindoh [1985]) is introduced. This is a refinement of the essential equilibrium concept which requires an equilibrium to change uniquely and continuously against perturbations in payoffs. Every such strongly stable equilibrium is isolated and strictly proper.

In Sect. 2.5, the concept of regular equilibria (Harsanyi [1973b]) is introduced and it is shown that every regular equilibrium is strongly stable.

The main result of Sect. 2.6 states that for almost all normal form games all equilibria are regular. This implies that for "nondegenerate" normal form games all equilibria possess all robustness properties one can hope for.

This chapter is based upon the references above and on van Damme [1981c].

2.1 Preliminaries

A *finite n-person normal form game* is a $2n$-tuple $\Gamma = (\Phi_1, \ldots, \Phi_n, R_1, \ldots, R_n)$, where Φ_i is a finite nonempty set and R_i is a mapping $R_i\colon \prod_{j=1}^{n} \Phi_j \to \mathbb{R}$, for

$i \in N(=\{1, \ldots, n\})$. The set Φ_i is the set of *pure strategies* of player i and R_i is the *payoff function* of this player.

Usually, we will just speak of a normal form game, rather than of a finite normal form game. Next, let us introduce the notation and terminology which will be used throughout Chaps. 2 – 5 with respect to these games. Let $\Gamma = (\Phi_1, \ldots, \Phi_n, R_1, \ldots, R_n)$ be fixed. We write:

$$m_i := |\Phi_i|\,, \quad m := \sum_{i=1}^{n} m_i\,, \quad m^* := \prod_{i=1}^{n} m_i\,. \tag{2.1.1}$$

A generic element of Φ_i will be denoted by φ_i. We assume that the elements of Φ_i are numbered and, consequently, we will speak about the kth pure strategy of player i. Therefore, and even more often, a generic element of Φ_i will also be denoted by k. A *mixed strategy* s_i of player i is a probability distribution on Φ_i. We denote the probability which s_i assigns to the pure strategy k of player i by s_i^k and we write S_i for the set of all mixed strategies of this player, hence

$$S_i := \left\{ s_i \in \mathscr{F}(\Phi_i, \mathbb{R});\ \sum_k s_i^k = 1,\ s_i^k \geqq 0 \quad \text{for all } k \in \Phi_i \right\}. \tag{2.1.2}$$

If $s_i \in S_i$, then $C(s_i)$ denotes the *Carrier* of s_i, i.e.

$$C(s_i) := \{k \in \Phi_i;\ s_i^k > 0\}\,. \tag{2.1.3}$$

s_i is said to be *completely mixed* if $C(s_i) = \Phi_i$. The pure strategy k of player i is identified with the mixed strategy which assigns probability 1 to k. We define the sets Φ and S by:

$$\Phi := \prod_{i=1}^{n} \Phi_i \quad \text{and} \quad S := \prod_{i=1}^{n} S_i\,. \tag{2.1.4}$$

Φ (resp. S) is the set of pure (resp. mixed) *strategy combinations* of Γ. A generic element of Φ (resp. S) is denoted by φ (resp. s). In a normal form game, the players make their choices independently of each other, therefore, the probability $s(\varphi)$ that $\varphi = (k_1, \ldots, k_n)$ occurs if $s = (s_1, \ldots, s_n)$ is played, is given by:

$$s(\varphi) := \prod_{i=1}^{n} s_i^{k_i}\,. \tag{2.1.5}$$

The *Carrier* of s, which is denoted by $C(s)$, is defined by:

$$C(s) := \{\varphi;\ s(\varphi) > 0\} = \prod_{i=1}^{n} C(s_i)\,, \tag{2.1.6}$$

and s is said to be *completely mixed*, if $C(s) = \Phi$.

If s is played, the *expected payoff* $R_i(s)$ for player i is given by:

$$R_i(s) := \sum_{\varphi} s(\varphi) R_i(\varphi)\,. \tag{2.1.7}$$

Let $s = (s_1, \ldots, s_n) \in S$ and let $\bar{s}_i \in S_i$. We denote by $s \backslash \bar{s}_i$ that strategy combination which results from s by replacing the strategy s_i of player i by the strategy $\bar{s}_i$ of this player. Hence, $s \backslash \bar{s}_i$ is the strategy combination $(s_1, \ldots, s_{i-1}, \bar{s}_i,$

$s_{i+1}, \ldots, s_n)$. We say that $\bar{s}_i$ is a *best reply* of player i against s if

$$R_i(s\backslash\bar{s}_i) = \max_{s_i' \in S_i} R_i(s\backslash s_i') . \tag{2.1.8}$$

The set of all *pure best replies* of player i against s (i.e. the best replies of player i which are in Φ_i) is denoted by $B_i(s)$. It is easily seen that $\bar{s}_i$ is a best reply against s if and only if we have

$$\text{if } R_i(s\backslash k) < R_i(s\backslash l), \text{ then } \bar{s}_i^k = 0 \quad \text{for all } k, l \in \Phi_i , \tag{2.1.9}$$

which is equivalent to

$$C(\bar{s}_i) \subset B_i(s) . \tag{2.1.10}$$

Hence, s_i is a best reply of player i against s if and only if $\bar{s}_i$ assigns a positive probability only to the pure best replies of player i against s.

Let $s, \bar{s} \in S$. We say that $\bar{s}$ is a *best reply* against s if $\bar{s}_i$ is a best reply against s for all i, and we denote the set of all *pure best replies* against s by $B(s)$, hence

$$B(s) := \prod_{i=1}^{n} B_i(s) . \tag{2.1.11}$$

A strategy combination s is a *Nash equilibrium* (Nash [1950, 1951]) of Γ if s is a best reply against itself. It follows from (2.1.10) that s is a Nash equilibrium if and only if

$$C(s) \subset B(s) . \tag{2.1.12}$$

Usually, we will speak of *equilibria* instead of Nash equilibria. We denote the set of equilibria of Γ by $E(\Gamma)$. Nash has shown that every finite normal form game possesses at least one equilibrium (Nash [1950, 1951]).

Equation (2.1.12) expresses that in an equilibrium each player only uses best replies. By requiring that each player should use all his best replies, a refinement of the Nash equilibrium concept, the so called *quasi-strict equilibrium** (Harsanyi [1973a]), is obtained. Hence, formally a *quasi-strict equilibrium* is defined as a strategy combination s which satisfies $C(s) = B(s)$. In Sect. 3.4 (Fig. 3.4.1), we show that not every game possesses a quasi-strict equilibrium. The game of Fig. 2.1.1 illustrates another drawback of the quasi-strictness concept: a quasi-strict equilibrium may involve dominated strategies, hence, it need not be 'reasonable'.

The set of equilibria of this game is $\{(s_1, 1); s_1 \in S_1\}$. All equilibria are quasi-strict, except for (1, 1), but this latter equilibrium is the most stable one since the first strategy of player 1 dominates all his other strategies.

In Harsanyi [1973a] the concept of quasi-strict equilibria is introduced as a generalization of the concept of *strict equilibria*. A *strict equilibrium* is a strategy combination s which is the only best reply against itself, i.e. it satisfies $\{s\} = B(s)$. Hence, a strict equilibrium is a quasi-strict equilibrium in pure strategies. This

* Harsanyi uses the term strong (resp. quasi-strong) instead of strict (resp. quasi-strict). We have changed his terminology in order to avoid confusion with the notion of strong equilibrium from Aumann [1959]

	1	2
1	1 / 1	1 / 0
2	1 / 1	0 / 0

Fig. 2.1.1. A quasi-strict equilibrium need not be 'reasonable' (the rows represent player 1's choices, the columns those of player 2, in each cell the upper left entry is the payoff to player 1, while the lower right entry is the payoff to player 2)

terminology might give the impression that quasi-strict equilibria possess similar properties as strict equilibria. This, however, is not true. As one can expect, strict equilibria possess all nice properties one can hope for, but, as the example in this section shows, quasi-strict equilibria need not be nice at all. In this chapter, it will be shown that regular equilibria (Sect. 2.5) possess similar robustness properties as strict equilibria.

Let $\mathscr{G}(\Phi_1,\ldots,\Phi_n)$ be the set of all games $\Gamma=(\Phi_1,\ldots,\Phi_n, R_1,\ldots,R_n)$ with pure strategy spaces $\Phi_1,\ldots,\Phi_n$. For $\Gamma\in\mathscr{G}(\Phi_1,\ldots,\Phi_n)$, let r_i be the vector of payoffs of player i associated with pure strategies in Γ, i.e.

$$r_i := \langle R_i(\varphi)\rangle_{\varphi\in\Phi}. \tag{2.1.13}$$

We can view r_i as an element of $\mathbb{R}^{m^*}$, where $m^*=|\Phi|$. Therefore r_i is called the *payoff vector* of player i in Γ. The *payoff vector* of the game Γ is the vector

$$r=(r_1,\ldots,r_n)\in\mathbb{R}^{nm^*}. \tag{2.1.14}$$

Hence, by imposing a fixed ordering on Φ, we obtain a one-to-one correspondence between $\mathscr{G}(\Phi_1,\ldots,\Phi_n)$ and $\mathbb{R}^{nm^*}$. We will denote the game $\Gamma\in\mathscr{G}(\Phi_1,\ldots,\Phi_n)$ which corresponds to $r\in\mathbb{R}^{nm^*}$ by $\Gamma(r)$. So $\mathscr{G}(\Phi_1,\ldots,\Phi_n)$ can be viewed as an nm^*-dimensional Euclidian space and we can speak about the (Euclidian) distance $\varrho(\Gamma,\Gamma')$ between two games and about the Lebesgue measure of a (measurable) set of games.

Let $\mathscr{S}$ be a statement about normal form games. We say that $\mathscr{S}$ is true for *almost all* games if for any $n\in\mathbb{N}$ and any n-tuple of finite nonempty sets $\Phi_1,\ldots,\Phi_n$, we have that

$$\lambda(\mathrm{cl}\{\Gamma\in\mathscr{G}(\Phi_1,\ldots,\Phi_n);\ \mathscr{S}\text{ is false}\})=0, \tag{2.1.15}$$

i.e. if the closure of the set of games for which $\mathscr{S}$ is false has Lebesgue measure zero. Notice that, if $\mathscr{S}$ is true for almost all games, then for any n-tuple $\Phi_1,\ldots,\Phi_n$, the set of games in $\mathscr{G}(\Phi_1,\ldots,\Phi_n)$ for which $\mathscr{S}$ is true contains an open and everywhere dense set (and, hence, the subset for which $\mathscr{S}$ is false is nowhere dense).

In this monograph, we are mainly interested in finite normal form games. However, at some instances such games will be approximated by infinite normal form games. An infinite normal form is a $2n$-tuple $G=(S_1,\ldots,S_n, R_1,\ldots,R_n)$, where S_i is an infinite set and R_i is a mapping from S to $\mathbb{R}$, for $i\in N$, where $S:=\prod_{i=1}^{n} S_i$. We will only be dealing with concave infinite games, i.e. infinite games which satisfy the conditions of Theorem 2.1.1 below. In such games, there is no need for the players to randomize and, consequently, an equilibrium of such a game is a strategy combination $s \in S$ which is a best reply against itself. We denote the set of equilibria of G by $E(G)$. Rosen has proved the following generalization of Nash's theorem on the existence of equilibria:

Theorem 2.1.1 (Rosen [1965]). *Let $G = (S_1, \ldots, S_n, R_1, \ldots, R_n)$ be an infinite n-person normal form game such that the following 3 conditions are satisfied for each $i \in N$:*

(i) *S_i is a nonempty, compact and convex subset of some finite dimensional Euclidean space,*
(ii) *the mapping R_i is continuous, and*
(iii) *for fixed $s \in S$, the mapping $s_i' \to R_i(s \backslash s_i')$ is concave.*

Then G possesses at least one equilibrium.

2.2 Perfect Equilibria

In Chap. 1 we saw that, to obtain sensible solutions for every game, the Nash equilibrium concept has to be refined. In this section we will consider one such refinement, the perfect equilibrium, which has been introduced in Selten [1975]. It will be shown that every normal form game possesses at least one perfect equilibrium and some properties of such equilibria will be derived.

The basic idea behind the *perfectness concept* is that each player with a small probability makes *mistakes*, which has the consequence that every pure strategy is chosen with a positive (although possibly small) probability. Mathematically, this idea is modelled via a *perturbed game*, i.e. a game in which each player is only allowed to use completely mixed strategies.

Definition 2.2.1. Let $\Gamma = (\Phi_1, \ldots, \Phi_n, R_1, \ldots, R_n)$ be an n-person normal form game. For $i \in N$, let η_i and $S_i(\eta_i)$ be defined by:

$$\eta_i \in \mathscr{F}(\Phi_i, \mathbb{R}) \text{ with } \eta_i^k > 0 \text{ for all } k \in \Phi_i \text{ and } \sum_k \eta_i^k < 1, \tag{2.2.1}$$

$$S_i(\eta_i) := \{s_i \in S_i;\ s_i^k \geqq \eta_i^k \quad \text{for all } k \in \Phi_i\}. \tag{2.2.2}$$

Furthermore, let $\eta = (\eta_1, \ldots, \eta_n)$ and $S(\eta) = \prod_{i=1}^{n} S_i(\eta_i)$. The *perturbed game* (Γ, η) is the infinite normal form game $(S_1(\eta_1), \ldots, S_n(\eta_n), R_1, \ldots, R_n)$.

It is easily seen that a perturbed game (Γ, η) satisfies the conditions of Theorem 2.1.1 and so such a game possesses at least one equilibrium. It is clear that in such an equilibrium a pure strategy which is not a best reply has to be chosen with a minimum probability. Therefore, we have [(cf. (2.1.9)].

Lemma 2.2.2. *A strategy combination $s \in S(\eta)$ is an equilibrium of (Γ, η) if and only if the following condition is satisfied:*

$$\text{if } R_i(s\backslash k) < R_i(s\backslash l), \text{ then } s_i^k = \eta_i^k \quad \text{for all } i, k, l. \tag{2.2.3}$$

The assumption that rational players will make mistakes only with a very small probability leads to the following definition:

Definition 2.2.3. Let Γ be a normal form game. An equilibrium s of Γ is a *perfect equilibrium* of Γ if s is a limit point of a sequence $\{s(\eta)\}_{\eta \downarrow 0}$ with $s(\eta) \in E(\Gamma, \eta)$ for all η, i.e. s is perfect if there exist sequences $\{s(t)\}_{t \in \mathbb{N}}$ and $\{\eta(t)\}_{t \in \mathbb{N}}$ with $s(t) \in E(\Gamma, \eta(t))$ for all $t \in \mathbb{N}$, and such that $s(t)$ converges to s and $\eta(t)$ converges to zero, as t tends to infinity.

Note that for an equilibrium s of Γ to be perfect it is sufficient that some perturbed games (Γ, η) with η close to zero possess an equilibrium close to s and that it is not required that all perturbed games (Γ, η) with η close to zero possess such an equilibrium s. Hence, requiring that an equilibrium is perfect is a weak requirement and, consequently, if an equilibrium fails to be perfect, it is very unstable. Let $\{(\Gamma, \eta(t))\}_{t \in \mathbb{N}}$ be a sequence of perturbed games for which $\eta(t)$ converges to zero as t tends to infinity. Since every game $(\Gamma, \eta(t))$ possesses at least one equilibrium $s(t)$, and since each $s(t)$ is an element of the compact set S, there exists at least one limit point of $\{s(t)\}_{t \to \infty}$. It easily follows from (2.2.3) that such a limit point is in fact an equilibrium of Γ. Hence, we have proved:

Theorem 2.2.4 (Selten [1975]). *Every normal form game possesses at least one perfect equilibrium.* [1]

By considering the game of Fig. 1.5.1, it is seen that not every Nash equilibrium is a perfect equilibrium (the equilibrium (R_1, R_2) of that game is not perfect), and so we have that the perfectness concept is a strict refinement of the Nash equilibrium concept.

Next, two characterizations of perfect equilibria will be derived. One of these characterizations uses the concept of ε-perfect equilibria which has been introduced in Myerson [1978]. Let $\Gamma = (\Phi_1, \dots, \Phi_n, R_1, \dots, R_n)$ be a normal form game and let $\varepsilon > 0$. A strategy combination $s \in S$ is an *ε-perfect equilibrium* of Γ if it is completely mixed and satisfies:

$$\text{if } R_i(s\backslash k) < R_i(s\backslash l), \text{ then } s_i^k \leqq \varepsilon \text{ for all } i, k, l. \tag{2.2.4}$$

An ε-perfect equilibrium of Γ need not be an equilibrium of Γ, but if ε is small, then an ε-perfect equilibrium is close to an equilibrium (Theorem 2.2.5). An ε-perfect equilibrium is another way of modelling the idea that rational players make mistakes (choose non-optimal strategies) only with a small probability (viz. a probability of at most ε) and, therefore, one expects that an equilibrium is perfect if and only if it is a limit point of ε-perfect equilibria. In Theorem 2.2.5 we prove that this is indeed the case. The second characterization of perfect equilibria given in Theorem 2.2.5 has first been obtained in Selten [1975]. This characterization is the most advantageous one from the viewpoint of mathematical simplicity.

Theorem 2.2.5. *Let $\Gamma = (\Phi_1, \dots, \Phi_n, R_1, \dots, R_n)$ be an n-person normal form game and let $s \in S$. The following assertions are equivalent:*

(i) *s is a perfect equilibrium of Γ,*
(ii) *s is a limit point of a sequence $\{s(\varepsilon)\}_{\varepsilon \downarrow 0}$, where $s(\varepsilon)$ is an ε-perfect equilibrium of Γ, for all ε, and*
(iii) *s is a limit point of a sequence $\{s(\varepsilon)\}_{\varepsilon \downarrow 0}$ of completely mixed strategy combinations with the property that s is a best reply against every element $s(\varepsilon)$ in this sequence.*

Proof. (i)→(ii): Let s be a limit point of a sequence $\{s(\eta)\}_{\eta \downarrow 0}$, with $s(\eta) \in E(\Gamma, \eta)$ for all η. Define $\varepsilon(\eta) \in \mathbb{R}_{++}$ by:

$$\varepsilon(\eta) := \max_{i,k} \eta_i^k .$$

Then $s(\eta)$ is an $\varepsilon(\eta)$-perfect equilibrium of Γ. This establishes (ii).
(ii)→(iii): let $\{s(\varepsilon)\}_{\varepsilon \downarrow 0}$ be as in (ii) and, without loss of generality, assume $s(\varepsilon)$ converges to s as ε tends to zero. It immediately follows from (2.2.4) that every element of $C(s)$ is a best reply against $s(\varepsilon)$ if ε is sufficiently small. Therefore, s is a best reply against $s(\varepsilon)$ if ε is sufficiently small.
(iii)→(i): Let $\{s(\varepsilon)\}_{\varepsilon \downarrow 0}$ be as in (iii) and without loss of generality assume that s is the unique limit of this sequence. Define $\eta(\varepsilon) \in \mathbb{R}^m_{++}$ by:

$$\eta_i^k(\varepsilon) = \begin{cases} s_i^k(\varepsilon) & \text{if } k \notin C(s_i) \\ \varepsilon & \text{otherwise} \end{cases} \quad \text{for all } i, k . \tag{2.2.5}$$

Then $\eta(\varepsilon)$ converges to zero as ε tends to zero and so, for sufficiently small ε, the perturbed game $(\Gamma, \eta(\varepsilon))$ is well-defined. For ε sufficiently small we have that $s(\varepsilon) \in S(\eta(\varepsilon))$ and it follows from (2.2.3) that in this case $s(\varepsilon)$ is actually an equilibrium of $(\Gamma, \eta(\varepsilon))$. □

In the examples of Sect. 1.5 we saw that the equilibria in dominated strategies were eliminated by the perfectness concept. For a normal form game $\Gamma = (\Phi_1, \dots, \Phi_n, R_1, \dots, R_n)$ we say that the strategy s_i' of player i is *dominated* by the strategy s_i'' if

$$R_i(s \backslash s_i') \leqq R_i(s \backslash s_i'') \quad \text{for all } s \in S, \text{ and}$$
$$R_i(s \backslash s_i') < R_i(s \backslash s_i'') \quad \text{for some } s \in S . \tag{2.2.6}$$

Notice that, to check whether s_i' is dominated by s_i'', it is sufficient to check whether (2.2.6) is satisfied for all $\varphi \in \Phi$, rather than for all $s \in S$. We say that s_i' is *strictly dominated* by s_i'' if all inequalities in (2.2.6) are strict, and we say that s_i' is *undominated*, if there is no strategy s_i'', which dominates it. Finally, a strategy combination $s \in S$ is said to be *undominated* if every component s_i of s is undominated.

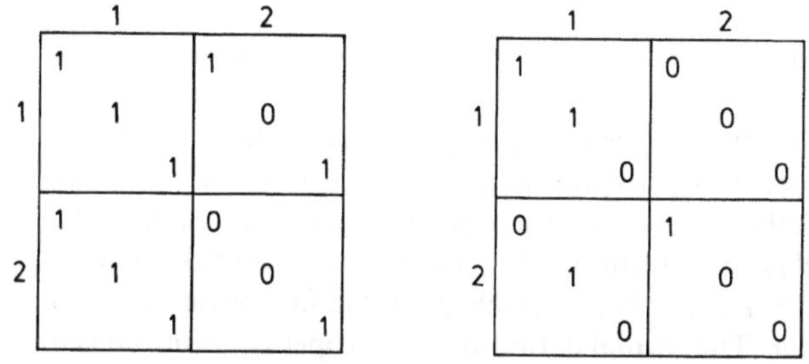

Fig. 2.2.1. An undominated equilibrium is not necessarily perfect. (Player 1 chooses a row, player 2 a column and player 3 a matrix; in each cell the upper left entry is the payoff to player 1, the entry in the middle is the payoff to player 2, etc.)

It easily follows from the characterization of perfect equilibria given in Theorem 2.2.5 (iii), that we have:

Corollary 2.2.6. *Every perfect equilibrium is undominated.*

The reader might wonder whether the converse of this corollary is true, i.e. whether an undominated equilibrium is perfect. We will prove that this is indeed correct for 2-person normal form games (Theorem 3.2.2). For games with more than 2 players, however, the converse is not true as the game of Fig. 2.2.1 shows. The strategy combination (2, 1, 1) is an undominated equilibrium of this game, which is not perfect. The only perfect equilibrium of this game is (1, 1, 1) as is easily verified.[2]

We close this section by giving the definition of a *strictly perfect equilibrium*. This concept has been introduced in Okada [1981a].

Definition 2.2.7. Let $\Gamma = (\Phi_1, \ldots, \Phi_n, R_1, \ldots, R_n)$ be an n-person normal form game. For $\hat{\eta} \in \mathbb{R}^m_{++}$, let $U_{\hat{\eta}} := \{\eta \in \mathbb{R}^m_{++}; \eta < \hat{\eta}\}$. s is a *strictly perfect equilibrium* of Γ if there exists some $\hat{\eta} \in \mathbb{R}^m_{++}$ and for each $\eta \in U_{\hat{\eta}}$ some $s(\eta) \in E(\Gamma, \eta)$ such that $\lim_{\eta \downarrow 0} s(\eta) = s$.

Obviously, each strictly perfect equilibrium is perfect, but, as we have seen in Sect. 1.5, there exist games without strictly perfect equilibria. In the next section we will consider another refinement of the perfectness concept: the properness concept. Contrary to the strict perfectness concept, this concept generates a nonempty set of solutions for every normal form game.

2.3 Proper Equilibria

In Chap. 1 (particularly in Sect. 1.5), we saw that a perfect equilibrium may be 'unreasonable'. In order to exclude these 'unreasonable' perfect equilibria, Myerson has introduced a refinement of the perfectness concept: the proper equilibrium (Myerson [1978]). In this section we will consider this concept, as well as the closely related concepts of weakly proper equilibria and strictly proper equilibria.[3]

The basic idea underlying the *properness concept* is that a player, although making mistakes, will try much harder to prevent the more costly mistakes than he will try to prevent the less costly ones, i.e. that there is some sort of *rationality in the mechanism of making mistakes.* As a result of this, a more costly mistake will (in Myerson's view) occur with a probability which is of smaller order than the probability of a less costly one. The formal definition of a proper equilibrium is in the same spirit as the characterization of perfect equilibria given in Theorem 2.2.5(ii):

Definition 2.3.1. Let $\Gamma = (\Phi_1, \ldots, \Phi_n, R_1, \ldots, R_n)$ be an n-person normal form game, let $\varepsilon \in \mathbb{R}_{++}$ and $s(\varepsilon) \in S$. We say that $s(\varepsilon)$ is an *ε-proper equilibrium* of Γ if $s(\varepsilon)$ is completely mixed and satisfies:

$$\textit{if } R_i(s(\varepsilon)\backslash k) < R_i(s(\varepsilon)\backslash l), \textit{ then } s_i^k(\varepsilon) \leqq \varepsilon s_i^l(\varepsilon) \quad \textit{for all } i, k, l. \tag{2.3.1}$$

$s \in S$ is a *proper equilibrium* of Γ if s is a limit point of a sequence $\{s(\varepsilon)\}_{\varepsilon \downarrow 0}$, where $s(\varepsilon)$ is an ε-proper equilibrium of Γ.

Notice that, if s is a proper equilibrium of Γ, then for every $\varepsilon > 0$ there exists some ε-proper equilibrium of Γ such that actually $s = \lim_{\varepsilon \downarrow 0} s(\varepsilon)$, due to the fact that an ε-proper equilibrium is ε'-proper for $\varepsilon' \geqq \varepsilon$. Furthermore, it is clear that a proper equilibrium is perfect, since an ε-proper equilibrium is ε-perfect.

From the proof of Theorem 2.2.5 we can deduce:

Lemma 2.3.2. *Let s be a proper equilibrium of a normal form game Γ and for $\varepsilon > 0$, let $s(\varepsilon)$ be an ε-proper equilibrium of Γ, such that $\lim_{\varepsilon \downarrow 0} s(\varepsilon) = s$. Then s is a best reply against $s(\varepsilon)$ for all ε which are sufficiently close to zero.*

By considering the game of Fig. 1.5.3, we see that the properness concept is a strict refinement of the perfectness concept (namely (R_1, R_2) is a perfect equilibrium of this game, which is not proper). Obviously we would like a refinement of the Nash equilibrium concept to generate a nonempty set of solutions for every normal form game. It will now be shown (as in Myerson [1978]) that this is indeed the case for the properness concept.

Theorem 2.3.3 (Myerson [1978]). *Every normal form game possesses at least one proper equilibrium.*[4]

Proof. Let $\Gamma = (\Phi_1, \ldots, \Phi_n, R_1, \ldots, R_n)$ be a normal form game. It suffices to show that, for $\varepsilon > 0$ sufficiently close to zero, there exists an ε-proper equilibrium of Γ. Let $\varepsilon \in (0, 1)$. For $i \in N$, define $\eta_i \in \mathscr{F}(\Phi_i, \mathbb{R})$ by:

$$\eta_i^k = \frac{\varepsilon^{m_i}}{m_i} \quad \text{for all } k \in \Phi_i.$$

Furthermore, let η, $S_i(\eta_i)$ and $S(\eta)$ be as in Def. 2.2.1. For $i \in N$, define the correspondence F_i from $S(\eta)$ to $S_i(\eta_i)$ by:

$$F_i(s) = \{\sigma_i \in S_i(\eta_i): \text{if } R_i(s\backslash k) < R_i(s\backslash l), \text{ then } \sigma_i^k \leqq \varepsilon\sigma_i^l, \text{ for all } k, l\} .$$

Then $F_i(s) \neq \varnothing$ for all $s \in S(\eta)$. Namely, let $s \in S(\eta)$ and define

$$\nu_i(s, k) = |\{l \in \Phi_i;\ R_i(s\backslash k) < R_i(s\backslash l)\}| \quad \text{for } k \in \Phi_i \text{, and}$$

$$\sigma_i^k = \varepsilon^{\nu_i(s,k)} / \sum_l \varepsilon^{\nu_i(s,l)} \qquad \text{for } k \in \Phi_i ,$$

Then $\sigma_i \in F_i(s)$. Furthermore we have that $F_i(s)$ is a closed and convex set for every $s \in S(\eta)$, and that the mapping F_i is upper semicontinuous. Let F be the n-tuple $(F_1, \ldots, F_n)$. Then F satisfies the conditions of the Kakutani Fixed Point Theorem (Kakutani [1941]) and, therefore F has a fixed point. Since every fixed point of F is an ε-proper equilibrium of Γ, the proof is complete. □

One of Myerson's motives for introducing the properness concept is that the perfectness concept has the drawback that adding strictly dominated strategies may enlarge the set of perfect equilibria. By means of the game of Fig. 2.3.1. we show that the properness concept suffers from the same drawback. In this game the second strategy of players 3 is strictly dominated and, therefore, one could consider this strategy as being strategically irrelevant. If one holds this view, then one can consider only (1, 1, 1) as being reasonable, since it is the unique perfect (proper) equilibrium of the game in which player 3 is restricted to his first strategy. However, this equilibrium is not the only proper equilibrium of the game of Fig. 2.3.1: also (2, 2, 1) is a proper equilibrium.

A second drawback of the properness concept is that the set of proper equilibria may change when a strategy that is already available as a mixed strategy is explicitly added as a pure strategy (see Fig. 10.1.2).

Another aspect of proper equilibrium which the reader might consider as being undesirable is that the properness concept requires a more costly mistake to be chosen with a probability which is of smaller order than the probability of a less costly one, even if this mistake is only a little bit more costly. To undo this latter

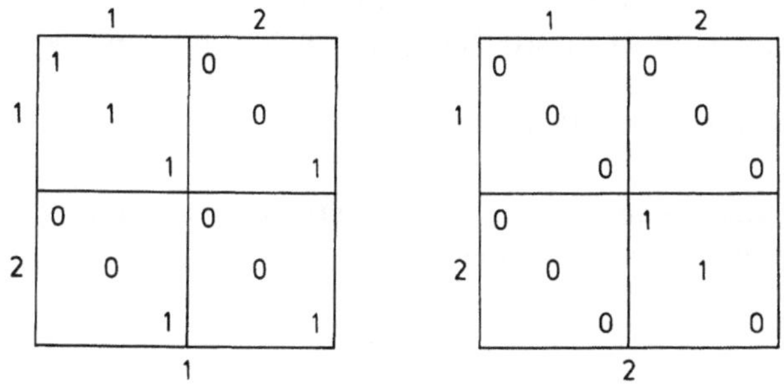

Fig. 2.3.1. Adding strictly dominated strategies may enlarge the set of the proper equilibria

	1	2	3
1	2, 2	1, 1	1, 1
2	2, 2	0, 1	3, 1
3	0, 2	0, 1	0, 0

Fig. 2.3.2. A weakly proper equilibrium need not be proper

drawback, we introduce the concept of *weakly proper equilibrium*, which requires only that a considerably more costly mistake should be chosen with a probability which is of smaller order.[5]

Definition 2.3.4. Let $\Gamma = (\Phi_1, \ldots, \Phi_n, R_1, \ldots, R_n)$ be an n-person normal form game and let $s \in S$. We say that s is a *weakly proper equilibrium* of Γ, if there exists a sequence $\{s(\varepsilon)\}_{\varepsilon \downarrow 0}$ of completely mixed strategy combinations with limit s, such that s is a best reply against every element in this sequence and such that

$$\text{if } R_i(s\backslash k) < R_i(s\backslash l), \text{ then } s_i^k(\varepsilon) \leqq \varepsilon s_i^l(\varepsilon) \text{ for all } i, k, l, \varepsilon. \tag{2.3.2}$$

From the characterization of perfect equilibria given in Theorem 2.2.5(iii), it is clear that every weakly proper equilibrium is perfect. The weakly properness concept is a strict refinement of the perfectness concept since the perfect equilibrium (R_1, R_2) of the game of Fig. 1.5.3. is not weakly proper. Furthermore, it is clear from Lemma 2.3.2 that every proper equilibrium is weakly proper. By means of the game of Fig. 2.3.2, we show that a weakly proper equilibrium is not necessarily proper. The unique proper equilibrium of this game is (1, 1) since properness requires that player 2 chooses his third strategy with an order smaller probability than his second one. According to the weakly properness concept, player 2 does not have to choose his third strategy with a much smaller probability than his second one since the third strategy is only a little bit worse (player 1 chooses his third strategy only with small probability). Consequently also the equilibrium (2, 1) is weakly proper.

From the above discussion it follows that we have:

Theorem 2.3.5. *Every proper equilibrium is weakly proper and every weakly proper equilibrium is perfect. The converses of both statements are false.*

Furthermore, we have:

Theorem 2.3.6. *Every strictly perfect equilibrium is weakly proper.*

Proof. Assume s is a strictly perfect equilibrium of a normal form game $\Gamma = (\Phi_1, \ldots, \Phi_n, R_1, \ldots, R_n)$. For $\varepsilon > 0$, define $\eta(\varepsilon)$ by

$$v_i(k) := |\{l \in \Phi_i;\ R_i(s \backslash k) \leqq R_i(s \backslash l)\}|$$
$$\eta_i^k(\varepsilon) := \varepsilon^{v_i(k)} \qquad \text{for } i \in N,\ k \in \Phi_i.$$

If ε is small, $\eta(\varepsilon)$ is close to zero, which implies that $(\Gamma, \eta(\varepsilon))$ has an equilibrium $s(\varepsilon)$ which is close to s. It follows from (2.2.3) that $s(\varepsilon)$ satisfies (2.3.2) and that s is a best reply against $s(\varepsilon)$ if ε is sufficiently small. Hence, s is a weakly proper equilibrium. □

The reader might conjecture that every strictly perfect equilibrium is even proper. In trying to prove this conjecture, the author has run into difficulties, caused by the fact that the correspondence $\eta \rightarrow E(\Gamma, \eta)$ may (possibly) be ill-behaved in the neighborhood of $\eta = 0$. To circumvent these difficulties, we will introduce a refinement of the strictly perfectness concept, *the strict properness concept* and we will prove that every strictly proper equilibrium is proper.[6]

Definition 2.3.7. Let $\Gamma = (\Phi_1, \ldots, \Phi_n, R_1, \ldots, R_n)$ be a normal form game. For $\hat{\eta} \in \mathbb{R}^m_{++}$, let $U_{\hat{\eta}}$ be as in Def. 2.2.7. s is a *strictly proper equilibrium* of Γ if there exists some $\hat{\eta} \in \mathbb{R}^m_{++}$ and a *continuous* map $\eta \rightarrow s(\eta)$ from $U_{\hat{\eta}}$ to S such that $s(\eta) \in E(\Gamma, \eta)$ for all η and $\lim_{\eta \downarrow 0} s(\eta) = s$.

Clearly, every strictly proper equilibrium is strictly perfect, but the author does not know whether the converse is also true. We will now show that the strict properness concept is indeed a refinement of the properness concept:

Theorem 2.3.8. *Every strictly proper equilibrium is proper.*

Proof. Let $\Gamma = (\Phi_1, \ldots, \Phi_n, R_1, \ldots, R_n)$ be an n-person normal form game and assume s is a strictly proper equilibrium of Γ. Let $\hat{\eta}$ and $s(\eta)$ (for $\eta \in U_{\hat{\eta}}$) be as in Definition 2.3.7. Let ε and V be given by:

$$0 < \varepsilon < \min_{i,\,k} \hat{\eta}_i^k,$$

$$V = \{\eta \in U_{\hat{\eta}};\ \varepsilon^m \leqq \eta_i^k \leqq \varepsilon \text{ for all } i, k\},$$

and define the correspondence F from S to V by:

$$F(s) = \{\eta \in V;\ \text{if } R_i(s \backslash k) < R_i(s \backslash l), \text{ then } \eta_i^k \leqq \varepsilon \eta_i^l \text{ for all } i, k, l\}.$$

As in the proof of Theorem 2.3.2, one can show that $F(s)$ is a nonempty, compact and convex set for every $s \in S$, and that F is upper semi-continuous. Define the correspondence G from V to V by $G(\eta) = F(s(\eta))$, where $s(\eta) \in E(\Gamma, \eta)$ is as in Def. 2.3.7. Then G satisfies the conditions of the Kakutani Fixed Point Theorem (Kakutani [1941]) and, therefore, G possesses a fixed point, Let $\eta(\varepsilon)$ be such a fixed point and write $s(\varepsilon)$ for $s(\eta(\varepsilon))$. We have

$$\text{if } R_i(s(\varepsilon) \backslash k) < R_i(s(\varepsilon) \backslash l), \text{ then } \eta_i^k(\varepsilon) \leqq \varepsilon \eta_i^l(\varepsilon) \text{ for all } i, k, l. \tag{2.3.3}$$

Since $s(\varepsilon)$ is an equilibrium of $(\Gamma, \eta(\varepsilon))$ we, moreover, have

$$s_i^l(\varepsilon) \geqq \eta_i^l(\varepsilon) \quad \text{for all } i, l, \text{ and} \tag{2.3.4}$$

$$\text{if } R_i(s(\varepsilon)\backslash k) < R_i(s(\varepsilon)\backslash l), \text{ then } s_i^k(\varepsilon) = \eta_i^k(\varepsilon) \quad \text{for all } i, k, l. \tag{2.3.5}$$

Hence, by combining Eqs. (2.3.3) – (2.3.5), we see that $s(\varepsilon)$ is an ε-proper equilibrium of Γ. Since ε can be chosen arbitrarily small and since $s(\varepsilon)$ is close to s if ε is close to zero, we have that s is a proper equilibrium of Γ. □

To conclude this section, we briefly consider the concept of *persistent equilibria* that has been introduced in Kalai and Samet [1984]. As we have already seen in Chap. 1, this concept has been introduced in order to eliminate "nonminimal" completely mixed equilibria (which by definition are strictly perfect).

Let $\Gamma = (\Phi_1, \ldots, \Phi_n, R_1, \ldots, R_n)$ be an n-person game. For $i \in N$, let $T_i \subset S_i$ be a nonempty compact and convex set of mixed strategies of player i. Then $T = \prod_{i=1}^{n} T_i$ is called a *retract*. A retract T is said to absorb a set of strategies T' if for every $s \in T'$ there exists a best reply against s that belongs to T. T is called a *Nash retract* if T absorbs itself; it is called an *essential retract* if it absorbs a neighborhood of T; and it is called a *persistent retract* if it is a minimal essential retract, i.e. if it does not properly contain an essential retract. (Minimality is needed since otherwise the set of all strategies would trivially qualify as a solution.) It is easy to see that every Nash retract contains a Nash equilibrium. An equilibrium that belongs to a persistent retract is said to be a *persistent equilibrium*. Clearly, every strict equilibrium is persistent.

The following theorem summarizes the main properties of persistent equilibria

Theorem 2.3.9 (Kalai and Samet [1984]).

(i) *Every game has a persistent retract (hence, every game has a persistent equilibrium).*
(ii) *Every persistent retract contains a proper equilibrium.*

The above theorem shows that every game has an equilibrium that is both proper and persistent. However, one cannot conclude that every proper equilibrium is persistent (cf. Fig. 1.5.6). Neither can one conclude that every persistent equilibrium is proper, or even perfect. The latter is demonstrated by the game of Fig. 3.4.1. In Sect. 3.4, it is shown that the strategy combination $(1, 1, 1)$, i.e. all players choose their first strategy, is the unique proper equilibrium of that game. Hence, if T is a persistent retract, then $(1, 1, 1) \in T$ (Theorem 2.3.9). Next, consider the strategy combination $s^i(\varepsilon)$ in which player i chooses his second strategy with probability ε and in which both other players choose 1. The unique best reply against $s^i(\varepsilon)$ has player $i+1 \,(\text{mod} 3)$ playing his second strategy with probability 1, which implies (by definition of an essential retract) that any essential retract contains both pure strategies of all players. Hence, convexity implies that the unique persistent retract is the space of all strategy combinations, so that, in particular, the imperfect equilibrium $(2, 2, 2)$ is persistent.

In the remainder of the monograph, persistent equilibria will not be considered any more. This is not because this is not an interesting concept; At this point in time just too few results are available. One interesting application should be noted, however: In repeated unanimity games, subgame consistent persistent equilibria lead to Pareto optimal outcomes (Kalai and Samet [1985]).

2.4 Essential Equilibria

In the previous sections of this chapter, we considered refinements of the Nash equilibrium concept which are based on the idea that a reasonable equilibrium should be stable against slight perturbations in the equilibrium strategies. In this section, we consider two refinements that are based on the idea that a reasonable equilibrium should be stable against perturbations in the payoffs of the game, viz. the essential equilibrium concept (Wu Wen-Tsün and Jiang Jia-He [1962]) and the concept of strongly stable equilibria (Kojima, Okada and Shindoh [1985]). The main result proved in this section is that every essential equilibrium is strictly perfect.

Definition 2.4.1. Let Γ be an n-person normal form game. An equilibrium s of Γ is an *essential equilibrium* of Γ if for every $\varepsilon > 0$ there exists some $\delta > 0$ such that for every game Γ' with $\varrho(\Gamma, \Gamma') < \delta$ there exists some $s' \in E(\Gamma')$ with $\varrho(s, s') < \varepsilon$.

In Chap. 1, we have already seen that there exist games without essential equilibria (the game of Fig. 1.5.5 is such a game). Wu Wen-Tsün and Jiang Jia-He, however, showed that such games are more or less exceptional. They proved:

Theorem 2.4.2. (Wu Wen-Tsün and Jiang Jia-He [1962], Theorems A and B).

(i) *For any n-tuple $\Phi_1, \ldots, \Phi_n$ of finite sets, the set of games in $\mathscr{G}(\Phi_1, \ldots, \Phi_n)$ for which all equilibria are essential is open and dense in $\mathscr{G}(\Phi_1, \ldots, \Phi_n)$.*

(ii) *If a game has finitely many equilibria, then it has at least one essential equilibrium.*

The proof of Theorem 2.4.2, given by Wu Wen-Tsün and Jiang Jia-He, is based on the theory of essential fixed points for continuous mappings (cf. Fort [1950]). For the special case of a bimatrix game a different proof of Theorem 2.4.2, only using game theoretic arguments, was given in Jansen [1981b]. Jansen's proof of part (ii) of the theorem can easily be generalized to the n-person case, but his proof of the first part essentially uses the 2-person character of the game. In Sect. 6 of this chapter, we will prove a slightly stronger assertion than the one of Theorem 2.4.2(i), by using properties of regular equilibria.

In the remainder of this section it is proved that stability against perturbations in payoffs implies stability against perturbations in strategies:

Theorem 2.4.3. *Every essential equilibrium is strictly perfect.*

Proof. Let $\Gamma = (\Phi_1, \ldots, \Phi_n, R_1, \ldots, R_n)$ be an n-person normal form game and assume $\tilde{s}$ is an essential equilibrium of Γ. Let $\eta = (\eta_1, \ldots, \eta_n)$ be such that (2.2.1) is satisfied. We will construct a normal form game Γ^η whose equilibria induce equilibria in (Γ, η). If η is small, Γ^η will be close to Γ and therefore Γ^η has an equilibrium close to $\tilde{s}$. In this case, the equilibrium induced in (Γ, η) will also be close to $\tilde{s}$ and this will establish the proof.

It will be convenient to denote a generic element of Φ_i by φ_i rather than by k. If $s_i \in S_i$, then we write $s_i(\varphi_i)$ for the probability which s_i assigns to φ_i. We also write $\eta_i(\varphi_i)$ for the minimum probability of φ_i in the perturbed game (Γ, η). For $i \in N$ and $s_i \in S_i$, define $\lambda_i \in (0, 1)$ and $s_i * \eta_i$ by

$$\lambda_i := \sum_{\varphi_i} \eta_i(\varphi_i) \quad \text{and} \quad s_i * \eta_i := (1 - \lambda_i) s_i + \eta_i .$$

Notice that $s_i * \eta_i$ can be viewed as a mixture of mixed strategies: s_i is chosen with probability $1 - \lambda_i$ and η_i / λ_i is chosen with probability λ_i. The interpretation of $s_i * \eta_i$ is: if the mistake probabilities of player i are determined by η_i, then this player will actually play $s_i * \eta_i$ if he intends to play s_i. For $s \in S$, let $s * \eta$ be defined by:

$$(s * \eta)_i = s_i * \eta_i \quad \text{for } i \in N ,$$

and let the game $\Gamma^\eta = (\Phi_1, \ldots, \Phi_n, R_1^\eta, \ldots, R_n^\eta)$ be defined by

$$R_i^\eta(\varphi) = R_i(\varphi * \eta) \quad \text{for } i \in N, \ \varphi \in \Phi . \tag{2.4.1}$$

Hence, $R_i^\eta(\varphi)$ is the expected payoff to player i if the players intend to play φ and make mistakes according to η. We claim that

$$R_i^\eta(s) = R_i(s * \eta) \quad \text{for all } i \in N \text{ and } s \in S . \tag{2.4.2}$$

Namely, for $i \in N$ and $s \in S$, we have

$$\begin{aligned} R_i^\eta(s) &= \sum_{\varphi} s(\varphi) R_i(\varphi * \eta) = \sum_{\varphi} s(\varphi) \sum_{\varphi'} (\varphi * \eta)(\varphi') R_i(\varphi') \\ &= \sum_{\varphi'} \left(\sum_{\varphi} s(\varphi)(\varphi * \eta)(\varphi') \right) R_i(\varphi') \overset{(*)}{=} \sum_{\varphi'} (s * \eta)(\varphi') R_i(\varphi') \\ &= R_i(s * \eta) , \end{aligned}$$

where the equality marked with $(*)$ follows from the fact that for all $i \in N$:

$$(s_i * \eta_i)(\varphi_i') = \sum_{\varphi_i} s_i(\varphi_i)(\varphi_i * \eta_i)(\varphi_i') \quad \text{for all } s_i \in S_i, \ \varphi_i' \in \Phi_i ,$$

and the fact that the players choose their strategies (and make their mistakes) independently. As a consequence of (2.4.2), we have that for all $i \in N$, $s \in S$, $\varphi_i \in \Phi_i$:

$$R_i^\eta(s \backslash \varphi_i') = (1 - \lambda_i) R_i(s * \eta \backslash \varphi_i') + \sum_{\varphi_i} \eta_i(\varphi_i) R_i(s * \eta \backslash \varphi_i)$$

which implies

$$R_i^\eta(s \backslash \varphi_i) < R_i^\eta(s \backslash \varphi_i') \quad \text{iff} \quad R_i(s * \eta \backslash \varphi_i) < R_i(s * \eta \backslash \varphi_i') \quad \text{for all } i, s, \varphi_i, \varphi_i' . \tag{2.4.3}$$

Next, let η be close to zero. For $\varphi\in\Phi$, we have that $\varphi*\eta$ is close to φ. This implies, since R_i is continuous, that $R_i^\eta(\varphi)$ is close to $R_i(\varphi)$. Hence, if η is close to zero, Γ^η is close to Γ. Therefore, in this case, Γ^η has an equilibrium $s^\eta=(s_1^\eta,\ldots,s_n^\eta)$ which is close to the essential equilibrium $\tilde{s}$ of Γ. We have

$$\text{if } R_i^\eta(s^\eta\backslash\varphi_i)<R_i^\eta(s^\eta\backslash\varphi_i'),\text{ then } s_i^\eta(\varphi_i)=0 \text{ for all } i\in N,\ \varphi_i,\ \varphi_i'\in\Phi_i,$$

from which it follows, by using (2.4.3), that for all $i\in N$ and φ_i, $\varphi_i'\in\Phi_i$,

$$\text{if } R_i(s^\eta*\eta\backslash\varphi_i)<R_i(s^\eta*\eta\backslash\varphi_i'),\text{ then } (s^\eta*\eta)_i(\varphi_i)=\eta_i(\varphi_i)\,.$$

This implies that $s^\eta*\eta$ is an equilibrium of (Γ,η). Since $s^\eta*\eta$ converges to $\tilde{s}$ as η tends to zero, we have that $\tilde{s}$ is a strictly perfect equilibrium of Γ. □

Next, we introduce a refinement of the essential equilibrium concept, viz. the concept of strongly stable equilibria introduced in Kojima, Okada and Shindoh [1985]. Loosely speaking it requires that an equilibrium changes uniquely and continuously against perturbations in payoffs.

Recall that, via (2.1.12) – (2.1.13), we identify a game $\Gamma\in\mathscr{G}(\Phi_1,\ldots,\Phi_n)$ with a vector $r=(r_1,\ldots,r_n)\in\mathbb{R}^{nm^*}$, where m^* is given by (2.1.1), so that we view $\mathscr{G}(\Phi_1,\ldots,\Phi_n)$ as an nm^*-dimensional Euclidean space.

Definition 2.4.4. Let $\tilde{\Gamma}=(\Phi_1,\ldots,\Phi_n,\tilde{R}_1,\ldots,\tilde{R}_n)$ be an n-person normal form game. An equilibrium $\tilde{s}$ is a strongly stable equilibrium of $\tilde{\Gamma}$ if there exist neighborhoods U of $\tilde{\Gamma}$ in $\mathbb{R}^{nm^*}$ and V of $\tilde{s}$ in $\mathbb{R}^m$, such that

(i) $|E(\Gamma)\cap V|=1$, for all $\Gamma\in U$, and
(ii) the mapping s: $U\to V$ defined by $\{s(\Gamma)\}=E(\Gamma)\cap V$ is continuous.

As an immediate consequence of Def. 2.4.4 we have

Corollary 2.4.5. *Every strongly stable equilibrium is essential.*

Kojima, Okada and Shindoh have given an example of an equilibrium that is essential though not strongly stable, hence, strong stability is a more stringent requirement than essentiality.

Let us call an equilibrium s of a game Γ *isolated* if there exists a neighborhood V of s such that $V\cap E(\Gamma)=\{s\}$. As a consequence of Def. 2.4.4, we have

Corollary 2.4.6. *Every strongly stable equilibrium is isolated*

Finally, by mimicking the proof of Theorem 2.4.3, we can show:

Theorem 2.4.7. *Every strongly stable equilibrium is strictly proper.*

2.5 Regular Equilibria

In this section we introduce the most stringent refinement of the Nash equilibrium concept which will be considered: the concept of *regular equilibria.* This concept has been introduced in Harsanyi [1973b]. It is shown that regular equilibria are strongly stable. A regular equilibrium need not be persistent, however, (cf. Fig. 1.5.6).

It should be noted that also in Jansen [1981b] a concept of regular equilibria has been introduced. Jansen's results, however, should not be confused with ours: what he calls a regular equilibrium is what we have called a quasi-strict equilibrium, and although every regular equilibrium is quasi-strict (Corollary 2.5.3), it is definitely not truc that every quasi-strict equilibrium is regular (as follows e.g. from the game of Fig. 2.1.1).

Before defining the concept of regular equilibria, let us introduce some notational conventions. Let $\Gamma = (\Phi_1, \ldots, \Phi_n, R_1, \ldots, R_n)$ be an n-person normal form game. For $i \in N$, we write X_i for $\mathscr{F}(\Phi_i, \mathbb{R})$, the set of all mappings from Φ_i to $\mathbb{R}$. We have $S_i \subset X_i$. A generic element of X_i is denoted by x_i and x_i^k denotes the value of x_i at k. We identify X_i with $\mathbb{R}^{m_i}$, where m_i is given by (2.1.1). X denotes $\prod_{i=1}^{n} X_i$ and a generic element of X is denoted by x. The set X can be identified with $\mathbb{R}^m$, where m is given by (2.1.1). If $x \in X$ and $x_i' \in X_i$, then $x \backslash x_i' := (x_1, \ldots, x_{i-1}, x_i', x_{i+1}, \ldots, x_n)$. By means of the Eqs. (2.1.5) and (2.1.7), we have extended R_i from Φ to S. We extend R_i from S to X by defining

$$R_i(x) := \sum_{\varphi \in \Phi} x(\varphi) R_i(\varphi); \quad x(\varphi) := \prod_{j=1}^{n} x_j^{k_j} \quad \text{if } \varphi = (k_1, \ldots, k_n). \tag{2.5.1}$$

In the next two sections it will be convenient to think of R_i as a mapping which is defined on $\mathbb{R}^m$. Notice that R_i is a polynomial and, hence, is infinitely often differentiable.

From (2.1.11) we see that a strategy combination $s \in S$ is an equilibrium of Γ if and only if the following condition is fulfilled:

$$\text{if } s_i^k > 0, \text{ then } R_i(s \backslash k) = \max_{l \in \Phi_i} R_i(s \backslash l) \text{ for all } i \in N,\ k \in \Phi_i.$$

Hence, $s \in \mathbb{R}_+^m$ is an equilibrium of Γ if and only if s is a solution to the following set of equations:

$$x_i^k \left[R_i(x \backslash k) - \max_{l \in \Phi_i} R_i(x \backslash l) \right] = 0 \quad \text{for all } i \in N,\ k \in \Phi_i, \text{ and} \tag{2.5.2}$$

$$\sum_k x_i^k - 1 = 0 \quad \text{for all } i \in N. \tag{2.5.3}$$

In (2.5.2) and (2.5.3) we have $m+n$ equations in m unknowns, but, since for each i at least one equation in (2.5.2) is trivially fulfilled, we actually have m equations in m unknowns. However, the system possesses the undesirable property that the mapping defined by the left hand side of (2.5.2) – (2.5.3) is not differentiable.

Therefore, we will consider a slightly different system of equations.[7] Namely, let $\varphi=(k_1,\ldots,k_n)\in\Phi$ be fixed, and consider the system:

$$x_i^k[R_i(x\backslash k)-R_i(x\backslash k_i)]=0 \quad \text{for all } i\in N,\ k\in\Phi_i,\ k\neq k_i, \tag{2.5.4}$$

$$\sum_k x_i^k-1=0 \quad \text{for all } i\in N. \tag{2.5.5}$$

If s is an equilibrium of Γ with $\varphi\in C(s)$, then s is a solution to the system (2.5.4) – (2.5.5) and, furthermore, the mapping defined by the left hand side of these equations is infinitely often differentiable. The price we have to pay for gaining this differentiability is that not every nonnegative solution of this system is an equilibrium and that not every equilibrium of Γ has to be a solution of the system. Nevertheless, it turns out to be more convenient to work with the system (2.5.4) – (2.5.5) than to work with the system (2.5.2) – (2.5.3). Let $F(\cdot|\varphi)$ be the mapping defined by the left hand side of (2.5.4) – (2.5.5). More precisely:

$$F_i^k(x|\varphi)=x_i^k[R_i(x\backslash k)-R_i(x\backslash k_i)] \quad \text{for } i\in N,\ k\in\Phi_i,\ k\neq k_i, \tag{2.5.6}$$

$$F_i^{k_i}(x|\varphi)=\sum_k x_i^k-1 \quad \text{for } i\in N. \tag{2.5.7}$$

Let $J(s|\varphi)$ be the Jacobian (i.e. the matrix of first order partial derivatives) of $F(\cdot|\varphi)$ evaluated at s, i.e.

$$J(s|\varphi)=\frac{\partial F(x|\varphi)}{\partial x}\Big|_{x=s}. \tag{2.5.8}$$

If s is an equilibrium of Γ with $\varphi\in C(s)$, then $F(s|\varphi)=0$. One can expect that an equilibrium s with $\varphi\in C(s)$ will have nice properties if $F(.|\varphi)$ is locally invertible at s, i.e. if $J(s|\varphi)$ is nonsingular. Therefore, we define:

Definition 2.5.1. An equilibrium s of Γ is a *regular equilibrium* if $J(s|\varphi)$ is nonsingular for some $\varphi\in C(s)$. An equilibrium is *irregular* if it is not regular.

By considering the game of Fig. 1.5.5, it can easily be verified that there exist games without regular equilibria (this also follows from Corollary 2.5.6). However, in Harsanyi [1973b], it has been proved that normal form games without regular equilibria are exceptional (also cf. Theorem 2.6.1). It should be noted that Def. 2.5.1 is slightly different from the definition of regular equilibria given by Harsanyi. Harsanyi defines a regular equilibrium as an equilibrium s for which the Jacobian $J(s|\hat{\varphi})$ is nonsingular, where $\hat{\varphi}$ is the strategy combination $(1,\ldots,1)$. To illustrate the difference between these definitions, consider the game of Fig. 2.5.1.

This game has two equilibria, viz. $\hat{\varphi}=(1,1)$ and $\tilde{\varphi}=(2,2)$. The equilibrium $\hat{\varphi}$ is not a reasonable one, but the equilibrium $\tilde{\varphi}$ possesses all nice properties one possibly could ask for. Therefore, the definition of a regular equilibrium should be such that $\hat{\varphi}$ is irregular and $\tilde{\varphi}$ is regular. Our definition of regular equilibria satisfies this condition. However, according to Harsanyi's definition both equilibria are irregular. This is the reason why we have modified Harsanyi's definition as in Def. 2.5.1. Actually, in the proofs in Harsanyi [1973b] it is

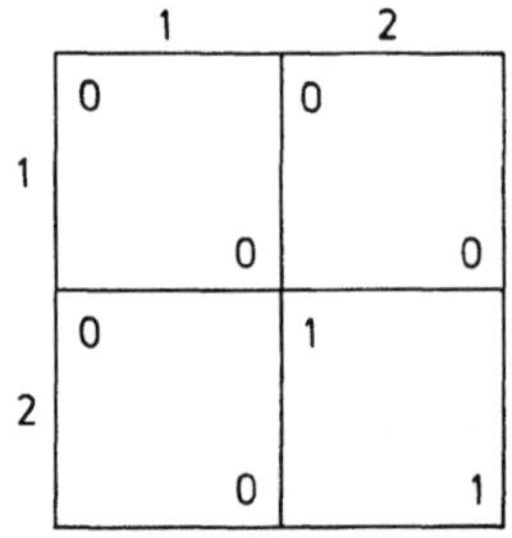

Fig. 2.5.1. An example to illustrate the role of the reference point φ in the definition of a regular equilibrium

implicitly assumed that every equilibrium s satisfies $\hat{\varphi} \in C(s)$ and so in his proofs Harsanyi actually works with our definition of regularity. Therefore, the results of Harsanyi [1973b] are correct if regularity is defined as in Def. 2.5.1.

Next, we will show that, if s is a regular equilibrium of Γ, then $J(s|\varphi)$ is nonsingular for all $\varphi \in C(s)$. First we prove:

Lemma 2.5.2. *Let $\Gamma = (\Phi_1, \ldots, \Phi_n, R_1, \ldots, R_n)$ be an n-person normal form game, let $s \in E(\Gamma)$ and let $\varphi = (k_1, \ldots, k_n) \in C(s)$. For $x \in \mathbb{R}^m$, let $F(x|\varphi)$ and $J(x|\varphi)$ be defined as before and let $\tilde{J}(x|\varphi)$ be the Jacobian which results from $J(s|\varphi)$ by crossing out the rows and columns corresponding to the pure strategies which do not belong to $C(s)$. Then the determinants of these Jacobians are related according to*

$$|J(s|\varphi)| = |\tilde{J}(s|\varphi)| \prod_{i \in N} \prod_{k \notin C(s_i)} [R_i(s \pm k) - R_i(s \pm k_i)] . \tag{2.5.9}$$

Proof. For $i \in N$ and $k \in \Phi_i \backslash C(s_i)$, we have

$$\frac{\partial F_i^k(x|\varphi)}{\partial x_j^l}\Big|_s = 0 \quad \text{for all } j \in N,\ l \in \Phi_j,\ l \neq k ,$$

$$\frac{\partial F_i^k(x|\varphi)}{\partial x_i^k}\Big|_s = R_i(s \backslash k) - R_i(s \backslash k_i) ,$$

which immediately implies (2.5.9). □

If s is a strict equilibrium of Γ, then the right hand side of (2.5.9) is nonzero and, hence, a strict equilibrium is regular. On the other hand, if an equilibrium is not quasi-strict, then the right hand side is zero, and so we have:

Corollary 2.5.3. *Every strict equilibrium is regular; every regular equilibrium is quasi-strict.*

Lemma 2.5.4. *Let $\Gamma = (\Phi_1, \ldots, \Phi_n, R_1, \ldots, R_n)$ be an n-person normal form game, let $s \in E(\Gamma)$ and let $\varphi = (k_1, \ldots, k_n)$ and $\varphi' = (l_1, \ldots, l_n)$ be in the carrier of s. Then $J(s|\varphi)$ is nonsingular if and only if $J(s|\varphi')$ is nonsingular.*

Proof. It follows from Lemma 2.5.2 that it suffices to show that $|\tilde{J}(s|\varphi)|=0$ if and only if $|\tilde{J}(s|\varphi')|=0$. By writing out the formulae, the reader can verify that for any $i \in N$:

$$\frac{\partial F_i^{k_i}(x|\varphi)}{\partial x_j^\alpha}\Big|_s = \frac{\partial F_i^{l_i}(x|\varphi')}{\partial x_j^\alpha}\Big|_s \qquad \text{for all } j \in N,\ \alpha \in C(s_j)\ ,$$

$$\frac{\partial F_i^k(x|\varphi)}{\partial x_i^\alpha}\Big|_s = \frac{\partial F_i^l(x|\varphi')}{\partial x_i^\alpha}\Big|_s \qquad \text{for all } k,\ l,\ \alpha \in C(s_i),\ k \neq k_i,\ l \neq l_i\ ,$$

$$\frac{\partial F_i^k(x|\varphi)}{\partial x_j^\alpha}\Big|_s = \frac{\partial F_i^k(x|\varphi')}{\partial x_j^\alpha}\Big|_s - \frac{s_i^k}{s_i^{k_i}}\frac{\partial F_i^{k_i}(x|\varphi')}{\partial x_j^\alpha}\Big|_s$$

$$\text{for all } k \in C(s_i),\ j \in N,\ \alpha \in C(s_j),\ k \neq k_i,\ j \neq i.$$

Therefore, $\tilde{J}(s|\varphi)$ results from $\tilde{J}(s|\varphi')$ by elementary algebraic operations which preserve the value of the determinant. □

We will now prove the main result of this section, viz. that every regular equilibrium is strongly stable.

Theorem 2.5.5. *Every regular equilibrium is strongly stable.*

Proof. Assume $\tilde{s}$ is a regular equilibrium of the n-person game $\tilde{\Gamma}=(\Phi_1,\ldots,\Phi_n, \tilde{R}_1,\ldots,\tilde{R}_n)$. Write $\tilde{r}$ for the payoff vector of $\tilde{\Gamma}$, i.e. $\tilde{r} \in \mathbb{R}^{nm^*}$ is as in (2.1.14) and let m be as in (2.1.1). Let $\varphi=(k_1,\ldots,k_n) \in C(\tilde{s})$ be fixed and define the mapping $F: \mathbb{R}^{nm^*} \times \mathbb{R}^m \to \mathbb{R}^m$ by:

$$F_i^k(r,x) = x_i^k[R_i(x\backslash k) - R_i(x\backslash k_i)] \quad \text{for } i \in N,\ k \in \Phi_i,\ k \neq k_i, \text{ and} \tag{2.5.10}$$

$$F_i^{k_i}(r,x) = \sum_k x_i^k - 1 \quad \text{for } i \in N. \tag{2.5.11}$$

(R_i denotes player i's payoff function in the game $\Gamma(r)$ associated with $r \in \mathbb{R}^{nm^*}$).

Since $\tilde{s}$ is an equilibrium of $\tilde{\Gamma}$ and $\varphi \in C(\tilde{s})$, we have $F(\tilde{r},\tilde{s})=0$. Furthermore, since $\tilde{s}$ is a regular equilibrium of $\tilde{\Gamma}$

$$J(\tilde{r},\tilde{s}) = \frac{\partial F(r,x)}{\partial x}\Big|_{(\tilde{r},\tilde{s})} \text{ is nonsingular} \tag{2.5.12}$$

By the implicit function theorem (Dieudonné [1960], p. 268) there exist neighborhoods U of $\tilde{r}$ and V of $\tilde{s}$ and a differentiable mapping $s: U \to V$ such that

$$\{(r,x) \in U \times V;\ F(r,x)=0\} = \{(r,s(r));\ r \in U\}\ . \tag{2.5.13}$$

By choosing U and V sufficiently small, it is guaranteed that for all $i \in N$ and $k \in \Phi_i$:

$$\text{if } \tilde{s}_i^k > 0, \text{ then } x_i^k > 0 \text{ for all } x \in V\ , \tag{2.5.14}$$

$$\text{if } \tilde{R}_i(\tilde{s}\backslash k) < \tilde{R}_i(\tilde{s}\backslash k_i), \text{ then } R_i(x\backslash k) < R_i(x\backslash k_i) \text{ for all } (r,x) \in U \times V\ . \tag{2.5.15}$$

From (2.5.13) – (2.5.15) and the fact that $\tilde{s}$ is a quasi-strict equilibrium of $\tilde{\Gamma}$ (Corollary 2.5.4), we can conclude that for all $i \in N$ and $r \in U$:

$$
\begin{array}{ll}
s_i^k(r) > 0 & \text{for all } k \in C(\tilde{s}_i), \\
R_i(s(r)\backslash k) = R_i(s(r)\backslash k_i) & \text{for all } k \in C(\tilde{s}_i), \\
R_i(s(r)\backslash k) < R_i(s(r)\backslash k_i) & \text{for all } k \notin C(\tilde{s}_i), \text{ and} \\
s_i^k(r) = 0 & \text{for all } k \notin C(\tilde{s}_i).
\end{array}
$$

Therefore, we have that $s(r) \in E(\Gamma(r))$, for all $r \in U$. Since every equilibrium s of $\Gamma(r)$ which is in V has to satisfy $F(s, r) = 0$, we have that $s(r)$ is the only equilibrium of $\Gamma(r)$ which is in V. This establishes condition (i) from Def. 2.4.4. Since the mapping $r \to s(r)$ is differentiable, also the second condition is satisfied. □

In Kojima, Okada and Shindoh [1985] a complete characterization of strongly stable equilibria has been derived in terms of the nonsingularity of the Jacobian of a piecewise continuously differentiable mapping (This mapping results from considering the nonlinear complementarity problem associated with the game). From this characterization one can draw the following conclusion.

Theorem 2.5.6 (Kojima, Okada and Shindoh [1985]). *An equilibrium is regular if and only if it is quasi-strict and strongly stable.*

In 2-person games, every strongly stable equilibrium is quasi-strict, hence, in this case regularity and strong stability are equivalent (Theorem 3.4.4), however, if there are at least 3 players, a strongly stable equilibrium need not be quasi-strict. The equilibrium (1, 1, 1) in the game of Fig. 3.4.1 is isolated and essential and, in fact, one can show that this is a strongly stable equilibrium. It is not quasi-strict, however. Hence, in general, regularity is a more stringent requirement than strong stability.

2.6 An "Almost all" Theorem

Intuitively, it will be clear that games with irregular equilibria are exceptional. Namely, the existence of an irregular equilibrium entails a special numerical relationship among the payoffs of the game and this relationship can be disturbed by slightly perturbing the payoffs of the game. In this section, we will prove that indeed for almost all normal form games all equilibria are regular. This result was first proved in Harsanyi [1973b] and the proof we will give essentially follows the same lines as the proof given in that paper. It is an application of Sard's Theorem (Sard [1942] or Milnor [1965]) in the way as initiated in Debreu [1970].

Theorem 2.6.1. *For almost all normal form games all equilibria are regular.*

Proof. Let $\Phi_1, \ldots, \Phi_n$ be an n-tuple of finite, nonempty sets and let $\mathscr{G}(\Phi_1, \ldots, \Phi_n)$ be the set of all games with pure strategy spaces $\Phi_1, \ldots, \Phi_n$. We identify $\mathscr{G}(\Phi_1, \ldots, \Phi_n)$ with $\mathbb{R}^{nm^*}$ via (2.1.13). Furthermore let $I(\Phi_1, \ldots, \Phi_n)$ be the set of all games in this class which have an irregular equilibrium. We will prove that $I(\Phi_1, \ldots, \Phi_n)$ is a closed set with Lebesgue measure zero.

We first prove that $I(\Phi_1, \ldots, \Phi_n)$ is closed. Let $F: \mathbb{R}^{nm^*} \times \mathbb{R}^m \to \mathbb{R}^m$ be the mapping defined by (2.5.10) – (2.5.11). Since this mapping depends on which reference point $\varphi \in \Phi$ is chosen, we will write $F(r, x \mid \varphi)$ for the image of (r, x) under this mapping. Let $J(r, x \mid \varphi)$ be the Jacobian of (2.5.12). If s is an equilibrium of $\Gamma(r)$ with $\varphi \in C(s)$, then $F(r, s \mid \varphi) = 0$ and, in this case, s is an irregular equilibrium of $\Gamma(r)$ if and only if $J(r, s \mid \varphi) = 0$. Let $\{\Gamma(r(t))\}_{t \in \mathbb{N}}$ be a sequence of games in $I(\Phi_1, \ldots, \Phi_n)$, such that $\lim_{t \to \infty} r(t) = r$. For $t \in \mathbb{N}$, let $s(t)$ be an irregular equilibrium of $\Gamma(r(t))$ and, without loss of generality, assume $s = \lim_{t \to \infty} s(t)$. Then s is an equilibrium of $\Gamma(r)$ since the correspondence which assigns to each game its set of equilibria is upper semi-continuous. We claim that s is an irregular equilibrium of $\Gamma(r)$. Namely, let $\varphi \in C(s)$. Then $\varphi \in C(s(t))$ for all t which are sufficiently large and therefore, $J(r(t), s(t) \mid \varphi) = 0$, for all sufficiently large t. Since F is infinitely often differentiable, J is continuous and so $J(r, s \mid \varphi) = 0$, which establishes our claim. Hence $\Gamma(r) \in I(\Phi_1, \ldots, \Phi_n)$ and so $I(\Phi_1, \ldots, \Phi_n)$ is closed.

Next, we will show that $\lambda(I(\Phi_1, \ldots, \Phi_n)) = 0$. For $i \in N$, let $C_i, B_i \subset \Phi_i$, and let $C = \prod_{i=1}^{n} C_i$ and $B = \prod_{i=1}^{n} B_i$. We write $G(C, B)$ (resp. $I(C, B)$) for the set of all games in $\mathscr{G}(\Phi_1, \ldots, \Phi_n)$ which have an equilibrium (resp. irregular equilibrium) s with $C(s) = C$ and $B(s) = B$. We have

$$\mathscr{G}(\Phi_1, \ldots, \Phi_n) = \bigcup_{C,\, B \subset \Phi} G(C, B); \qquad I(\Phi_1, \ldots, \Phi_n) = \bigcup_{C,\, B \subset \Phi} I(C, B),$$

hence, since Φ is finite, it suffices to show that for all $C, B \subset \Phi$:

$$\lambda(I(C, B)) = 0. \tag{2.6.1}$$

Hence, let $C, B \subset \Phi$ be fixed. If $C = \varnothing$ or $C \not\subset B$, then (2.6.1) is trivially fulfilled, so assume $C \neq \varnothing$, $C \subset B$. Let $\hat{\varphi} = (k_1, \ldots, k_n) \in C$ be fixed. For $i \in N$, let $\Phi(i) \subset \Phi$ be defined by

$$\Phi(i) = \{\varphi \in \Phi;\ \varphi = \hat{\varphi} \backslash k \quad \text{for some } k \in B_i \backslash \{k_i\}\}.$$

Assume that we have given r and s such that:

$$\Gamma(r) \in G(C, B), \quad s \in E(\Gamma(r)), \quad C(s) = C \quad \text{and} \quad B(s) = B. \tag{2.6.2}$$

Then, to be able to compute the complete payoff vector r, we actually only have to know the collection of payoffs

$$\{R_i(\varphi);\ \varphi \in \Phi \backslash \Phi(i),\ i \in N\}. \tag{2.6.3}$$

Namely, if $k \in B_i \backslash \{k_i\}$, then the payoff $R_i(\hat{\varphi} \backslash k)$ can be computed from the equation

$$R_i(s \backslash k) = R_i(s \backslash k_i),$$

since this equation [once we know (2.6.3)] only contains $R_i(\phi\backslash k)$ as an unknown variable, and since this variable occurs in this equation with a positive coefficient. Let us denote by H the mapping by means of which the complete payoff vector r can be computed from the data in (2.6.3). To be more precise, let

$$P := \{\varrho = (\varrho_1, \ldots, \varrho_n);\ \varrho_i\colon \Phi\backslash\Phi(i) \to \mathbb{R}\}, \text{ and}$$

$$S(C) := \{s \in S;\ s = (s_1, \ldots, s_n),\ C(s_i) = C_i\},$$

and for $\varrho \in P$, $s \in S(C)$ define $H(\varrho, s)$ as the unique vector $r \in \mathbb{R}^{nm^*}$ which satisfies

$$R_i(\varphi) = \varrho_i(\varphi) \qquad \text{for } i \in N,\ \varphi \in \Phi\backslash\Phi(i), \text{ and} \tag{2.6.4}$$

$$R_i(s\backslash k) = R_i(s\backslash k_i) \quad \text{for } i \in N,\ k \in B_i\backslash\{k_i\}. \tag{2.6.5}$$

Notice that there is indeed a unique vector r satisfying (2.6.4) and (2.6.5), since for fixed $i \in N$, $k \in B_i\backslash\{k_i\}$, Eq. (2.6.5) only contains as an unknown $R_i(\phi\backslash k)$ and this occurs in the equation with the positive coefficient $s_{-i}(\phi_{-i}) := s(\phi)/s_i(\phi_i)$. Furthermore, since $R_i(\phi\backslash k)$ is computed by using multiplications, additions and subtractions and only dividing by $s_{-i}(\phi_{-i})$, we have that H is infinitely often differentiable on $P \times S(C)$.

For $r = (r_1, \ldots, r_n) \in \mathbb{R}^{nm^*}$, let $\varrho(r) \in P$ be the vector $(\varrho_1(r_1), \ldots, \varrho_n(r_n))$ where $\varrho_i(r_i)$ is the restriction of r_i to $\Phi\backslash\Phi(i)$, for $i \in N$. The mapping H is constructed in such a way that $H(\varrho(r), s) = r$, if r and s satisfy (2.6.2). Therefore, we have:

$$G(C, B) \subset \{\Gamma(r);\ \exists_{(\bar\varrho, \bar s) \in P \times S(C)}[H(\bar\varrho, \bar s) = \bar r]\}, \tag{2.6.6}$$

from which we can conclude that (2.6.1) is true in the special case where $C \neq B$. Namely, in this case

$$\dim(P) = nm^* - |B| + n, \quad \dim(S(C)) = |C| - n,$$

and, therefore, $\dim(P \times S(C)) < nm^*$ if $|C| < |B|$. This implies, by means of (2.6.6), that $\lambda(G(C, B)) = 0$ if $C \neq B$ (recall we assumed that $C \subset B$). Hence, (2.6.1) is true if $C \neq B$.

Next, assume $C = B$. In this case, it easily follows from the definition of H that, if r, s are such that (2.6.2) is satisfied, then

$$\frac{\partial H(\varrho, \sigma)}{\partial(\varrho, \sigma)}\Big|_{(\varrho(r),\, s)} \text{ is singular iff } \frac{\partial F(r, x|\phi)}{\partial x}\Big|_{(r,\, s)} \text{ is singular.} \tag{2.6.7}$$

From (2.6.6) and (2.6.7) we see that $I(C, C)$ is a subset of

$$\left\{\Gamma(\bar r);\ \exists_{\bar\varrho, \bar s}\left[H(\bar\varrho, \bar s) = \bar r \text{ and } \frac{\partial H(\varrho, \sigma)}{\partial(\varrho, \sigma)}\Big|_{(\bar\varrho,\, \bar s)} \text{ is singular}\right]\right\}. \tag{2.6.8}$$

Sard's theorem (Milnor [1965], p. 10) assures us that Lebesgue measure of the set in (2.6.8) is zero and, therefore, we have $\lambda(I(C, C)) = 0$. □

If all equilibria of a normal form game Γ are regular, then this game has a finite number of equilibria (Corollary 2.5.7). Harsanyi has sharpened this result and proved that a game for which all equilibria are regular, in fact has an odd number of equilibria (Harsanyi [1973b, Theorem 1]; for alternative proofs that almost all games have an odd number of equilibria, see Rosenmüller [1971] or Wilson [1971]). Therefore, by combining the Theorems 2.3.5, 2.3.7, 2.5.8, 2.6.1 and the Corollaries 2.5.3, 2.5.6, we can draw our main conclusion for normal form games:

Theorem 2.6.2. *Almost all normal form games possess an odd number of equilibria, which are all regular (hence, quasi-strict, strongly stable, essential, isolated, strictly proper, proper and perfect).*

It should be stressed that Theorem 2.6.2 is of limited value for the study of extensive form games as any nontrivial such game gives rise to a nongeneric normal form.

Notes

1. An algorithm to find perfect equilibria in n-person normal form games has been given in Van den Elzen [1991, Chap. 8].
2. Corollary 2.2.6 has been generalized in Okada [1988]. Okada introduces a concept of lexicographic dominance and he shows that a perfect equilibrium is lexicographically undominated and that a lexicographically undominated strategy combination is undominated. Okada also gives examples to illustrate that in n-player games with $n \geq 3$ the reverse implications are not correct.
3. The concepts of perfect and proper equilibria are defined only for finite games. For generalizations of these concepts to infinite games, see Simon [1987] and Simon and Stinchcombe [1990]. For applications see Levin and Harstad [1984] and Chatterjee and Samuelson [1990].
4. A path following algorithm to find a proper equilibrium of an n-person normal form game has been given in Yamamoto [1990].
5. In Blume, Brandenburger and Dekel [1991 a] refinements of Nash equilibrium are rationalized by the assumption that players have lexicographic preferences. Nash, perfect and proper equilibria can be characterized by means of different versions of lexicographic expected utility maximization.
6. Garcia Jurado and Prada Sànchez [1990] introduce a refinement of properness (called strongly proper equilibrium) that requires, in addition to (2.3.1), that strategies that are equally good are chosen with the same probability.
7. It would have been more convenient to define regularity by means of the system that is obtained by replacing (2.5.4) with the equation $x_i^k[R_i(x\backslash k) - R_i(x)] = 0$. This definition is proposed in Ritzberger and Vogelsberger [1989]. The modified definition of regularity has the same properties as the one used in the book, but the smoothness of the modified system allows some new insights into the structure of the Nash equilibrium correspondence. In Jansen [1987] an alternative definition of regularity is given for bimatrix games. Jansen exploits the fact that finding the equilibria of a bimatrix game is equivalent to solving a linear complimentarity problem.

3 Matrix and Bimatrix Games

In this chapter, we study 2-person normal form games, zero-sum games (matrix games) as well as nonzero-sum games (bimatrix games). It is our objective to investigate whether for this special class of games the results of the previous chapter can be refined and specialized.

In Sect. 3.1, some notation and terminology is introduced, something is said about the structure of the set of equilibria of a bimatrix game and some well-known results concerning matrix games are stated.

In Sect. 3.2, perfect equilibria are studied. The main result is that an equilibrium of a bimatrix game is perfect if and only if it is undominated. It follows that one can verify whether an equilibrium of a bimatrix game is perfect by solving a linear programming problem.

Regular equilibria are studied in Sect. 3.3. A first characterization of such equilibria is derived and it is shown that, for a bimatrix game which is nondegenerate in the sense of Lemke and Howson [1964], all equilibria are regular.

In Sect. 3.4, further characterizations of regular equilibria are derived. The main result is that an equilibrium s of a bimatrix game is regular if and only if s is isolated and perfect in the s-restriction of Γ. From this result several other characterizations follow easily.

In the last section of the chapter zero-sum games are studied. Attention is focused on proper equilibria, and it is shown that an equilibrium of a matrix game is proper if and only if it is a pair of best strategies in the sense of Dresher [1961]. Therefore, it is easy to check whether an equilibrium of a matrix game is proper. The results in this chapter are based upon van Damme [1980b, 1981b].

3.1 Preliminaries

A 2-person normal form game $\Gamma = (\Phi_1, \Phi_2, R_1, R_2)$ is also called a *bimatrix game*, for reasons which will be clear. With respect to such a game, we will use the same notation and terminology as the one we used for general n-person normal form games (cf. Sect. 2.1). Hence, S_i denotes the set of all mixed strategies of player i in Γ and $S = S_1 \times S_2$. For $s \in S$, the expected payoff $R_i(s)$ to player i if s is played is defined as in (2.1.7). Furthermore, as in section 2.5, we write X_i for $\mathcal{F}(\Phi_i, \mathbb{R})$ (the set of all functions from Φ_i to $\mathbb{R}$) and $X = X_1 \times X_2$. R_i is extended from S to X as in (2.5.1). Note that R_i is bilinear. Hence, if $x_1 \in X_1$, then R_i and x_1 determine a linear mapping from X_2 to $\mathbb{R}$, which we denote by $R_i(x_1)$. Hence, we have $(R_i(x_1))(x_2)=R_i(x_1,x_2)$ (for $x_2\in X_2$). Similarly, R_i and $x_2\in X_2$ determine a linear mapping $R_i(x_2)$ from X_1 to $\mathbb{R}$. The mappings $R_i(x_1)$ and $R_i(x_2)$ cannot be mixed up, since the index will always point out which mapping is meant. We say that R_i is *row-regular*, if $R_i(x_1)\neq 0$ for all $x_1\neq 0$, R_i is *column-regular*, if $R_i(x_2)\neq 0$ for all $x_2\neq 0$, and R_i is *regular* (or nonsingular) if R_i is both row-regular and column-regular.

Let $s=(s_1,s_2)\in S$. If player 1 knows that player 2 plays s_2, then he will assign a positive probability only to the pure strategies belonging to $B_1(s)$, and a similar remark applies to player 2. Hence, if s is played, the most relevant payoffs are the payoffs $R_i(\varphi)$ with $\varphi\in B(s)$. Therefore, we define the *s-restriction* of Γ as the game $\Gamma^s=(\Phi_1^s,\Phi_2^s,R_1^s,R_2^s)$, where $\Phi_i^s:=B_i(s)$ and R_i^s is the restriction of R_i to $\Phi^s:=\Phi_1^s\times\Phi_2^s$. R_i^s is called the *s-restriction* of the payoff matrix R_i. We write X_i^s for $\mathscr{F}(\Phi_i^s,\mathbb{R})$ and we extend R_i^s bilinearly from Φ^s to $X^s:=X_1^s\times X_2^s$. A generic element of X_i^s is denoted by x_i^s and the *extension* of x_i^s to Φ_i is the element $x_i\in\mathscr{F}(\Phi_i,\mathbb{R})$, defined by

$$x_i(\varphi_i)=\begin{cases} x_i^s(\varphi_i) & \text{if } \varphi_i\in\Phi_i^s,\\ 0 & \text{otherwise}.\end{cases}$$

When no confusion can result, x_i and x_i^s are identified and, consequently, sometimes x_i is used to denote a generic element of X_i^s.

Next, let us briefly say something about the structure of the set of equilibria of a bimatrix game. If s, s' are equilibria of a bimatrix game Γ, then s and s' are said to be *interchangeable* if (s_1,s_2') and (s_1',s_2) are also equilibria of Γ. A set of equilibria is said to be a *maximal Nash subset* (Millham [1974], Heuer and Millham [1976]) or *subsolution* (Nash [1951]) if it is a maximal set with the property that all its elements are interchangeable. Theorem 2.1.1, of which a proof can be found in Jansen [1981a], shows that maximal Nash subsets are important in the study of the structure of the set of equilibria:

Theorem 3.1.1. *Let Γ be a bimatrix game. Then*

(i) *every maximal Nash subset is a closed and convex polyhedral set,*
(ii) *the set of equilibria of Γ is the (not necessarily disjoint) union of all maximal Nash subsets,*
(iii) *there are only finitely many maximal Nash subsets.*

By using Theorem 3.1.1, it is straightforward to prove

Corollary 3.1.2. *An equilibrium s of a bimatrix game is isolated if and only if $\{s\}$ is a maximal Nash subset.*

For general n-person games, also interchangeability of equilibria and maximal Nash subsets can be defined. If $n\geqq 3$, however, there may exist an uncountable number of maximal Nash subsets (Chin, Parthasarathy and Raghavan [1974], p. 3). Furthermore, if $n\geqq 3$ and $\{s\}$ is a maximal Nash subset, then s is not necessarily isolated.

To conclude our preliminary discussion on bimatrix games, let us remark that we can assume, whenever it is convenient to do so, that all payoffs in a bimatrix game are positive. Namely, let $\Gamma = (\Phi_1, \Phi_2, R_1, R_2)$ be an arbitrary bimatrix game and let Γ' result from Γ by adding a fixed amount to all payoffs in Γ, such that all payoffs in Γ' are positive. Then the strategic situation in Γ' is the same as the one in Γ, and so the equilibria of these games are the same and, hence, in order to analyse Γ, we can just as well analyse Γ'.

In the remaining part of this section, we consider *matrix games*, i.e. bimatrix games $(\Phi_1, \Phi_2, R_1, R_2)$ with $R_2 = -R_1$. We denote the matrix game in which the payoff function of player 1 is R by (Φ_1, Φ_2, R). For a matrix game $\Gamma = (\Phi_1, \Phi_2, R)$ all equilibria are interchangeable, i.e. $E(\Gamma)$ is a maximal Nash subset, and all equilibria yield the same payoff to player 1. This payoff is called the *value* of the game Γ and is denoted by $v(\Gamma)$. It is easily seen, that $E(\Gamma) = O_1(\Gamma) \times O_2(\Gamma)$, where

$$O_1(\Gamma) = \{s_1 \in S_1;\ R(s_1, l) \geqq v(\Gamma) \text{ for all } l \in \Phi_2\}, \text{ and} \tag{3.1.1}$$

$$O_2(\Gamma) = \{s_2 \in S_2;\ R(k, s_2) \leqq v(\Gamma) \text{ for all } k \in \Phi_1\}. \tag{3.1.2}$$

$O_i(\Gamma)$ is a closed, convex polyhedral set, the elements of which are called the *optimal strategies* of player i in Γ. It is easily seen that $s_1 \in O_1(\Gamma)$ if and only if

$$\min_{s_2 \in S_2} R(s_1, s_2) = \max_{s'_1 \in S_1} \min_{s_2 \in S_2} R(s'_1, s_2),$$

and so in a matrix game a strategy of player 1 is an equilibrium strategy if and only if it is a so called *maximin strategy*. (As we know from the game of Fig. 1.6.2, this is not true for nonzero-sum games). A similar property holds for the optimal strategies of player 2. From (3.1.1) and (3.1.2) it follows that $O_1(\Gamma)$ and $v(\Gamma)$ can be determined by solving the linear programming problem (3.1.3):

$$\begin{aligned}&\text{maximize } v\\ &\text{subject to } R(s_1, l) \geqq v \quad \text{for all } l \in \Phi_2;\ s_1 \in S_1.\end{aligned} \tag{3.1.3}$$

To be more precise: we have that $s_1 \in O_1(\Gamma)$ and $v = v(\Gamma)$ if and only if (s_1, v) solves (3.1.3), (where "solves" means "is an optimal solution of"). Therefore, $v(\Gamma)$ and $O_1(\Gamma)$ can be determined easily (cf. Balinski [1961]). An important property of matrix games, which first was proved in Gale and Sherman [1950] and, independently, in Bohnenblust, Karlin and Shapley [1950], is that a pure strategy is used by some optimal strategy with a positive probability if and only if this strategy is a best reply against all optimal strategies of the opponent. Hence, for a matrix game Γ:

$$\bigcup_{s_i \in O_i(\Gamma)} C(s_i) = \bigcap_{s_j \in O_j(\Gamma)} B_i(s) \quad i, j \in \{1, 2\}, i \neq j, \tag{3.1.4}$$

or, to put it differently

$$\bigcup_{s \in E(\Gamma)} C(s) = \bigcap_{s \in E(\Gamma)} B(s). \tag{3.1.5}$$

We denote the set of pure strategies which are in the Carrier of some equilibrium strategy of player i in Γ by $C_i(\Gamma)$ (hence $C_i(\Gamma)$ is the set of (3.1.4)) and $C(\Gamma)$ denotes the set $C_1(\Gamma) \times C_2(\Gamma)$.

Finally we remark that, just as we can assume that in a bimatrix game all payoffs are positive, we can assume, whenever it is convenient to do so, that all payoffs to player 1 in a matrix game are positive (and, hence, that all payoffs to player 2 are negative).

3.2 Perfect Equilibria

In Chap. 2 (Corollary 2.2.6), we have seen that a perfect equilibrium is always undominated, but that an undominated equilibrium is not necessarily perfect. It is our objective in this section to show that in a bimatrix game, however, every undominated equilibrium is perfect. Furthermore, it is shown that in this case one can verify whether an equilibrium is perfect by solving a linear programming problem. The proof of the main result of the section is based on the following lemma:

Lemma 3.2.1. *Let $\Gamma = (\Phi_1, \Phi_2, R_1, R_2)$ be a bimatrix game and let $s_1 \in S_1$. Define the matrix game $\Gamma(s_1) = (\Phi_1, \Phi_2, R)$ by:*

$$R(k, l) = R_1(k, l) - R_1(s_1, l) \quad \text{for } k \in \Phi_1 \text{ and } l \in \Phi_2 .$$

Then s_1 is undominated in Γ if and only if $v(\Gamma(s_1)) = 0$ and $C_2(\Gamma(s_1)) = \Phi_2$.

Proof. It is easily verified that the assertions (3.2.1) – (3.2.4) are all equivalent:

$$s_1 \text{ is dominated by } s_1' \text{ in } \Gamma, \text{ for some } s_1', \tag{3.2.1}$$

$$R_1(s_1', l) \geqq R_1(s_1, l) \text{ for all } l \text{ and } R(s_1', l) > R_1(s_1, l) \text{ for some } l, \text{ for some } s_1', \tag{3.2.2}$$

$$R(s_1', l) \geqq 0 \text{ for all } l \text{ and } R(s_1', l) > 0 \text{ for some } l, \text{ for some } s_1', \tag{3.2.3}$$

$$v(\Gamma(s_1)) \geqq 0 \text{ and, if } v(\Gamma(s_1)) = 0, \text{ then } C_2(\Gamma(s_1)) \neq \Phi_2 . \tag{3.2.4}$$

Since in $\Gamma(s_1)$ player 1 can guarantee himself a payoff 0 by playing s_1, we have $v(\Gamma(s_1)) \geqq 0$, hence, (3.2.4) is equivalent to

$$\text{if } v(\Gamma(s_1)) = 0, \text{ then } C_2(\Gamma(s_1)) \neq \Phi_2 . \tag{3.2.5}$$

The equivalence of (3.2.1) and (3.2.5) establishes the lemma. □

Theorem 3.2.2. *An equilibrium of a bimatrix game is perfect if and only if it is undominated.*

Proof. Let $s=(s_1, s_2)$ be an equilibrium of the bimatrix game $\Gamma=(\Phi_1, \Phi_2, R_1, R_2)$. In view of Corollary 2.2.6 it suffices to show that s is perfect if s is undominated. Assume s is undominated and let $\Gamma(s_1)$ be as in Lemma 3.2.1. Since $v(\Gamma(s_1))=0$, we have $s_1 \in O_1(\Gamma(s_1))$. Furthermore, let s_2' be a completely mixed element of $O_2(\Gamma(s_1))$, the existence of which is guaranteed by $C_2(\Gamma(s_1))=\Phi_2$ and Eq. (3.1.4.). Then s_1 is a best reply against s_2' in $\Gamma(s_1)$, which implies that s_1 is also a best reply against s_2' in Γ. For $\varepsilon>0$, let $s_2(\varepsilon)=(1-\varepsilon)s_2+\varepsilon s_2'$. Then $s_2(\varepsilon)$ is completely mixed, s_1 is a best reply against $s_2(\varepsilon)$ and $s_2(\varepsilon)$ converges to s_2 as ε tends to zero. Since undominated strategies of player 2 can be characterized similarly as undominated strategies of player 1, we can follow the same procedure for this player and, hence, we can construct, for $\varepsilon>0$, a completely mixed strategy $s_1(\varepsilon)\in S_1$, such that s_2 is a best reply against $s_1(\varepsilon)$ and such that $s_1(\varepsilon)$ converges to s_1 as ε tends to zero. Let $s(\varepsilon)=(s_1(\varepsilon), s_2(\varepsilon))$. Then $\{s(\varepsilon)\}_{\varepsilon\downarrow 0}$ satisfies the condition of Theorem 2.2.5(iii), which shows that s is a perfect equilibrium of Γ. □

In view of the theorem above, in order to verify whether an equilibrium of a bimatrix game is perfect, it suffices to check whether both equilibrium strategies are undominated. We will now demonstrate that this is an easy task. Let us confine ourselves to the equilibrium strategy of player 1. It follows from Lemma 3.2.1, that it suffices to show that the value of a matrix game $\Gamma=(\Phi_1, \Phi_2, R)$ can be easily determined and that it is easy to verify whether $C_2(\Gamma)=\Phi_2$. Without loss of generality, we can restrict ourselves to the case where $R(\varphi)>0$ for all $\varphi\in\Phi$.

The value of Γ can be determined by solving the linear programming problem (3.1.3). It is convenient to make a change of variables and to write x_1 for s_1/v (this causes no trouble, since $v(\Gamma)>0$). By this change of variables, (3.1.3) is transformed into (3.2.6), in which $x_1(\Phi_1)$ denotes the sum of all x_1^k with $k\in\Phi_1$.

$$\begin{aligned}&\text{minimize } x_1(\Phi_1),\\ &\text{subject to } R(x_1, l)\geqq 1 \text{ for all } l\in\Phi_2;\ x_1\in X_1;\ x_1\geqq 0.\end{aligned} \tag{3.2.6}$$

It is easily seen that:

if (s_1, v) solves (3.1.3), then s_1/v solves (3.2.6), and

if x_1 solves (3.2.6), then $(x_1(\Phi_1))^{-1}(x_1, 1)$ solves (3.1.3),

which implies that

$$l\in C_2(\Gamma) \text{ iff } R(x_1, l)=1 \text{ for all } x_1 \text{ which solve (3.2.6).} \tag{3.2.7}$$

Let us consider (3.2.6) in conjunction with its dual. We have that x_1 solves (3.2.6) and x_2 solves the dual of this problem if and only if (x_1, x_2) is a solution to the following system of inequalities:

$$\begin{aligned}&x_1(\Phi_1)=x_2(\Phi_2),\\ &R(x_1, l)\geqq 1 \qquad \text{for all } l\in\Phi_2,\\ &R(k, x_2)\leqq 1 \qquad \text{for all } k\in\Phi_1, \text{ and}\\ &x_1\in X_1,\ x_2\in X_2,\ x_1\geqq 0,\ x_2\geqq 0.\end{aligned} \tag{3.2.8}$$

(3.2.7) leads us to the following linear programming problem, in which $R(x_1, \Phi_2)$ denotes the sum of all $R(x_1, l)$, where l ranges over Φ_2:

$$\begin{aligned} &\text{maximize} && R(x_1, \Phi_2)\,, \\ &\text{subject to} && x_1(\Phi_1) = x_2(\Phi_2)\,, \\ & && R(x_1, l) \geqq 1 \quad \text{for all } l \in \Phi_2\,, \\ & && R(k, x_2) \leqq 1 \quad \text{for all } k \in \Phi_1\,, \text{ and} \\ & && x_1 \in X_1,\ x_2 \in X_2,\ x_1 \geqq 0,\ x_2 \geqq 0\,. \end{aligned} \tag{3.2.9}$$

From (3.2.7) one can conclude that $C_2(\Gamma) = \Phi_2$ if and only if the value of the linear programming problem (3.2.9) is $m_2(=|\Phi_2|)$. Furthermore, if (x_1, x_2) solves (3.2.9), then $v(\Gamma) = (x_1(\Phi_1))^{-1}$ and so, we have proved:

Theorem 3.2.3. *Let $\Gamma = (\Phi_1, \Phi_2, R)$ be a matrix game with $R(\varphi) > 0$ for all $\varphi \in \Phi$. Then*

(i) $v(\Gamma) = (x_1(\Phi_1))^{-1}$, *where* x_1 is *such that* (x_1, x_2) *solves* (3.2.9),
(ii) $C_2(\Gamma) = \Phi_2$ *if and only if the value of* (3.2.9) *is* m_2.

The linear programming problem (3.2.9) can easily be solved and so, in a bimatrix game, it can easily be checked whether a strategy of player 1 is dominated. The same holds for player 2 and, therefore, it is easy to verify whether an equilibrium of a bimatrix game is perfect.[1]

3.3 Regular Equilibria

In this section, a first characterization of regular equilibria in a bimatrix game is derived. Furthermore, it is shown that for a game which is nondegenerate in the sense of Lemke and Howson [1964] or Shapley [1974], all equilibria are regular.

Let $s = (s_1, s_2)$ be an equilibrium of a bimatrix game $\Gamma = (\Phi_1, \Phi_2, R_1, R_2)$. From Corollary 2.5.3 we know that s has to be quasi-strict in order to be regular.

Let us investigate under which conditions a quasi-strict equilibrium s is regular. Without loss of generality, assume that $\phi = (1, 1) \in \Phi^s$, where $\Phi^s = C(s) = B(s)$. Define the $|C(s_1)| \times |C(s_2)|$ matrices A_1 and A_2 by (3.3.1) – (3.3.4).

$$A_1(1, l) = 1 \qquad \text{for } l \in \Phi_2^s\,, \tag{3.3.1}$$

$$A_1(k, l) = R_1(k, l) - R_1(1, l) \quad \text{for } k \in \Phi_1^s,\ l \in \Phi_2^s,\ k \neq 1\,, \tag{3.3.2}$$

$$A_2(k, 1) = 1 \qquad \text{for } k \in \Phi_1^s\,, \tag{3.3.3}$$

$$A_2(k, l) = R_2(k, l) - R_2(k, 1) \quad \text{for } k \in \Phi_1^s,\ l \in \Phi_2^s,\ l \neq 1\,. \tag{3.3.4}$$

Let the Jacobian $\tilde{J}(s|\phi)$ be as in Lemma 2.5.2. Since s is quasi-strict, it follows that $\tilde{J}(s|\phi)$ is nonsingular if and only if the matrix $\begin{pmatrix} \varnothing & A_1 \\ A_2^T & \varnothing \end{pmatrix}$ is nonsingular.

Obviously, this matrix is nonsingular, if and only if A_1 and A_2 are both nonsingular, which can be the case only if $|C(s_1)|=|C(s_2)|$. We have proved:

Lemma 3.3.1. *An equilibrium s of a bimatrix game is regular if and only if it is a quasi-strict equilibrium with* $|C(s_1)|=|C(s_2)|$ *for which the matrices* A_1 *and* A_2 *as defined in* (3.3.1) – (3.3.4) *are nonsingular.*

Next, assume $\Gamma=(\Phi_1, \Phi_2, R_1, R_2)$ is a bimatrix game in which all payoffs are positive and assume s is a quasi-strict equilibrium of Γ with $|C(s_1)|=|C(s_2)|$. We claim that, in this case, A_i is nonsingular if and only if the s-restricted payoff matrix R_i^s is nonsingular ($i\in\{1,2\}$). Let us demonstrate this fact for $i=1$. Assume A_1 is singular and let $x_2\in X_2^s$ be such that $x_2\neq 0$ and $A_1(x_2)=0$. Then

if $R_1^s(1,x_2)=0$, then $R_1^s(x_2)=0$, and
if $R_1^s(1,x_2)\neq 0$, then $R_1^s(x_2)$ is a scalar multiple of $R_1^s(s_2)$,

whereas x_2 is not a scalar multiple of s_2 (since $x_2(\Phi_2^s)=0$ and $s_2(\Phi_2^s)=1$).

In both cases, we can conclude that R_1^s is singular.

Next, assume R_1^s is singular and let $x_2\in X_2^s$ be such that $x_2\neq 0$ and $R_1^s(x_2)=0$. Then

if $x_2(\Phi_2^s)=0$, then $A_1(x_2)=0$, and
if $x_2(\Phi_2^s)\neq 0$, then $x_2(\Phi_2^s)s_2-x_2$ is in the null space of A_1,

whereas $x_2(\Phi_2^s)s_2\neq x_2$ (since $R_1^s(x_2)=0$ and $R_1^s(s_s)>0$ since R_1 is positive).

In both cases, one can conclude that A_1 is singular, Similarly, it can be proved that A_2 is singular if and only if R_2^s is singular and so we have:

Theorem 3.3.2. *Let* $\Gamma=(\Phi_1, \Phi_2, R_1, R_2)$ *be a bimatrix game with positive payoffs. Then s is a regular equilibrium of* Γ, *if and only if s is a quasi-strict equilibrium with* $|C(s_1)|=|C(s_2)|$, *for which the matrices* R_1^s *and* R_2^s *are nonsingular.*

Up to now, nothing has been said about the actual computation of equilibria. Algorithms to compute the set of *all* equilibria of a bimatrix game have been given in Vorob'ev [1958], Mills [1960], Kuhn [1961] and Winkels [1979], but all these are more of theoretical than of practical interest. An efficient algorithm to compute *one* equilibrium of a bimatrix game, has been proposed in Lemke and Howson [1964][2]. This algorithm is based on path following and can be applied to *nondegenerate games* (see below). To compute an equilibrium of a degenerate game, one first has to perturb the game slightly (e.g. by means of the scheme proposed by Lemke and Howson) in order to get a nondegenerate game to which the algorithm can be applied. In this monograph, we will not consider the question how to compute equilibria, but we will show that a bimatrix game which satisfies the nondegeneracy condition of Lemke and Howson possesses only regular equilibria. The following lemma is essential:

Lemma 3.3.3. *For a bimatrix game* Γ *which is such that* $|B(s)|\leqq|C(s)|$ *for all* $s\in S$, *all equilibria are regular.*

Proof. Assume Γ satisfies the condition of the lemma and let s be an equilibrium of Γ. Then $C(s) \subset B(s)$ and, therefore $C(s) = B(s)$. Hence, s is a quasi-strict equilibrium. From Lemma 3.3.1 it follows that it suffices to show that the matrices A_1 and A_2, as defined in (3.3.1) – (3.3.4) are nonsingular. Assume A_1 is column-singular and let $x_2^s \in X_2^s$ be such that $x^s{}_2 \neq 0$ and $A_1(x_2^s) = 0$. Let x_2 be the extension from $x^s{}_2$ to X_2 and, for $\varepsilon > 0$, let $s_2(\varepsilon) = s_2 + \varepsilon x_2$ and $s(\varepsilon) = (s_1, s_2(\varepsilon))$. If ε is sufficiently small, then $s(\varepsilon) \in S$, $C(s(\varepsilon)) = C(s)$ and $B(s(\varepsilon)) = B(s)$. Define $\tilde{\varepsilon}$ by

$$\tilde{\varepsilon} := \sup\{\varepsilon > 0;\ s(\varepsilon) \in S,\ C(s(\varepsilon)) = C(s),\ B(s(\varepsilon)) = B(s)\}.$$

Notice that $\tilde{\varepsilon}$ is finite since x_2 has at least one negative component. Since s is an equilibrium of Γ, we can conclude from the definition of $\tilde{\varepsilon}$

$$C(s(\tilde{\varepsilon})) \subset C(s) = B(s) \subset B(s(\tilde{\varepsilon})), \text{ where at least one inclusion is strict.}$$

This contradicts the condition of the lemma and so A_1 is column-regular, which implies $|C(s_1)| \geqq |C(s_2)|$. Similarly, it can be shown that A_2 is row-regular, and so $|C(s_1)| \leqq |C(s_2)|$. Hence, $|C(s_1)| = |C(s_2)|$ and A_1 and A_2 are both nonsingular. This implies that s is a regular equilibrium of Γ. □

Next, let us turn to the nondegeneracy assumption which is imposed in Lemke and Howson [1964]. We restrict ourselves to bimatrix games $(\Phi_1, \Phi_2, R_1, R_2)$ in which all payoffs are positive. For $s \in S$, define matrices $\tilde{R}_1^s$ and $\tilde{R}_2^s$ by

$$\tilde{R}_1^s := [R_1(k,l)]_{k,l} \quad (k \in B_1(s),\ l \in C(s_2)), \text{ and}$$

$$\tilde{R}_2^s := [R_2(k,l)]_{k,l} \quad (k \in C(s_1),\ l \in B_2(s)).$$

The *nondegeneracy assumption* which Lemke and Howson impose is:

the rows of $\tilde{R}_1^s$ are independent, for all $s \in S$, and

the columns of $\tilde{R}_2^s$ are independent, for all $s \in S$.

Obviously if this condition is satisfied, then for every $s \in S$:

$$|C(s_2)| \geqq \operatorname{rank}(\tilde{R}_1^s) = |B_1(s)|,$$

$$|C(s_1)| \geqq \operatorname{rank}(\tilde{R}_2^s) = |B_2(s)|,$$

from which it follows that

$$|C(s)| \geqq |B(s)| \quad \text{for every } s \in S.$$

Hence, if the nondegeneracy condition of Lemke and Howson is satisfied, then the condition of Lemma 3.3.1 is fulfilled and all equilibria are regular. In Shapley [1974] a similar nondegeneracy condition is imposed and, therefore, we also have that all equilibria are regular if that condition is fulfilled.

Theorem 3.3.4. *For a bimatrix game which satisfies the nondegeneracy condition of Lemke and Howson* [1964] (*or Shapley* [1974]) *all equilibria are regular.*

3.4 Characterizations of Regular Equilibria

In this section several characterizations of regular equilibria are derived[3]. The fundamental result of the section is that an equilibrium s of a bimatrix game Γ is regular if and only if s is an isolated equilibrium of Γ which is perfect in the s-restriction of Γ. Once this result has been established, several other characterizations of regular equilibria follow easily. At the end of the section, an example is given to demonstrate that the results of this section cannot be generalized to games with more than two players.

We first prove our main result:

Theorem 3.4.1. *s is a regular equilibrium of the bimatrix game $\Gamma = (\Phi_1, \Phi_2, R_1, R_2)$ if and only if s is an isolated equilibrium of Γ, which is perfect in the s-restriction Γ^s of Γ.*

Proof. If s is a regular equilibrium of Γ, then s is isolated (Theorem 2.5.5 and Corollary 2.4.6). Furthermore, if s is regular in Γ, then s is regular in Γ^s as follows from Lemma 2.5.2 and, therefore, s is a perfect equilibrium of Γ^s (Theorem 2.4.7). Next, assume $s = (s_1, s_2)$ is an isolated equilibrium of Γ, which is perfect in Γ^s. Since s is perfect in Γ^s there exists a sequence $\{s(\varepsilon)\}_{\varepsilon \downarrow 0}$ which converges to s, such that $C(s(\varepsilon)) = B(s)$ and such that s is a best reply against $s(\varepsilon)$ for every $\varepsilon > 0$. In this case, we have that $(s_1, s_2(\varepsilon))$ and $(s_1(\varepsilon), s_2)$ are both equilibria of Γ, for any $\varepsilon > 0$. Since s is isolated, we therefore have $s_1(\varepsilon) = s_1$ and $s_2(\varepsilon) = s_2$, for all ε sufficiently close to zero, which implies that $C(s) = B(s)$. Hence, s is a quasi-strict equilibrium of Γ. In view of Lemma 3.3.1, it remains to be shown that the matrices A_1 and A_2, as defined in (3.3.1) – (3.3.4) are nonsingular. Assume $x_2^s \in X_2^s$ is such that $A_1(x_2^s) = 0$. Let x_2 be the extension of x_2^s to X_2 and, for $\varepsilon > 0$, let $\bar{s}_2(\varepsilon) = s_2 + \varepsilon x_2$. For ε sufficiently close to zero, we have that $\bar{s}_2(\varepsilon) \in S_2$ and that $(s_1, \bar{s}_2(\varepsilon))$ is an equilibrium of Γ. Since s is isolated, we have $\bar{s}_2(\varepsilon) = s_2$ for ε close to zero and, therefore, $x_2 = 0$. Hence, A_1 is column-regular. Similarly, it can be shown that A_2 is row-regular, from which it follows that both matrices are square and nonsingular. □

If s is a regular equilibrium, then s is quasi-strict, as we know from Corollary 2.5.3. Furthermore, if s is a quasi-strict equilibrium of a bimatrix game Γ, then s is a completely mixed equilibrium of Γ^s. Since every completely mixed equilibrium is perfect, we immediately deduce from Theorem 3.4.1:

Corollary 3.4.2. *An equilibrium of a bimatrix game is regular if and only if it is isolated and quasi-strict.*

The game of Fig. 1.5.3 shows that a perfect equilibrium s of a bimatrix game Γ is not necessarily a perfect equilibrium of the s-restriction Γ^s of Γ (namely take $s = (M_1, M_2)$). Consequently, a perfect and isolated equilibrium of a bimatrix game need not be regular (it is easily verified, that (M_1, M_2) is an isolated, though not a regular equilibrium of the game of Fig. 1.5.3). Next, we will show, that a weakly proper equilibrium of Γ has the property that it is perfect in Γ^s, from which it can be concluded that a weakly proper and isolated equilibrium is regular.

Lemma 3.4.3. *Let s be a weakly proper equilibrium of the bimatrix game* $\Gamma=(\Phi_1, \Phi_2, R_1, R_2)$. *Then s is a perfect equilibrium of the s-restriction* Γ^s *of* Γ.

Proof. In view of Theorem 3.2.2, it suffices to show that s is undominated in Γ^s. Assume s_1 is dominated by s_1' in Γ^s and let $l' \in \Phi_2^s(=B_2(s))$ be such that

$$R_1(s_1', l') > R_1(s_1, l').$$

Let the sequence $\{s(\varepsilon)\}_{\varepsilon \downarrow 0}$ be as in Def. 2.3.4. Notice that, if $l \in \Phi_2 \backslash B_2(s)$, then

$$s_2^l(\varepsilon) \leqq \varepsilon s_2^{l'}(\varepsilon).$$

Let M be defined by

$$M=\max\{R_1(k, l);\ k \in \Phi_1,\ l \in \Phi_2\}.$$

Then we have, for ε sufficiently close to zero

$$\begin{aligned}
&R_1(s_1', s_2(\varepsilon)) - R_1(s_1, s_2(\varepsilon)) = R_1(s_1' - s_1, s_2(\varepsilon)) \\
&= \sum_{l \in B_2(s)} R_1(s_1' - s_1, l) s_2^l(\varepsilon) + \sum_{l \notin B_2(s)} R_1(s_1' - s_1, l) s_2^l(\varepsilon) \\
&\geqq R_1(s_1' - s_1, l') s_2^{l'}(\varepsilon) - 2M \sum_{l \notin B_2(s)} s_2^l(\varepsilon) \\
&\geqq R_1(s_1' - s_1, l') s_2^{l'}(\varepsilon) - 2M m_2 \varepsilon s_2^{l'}(\varepsilon) > 0,
\end{aligned}$$

which contradicts the definition of $\{s(\varepsilon)\}_{\varepsilon \downarrow 0}$. Hence, s_1 is undominated in Γ^s. Similarly, it can be shown that s_2 is undominated in Γ^s. Hence, s is a perfect equilibrium of Γ^s. □

By combining Theorems 2.3.6, 2.4.3, 2.5.5 and Corollaries 2.4.5, 2.4.6, 3.4.2 with Lemma 3.4.3, we see that regular equilibria can be characterized as follows:

Theorem 3.4.4. *For a bimatrix game* Γ, *the following assertions are equivalent:*

(i) *s is a regular equilibrium of* Γ,
(ii) *s is a strongly stable equilibrium.*
(iii) *s is an isolated and essential equilibrium of* Γ,
(iv) *s is an isolated and strictly perfect equilibrium of* Γ,
(v) *s is an isolated and weakly proper equilibrium of* Γ, *and*
(vi) *s is an isolated and quasi-strict equilibrium of* Γ.

The final characterization of regular equilibria specializes Theorem 2.5.6 to the bimatrix context:

Theorem 3.4.5 (cf. Jansen [1981b], Theorem 7.4). *An equilibrium of a bimatrix game is regular if and only if it is essential and quasi-strict.*

Proof. The only if part of the theorem follows from the Corollaries 2.4.5, 2.5.3 and 2.5.5. Next, assume $s=(s_1, s_2)$ is an essential and quasi-strict equilibrium of the bimatrix game $\Gamma=(\Phi_1, \Phi_2, R_1, R_2)$. Without loss of generality, assume that all payoffs in Γ are positive. In view of Theorem 3.3.2 it suffices to show that R_1^s and R_2^s are nonsingular. Assume R_1^s is row-singular and let $x_1 \in X_1^s$ be such that $x_1 \neq 0$ and $R_1^s(x_1)=0$. Without loss of generality assume $x_1^1 \neq 0$ and, for $\varepsilon>0$, define the bimatrix game $\Gamma^{s,\varepsilon}=(\Phi_1^s, \Phi_2^s, R_1^{s,\varepsilon}, R_2^s)$ by:

$$R_1^{s,\varepsilon}(1, l)=R_1^s(1, l)+\varepsilon \quad \text{for } l \in \Phi_2^s \text{, and}$$
$$R_1^{s,\varepsilon}(k, l)=R_1^s(k, l) \quad \text{for } k \in \Phi_1^s,\ l \in \Phi_2^s,\ k \neq 1.$$

Since s is an essential equilibrium of Γ, we have that s is an essential equilibrium of Γ^s and, therefore, if ε is sufficiently close to zero, there exists an equilibrium $s(\varepsilon)$ of $\Gamma^{s,\varepsilon}$ which is close to s. In this case we have:

$$\varepsilon x_1^1=\varepsilon x_1^1+R_1(x_1, s_2(\varepsilon))=R_1^{s,\varepsilon}(x_1, s_2(\varepsilon))=x_1(\Phi_1^s)R_1^{s,\varepsilon}(s(\varepsilon)).$$

Therefore, $x_1(\Phi_1^s) \neq 0$ and $R_1(s)=\lim_{\varepsilon \to 0} R_1^{s,\varepsilon}(s(\varepsilon))=0$, but this contradicts our assumption that all payoffs in Γ are positive. Hence, R_1^s is row-regular. Similarly, it can be shown that R_2^s is column-regular and, therefore, R_1^s and R_2^s are both nonsingular, which implies that s is a regular equilibrium of Γ. □

We conclude this section by presenting an example which illustrates that the results obtained in this section are not true for games with more than two players. Consider the 3-person game of Fig. 3.4.1. The reader can verify that this game has exactly two equilibria, viz. $\varphi=(\varphi_1, \varphi_2, \varphi_3)=(1, 1, 1)$ and $\bar{\varphi}=(\bar{\varphi}_1, \bar{\varphi}_2, \bar{\varphi}_3)=(2, 2, 2)$. We will show that the second equilibrium is not perfect. Let $s=(s_1, s_2, s_3)$ be a mixed strategy combination. Then, we have:

$\bar{\varphi}_1$ is a best reply against s if and only if $2s_2^2 s_3^1 \leqq s_2^1 s_3^2$,

$\bar{\varphi}_2$ is a best reply against s if and only if $s_1^1 s_3^2 \leqq 2s_1^2 s_3^1$, and

$\bar{\varphi}_3$ is a best reply against s if and only if $2s_1^2 s_2^1 \leqq s_1^1 s_2^2$,

from which it follows that

if $\bar{\varphi}$ is a best reply against s, then $s_2^1=0$ or $s_2^2=0$.

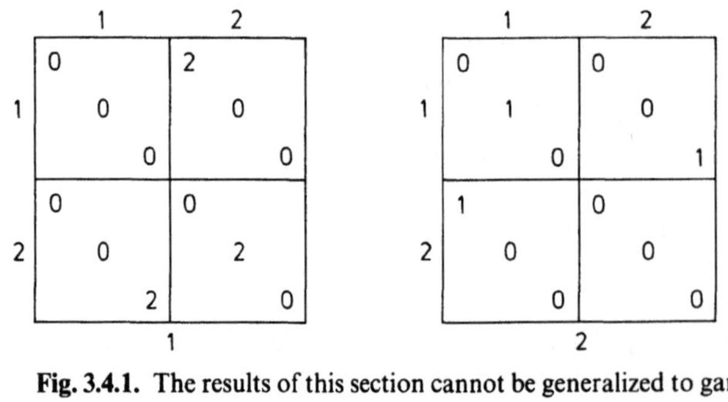

Fig. 3.4.1. The results of this section cannot be generalized to games with more than two players

Hence, there does not exist a completely mixed strategy combination s such that $\bar{\varphi}$ is a best reply against s, and so $\bar{\varphi}$ is not perfect. (However, $\bar{\varphi}$ is persistent (cf. the discussion at the end of Sect. 2.3).) Since this game possesses at least one essential equilibrium (by Theorem 2.4.2) and since $\bar{\varphi}$ is not essential (Theorem 2.4.3), we have that φ is essential. Hence, φ is also strictly perfect and weakly proper. Moreover, φ is isolated, and it can be shown that φ changes continuously and uniquely when payoffs are perturbed, so that φ is strongly stable. However, φ is not regular, since it is not quasi-strict.

Notice that this example shows that a game with finitely many equilibria need not possess a regular equilibrium and that there exist games without quasi-strict equilibria (the latter phenomenon has also been observed in Okada [1982]).

3.5 Matrix Games

In this section, we consider matrix games, i.e. 2-person zero-sum games in normal form. We concentrate on proper equilibria and show that an equilibrium is proper if and only if both equilibrium strategies are optimal in the sense of Dresher [1961].

Let us first briefly consider the question which conditions an equilibrium has to satisfy in order to be regular (or, equivalently, strongly stable). In Sect. 3.1, we have seen that the set of equilibria of a matrix game is convex and, therefore, we can conclude from Corollary 2.5.7 that, if s is a regular equilibrium of a matrix game Γ, s is the unique equilibrium of Γ. Conversely, if s is the unique equilibrium of Γ, then s is isolated and proper (since every game has a proper equilibrium) and, therefore, in view of Theorem 3.4.4, regular. So, we have proved:

Theorem 3.5.1. *s is a regular equilibrium of a matrix game Γ if and only if s is the unique equilibrium of Γ.*

In Bohnenblust, Karlin and Shapley [1950] it is shown that the set of all matrix games with a unique equilibrium is open and dense in the set of all matrix games. By similar methods as the ones used in the proof of Theorem 2.6.1, we can prove the slightly stronger result that, within the set of all matrix games, the set of games with an irregular equilibrium is closed and has Lebesque measure zero. Hence, we have:

Theorem 3.5.2. *Almost all matrix games have a unique equilibrium.*

Hence, if a matrix game is "nondegenerate", it has a unique equilibrium. However, almost any matrix game arising from a nontrivial extensive form will be "degenerate" and have more than one equilibrium so that the result of Theorem 3.5.2 is not as strong as it might look at first sight. Therefore, in the remainder of this section, we will consider matrix games with more than one equilibrium and we will investigate whether there exists an equilibrium which should be preferred to all others.

	1	2	3
1	1	2	3
2	1	1	4

Fig. 3.5.1. Which strategy should be chosen by player 1 in order to exploit the mistakes of player 2 optimally?

If, in a matrix game Γ, player 1 has more than one optimal (i.e. equilibrium, maximin) strategy, then by using any of these he can guarantee himself a payoff of at least $v(\Gamma)$. If, moreover, his opponent also uses an optimal strategy, then every optimal strategy of player 1 yields exactly $v(\Gamma)$ and so, if both players play optimally, there is no reason to prefer one optimal strategy to another. However, if player 1 considers the possibility that his opponent may make a mistake and may, therefore, fail to choose an optimal strategy, then by playing a specific optimal strategy he can perhaps take maximal advantage of such a mistake. Hence, the question which has to be considered is: which strategy should be chosen in order to exploit the potential mistakes of the opponent optimally?* One approach in trying to solve this problem, is to restrict oneself to perfect equilibria. The game of Fig. 3.5.1 shows that this approach does not always lead to a definite answer.

In this game all equilibria are perfect: if player 1 expects that player 2 will choose his second pure strategy with a greater probability than his third, then he should play his first strategy; if he expects the third pure strategy of player 2 to occur with the greater probability, then he should play his second strategy; and if he expects both mistakes to occur with the same probability, then all his strategies are equally good. Hence, if one follows this approach, one has to know how the opponent makes his mistakes, in order to obtain a definite answer.

If one expects that the opponent will make a more costly mistake with a much smaller probability than a less costly one and, hence, that the opponent makes his mistakes in a more or less rational way, one is led to proper equilibria. According to the properness concept, in the game of Fig. 3.5.1, player 2 will mistakenly choose his second strategy with a much greater probability than his third strategy and, therefore, player 1 should play his first strategy. Hence, in this example, the properness concept leads to a definite answer. Below, we will prove that the same is the case for any matrix game. To be more precise, we will prove that proper equilibria of a matrix game Γ are *equivalent*, where two strategy pairs s and s' are said to be *equivalent* in Γ if

$$R(s_1, l) = R(s'_1, l) \quad \text{for all} \quad l \in \Phi_2, \text{ and}$$

$$R(k, s_2) = R(k, s'_2) \quad \text{for all} \quad k \in \Phi_1 .$$

A slightly different approach to solve the problem has been proposed in Dresher [1961]. This approach amounts to a lexicographic application of the maximin

* This question has also been considered in Ponssard [1976].

criterion. The idea underlying it is that, since one does not know beforehand which mistake will be made by the opponent, one should follow a conservative plan of action and maximize the minimum gain resulting from the opponent's mistakes. If, in the game of Fig. 3.5.1, player 1 behaves in this way, then he will play his first strategy since this strategy guarantees a payoff 2 if player 2 makes a mistake, whereas his second strategy guarantees only 1 in this case. Hence, in this example, Dresher's approach yields a definite answer, which is the same as the one given by the properness concept (which is no coincidence, as we will see in Theorem 3.5.5).

For a matrix game $\Gamma = (\Phi_1, \Phi_2, R)$ *Dresher's procedure* to select a particular optimal strategy of player 1 is described as follows:

(i) Set $t := 0$, write $\Phi_1^t := \Phi_1$, $\Phi_2^t := \Phi_2$ and $\Gamma^t := (\Phi_1^t, \Phi_2^t, R)$. Compute $O_1(\Gamma^t)$, i.e. the set of optimal strategies of player 1 in the game Γ^t.

(ii) If all elements of $O_1(\Gamma^t)$ are equivalent in Γ^t, then go to (v), otherwise go to (iii).

(iii) Assume that player 2 makes a mistake in Γ^t, i.e. that he assigns a positive probability only to the pure strategies which yield player 1 a payoff greater than $v(\Gamma^t)$. Hence, restrict player 2's pure strategy set to $\Phi_2^t \backslash C_2(\Gamma^t)$.

(iv) Determine the optimal strategies of player 1 which maximize the minimum gain resulting from mistakes of player 2. Hence, compute the optimal strategies of player 1 in the game $\Gamma^{t+1} := (\Phi_1^{t+1}, \Phi_2^{t+1}, R)$, where $\Phi_1^{t+1} := \text{ext}\, O_1(\Gamma^t)$ is the (finite) set of extreme optimal strategies of player 1 in Γ^t and $\Phi_2^{t+1} := \Phi_2^t \backslash C_2(\Gamma^t)$. Replace t by $t+1$ and repeat step (ii).

(v) The set of *Dresher-optimal* (or shortly *D-optimal strategies*) of player 1 in Γ is the set $D_1(\Gamma) := O_1(\Gamma^t)$.

Note that $D_1(\Gamma)$ is well-defined since in each iteration the number of permissible pure strategies of player 2 decreases with at least one, such that, eventually, all remaining optimal strategies of player 1 must be equivalent in Γ^t. We claim that all D-optimal strategies of player 1 in Γ are, in fact, equivalent in Γ. Let $s_1, s_1' \in D_1(\Gamma)$ and let $\Gamma^0, \ldots, \Gamma^\tau$ be the sequence of games generated by the above procedure. Then s_1, s_1' are equivalent in Γ^τ by definition of $D_1(\Gamma)$. But then s_1 and s_1' are also equivalent in $\Gamma^{\tau-1}$ since every element l of $\Phi_2^{\tau-1}$ is either an element of $C_2(\Gamma^{\tau-1})$, in which case s_1 and s_1' both yield $v(\Gamma^{\tau-1})$ against l, or an element of Φ_2^τ, in which case s_1 and s_1' yield the same against l since they are equivalent in Γ^τ. Hence, inductively it can be proved that s_1 and s_1' are equivalent in Γ^t for all $t \in \{0, \ldots, \tau\}$, which shows that s_1 and s_1' are equivalent in Γ.

It will be clear that by reversing the roles of the players one obtains a procedure for selecting a particular optimal strategy (or more precisely a particular class of equivalent optimal strategies) of player 2. The set of all D-optimal strategies of player 2 in Γ will be denoted by $D_2(\Gamma)$, and the product $D_1(\Gamma) \times D_2(\Gamma)$ will be denoted by $D(\Gamma)$. We have already seen:

Lemma 3.5.3. *If $s, s' \in D(\Gamma)$, then s and s' are equivalent.*

From the description of Dresher's procedure it will be clear that a D-optimal strategy cannot be dominated. Hence, by Theorem 3.2.2, we have

Theorem 3.5.4. *A pair of D-optimal strategies constitutes a perfect equilibrium.*

We have already seen that the converse of Theorem 3.5.4 is false: in the game of Fig. 3.5.1 only (1, 1) is D-optimal, whereas all equilibria are perfect. This is not really surprising since the perfectness concept allows all kinds of mistakes, whereas player 1 assumes, if he plays a D-optimal strategy, that his opponent makes his mistakes as if he actually wishes to minimize player 1's gain resulting from his mistakes. Hence, by playing a D-optimal strategy you optimally exploit the mistakes of your opponent only if he makes his mistakes in a rational way.Based on this observation, one might conjecture that D-optimal strategies are related to proper equilibria. In the following theorem, we prove that this conjecture is correct.

Theorem 3.5.5. *For matrix games, the following assertions are equivalent:*

(i) *s is a proper equilibrium,*
(ii) *s is a weakly proper equililibrium,*
(iii) *s is a D-optimal strategy pair.*

Proof. Let $\Gamma = (\Phi_1, \Phi_2, R)$ be a matrix game. It suffices to show that (ii) implies (iii) and that (iii) implies (i).
(ii)→(iii): Let $s=(s_1, s_2)$ be a weakly proper equilibrium of Γ and let $\{s(\varepsilon)\}_{\varepsilon \downarrow 0}$ be as in Def. 2.3.4. We will show that $s_1 \in D_1(\Gamma)$. Let $\Gamma^0, \ldots, \Gamma^\tau$ be the sequence of games generated by Dresher's procedure for player 1. We have that $s_1 \in D_1(\Gamma)$ if and only if $s_1 \in O_1(\Gamma^t)$ for all $t \in \{0, \ldots, \tau\}$. We will show, by using induction with respect to t, that s_1 has the latter property.

First of all, $s_1 \in O_1(\Gamma^0)$, since s is an equilibrium of Γ.

Next, assume $t \in \{1, \ldots, \tau\}$ is such that

$$s_1 \in O_1(\Gamma^{t'}) \quad \text{for all } t' \in \{0, \ldots, t-1\},\ s_1 \notin O_1(\Gamma^t).$$

Let $\bar{s}_1 \in O_1(\Gamma^t)$. Then $\bar{s}_1 \in O_1(\Gamma^{t'})$ for all $t' \in \{0, \ldots, t\}$ and, therefore

$$R(s_1, l) = R(\bar{s}_1, l) \quad \text{for all } l \in \Phi_2 \backslash \Phi_2^t. \tag{3.5.1}$$

Since $s_1 \notin O_1(\Gamma^t)$, there exists some $l \in \Phi_2^t$ such that

$$R(s_1, l) < v(\Gamma^t). \tag{3.5.2}$$

Let Ψ_2^t be the set of all $l \in \Phi_2^t$ for which (3.5.2) is satisfied. Since $\bar{s}_1 \in O_1(\Gamma^t)$ we have

$$R(s_1, l) < v(\Gamma^t) \leqq R(\bar{s}_1, l) \quad \text{for all } l \in \Psi_2^t. \tag{3.5.3}$$

Furthermore, if $k \in \Phi_2^t \backslash \Psi_2^t$ and $l \in \Psi_2^t$, then

$$R(s_1, l) < v(\Gamma^t) \leqq R(s_1, k).$$

from which it follows, by definition of $s(\varepsilon)$, that

$$s_2^k(\varepsilon) \leqq \varepsilon s_2^l(\varepsilon) \quad \text{for all } k \in \Phi_2^t \backslash \Psi_2^t \text{ and } l \in \Psi_2^t. \tag{3.5.4}$$

By using (3.5.1), (3.5.3) and (3.5.4) one can show, similarly as in the proof of Lemma 3.4.3, that

$$R(\bar{s}_1, s_2(\varepsilon)) > R(s_1, s_2(\varepsilon)) \quad \text{for all } \varepsilon \text{ sufficiently close to zero,}$$

but this contradicts the definition of $\{s(\varepsilon)\}_{\varepsilon\downarrow 0}$. This establishes the induction step, and so $s_1 \in O_1(\Gamma^t)$, for all $t \in \{0, \ldots, \tau\}$. Hence, $s_1 \in D_1(\Gamma)$. Similarly, it can be shown that $s_2 \in D_2(\Gamma)$.
(iii) → (i): Assume $s = (s_1, s_2) \in D(\Gamma)$ and let $\bar{s} = (\bar{s}_1, \bar{s}_2)$ be a proper equilibrium of Γ. From the first part of the theorem, it follows that $\bar{s} \in D(\Gamma)$, hence, by Lemma 3.5.3, we have that s and $\bar{s}$ are equivalent. For $\varepsilon > 0$, let $\bar{s}(\varepsilon)$ be an ε-proper equilibrium of Γ such that $\bar{s}(\varepsilon)$ converges to $\bar{s}$ as ε tends to zero. Define $s(\varepsilon)$ by:

$$s_i(\varepsilon) = (1-\varepsilon)s_i + \varepsilon \bar{s}_i(\varepsilon) \quad \text{for } i \in \{1, 2\}.$$

We will show that $s(\varepsilon)$ is an ε-proper equilibrium of Γ if ε is sufficiently close to zero. It is clear that $s(\varepsilon)$ is completely mixed. Assume $k, l \in \Phi_1$ and $\varepsilon > 0$ are such that

$$R(k, s_2(\varepsilon)) < R(l, s_2(\varepsilon)). \tag{3.5.5}$$

Then we have, if ε is sufficiently close to zero

$$\begin{aligned} &R(k, s_2) < R(l, s_2) \quad \text{or} \\ &R(k, s_2) = R(l, s_2) \quad \text{and} \quad R(k, \bar{s}_2(\varepsilon)) < R(l, \bar{s}_2(\varepsilon)). \end{aligned} \tag{3.5.6}$$

Since s_2 is equivalent to $\bar{s}_2$, Eq. (3.5.6) is equivalent to

$$\begin{aligned} &R(k, \bar{s}_2) < R(l, \bar{s}_2) \quad \text{or} \\ &R(k, \bar{s}_2) = R(l, \bar{s}_2) \quad \text{and} \\ &R(k, \bar{s}_2(\varepsilon)) < R(l, \bar{s}_2(\varepsilon)). \end{aligned} \tag{3.5.7}$$

If ε is sufficiently close to zero, (3.5.7) implies

$$R(k, \bar{s}_2(\varepsilon)) < R(l, \bar{s}_2(\varepsilon)), \tag{3.5.8}$$

and, therefore, if (3.5.5) is satisfied, we have

$$\bar{s}_1^k(\varepsilon) \leqq \varepsilon \bar{s}_1^l(\varepsilon)$$

since $\bar{s}(\varepsilon)$ is an ε-proper equilibrium of Γ. From Lemma 2.3.2 we know that $\bar{s}_1$ is a best reply against $\bar{s}_2(\varepsilon)$ if ε is small. This implies, since s_1 is equivalent to $\bar{s}_1$, that s_1 is a best reply against $\bar{s}_2(\varepsilon)$ if ε is small. Hence, if (3.5.8) is satisfied and ε is small, then $s_1^k = 0$. Therefore, if ε is sufficiently small and if (3.5.5) is satisfied

$$s_1^k(\varepsilon) = (1-\varepsilon)s_1^k + \varepsilon \bar{s}_1^k(\varepsilon) = \varepsilon \bar{s}_1^k(\varepsilon) \leqq \varepsilon^2 \bar{s}_1^l(\varepsilon) \leqq \varepsilon s_1^l(\varepsilon).$$

Similarly, we can prove, that for all ε sufficiently close to zero

$$\text{if } R(s_1(\varepsilon), k) > R(s_1(\varepsilon), l), \text{ then } s_2^k(\varepsilon) \leqq \varepsilon s_2^l(\varepsilon) \text{ for all } k, l \in \Phi_2,$$

and, therefore, $s(\varepsilon)$ is an ε-proper equilibrium of Γ if ε is sufficiently close to zero. Hence, s is a proper equilibrium of Γ. □

Since the set of D-optimal strategies of a matrix game can easily be determined (the only thing one has to do is to compute the optimal strategy sets of a finite number of matrix games, which is an easy task, cf. Balinski [1961]), we have that a proper equilibrium of a matrix game can be computed easily. Furthermore, as a consequence of Lemma 3.5.3 and Theorem 3.5.5 we have

Corollary 3.5.6. *If s, s' are weakly proper equilibria of a matrix game, then s and s' are equivalent.*

So, a matrix game has essentially one (weakly) proper equilibrium and therefore if one expects the opponent to make his mistakes in a rational way, there is a unique way to exploit these mistakes optimally.

Notes

1. An algorithm to find a perfect equilibrium in a bimatrix game has been described in Van den Elzen and Talman [1991]. The structure of the set of perfect equilibria of bimatrix games is investigated in Borm et al. [1988]. It is shown in that paper that the structure of this set resembles that of the Nash equilibrium set (see Theorem 3.3.1). Borm [1990] studies the class of $2 \times n$ games in detail. In addition to showing how one can find all perfect and all proper equilibria, he characterizes stable sets and persistent retracts. He finds that stable sets consist of either one or two perfect equilibria, that each stable component contains a proper equilibrium, and that each persistent equilibrium is perfect.
2. Also see Krohn et al. [1989].
3. An alternative (equivalent) definition of regularity in bimatrix games is given in Jansen [1987].

4 Control Costs

In this chapter, games with control costs* are studied. These are normal form games in which each player, in addition to receiving his payoff from the game, incurs costs depending on how well he chooses to control his actions. Such a game models the idea that a player can reduce the probability of making mistakes, but that he can only do so by being extra prudent, hence, by spending an extra effort, which involves some costs.[1] The goal of the chapter is to investigate what the consequences are of viewing an ordinary normal form game as a limiting case of a game with control costs, i.e. it is examined which equilibria are still viable when infinitesimal control costs are incorporated into the analysis of normal form games.[2]

In Sect. 4.1, the motives for studying games with control costs are discussed and the formal definition of such a game is given, while it is shown in Sect. 4.2 that a game with control costs always possesses at least one equilibrium. In this section, also a characterization of equilibria of such a game is derived.

In Sect. 4.3, a first analysis is performed on the question of which equilibria of an ordinary normal form game are still viable when control costs are taken into account, i.e. it is investigated which equilibria can be approximated by equilibria of games with control costs as these costs go to zero. It is shown that the set of equilibria which are approximated may depend on the control costs and that there exist equilibria (even perfect ones) which cannot be approximated.

In Sect. 4.4, it is shown that the basic assumption underlying the properness concept, viz. that a more costly mistake is made with a probability which is of smaller order, cannot be justified by the control cost approach. Namely, if the players incur control costs, then a non-optimal strategy is chosen with a probability which is approximately inversely proportional to the loss which is incurred if this strategy is played.

In Sect. 4.5, it is investigated whether control costs result in playing a perfect equilibrium. It is shown that this is the case only if the control costs are sufficiently high. For the special case of a bimatrix game, an exact formulation of the control costs being sufficiently high is given.

Finally, it is shown in Sect. 4.6, that a regular equilibrium is always approximated by equilibria of games with control costs as these costs go to zero.

The material in this chapter is based on van Damme [1982].

---

* The idea to study games was raised by R. Selten.

4.1 Introduction

Consider the 2-person normal form game of Fig. 4.1.1, in which M is some positive real number. The equilibria of this game are the strategy pairs in which player 2 plays his first strategy. What player 1 should do depends on how player 2 makes his mistakes. If we write p_k for the probability that player 2 mistakenly chooses $k(k \in \{2, 3\})$, then

if $p_2 > Mp_3$, player 1 should play his first strategy,

if $p_2 = Mp_3$, player 1 can play anything, and

if $p_2 < Mp_3$, player 1 should play his second strategy.

The perfectness concept does not give an opinion on which of these cases will prevail and, consequently, all equilibria are perfect. If player 1 believes in the properness concept (or weakly properness concept), however, then he should always (no matter what the value of M is) choose his first strategy since, according to this concept, p_3 is of smaller order than p_2.

In this game both the perfectness concept and the properness concept yield unsatisfactory answers: the perfectness concept does not discriminate between the Nash equilibria and the properness concept yields a solution which is independent of M, whereas, intuitively, one would say that player 1 should choose his first strategy for small values of M and his second strategy for large values of M.

The game of Fig. 4.1.1 illustrates a phenomenon which occurs for many games: Since the perfectness concept allows for all kinds of mistakes, there exists a plethora of perfect equilibria, some of which are unreasonable. Contrary to this, the properness concept severely restricts the possibilities for making mistakes, such that some unreasonable perfect equilibria are excluded. However, it is unclear whether it is appropriate to restrict oneself to proper equilibria since the basic assumption underlying this concept (viz. that more costly mistakes occur with a probability which is of smaller order) is questionable. In this chapter, as well as in the next one, we investigate whether this assumption can be justified. It will be clear that in order to judge the reasonableness of this assumption and in order to be able to refine the perfectness concept in a well-founded way, one needs a model in which the mistake probabilities are endogenously determined, rather than a model in which these probabilities can be chosen in some ad hoc way. In this

	1	2	3
1	2, 2	1, 1	0, 0
2	2, 2	0, 1	M, 0

Fig. 4.1.1. A bimatrix game to illustrate undesirable properties of the perfectness concept and the properness concept

chapter, one such model is developed. A second model will be considered in the next chapter.

Obviously, the reason that players make mistakes lies in the fact that they are not as careful as is needed to prevent these mistakes. In this chapter, we elaborate on the idea that the players are not as careful as is needed simply because it is too costly to be so careful. The basic idea behind the approach of this chapter is that a player can reduce his probability of making mistakes, but that this can be done only by being extra careful. If a player wants to be extra careful, this means that he actually has to do something, i.e. that he has to spend an extra effort which involves some costs. Hence, in this chapter, we assume that a player can reduce his probability of making mistakes, but that he incurs costs in doing so. Obviously, the more a player wants to reduce his mistake probability, the more effort he has to spend and, hence, the higher the cost he incurs. Furthermore, it will be clear that, if it becomes increasingly difficult (i.e. increasingly costly) to reduce the mistake probability further and further, then a player will choose to prevent mistakes only up to a certain level and, hence, each player will make mistakes with a positive probability. In this chapter, this is viewed as the basic reason why mistakes occur and it is investigated what the consequences of this view are, i.e. it is investigated whether this view leads to a refinement of the perfectness concept, whether it leads to a justification of the properness concept, etc.

Next, let us state the ideas outlined above more precisely. For the remainder of this section, let an n-person normal form game $\Gamma = (\Phi_1, \ldots, \Phi_n, R_1, \ldots, R_n)$ be fixed. Let s_i be a mixed strategy of player i in Γ. We will assume that player i incurs *control costs* $\sum_k f_i^k(s_i^k)$ if he wants to play s_i in Γ, where f_i^k represents the cost player i incurs due to his special attention to the strategy k. We will assume that player i can control all his pure strategies equally good (or equally bad), which implies that all f_i^k are equal to some function f_i. Hence, we assume

Assumption 4.1.1. $f_i^1 = \ldots = f_i^k \ldots = f_i^{m_i} =: f_i$.

The function f_i is called the *control cost function* of player i.

The following assumption has been motivated above.

Assumption 4.1.2. f_i: $[0, 1] \to \mathbb{R} \cup \{\infty\}$ is a decreasing function with $f_i(0) = \infty$.

Notice that we allow a control cost function to take the value ∞. However, from the fact that f_i is decreasing it follows that $f_i(x) < \infty$ if $x > 0$. The conventions which will be used with respect to $-\infty$ and ∞ are the same as in Rockefellar [1970]. The exact formulation of the assumption that it becomes increasingly difficult for player i to decrease his mistake probability further and further is

$$\text{if } 0 < \varepsilon < x < y, \text{ then } f_i(x - \varepsilon) - f_i(x) > f_i(y - \varepsilon) - f_i(y),$$

which is equivalent to saying that f_i is strictly convex.

Assumption 4.1.3. f_i is a strictly convex function.

Finally, we make an additional assumption for mathematical convenience.

Assumption 4.1.4. f_i is at least twice differentiable.

From now on, a *control cost function* is a function for which Assumptions 4.1.2–4.1.4 are satisfied. Let $f = (f_1, \ldots, f_n)$ be an n-tuple of such functions. If, in the game Γ, each player i, in addition to receiving a payoff as described by R_i, incurs a cost as determined by f_i, then the strategic situation is more adequately described by the infinite normal form game $\Gamma^f = (S_1, \ldots, S_n, R_1^f, \ldots, R_n^f)$, where

$$R_i^f(s) := R_i(s) - \sum_k f_i(s_i^k) \quad \text{for } i \in N, s \in S.$$

The game Γ^f is called a *game with control costs*. Note that in such a game a payoff of $-\infty$ can occur. Strategically, however, this is irrelevant since each player can guarantee a finite payoff by playing a completely mixed strategy.

If control costs are present in an ordinary normal form game, these costs will be of much less importance than the ordinary payoffs of the game. This can be modelled by considering infinitesimal control costs, i.e. by approximating an ordinary normal form game Γ with games $\Gamma^{\varepsilon f}$ by letting ε go to zero. In this chapter, it is investigated what the consequences are of doing so. Hence, we examine which equilibria of Γ are approximated by equilibria of $\Gamma^{\varepsilon f}$ as ε goes to zero. Such equilibria are called *f-approachable equilibria*, and we are particularly interested in aspects as:

(i) How does the set of f-approachable equilibria depend on f?
(ii) Is an f-approachable equilibrium perfect (resp. proper)?
(iii) Is a perfect (resp. proper) equilibrium f-approachable?

Before considering these questions, let us first study games with control cost in somewhat greater detail.

4.2 Games with Control Costs

In this section, we prove some elementary properties of games with control costs. In particular it is shown that any such game possesses at least one equilibrium. Furthermore, a condition which is necessary and sufficient for a strategy combination to be an equilibrium is derived.

Throughout the section we consider a fixed n-person normal form game $\Gamma = (\Phi_1, \ldots, \Phi_n, R_1, \ldots, R_n)$ and a fixed n-tuple of control cost functions $f = (f_1, \ldots, f_n)$. We write S_i for the set of mixed strategies of player i and, if $\varepsilon > 0$, then

$$S_i(\varepsilon) := \{s_i \in S_i;\ s_i^k \geqq \varepsilon \text{ for all } k\}, \quad S(\varepsilon) = \prod_{i=n}^{n} S_i(\varepsilon). \tag{4.2.1}$$

The object of study is the game Γ^f, defined as in the previous section. Throughout the section, when we speak of a best reply, we will mean a best reply in Γ^f.

Lemma 4.2.1. *There exists some $\varepsilon > 0$, such that for all $\bar{s} \in S$: if s is a best reply against $\bar{s}$, then $s \in S(\varepsilon)$.*

Proof. Let s' be an arbitrary completely mixed strategy combination. Since $f_i(0) = \infty$, we have for each $i \in N$ and $k \in \Phi_i$,

$$\lim_{s_i^k \to 0} R_i^f(\bar{s} \backslash s_i) < R_i^f(\bar{s} \backslash s_i'),$$

which proves the lemma. □

Lemma 4.2.2. *Every strategy combination has a unique best reply. The mapping which assigns to each strategy combination its best reply is continuous.*

Proof. Let $\bar{s} \in S$, let $\varepsilon > 0$ be as in Lemma 4.2.1 and let $i \in N$. From Lemma 4.2.1, we know that every best reply of player i against $\bar{s}$ is an element of $S_i(\varepsilon)$. Since $S_i(\varepsilon)$ is compact and since $s_i \to R_i^f(\bar{s} \backslash s_i)$ is continuous on $S_i(\varepsilon)$, we have that a best reply of player i against $\bar{s}$ exists. The best reply of player i must be unique since the mapping $s_i \to R_i^f(\bar{s} \backslash s_i)$ is strictly concave. Continuity follows in a standard way. □

The unique best reply against $\bar{s} \in S$ in the game Γ^f will be denoted by $b(\bar{s}, f) = (b_1(\bar{s}, f), \ldots, b_n(\bar{s}, f))$. We have

Lemma 4.2.3. *If $s, \bar{s} \in S$, then $s = b(\bar{s}, f)$ if and only if s is a solution to the following set of equations:*

$$R_i(\bar{s} \backslash k) - R_i(\bar{s} \backslash l) = f_i'(s_i^k) - f_i'(s_i^l) \quad \text{for all } i \in N,\ k,\ l \in \Phi_i. \tag{4.2.2}$$

Proof. $s = b(\bar{s}, f)$ if and only if for each $i \in N$ we have that s_i is a solution of the convex programming problem.

$$\text{maximize } R_i^f(\bar{s} \backslash s_i), \text{ subject to } s_i \in S_i. \tag{4.2.3}$$

Since every solution of (4.2.3) is in the interior of S_i, we have that $s_i \in S_i$ solves (4.2.3) if and only if

$$\frac{\partial R_i^f(\bar{s} \backslash s_i)}{\partial s_i^k} = \lambda_i \quad \text{for all } k \in \Phi_i, \tag{4.2.4}$$

where $\lambda_i \in \mathbb{R}$ is a Lagrange multiplier. This completes the proof since (4.2.2) is nothing else than a restatement of (4.2.4). □

Let $\bar{s} \in S$ and let $s = b(\bar{s}, f)$. Equation (4.2.2) shows that

$$\text{if } R_i(\bar{s} \backslash k) < R_i(\bar{s} \backslash l), \text{ then } f_i'(s_i^k) < f_i'(s_i^l) \text{ for all } i \in N,\ k,\ l \in \Phi_i.$$

Since f_i is convex, f_i' is increasing and, therefore, we have

Corollary 4.2.4. *Let $\bar{s} \in S$ and $s = b(\bar{s}, f)$. Then*

$$\textit{if } R_i(\bar{s} \backslash k) < R_i(\bar{s}, l), \textit{ then } s_i^k < s_i^l \textit{ for all } i \in N,\ k,\ l \in \Phi_i.$$

The following corollary of Lemma 4.2.3, will turn out to be useful in Sect. 4.6:

Lemma 4.2.5. *Let $\bar{s}\in S$, $i\in N$, $\Psi_i\subset\Phi_i$ and $s_i\in S_i$. If s_i is such that*

$$R_i(\bar{s}\backslash k)-R_i(\bar{s}\backslash l)=f_i'(s_i^k)-f_i'(s_i^l) \quad \text{for all } k,\ l\in\Psi_i, \text{ and} \tag{4.2.5}$$

$$s_i^k=b_i^k(\bar{s},f) \quad \text{for all } k\notin\Psi_i, \tag{4.2.6}$$

then $s_i=b_i(\bar{s},f)$.

Proof. From Lemma 4.2.3 we see that for all $k,\ l\in\Psi_i$:

$$f_i'(s_i^k)-f_i'(s_i^l)=R_i(\bar{s}\backslash k)-R_i(\bar{s}\backslash l)=f_i'(b_i^k(\bar{s},f))-f_i'(b_i^l(\bar{s},f)),$$

from which it follows, by using the monotonicity of f_i', that

$$\text{if } s_i^k<b_i^k(\bar{s},f) \text{ for some } k\in\Psi_i, \text{ then } s_i^k<b_i^k(\bar{s},f) \text{ for all } k\in\Psi_i. \tag{4.2.7}$$

Equation (4.2.6) has the consequence that

$$\sum_{k\in\Psi_i} s_i^k=\sum_{k\in\Psi_i} b_i^k(\bar{s},f), \tag{4.2.8}$$

and so, by combining (4.2.7) and (4.2.8), we see that $s_i^k=b_i^k(\bar{s},f)$ for all $k\in\Psi_i$. □

By combining the results of the Lemmas 4.2.1 and 4.2.3, we obtain the main result of this section:

Theorem 4.2.6. *The game Γ^f possesses at least one equilibrium. All equilibria of Γ^f are completely mixed. A strategy combination $s\in S$ is an equilibrium of Γ^f if and only if s is a solution to the following set of equations*

$$R_i(s\backslash k)-R_i(s\backslash l)=f_i'(s_i^k)-f_i'(s_i^l) \quad \text{for all } i\in N,\ k,\ l\in\Phi_i. \tag{4.2.9}$$

Proof. Let ε be as in Lemma 4.2.1. It follows from that lemma that s is an equilibrium of Γ^f if and only if s is an equilibrium of the game $(S_1(\varepsilon),\dots,S_n(\varepsilon), R_1^f,\dots,R_n^f)$. In view Theorem 2.1.1, the latter game has an equilibrium and, hence, also Γ^f has an equilibrium. The other assertions of the theorem follow directly from Lemma 4.2.1 and Lemma 4.2.3. □

4.3 Approachable Equilibria

In this section, we start investigating the consequences of viewing an ordinary normal form game as a limiting case of a game with control costs. Hence, for an n-person normal form game Γ and an n-tuple of control cost functions f, we consider the question which strategy combinations of Γ are approached by equilibria of $\Gamma^{\varepsilon f}$ as ε tends to zero. Theorem 4.3.1 shows that only equilibria of Γ can be approached in this way.

Theorem 4.3.1. *Let $\Gamma = (\Phi_1, \dots, \Phi_n, R_1, \dots, R_n)$ be an n-person normal form game and let $f = (f_1, \dots, f_n)$ be an n-tuple of control cost functions. If $s \in S$ is the limit of a sequence $\{s(\varepsilon)\}_{\varepsilon \downarrow 0}$ with $s(\varepsilon) \in E(\Gamma^{\varepsilon f})$ for all ε, then s is an equilibrium of Γ.*

Proof. Assume s is the limit of such a sequence $\{s(\varepsilon)\}_{\varepsilon \downarrow 0}$. Let $i \in N$ and $k \in \Phi_i$. We have to prove that $s_i^k = 0$ if k is not a best reply against s in Γ. For $\varepsilon > 0$ and $l \in \Phi_i$, we have, by (4.2.9)

$$R_i(s(\varepsilon)\backslash k) - R_i(s(\varepsilon)\backslash l) = \varepsilon\{f_i'(s_i^k(\varepsilon)) - f_i'(s_i^l(\varepsilon))\}. \tag{4.3.1}$$

If k is not a best reply against s in Γ, while l is a best reply against s, then the limit (as ε tends to zero) of the left hand side of (4.3.1) is negative, hence, in this case we have

$$\lim_{\varepsilon \downarrow 0} \{f_i'(s_i^k(\varepsilon)) - f_i'(s_i^l(\varepsilon))\} = -\infty,$$

which implies, since f_i is decreasing, that $s_i^k(\varepsilon)$ converges to zero as ε tends to zero. Hence, $s_i^k = 0$ if k is not a best reply against s in Γ. □

If one holds the view, that the players in a normal form game always incur some infinitesimal control costs, then one can consider as being reasonably only those equilibria which can be obtained as a limit in the way of Theorem 4.3.1. The fact that, in general, such a limit need not exist (see below), motivates the different approachability concepts introduced in Def. 4.3.2.

Definition 4.3.2. Let Γ be an n-person normal form game and let $f = (f_1, \dots, f_n)$ be an n-tuple of control cost functions. A strategy combination $s \in S$ is said to be *weakly f-approachable*, if s is a limit point of a collection $\{s(\varepsilon)\}_{\varepsilon \downarrow 0}$ with $s(\varepsilon) \in E(\Gamma^{\varepsilon f})$ for all ε; if s is actually the limit of such collection, then s is said to be *f-approachable*, and s is said to be *continuously f-approachable* if s is the limit of such a collection $\{s(\varepsilon)\}_{\varepsilon \downarrow 0}$ with the property that the mapping $\varepsilon \rightarrow s(\varepsilon)$ is continuous in a neighborhood of zero.

We have the following:

Theorem 4.3.3. *Let Γ and f be as in* Def. 4.3.2. *Then*

(i) *if s is weakly f-approachable, then $s \in E(\Gamma)$,*
(ii) *there exists at least one weakly f-approachable equilibrium of Γ,*
(iii) *if f is such that the equations in* (4.2.9) *are algebraic, then there exists at least one continuously f-approachable equilibrium of Γ.*

Proof. (i) follows in the same way as in the proof of Theorem 4.3.1, and (ii) follows from Theorem 4.2.6 and the compactness of S. The assertion in (iii) can be proved in the same way as in Harsanyi [1973b, Lemmas 2, 3, and 4]. □

In the remainder of this section, we elucidate the concepts introduced above by computing the set of approachable equilibria in a simple example (the game of

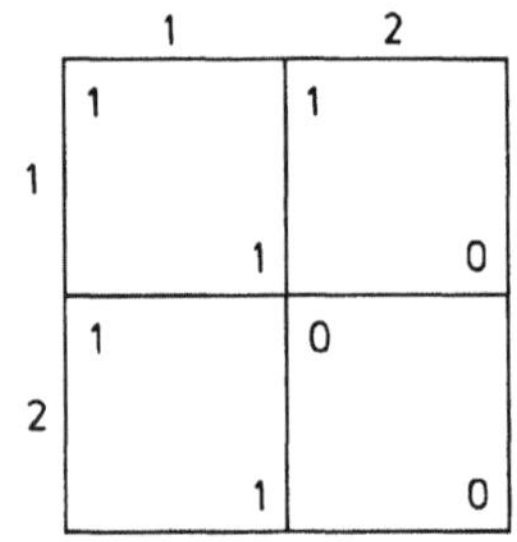

Fig. 4.3.1. A bimatrix game to illustrate the approachability concepts

Fig. 4.3.1). This example will serve to demonstrate the following properties of the approachability concepts:

(i) the set of weakly f-approachable equilibria may depend upon the choice of f,
(ii) there exist equilibria which fail to be weakly f-approachable for any choice of f, and
(iii) f-approachable equilibria need not exist.

The equilibria of the game Γ of Fig. 4.3.1 are the strategy pairs in which player 2 plays his first strategy. Let $f = (f_1, f_2)$ be a pair of control cost functions and, for $\varepsilon > 0$, let $(s_1(\varepsilon), s_2(\varepsilon))$ be an equilibrium of $\Gamma^{\varepsilon f}$. Because of Corollary 4.2.4, we have $s_1^1(\varepsilon) > \frac{1}{2}$ for all $\varepsilon > 0$ and, therefore, the equilibria with $s_1^1 < \frac{1}{2}$ fail to be weakly f-approachable for any choice of f.

Next, let us show that every equilibrium with $s_1^1 \geqq \frac{1}{2}$ is continuously f-approachable for some choice of f. Let us write $p(\varepsilon)$ for $s_1^2(\varepsilon)$ and $q(\varepsilon)$ for $s_2^2(\varepsilon)$. From (4.2.9) we know that $p(\varepsilon)$ and $q(\varepsilon)$ satisfy:

$$f_1'(1-p(\varepsilon)) - f_1'(p(\varepsilon)) = q(\varepsilon)/\varepsilon, \tag{4.3.2}$$

$$f_2'(1-q(\varepsilon)) - f_2'(q(\varepsilon)) = 1/\varepsilon. \tag{4.3.3}$$

For $i \in \{1, 2\}$, define the mapping g_i: $[0, \frac{1}{2}] \to [0, \infty]$ by

$$g_i(x) = f_i'(1-x) - f_i'(x).$$

Then (4.3.2) and (4.3.3) are equivalent to

$$g_1(p(\varepsilon)) = q(\varepsilon)/\varepsilon, \quad g_2(q(\varepsilon)) = 1/\varepsilon. \tag{4.3.4}$$

Since f_i is convex, g_i is decreasing and, hence, g_i has an inverse h_i: $[0, \infty] \to [0, \frac{1}{2}]$ which is continuous (as follows e.g. from the implicit function theorem). Equation (4.3.4) is equivalent to

$$p(\varepsilon) = h_1(q(\varepsilon)/\varepsilon), \quad q(\varepsilon) = h_2(1/\varepsilon), \tag{4.3.5}$$

from which it follows that $\Gamma^{\varepsilon f}$ has a unique equilibrium and that this equilibrium depends continuously on ε. Substituting the expression for $1/\varepsilon$ of (4.3.4) in the expression for $p(\varepsilon)$ in (4.3.5), yields

$$p(\varepsilon) = h_1[q(\varepsilon) g_2(q(\varepsilon))]. \tag{4.3.6}$$

Next, assume f_2 is such that

$$L(f_2) := \lim_{x \downarrow 0} x f'_2(x) \text{ exists,}$$

where we allow the possibility that $L(f_2) = -\infty$. Since $q(\varepsilon)$ converges to zero as ε tends to zero, we have

$$s_1^2 = \lim_{\varepsilon \downarrow 0} p(\varepsilon) = h_1(-L(f_2)),$$

from which it follows that every equilibrium with $s_1^2 \leqq \frac{1}{2}$ is continuously f-approachable for some f. Namely, choosing

$f_2(x) = 1/x$,	yields $L(f_2) = -\infty$	and, hence, $s_1^2 = 0$,
$f_2(x) = a \log x$ with $a < 0$,	yields $L(f_2) = a$	and, hence, $0 < s_1^2 < \frac{1}{2}$,
$f_2(x) = \log\lvert\log x\rvert$,	yields $L(f_2) = 0$	and, hence, $s_1^2 = \frac{1}{2}$.

Furthermore, it follows from (4.3.6) that f-approachable equilibria may not exist if f_2 is not nice. Namely, if f_2 is such that

$$\liminf_{x \downarrow 0} x f'_2(x) = -\infty \quad \text{and} \quad \limsup_{x \downarrow 0} x f'_2(x) = 0$$

(indeed some such f_2 which also satisfies Assumptions 4.1.1 – 4.1.4 can be constructed), then all equilibria with $s_1^2 \leqq \frac{1}{2}$ will be weakly f-approachable, but none of them is f-approachable.

Notice that, in the game of Fig. 4.3.1, the perfect equilibrium is obtained only if $L(f_2) = -\infty$. Intuitively this can be explained as follows. If this condition is fulfilled, the costs of player 2 increase very fast if this player tries to reduce his mistake probability, which means that player 2 has great trouble in controlling his actions. In this case, player 2 will play his second strategy with a relatively large probability, with the consequence that the second strategy of player 1 is much worse than his first strategy, which in turn implies that player 1 will choose his second strategy with a small probability. If, however, $L(f_2) > -\infty$, then player 2 is capable of choosing his second strategy with a relatively small probability, which implies that the second strategy of player 1 is only a little bit worse than his first one, in which case the control costs force player 1 to choose his second strategy with a relatively large probability.

In Sect. 4.5 we will see that, if both players in a bimatrix game have great trouble in controlling their actions, then the control costs will result in a perfect equilibrium being played (Theorem 4.5.1).

We conclude this section by presenting an example which shows that there exist perfect equilibria which fail to be weakly f-approachable for any choice of f. Namely, consider the game of Fig. 4.1.1 with $M = 1$. All equilibria of this game are perfect, however, the equilibria in which player 1 assigns a probability greater than $\frac{1}{2}$ to his second strategy can never be weakly f-approachable, as follows from Corollary 4.2.4.

As a consequence of Corollary 4.2.4, the unreasonable perfect equilibria in which a player chooses a more costly mistake with a greater probability than a less costly one are excluded by the approachability concept. Notice, however, that Corollary 4.2.4 essentially uses Assumption 4.1.1. If this assumption would not be satisfied and if, for instance, in the game of Fig. 4.1.1 (with $M=1$) player 2 could control his second strategy much better than his third, then it might be optimal for player 1 to choose his second strategy. Hence, to reach the conclusion that there exist perfect equilibria which fail to be weakly f-approachable for any choice of f, one essentially needs the assumption that a player can control all his strategies equally good. However, even with non-identical control costs some Nash equilibria will still be unapproachable. Namely, the equilibrium (R_1, R_2) of the game of Fig. 1.5.1 fails to be weakly f-approachable for any choice of f, even if Assumption 4.1.1 is dropped.

4.4 Proper Equilibria

The basic assumption underlying the concepts of proper and weakly proper equilibria is that, if a pure strategy k is worse than a pure strategy l, then k will be mistakenly chosen with a probability which is an order smaller than the probability with which l is mistakenly chosen. Our objective in this section is to show that this assumption cannot be justified by means of the control cost approach.

To make the above statement more precise, consider an n-person normal form game $\Gamma = (\Phi_1, \ldots, \Phi_n, R_1, \ldots, R_n)$. If s is a weakly proper equilibrium of Γ and if $\{s(\varepsilon)\}_{\varepsilon \downarrow 0}$ is as in Def. 2.3.4, then for all i, k, l

$$\text{if } R_i(s \backslash k) < R_i(s \backslash l), \text{ then } \lim_{\varepsilon \downarrow 0} s_i^k(\varepsilon)/s_i^l(\varepsilon) = 0. \tag{4.4.1}$$

In this section, however, it will be shown that if f is an n-tuple of control cost functions and if the equilibrium s of Γ is continuously approached by $s(\varepsilon) \in E(\Gamma^{\varepsilon f})$, then for all i, k, l

$$\text{if } R_i(s \backslash k) < R_i(s \backslash l) < \max_{k' \in \Phi_i} R_i(s \backslash k'), \text{ then } \limsup_{\varepsilon \downarrow 0} s_i^k(\varepsilon)/s_i^l(\varepsilon) > 0. \tag{4.4.2}$$

Hence, if a player incurs control costs, then he will not make a more costly mistake with a probability which is an order smaller than the probability of a less costly one.

This section is devoted to the proof of (4.4.2), as well as several specializations of it for cases in which the control cost functions satisfy some extra conditions. For the remainder of the section, let an n-person normal form game $\Gamma = (\Phi_1, \ldots, \Phi_n, R_1, \ldots, R_n)$ and an n-tuple of control cost functions $f = (f_1, \ldots, f_n)$ be fixed. Furthermore, assume $s \in E(\Gamma)$ is continuously approached by $s(\varepsilon) \in (\Gamma^{\varepsilon f})$ as ε tends to 0. Let $i \in N$ and assume, without loss of generality, that $s_i^1 > 0$. Finally, suppose $k, l \in \Phi_i$ are such that

$$R_i(s \backslash k) < R_i(s \backslash l) < R_i(s \backslash 1) \left(= \max_{k' \in \Phi_i} R_i(s \backslash k') \right). \tag{4.4.3}$$

To simplify notation, we write,

$$\varkappa(\varepsilon) := s_i^k(\varepsilon), \qquad \lambda(\varepsilon) := s_i^l(\varepsilon), \qquad \mu(\varepsilon) := s_i^1(\varepsilon). \tag{4.4.4}$$

The proof of (4.4.2) is based on Eq. (4.2.9). From that equation we see that for every $\varepsilon>0$

$$\frac{f_i'(\lambda(\varepsilon))}{f_i'(\varkappa(\varepsilon))}=\frac{R_i(s(\varepsilon)\backslash 1)-R_i(s(\varepsilon)\backslash l)-\varepsilon f_i'(\mu(\varepsilon))}{R_i(s(\varepsilon)\backslash 1)-R_i(s(\varepsilon)\backslash k)-\varepsilon f_i'(\mu(\varepsilon))}, \tag{4.4.5}$$

from which it follows that for every $\varepsilon>0$

$$\frac{\varkappa(\varepsilon)}{\lambda(\varepsilon)}=\frac{\varkappa(\varepsilon)f_i'(\varkappa(\varepsilon))[R_i(s(\varepsilon)\backslash 1)-R_i(s(\varepsilon)\backslash l)-\varepsilon f_i'(\mu(\varepsilon))]}{\lambda(\varepsilon)f_i'(\lambda(\varepsilon))[R_i(s(\varepsilon)\backslash 1)-R_i(s(\varepsilon)\backslash k)-\varepsilon f_i'(\mu(\varepsilon))]}. \tag{4.4.6}$$

Since both k and l are not a best reply against s, we have that $\varkappa(\varepsilon)$ and $\lambda(\varepsilon)$ converge to 0 as ε tends to 0. Since, furthermore, $\mu(\varepsilon)$ tends to $s_i^1>0$, we immediately deduce from (4.4.6)

Theorem 4.4.1. *If* $\lim_{x\downarrow 0} xf_i'(x)$ *exists and is finite and unequal to* 0, *then*

$$\lim_{\varepsilon\downarrow 0}\frac{s_i^k(\varepsilon)}{s_i^l(\varepsilon)}=\frac{R_i(s\backslash 1)-R_i(s\backslash l)}{R_i(s\backslash 1)-R_i(s\backslash k)}.$$

Hence if the condition of Theorem 4.4.1 is satisfied, a pure strategy which is not a best reply is chosen with a probability which is inversely proportional to the loss the player incurs when he plays this strategy. In Theorem 4.4.2 we show that a similar property holds in case $\lim_{x\downarrow 0} xf_i'(x)=-\infty$.

Theorem 4.4.2. *If* $\lim_{x\downarrow 0} xf_i'(x)=-\infty$, *then*

$$\limsup_{\varepsilon\downarrow 0}\frac{s_i^k(\varepsilon)}{s_i^l(\varepsilon)}\geq\frac{R_i(s\backslash 1)-R_i(s\backslash l)}{R_i(s\backslash 1)-R_i(s\backslash k)}.$$

Proof. The theorem immediately follows from (4.4.6) if there exists a sequence $\{\varepsilon_t\}_{t\in\mathbb{N}}$ converging to 0 such that

$$\varkappa(\varepsilon_t)f_i'(\varkappa(\varepsilon_t))\leqq\lambda(\varepsilon_t)f_i'(\lambda(\varepsilon_t))\quad\text{for all } t.$$

Assume such a sequence does not exist and let $\varepsilon_1>0$ be such that

$$\varkappa(\varepsilon)f_i'(\varkappa(\varepsilon))>\lambda(\varepsilon)f_i'(\lambda(\varepsilon))\text{ for all }\varepsilon\in(0,\varepsilon_1]\text{, and} \tag{4.4.7}$$

$$\varkappa(\varepsilon)<\lambda(\varepsilon)\qquad\text{for all }\varepsilon\in(0,\varepsilon_1]. \tag{4.4.8}$$

Note that such ε_1 can be found because of Corollary 4.2.4. Since $\lambda(\varepsilon)$ depends continuously on ε and converges to 0 as ε tends to 0, we can define a sequence $\{\varepsilon_t\}_{t\in\mathbb{N}}$ by

$$\varepsilon_{t+1}:=\min\{\varepsilon\in(0,\varepsilon_t);\ \lambda(\varepsilon)=\varkappa(\varepsilon_t)\}. \tag{4.4.9}$$

Let us write $\varkappa_t$ (resp. λ_t) for $\varkappa(\varepsilon_t)$ (resp. $\lambda(\varepsilon_t)$). Since ε_t converges to 0 as t tends to infinity, λ_t converges also to 0. Furthermore, we have

$$\lambda_{t+1}f_i'(\lambda_{t+1})=\varkappa_t f_i'(\varkappa_t)>\lambda_t f_i'(\lambda_t)\quad\text{for all } t\in\mathbb{N},$$

which contradicts the condition of the theorem. □

Before proving (4.4.2), we consider another special case in which a similar result as in Theorem 4.4.2 holds.

Theorem 4.4.3. *If the limit in* Eq. (4.4.10) *exists, then*

$$\lim_{\varepsilon \downarrow 0} \frac{s_i^k(\varepsilon)}{s_i^l(\varepsilon)} \geqq \frac{R_i(s\backslash 1) - R_i(s\backslash l)}{R_i(s\backslash 1) - R_i(s\backslash k)} \,. \tag{4.4.10}$$

Proof. Assume that the limit in (4.4.10) exists, but that the inequality is not valid. Let ε_1 be such that (4.4.8) is satisfied, define $\{\varepsilon_t\}_{t \in \mathbb{N}}$ as in (4.4.9) and let $\varkappa_t$ and $\lambda_t (t \in \mathbb{N})$ be defined as above. Our assumptions imply that

$$\lim_{t \to \infty} \frac{\lambda_t - \lambda_{t+1}}{\lambda_{t-1} - \lambda_t} = \lim_{t \to \infty} \frac{\lambda_t}{\lambda_{t-1}} < \frac{R_i(s\backslash 1) - R_i(s\backslash l)}{R_i(s\backslash 1) - R_i(s\backslash k)} \,,$$

which, when combined with (4.4.5), yields

$$\lim_{t \to \infty} \frac{(\lambda_t - \lambda_{t+1}) f_i'(\lambda_{t+1})}{(\lambda_{t-1} - \lambda_t) f_i'(\lambda_t)} < 1 \,,$$

which in turn implies that

$$\sum_t (\lambda_t - \lambda_{t+1}) f_i'(\lambda_{t+1}) > -\infty \,.$$

Therefore we have, since f_i' is increasing

$$\int_0^{\lambda_1} f_i'(x)\,dx = \sum_t \int_{\lambda_{t+1}}^{\lambda_t} f_i'(x)\,dx \geqq \sum_t (\lambda_t - \lambda_{t+1}) f_i'(\lambda_{t+1}) > -\infty \,,$$

but this contradicts the fact that $f_i(0) = \infty$. □

As an immediate consequence of Theorem 4.4.3, we have

Corollary 4.4.4. *The basic assumption underlying the (weakly) properness concept cannot be justified by the control cost approach* (i.e. *the statement in* (4.4.2) *is correct*).

4.5 Perfect Equilibria

In Sect. 4.3, we saw that f-approachable equilibria need not be perfect: the perfect equilibrium of the game of Fig. 4.3.1 is f-approachable only if the control costs have considerable influence on the strategy choice of player 2, i.e. only if $\lim_{x \downarrow 0} x f_2'(x) = -\infty$. By reversing the roles of the players it is clear that also $\lim_{x \downarrow 0} x f_1'(x) = -\infty$ is necessary to ensure that for every bimatrix game only perfect equilibria will be f-approachable. In the next theorem we show that these conditions are also sufficient to guarantee perfectness for bimatrix games.

Theorem 4.5.1. *Let $f=(f_1, f_2)$ be a pair of control cost functions such that*

$$\lim_{x \downarrow 0} x f'_i(x) = -\infty \quad \text{for } i=1, 2.$$

Then, for every bimatrix game, every weakly f-approachable equilibrium is perfect.

Proof. Let $\Gamma=(\Phi_1, \Phi_2, R_1, R_2)$ be a bimatrix game. We will prove that every f-approachable equilibrium is perfect (the general case can be proved by the same methods). Hence, let s be an f-approachable equilibrium of Γ and, for $\varepsilon>0$, let $s(\varepsilon) \in E(\Gamma^{\varepsilon f})$ be such that $s(\varepsilon)$ converges to s as ε tends to 0. In view of Theorem 3.2.2, if suffices to show that s is undominated. Assume, to the contrary, that s_1 is dominated by $\bar{s}_1$. We distinguish two cases:
Case I: $C(\bar{s}_1) \not\subset C(s_1)$. Since every $s_2(\varepsilon)$ is completely mixed, we have

$$R_1(\bar{s}_1, s_2(\varepsilon)) > R_1(s_1, s_2(\varepsilon)) \quad \text{for all } \varepsilon>0$$

and, therefore, for every $\varepsilon>0$, there exist $k_\varepsilon \in C(s_1)$ and $\bar{k}_\varepsilon \in C(\bar{s}_1) \backslash C(s_1)$, such that

$$R_1(\bar{k}_\varepsilon, s_2(\varepsilon)) > R_1(k_\varepsilon, s_2(\varepsilon)).$$

Corollary 4.2.4 leads to the conclusion that

$$0 = \lim_{\varepsilon \downarrow 0} s_1^{\bar{k}_\varepsilon}(\varepsilon) \geqq \liminf_{\varepsilon \downarrow 0} s_1^{k_\varepsilon}(\varepsilon) \geqq \min_{k \in C(s_1)} s_1^k > 0.$$

The contradiction shows that Case I cannot occur.
Case II: $C(\bar{s}_1) \subset C(s_1)$. Since $\frac{1}{2}(s_1+\bar{s}_1)$ dominates s_1, we can assume $C(\bar{s}_1)=C(s_1)$. Let $l \in \Phi_2$ be such that

$$R_1(\bar{s}_1, l) > R_1(s_1, l). \tag{4.5.1}$$

Then $l \notin C(s_2)$. Without loss of generality, assume $1 \in C(s_2)$. From (4.2.9) we obtain that for every $\varepsilon>0$

$$[R_2(s_1(\varepsilon), 1) - R_2(s_1(\varepsilon), l)]/\varepsilon = f'_2(s_2^1(\varepsilon)) - f'_2(s_2^l(\varepsilon)). \tag{4.5.2}$$

Multiplying both sides of (4.5.2) with $s_2^l(\varepsilon)$ and using the condition of the theorem together with the fact that $s_2^l(\varepsilon)$ converges to 0 as ε tends to 0, yields

$$\lim_{\varepsilon \downarrow 0} s_2^l(\varepsilon)/\varepsilon = \infty. \tag{4.5.3}$$

Applying (4.2.9) with respect to player 1 yields that for all $k, \bar{k} \in C(s_1)$ and all $\varepsilon>0$

$$[R_1(\bar{k}, s_2(\varepsilon)) - R_1(k, s_2(\varepsilon))]/\varepsilon = f'_1(s_1^{\bar{k}}(\varepsilon)) - f'_1(s_1^k(\varepsilon)),$$

from which it follows, by multiplying both sides with $\bar{s}_1^{\bar{k}} s_1^k$ and summing over all $\bar{k}$, $k \in C(s_1)$, that

$$\begin{aligned}&[R_1(\bar{s}_1, s_2(\varepsilon)) - R_1(s_1, s_2(\varepsilon))]/\varepsilon \\ &\quad = \sum_{\bar{k} \in C(s_1)} \bar{s}_1^{\bar{k}} f'_1(s_1^{\bar{k}}(\varepsilon)) - \sum_{k \in C(s_1)} s_1^k f'_1(s_1^k(\varepsilon)).\end{aligned} \tag{4.5.4}$$

The limit, as ε goes to 0, of the right hand side of (4.5.4) is finite. For the left hand side we have, since $\bar{s}_1$ dominates s_1 and because of (4.5.1) and (4.5.3):

$$\lim_{\varepsilon \downarrow 0} [R_1(\bar{s}_1, s_2(\varepsilon)) - R_1(s_1, s_2(\varepsilon))]/\varepsilon$$
$$\geqq \lim_{\varepsilon \downarrow 0} [R_1(\bar{s}_1, l) - R_1(s_1, l)] s_2^l(\varepsilon)/\varepsilon = \infty ,$$

The contradiction shows that s_1 cannot be dominated. Similarly it can be shown that s_2 is undominated and, therefore, s is a perfect equilibrium of Γ. □

The reader might wonder whether Theorem 4.5.1 can be generalized to n-person games. A simple example can serve to show that, in order to guarantee perfectness in this case, the control costs have to satisfy more stringent conditions. Namely, consider the game Γ described by the following rules:

(i) There are n players, every one of them having 2 pure strategies.
(ii) Each player $i \in \{2, \ldots, n\}$ receives 1 if he plays his first strategy and 0 if he plays his second strategy.
(iii) The first strategy of player 1 yields him 1 if all players play 1 or if all other players play 2 and 0 otherwise; his second strategy yields 1 if all other players play 1 and 0 otherwise.

Γ can be considered as an n-person analogon of the game of Fig. 4.3.1.

Let $f = (f_1, \ldots, f_1)$ be an n-tuple of identical control cost functions. Then, in the same way as in Sect. 4.3, it is seen that the perfect equilibrium of Γ (in which all players choose their first strategy) is f-approachable if and only if $\lim_{x \downarrow 0} x^{n-1} f'_1(x) = -\infty$. Hence, in order to obtain perfectness for n-person games, the control costs have to to infinity very fast. It is unknown to the author whether, for an n-tuple of control cost functions $f = (f_1, \ldots, f_n)$ which is such that $\lim_{x \downarrow 0} x^{n-1} f'_i(x) = -\infty$ for all $i \in N$, only perfect equilibria of an n-person game are f-approachable.

The results obtained in this section are of some relevance with respect to the Harsanyi/Selten solution theory for noncooperative games (Harsanyi [1978], Harsanyi and Selten [1988]). An essential element in this theory is the *tracing procedure* (or more precisely the logarithmic tracing procedure, (Harsanyi [1975]), which is a mathematical procedure[3] to determine a unique solution of a noncooperative game, once one has given for each player i a probability distribution p_i, representing the other players' initial expectations about player i's likely strategy choice (how these p_i's should be determined is another important element of the theory). To determine the Harsanyi/Selten solution of a noncooperative game Γ, however, one cannot apply the theory directly to Γ, but rather one has apply the theory to a sequence of perturbed games $\{(\Gamma, \eta)\}_{\eta \downarrow 0}$ (this is necessary to ensure that the solution is a perfect equilibrium). If it would be the case that the tracing procedure would always (no matter which prior is chosen) end up with a perfect equilibrium, then one could circumvent this roundabout way and apply the theory directly to Γ. (This would simplify the theory considerably.) Unfortunately, the tracing procedure may yield a non-perfect equilibrium for some priors. Namely, the logarithmic tracing procedure involves approximating a normal form game with games with logarithmic control costs and since logarithmic functions do not satisfy the condition of Theorem 4.5.1, one may expect non-perfect equilibria. A simple example where this occurs is the game of Fig. 4.3.1, but this example is not really convincing, since Harsanyi argues that one should first eliminate all dominated pure strategies, before applying the tracing procedure (Harsanyi [1975], p. 69). However, also examples without dominated pure strategies, in which, nevertheless, the tracing procedure yields a non-perfect equilibrium can be constructed.

4.6 Regular Equilibria

In Sect. 4.3, we saw that, for a given game Γ, the set of equilibria of Γ which are weakly f-approachable may depend upon the choice of f and, that there does not necessarily exist an equilibrium of Γ which is weakly f-approachable for every possible choice of f. This raises the question which conditions an equilibrium has to satisfy in order to be weakly f-approachable for every possible choice of the control cost functions f. This question is answered in Theorem 4.6.1, the proof of which is the subject of this section.

Theorem 4.6.1. *A regular equilibrium is f-approachable for every possible choice of f.*

Proof. Assume s is a regular equilibrium of an n-person normal form game $\Gamma = (\Phi_1, \ldots, \Phi_n, R_1, \ldots, R_n)$ and let $f = (f_1, \ldots, f_n)$ be an n-tuple of control cost functions. We have to construct, for every sufficiently small $\varepsilon > 0$, an equilibrium $s(\varepsilon)$ of $\Gamma^{\varepsilon f}$, such that $s(\varepsilon)$ converges to s as ε tends to 0. This desired $s(\varepsilon)$ will be constructed by using the implicit function theorem in combination with Brouwer's Fixed Point Theorem.

Let us first fix our notation. We write

$$\Psi_i := \Phi_i \backslash C(s_i) \qquad \Psi := \prod_{i=1}^{n} \Psi_i,$$

$$X_i := \mathscr{F}(\Phi_i, \mathbb{R}) \qquad X := \prod_{i=1}^{n} X_i,$$

$$Y_i := \mathscr{F}(C(s_i), \mathbb{R}) \qquad Y := \prod_{i=1}^{n} Y_i,$$

$$Z_i := \mathscr{F}(\Psi_i, \mathbb{R}) \qquad Z := \prod_{i=1}^{n} Z_i.$$

A generic element of X_i is denoted by x_i, etc. If $x \in X$, then we write $x = (y, z)$ where $y \in Y$ and $z \in Z$. The equilibrium s is viewed both as an element of X and as an element of Y. The restriction of $x \in X$ to Z is denoted by ζx and $b(x, \varepsilon)$ denotes the best reply against x in the game $\Gamma^{\varepsilon f}$. Finally, without loss of generality, it is assumed that $(1, \ldots, 1) \in C(s)$.

Since s is a quasi-strict equilibrium of Γ (Corollary 2.5.3), there exists a neighborhood $\mathscr{X}$ of s in X, such that

$$x_i^k > 0 \qquad \text{for all } x \in \mathscr{X},\ i \in N \text{ and } k \in C(s_i),$$

$$R_i(x \backslash k) < R_i(x \backslash 1) \quad \text{for all } x \in \mathscr{X},\ i \in N \text{ and } k \in \Psi_i,$$

from which it follows, as in the proof of Theorem 4.3.1, that

$$\lim_{\varepsilon \downarrow 0} \zeta b(x, \varepsilon) = 0 \quad \text{for all } x \in \mathscr{X}. \tag{4.6.1}$$

Next, consider the mapping $F\colon \mathscr{X} \times \mathbb{R} \to Y$ defined by

$$F_i^k(x, \varepsilon) := R_i(x\backslash k) - R_i(x\backslash 1) - \varepsilon[f_i'(x_i^k) - f_i'(x_i^1)] \quad \text{for } i \in N,\ k \in C(s_i),\ k \neq 1, \tag{4.6.2}$$

$$F_i^{k_i}(x, \varepsilon) := \sum_k x_i^k - 1 \quad \text{for } i \in N. \tag{4.6.3}$$

F is a differentiable mapping and since s is a regular equilibrium of Γ we have

$$\frac{\partial F(x, \varepsilon)}{\partial y}\Big|_{(s,0)} \text{ is nonsingular.}$$

Hence, by the Implicit Function Theorem (Dieudonné [1960], p. 268) there exist neighborhoods $\mathscr{Y}$ of s in Y, $\mathscr{Z}$ of $0 = \zeta s$ in Z and $\mathscr{E}$ of 0 in $\mathbb{R}_+$, as well as a differentiable mapping $\boldsymbol{y}\colon \mathscr{Z} \times \mathscr{E} \to \mathscr{Y}$, such that

$$\{(y, z, \varepsilon) \in \mathscr{Y} \times \mathscr{Z} \times \mathscr{E}; F(y, z, \varepsilon) = 0\} = \{(\boldsymbol{y}(z, \varepsilon), z, \varepsilon); z \in \mathscr{Z}, \varepsilon \in \mathscr{E}\}. \tag{4.6.4}$$

Since every open neighborhood contains a closed neighborhood (i.e. closed set with nonempty interior), we can choose $\mathscr{Y}$ and $\mathscr{Z}$ such that

$$\mathscr{Y} \times \mathscr{Z} \subset \mathscr{X} \text{ and } \mathscr{Z} \text{ is compact and convex.} \tag{4.6.5}$$

Define mappings $\boldsymbol{x}$ and $\boldsymbol{z}$ with domain $\mathscr{Z} \times \mathscr{E}$, by

$$\boldsymbol{x}(z, \varepsilon) := (\boldsymbol{y}(z, \varepsilon), z), \quad \boldsymbol{z}(z, \varepsilon) := \zeta b(\boldsymbol{x}(z, \varepsilon), \varepsilon).$$

Note that both mappings are continuous, because of Lemma 4.2.2. In view of (4.6.1) and (4.6.3), we choose $\mathscr{E}$ so small that

$$\boldsymbol{z}(z, \varepsilon) \in \mathscr{Z} \quad \text{for all } z \in \mathscr{Z} \text{ and } \varepsilon \in \mathscr{E}. \tag{4.6.6}$$

Let $\mathscr{E}$ be such that (4.6.6) is satisfied and let $\varepsilon \in \mathscr{E}$. The mapping $z \to \boldsymbol{z}(z, \varepsilon)$ is a continuous mapping from the nonempty compact and convex set $\mathscr{Z}$ into itself and, hence, by Brouwer's Fixed Point Theorem (Lefschetz [1949], p. 117), it has a fixed point. Let $z(\varepsilon)$ be a fixed point of this mapping and let us write $x(\varepsilon)$ for $\boldsymbol{x}(z(\varepsilon), \varepsilon)$. From (4.6.4) we can conclude that

$$R_i(x(\varepsilon)\backslash k) - R_i(x(\varepsilon)\backslash 1) = \varepsilon[f_i'(x_i^k(\varepsilon)) - f_i'(x_i^1(\varepsilon))] \quad \text{for } i \in N, k \in C(s_i) \tag{4.6.7}$$

and

$$\sum_k x_i^k(\varepsilon) = 1 \quad \text{for } i \in N. \tag{4.6.8}$$

Furthermore, since $z(\varepsilon)$ is a fixed point of the mapping $z \to \zeta b(\boldsymbol{x}(z, \varepsilon), \varepsilon)$, we have

$$\zeta x(\varepsilon) = z(\varepsilon) = \zeta b(x(\varepsilon), \varepsilon). \tag{4.6.9}$$

From (4.6.7) – (4.6.9) it follows, by applying Lemma 4.2.5, that $x(\varepsilon)$ is an equilibrium of $\Gamma^{\varepsilon f}$. This completes the proof since $x(\varepsilon)$ converges to s as ε tends to 0. □

Notes

1. In Stahl [1990a] an alternative way of introducing control costs is proposed and analyzed. In Stahl's model a player plays all pure strategies with the same probability if he doesn't exert any effort. If a player wants to play a different strategy he incurs control costs that depend on the entropy reduction that he wants to obtain. Another model in which a player has to choose both target strategies and 'desired' error probabilities is described in Beja [1990].
2. There is a close relation between the theory developed in this chapter and the theory of boundedly rational behavior developed in Rosenthal [1989].
3. Some of the statements in Harsanyi [1975] concerning the properties of the tracing procedure are imprecise. For further details and mathematically correct formulations, see Schanuel, Simon and Zame [1991].

5 Incomplete Information

Games with incomplete information are games in which some of the data are unknown to some of the players. In this chapter, a particular class of games with incomplete information, the class of disturbed games, is studied. A disturbed game is a normal form game in which each player, although knowing his own payoff function exactly, has only imprecise information about the payoff functions of his opponents. We study such games since we feel that it is more realistic to assume that each player always has some slight uncertainty about the payoffs of his opponents rather than to assume that he knows these payoffs exactly. Our objective in this chapter is to study what the consequences are of this more realistic point of view.[1]

Disturbed games are introduced informally in Sect. 5.1 and formally in Sect. 5.2. In this latter section it is also shown that every disturbed game possesses an equilibrium, provided that some regularity condition is satisfied.

In Sect. 5.3, it is shown that equilibria of disturbed games converge to equilibria of undisturbed games as the disturbances go to 0. Equilibria which can be approximated in this way are called $\mathscr{P}$-firm equilibria, where $\mathscr{P}$ summerizes the characteristics of the disturbances (i.e. of the uncertainty the players have). Since there exist equilibria which fail to be $\mathscr{P}$-firm whatever $\mathscr{P}$ is, not every equilibrium of a normal form game is viable when the slight uncertainty which each player has about his opponents' payoffs is taken into account.

The main result of Sect. 5.4 states that, if the uncertainty about the payoffs is a special kind, this uncertainty will force the players to play a perfect equilibrium, hence, under some conditions on $\mathscr{P}$, every $\mathscr{P}$-firm equilibrium is perfect. However, there exist perfect equilibria which fail to be $\mathscr{P}$-firm for every choice of $\mathscr{P}$.

If the uncertainty about the payoffs is of a very special kind, the players will be forced to play a weakly proper equilibrium, as is shown in Sect. 5.5. This implies that the assumption that considerably worse mistakes are chosen with probabilities that are of smaller order is justifiable. The properness concept, however, cannot be justified by the approach of this chapter.

In general, the set of equilibria which are $\mathscr{P}$-firm may depend upon the choice of $\mathscr{P}$ (i.e. upon the exact characteristics of the disturbances). However, in Sect. 5.6, it is shown that every strictly proper equilibrium of a normal form game Γ is $\mathscr{P}$-firm for all disturbances $\mathscr{P}$ which occur only with a small probability and that every regular equilibrium is $\mathscr{P}$-firm for all disturbances $\mathscr{P}$.

Finally, Sect. 5.7 is devoted to the (technical) proofs of the results of Sect. 5.5.

5.1 Introduction

In the preceding chapters we have assumed that each participating player knows the payoff functions (utility functions) of the other players exactly. This assumption, however, is questionable, due to the subjective character of the utility concept. It is more realistic to assume that each player, although knowing his own payoff function exactly, has only somewhat imprecise information about the payoffs of the other players. In this chapter, the consequences of this more realistic point of view will be investigated. Hence, we will examine which influence inexact information about the payoffs has on the strategy choices in a normal form game.

To get a feeling for what these consequences might be, consider the game Γ of Fig. 5.1.1. Γ has two equilibria, viz. $(1, 1)$ and $(2, 2)$. Next, let us suppose that each player only knows that his own payoffs are as in Γ and that the payoffs of his opponent are approximately as described by Γ. In this case, is it still sensible for player 1 to play his second strategy? It is only sensible for him to do so, if he is absolutely sure that player 2 will play his second strategy. However, since he is not sure that the payoffs of player 2 are actually as described by Γ, he cannot be sure of this. Namely, the actual payoffs of player 2 might be such that his first strategy strictly dominates his second in which case player 2 will certainly play his first strategy. Hence, there is a positive probability that player 2 will play his first strategy, which implies that the only rational choice for player 1 is to play his first strategy. Similarly it is seen that also player 2 has to play his first strategy and, hence, only the equilibrium $(1, 1)$ is viable when each player has somewhat imprecise information about the other player's payoffs.

The example shows that there exist equilibria which are not viable when slight uncertainty about payoffs is taken into account. Our aim in this chapter is to investigate which equilibria are still viable in this case. To model the situation in which each player is uncertain about the payoffs of the other players, we will follow the approach as proposed in Harsanyi [1968, 1973a]. The basic assumption underlying this approach is that this uncertainty is caused by the fact that each player's utility function is subject to small random fluctuations as a result of changes in this player's mood or taste of which the precise effects are known only to this player himself. Hence, we will consider games with randomly fluctuating payoffs (which will be called *disturbed games*) rather than games with a priori fixed and constant payoffs. According to this model, a game in which each

	1	2
1	1, 1	0, 0
2	0, 0	0, 0

Fig. 5.1.1. The influences of incomplete information on the strategy choices

player has exact knowledge of the payoff functions of the other players corresponds to the limiting situation in which the random disturbances are 0. Therefore, to model the situation in which each player has slight uncertainty, we will consider sequences of disturbed games in which the disturbances go to 0 and we will investigate which equilibria are approximated by equilibria of such disturbed games. These equilibria will be called *firm equilibria.*

In the game of Fig. 5.1.1, only the perfect equilibrium (1, 1) is firm. This is not really surprising since for each player the situation in which he is slightly uncertain about the payoffs of his opponent is equivalent to the situation in which he knows that his opponent with a small probability makes mistakes. Since something similar is true for an arbitrary game (at least if Assumption 5.3.3. is satisfied), the disturbed game model can be viewed as a model in which the mistake probabilities are endogenously determined. This makes it interesting to look at the relations between firm equilibria on the one side and perfect (resp. weakly proper, resp. proper) equilibria on the other side. We will see that, under some conditions on the disturbances, a firm equilibrium is perfect, but that the converse is not true. Hence, the approach of this chapter yields (for normal form games) a refinement of the perfectness concept. Moreover, we will see that, under stronger conditions on the disturbances, every firm equilibrium is weakly proper. This implies that the assumption of a considerably worse mistake being chosen with an order smaller probability than a less costly mistake is justifiable. However, since firm equilibria possess more monotonicity properties than arbitrary weakly proper equilibria, not every weakly proper equilibrium is firm and so our approach yields a refinement of the weak properness concept. Finally, we will see that the incomplete information model cannot justify the assumption that also nonserious mistakes are chosen with a probability that is of smaller order, hence, a firm equilibrium need not be proper.

To conclude this section, let us say something about the terminology which is used throughout the chapter.

We consider a fixed class of n-person normal form games $\mathscr{G}(\Phi_1, \ldots, \Phi_n)$. A game in this class is completely determined by its payoff vector $r = (r_1, \ldots, r_n)$, where $r_i \in \mathbb{R}^{m^*}$ is given by (2.1.13). We assume $\mathbb{R}^{m^*}$ is endowed with its Borel σ-field $\mathscr{B}$ and whenever we speak of measurable, we mean Borel measurable. In order to have a consistent framework for our probability calculations, we assume a basic probability space $(\Omega, \mathscr{A}, \mathbb{P})$ is given. All random variables to be considered are defined on this space. If X_i is a random vector with values in $\mathbb{R}^{m^*}$, then $X_i(\varphi)$ is the component of X_i corresponding to $\varphi \in \Phi$. Furthermore, for $s \in S$, the random variable $X_i(s)$ is defined by $X_i(s) := \sum_{\varphi} s(\varphi) X_i(\varphi)$.

We use standard measure theoretic terminology. Standard results from measure theory will be used without giving references. For proofs as well as for definitions of measure theoretic concepts, we refer to Halmos [1950] or Kingman and Taylor [1966].

5.2 Disturbed Games

In this section, we introduce the model (the disturbed game $\Gamma(\mu)$) by means of which we will investigate what the consequences are of each player not knowing the payoff functions of the other players exactly. This model is a generalization of the model introduced in Harsanyi [1973a]. Furthermore, it is shown that every disturbed game which satisfies some continuity condition possesses at least one equilibrium.

Definition 5.2.1. Let $\Gamma=(\Phi_1,\ldots,\Phi_n,R_1,\ldots,R_n)$ be an n-person normal form game and let $\mu=(\mu_1,\ldots,\mu_n)$ be an n-tuple of probability measures on $\mathbb{R}^{m^*}$. For $i\in N$, let X_i be a random vector with distribution μ_i. The *disturbed game* $\Gamma(\mu)$ is described by the following rules:

(i) Chance chooses an outcome x_i of X_i for each player i (independently).
(ii) Player $i(i\in N)$ gets to hear the outcome x_i of X_i (and nothing more).
(iii) Player $i(i\in N)$ chooses an element $s_i\in S_i$.
(iv) If the outcome $x=(x_1,\ldots,x_n)$ resulted in (i) and if $s=(s_1,\ldots,s_n)$ has been chosen in (iii), then player i receives the expected payoff $R_i^{x_i}(s):=R_i(s)+x_i(s)$.

It is assumed that the basic characteristics of $\Gamma(\mu)$, i.e. the game Γ itself and the distributions $(\mu_1,\ldots,\mu_n)$ are known to all players; the outcome of X_i, however, is only known to player i. The probability distribution μ_i represents the information which every player different from i has about the disturbances in the payoffs of player i; the precise effects of the disturbances are known only to the player himself.

The disturbed game $\Gamma(\mu)$ is a (possibly infinite) extensive form game with perfect recall (cf. Assumption 6.1.7) and, therefore, the players can restrict themselves to behavior strategies in $\Gamma(\mu)$ (Aumann [1964]), a *behavior strategy* of player i being a measurable function σ_i from $\mathbb{R}^{m^*}$ to S_i. Two behavior strategies of player i are said to be *equivalent* if they differ only at a set of μ_i-measure zero. If σ_i is a behavior strategy of player i, then σ_i induces an element s_i of S_i defined by $s_i:=\int\sigma_i d\mu_i$. We call s_i the *aggregate* of σ_i. If player i plays σ_i, then to an outside observer (and to a player $j\neq i$) it will look as if i plays the aggregate s_i of σ_i. For $i\in N$, $k\in\Phi_i$, $x_i\in\mathbb{R}^{m^*}$ and $s\in S$, we define

$$B_i(s|x_i) := \left\{k\in\Phi_i;\ R_i^{x_i}(s\backslash k)=\max_{l\in\Phi_i} R_i^{x_i}(s\backslash l)\right\}, \text{ and} \tag{5.2.1}$$

$$\mathscr{X}_i^k(s) := \{x_i\in\mathbb{R}^{m^*};\ k\in B_i(s|x_i)\}. \tag{5.2.2}$$

$B_i(s|x_i)$ is the set of all pure best replies of player i against a behavior strategy combination with aggregate s if the realization of X_i is x_i. We say that a behavior strategy combination σ is an equilibrium of $\Gamma(\mu)$ if each player i always (i.e. for all realizations of X_i) chooses a best reply. Hence, a strategy combination σ with aggregate s is an *equilibrium* of $\Gamma(\mu)$ if we have

$$\text{if } \sigma_i^k(x_i)>0, \text{ then } k\in B_i(s|x_i) \text{ for all } i\in N,\ k\in\Phi_i \text{ and } x_i\in\mathbb{R}^{m^*}, \tag{5.2.3}$$

or equivalently

$$\text{if } \sigma_i^k(x_i)>0, \text{ then } x_i\in\mathscr{X}_i^k(s) \text{ for all } i\in N,\ k\in\Phi_i \text{ and } x_i\in\mathbb{R}^{m^*}. \tag{5.2.4}$$

The disturbed game $\Gamma(\mu)$ can be expected to possess an equilibrium only if some regularity conditions are satisfied. Therefore, *throughout the chapter, we will assume*:

Assumption 5.2.2. Every μ_i can be written as $\mu_i=\alpha_i\mu_i^d+(1-\alpha_i)\mu_i^{\mathrm{ac}}$ with $\alpha_i\in[0,1)$, where μ_i^d is a discrete probability measure with only finitely many atoms and where μ_i^{ac} is a probability measure which is absolutely continuous with respect to Lebesgue measure, such that the associated density f_i is continuous.

Next, we will show that the disturbed game $\Gamma(\mu)$ possesses an equilibrium if μ satisfies Assumption 5.2.2. For related (and, in fact, stronger) existence results, we refer to Milgrom and Weber [1981] and to Radner and Rosenthal [1982]. Our proof, which follows the ideas outlined in Harsanyi [1973a], proceeds by constructing a correspondence B from S to S whose fixed points induce equilibria in $\Gamma(\mu)$. Let us first show how B comes about. Let $\Gamma(\mu)$ be such that Assumption 5.2.2 is satisfied and, for $i\in N$ and $s\in S$, define $b_i^{\mathrm{ac}}(s)$ as the vector of which the kth component is given by

$$b_i^{\mathrm{ac},k}(s):=\mu_i^{\mathrm{ac}}(\mathscr{X}_i^k(s)) \quad \text{for } k\in\Phi_i. \tag{5.2.5}$$

As a consequence of the fact that

$$\mathscr{X}_i^k(s)\cap\mathscr{X}_i^l(s)\subset\{x_i\in\mathbb{R}^{m^*};\ R_i^{x_i}(s\backslash k)=R_i^{x_i}(s\backslash l)\}, \tag{5.2.6}$$

and since, if $k\neq l$, the set in the right hand side of (5.2.6) is a hyperplane with Lebesgue measure 0, we have

$$\mu_i^{\mathrm{ac}}(\mathscr{X}_i^k(s)\cap\mathscr{X}_i^l(s))=0 \quad \text{if } k\neq l. \tag{5.2.7}$$

Equation (5.2.7) implies that $b_i^{\mathrm{ac}}(s)\in S_i$ for all $i\in N$, $s\in S$. Another important consequence of (5.2.7) is that $s\in S$ is the aggregate of an equilibrium σ of $\Gamma(\mu)$ if and only if every component s_i of s can be written as

$$s_i=\alpha_i\sum_{x_i\in A_i}\mu_i^d(x_i)b_i(s|x_i)+(1-\alpha_i)b_i^{\mathrm{ac}}(s) \quad \text{with } b_i(s|x_i)\in\mathrm{conv}B_i(s|x_i) \text{ for all } x_i\in A_i, \tag{5.2.8}$$

where A_i denotes the set of atoms of μ_i and where "conv" stands for "convex hull", hence $\mathrm{conv}B_i(s|x_i)$ is the set of all mixed strategies of player i with Carrier $B_i(s|x_i)$. Namely, it immediately follows from (5.2.3) – (5.2.4) that the aggregate s of an equilibrium σ of $\Gamma(\mu)$ satisfies (5.2.8) and, conversely, if $s\in S$ satisfies (5.2.8), then every behavior strategy combination σ, defined by

$$\left.\begin{array}{ll}\sigma_i(x_i)=b_i(s|x_i) & \text{if } x_i\in A_i,\\ \sigma_i(x_i)\in\mathrm{conv}\,B_i(s|x_i) & \text{otherwise,}\end{array}\right\} \text{ for } i\in N,$$

is an equilibrium of $\Gamma(\mu)$. Hence, to prove that $\Gamma(\mu)$ possesses an equilibrium, it suffices to show that there exists some $s \in S$ for which (5.2.8) is satisfied for every $i \in N$. Now, for $i \in N$ and $s \in S$, define the subset $B_i(s)$ of S_i by

$$B_i(s) := \alpha_i \sum_{x_i \in A_i} \mu_i^d(x_i) \operatorname{conv} B_i(s|x_i) + (1-\alpha_i)\{b_i^{ac}(s)\},$$

and let $B(s) := (B_1(s), \ldots, B_n(s))$. Then the fixed points of B correspond to the solutions of (5.2.8), hence, $\Gamma(\mu)$ has an equilibrium if B possesses a fixed point. The latter, however, follows from the Kakutani Fixed Point Theorem (Kakutani [1941]): Since A_i is finite, $B_i(s)$ is nonempty compact and convex, for every $s \in S$, while the finiteness of A_i and the continuity of the density of μ_i^{ac} imply that the correspondence B_i is upper semi-continuous. Hence, we have proved.

Theorem 5.2.3. *Every disturbed game possesses at least one equilibrium.*

Note that it follows from (5.2.7) that, if every μ_i is atomless (the case which is considered in Harsanyi [1973a]), then, in every equilibrium of $\Gamma(\mu)$, each player will choose a pure strategy (an element of Φ_i) almost everywhere. Moreover, in this case, there exists for every equilibrium σ of $\Gamma(\mu)$ another equilibrium σ' of $\Gamma(\mu)$ such that σ' is equivalent to σ and such that $\sigma_i'(x_i) \in \Phi_i$ for all $i \in N$ and $x_i \in \mathbb{R}^{m^*}$. Such σ' is called a *purification* of σ. Hence, we have

Theorem 5.2.4 (Harsanyi [1973a]). *If every μ_i is atomless, then every equilibrium of $\Gamma(\mu)$ has a purification.*

For more results on purification we refer to Milgrom and Weber [1985], Aumann et al. [1983] and to Radner and Rosenthal [1982]. We will return to this subject in Sect. 5.6.

Above, we have seen that there is a one-to-one correspondence between equivalence classes of equilibria of $\Gamma(\mu)$ and elements $s \in S$ for which (5.2.8) is satisfied. Therefore, in the remainder of this chapter, when the speak of an *equilibrium* of $\Gamma(\mu)$, we will mean an element $s \in S$ for which (5.2.8) is satisfied. The set of equilibria of $\Gamma(\mu)$ will be denoted by $E(\Gamma(\mu))$

5.3 Firm Equilibria

In order to investigate which equilibria of an ordinary normal form game are still viable in the case where each player has some slight uncertainty about the exact payoffs of the other players, we will approximate a normal form game Γ with disturbed games $\{\Gamma(\mu^\varepsilon)\}_{\varepsilon \downarrow 0}$ in which the disturbances go to 0 as ε tends to 0.

We will first consider the case in which μ^ε *converges weakly* to 0 as ε tends to 0 or more precisely the case in which μ^ε converges weakly to the probability distribution which assigns all mass to 0 (cf. Billingsley [1968]). We say that μ^ε *converges weakly* to 0 as ε tends to 0 (which we denote by $\mu^\varepsilon \xrightarrow{w} 0(\varepsilon \to 0)$) if

$$\lim_{\varepsilon \downarrow 0} \mu_i^\varepsilon(A) = 1 \quad \text{for all neighborhoods } A \text{ of } 0 \text{ in } \mathbb{R}^{m^*} \text{ and all } i \in N.$$

If $\mu^\varepsilon \xrightarrow{w} 0\,(\varepsilon \to 0)$, then for small ε the players are almost sure that the payoffs in $\Gamma(\mu^\varepsilon)$ are very close to the payoffs in Γ. Note that, if μ^ε is the distribution of the random vector X^ε, then $\mu^\varepsilon \xrightarrow{w} 0\,(\varepsilon \to 0)$ corresponds to X^ε *converges in probability* to 0 as ε tends to 0.

Theorem 5.3.1. *Let $\Gamma = (\Phi_1, \ldots, \Phi_n, R_1, \ldots, R_n)$ be a normal form game and, for $\varepsilon > 0$, let $\mu^\varepsilon = (\mu_1^\varepsilon, \ldots, \mu_n^\varepsilon)$ be an n-tuple of probability distributions such that $\mu^\varepsilon \xrightarrow{w} 0$ $(\varepsilon \to 0)$. Let $s \in S$ and assume that, for $\varepsilon > 0$, there exists $s(\varepsilon) \in E(\Gamma(\mu^\varepsilon))$ such that $s = \lim_{\varepsilon \downarrow 0} s(\varepsilon)$. Then $s \in E(\Gamma)$.*

Proof. Let $i \in N$ and assume $k, l \in \Phi_i$ are such that $R_i(s\backslash k) < R_i(s\backslash l)$. We have to show that $s_i^k = \lim_{\varepsilon \downarrow 0} s_i^k(\varepsilon) = 0$. The assumption implies that there exist neighborhoods U of s in $\mathbb{R}^m$ and V of 0 in $\mathbb{R}^{m^*}$ such that, for all $\bar{s} \in U$ and $x_i \in V$

$$R_i^{x_i}(\bar{s}\backslash k) < R_i^{x_i}(\bar{s}\backslash l).$$

If ε is sufficiently small, then $s(\varepsilon) \in U$ and $\mathscr{X}_i^k(s(\varepsilon)) \cap V = \varnothing$, which implies that

$$s_i^k(\varepsilon) \leqq \mu_i^\varepsilon[\mathscr{X}_i^k(s(\varepsilon))] \leqq \mu_i^\varepsilon[\mathbb{R}^{m^*} \backslash V],$$

from which we see that indeed $\lim_{\varepsilon \downarrow 0} s_i^k(\varepsilon) = 0$. □

In the introduction of the chapter we have seen that in general not all equilibria of a normal form game can be obtained as a limit in the way of Theorem 5.3.1. This observation motivates the following definition.

Definition 5.3.2. Let Γ be a normal form game and let $\mathscr{P} = \{\mu^\varepsilon; \varepsilon > 0\}$ be a collection of probability distributions such that $\mu^\varepsilon \xrightarrow{w} 0\,(\varepsilon \to 0)$. An equilibrium s of Γ is said to be *$\mathscr{P}$-firm* if $s = \lim_{\varepsilon \downarrow 0} s(\varepsilon)$, where $s(\varepsilon) \in E(\Gamma(\mu^\varepsilon))$ for all $\varepsilon > 0$.

Similarly as in the previous chapter (Def. 4.3.2), one can also define the notions of weakly $\mathscr{P}$-firm and continuously $\mathscr{P}$-firm equilibria. In this chapter, however, we will restrict our attention to $\mathscr{P}$-firm equilibria (although various results can be generalized either to weakly firm equilibria or to continuously firm equilibria).

A main topic in this chapter is to investigate whether a $\mathscr{P}$-firm equilibrium is perfect. A simple example can already show that, to establish perfectness, $\mathscr{P}$ has to have some additional properties. Namely, consider the game Γ of Fig. 4.3.1 and, for $\varepsilon > 0$, let $\mu^\varepsilon = (\mu_1^\varepsilon, \mu_2^\varepsilon)$ be such that $\mu_1^\varepsilon = \mu_2^\varepsilon$ is the uniform distribution on the ball (in $\mathbb{R}^4$) with radius ε and centre 0. If ε is small, player 2 will play his first strategy with probability 1 in $\Gamma(\mu^\varepsilon)$, hence, by symmetry, player 1 will play both his pure strategies with probability $\frac{1}{2}$. This shows that, if $\mathscr{P} := \{\mu^\varepsilon; \varepsilon > 0\}$, then only the non-perfect equilibrium in which player 1 chooses both strategies with the same probability is $\mathscr{P}$-firm.

Obviously, the reason that in this example the perfect equilibrium is not obtained, is the fact that μ^ε has a bounded support. One can reasonably expect every $\mathscr{P}$-firm equilibrium to be perfect only in the case in which $\mathscr{P} = \{\mu^\varepsilon : \varepsilon > 0\}$ is such that every equilibrium of $\Gamma(\mu^\varepsilon)$ is completely mixed. This is assured by the following assumption, which *we assume throughout the remainder of the chapter.*

Assumption 5.3.3. For every $i \in N$ we have that Lebesgue measure λ on $\mathbb{R}^{m^*}$ is absolutely continuous with respect to μ_i (i.e. if $\lambda(A) > 0$, then $\mu_i(A) > 0$ for every $A \in \mathscr{B}$).

Let $\Gamma(\mu)$ be a disturbed game. Since the set $\mathscr{X}_i^k(s)$ has positive Lebesgue measure for all i, k and s, this set also has positive μ_i-measure (and, hence, positive μ_i^{ac}-measure) if Assumption 5.3.3 is satisfied. Hence, from (5.2.8) and Assumption 5.2.2, we see

Corollary 5.3.4. *If* Assumption 5.3.3 *is satisfied, then every equilibrium of* $\Gamma(\mu)$ *is completely mixed.*

We will also assume that a player does not know one component of another player's payoff vector better than any other component. Hence, *throughout the remainder of the chapter, we assume*:

Assumption 5.3.5. For every $i \in N$, the distribution μ_i is invariant with respect to coordinate permutations of $\mathbb{R}^{m^*}$, i.e. for every permutation π of $\{1, \ldots, m^*\}$ and for every $A \in \mathscr{B}$, we have $\mu_i(A) = \mu_i(\tilde{\pi} A)$, where $\tilde{\pi}$ is the transformation of $\mathbb{R}^{m^*}$ which permutes the coordinates according to π.

As a consequence of this symmetry assumption, equilibria of $\Gamma(\mu)$ possess a natural monotonicity property:

Theorem 5.3.6. *Let* $\Gamma = (\Phi_1, \ldots, \Phi_n, R_1, \ldots, R_n)$ *be a normal form game and let* $\mu = (\mu_1, \ldots, \mu_n)$ *be such that* Assumption 5.3.5 *is satisfied. Then, for* $s \in E(\Gamma(\mu))$, $i \in N$ *and* k, $l \in \Phi_i$, *we have*

$$\text{if } R_i(s\backslash k) < R_i(s\backslash l), \text{ then } s_i^k < s_i^l. \tag{5.3.1}$$

Proof. Assume $s \in E(\Gamma(\mu))$ is such that the condition of (5.3.1) is satisfied. Let $\text{int}\mathscr{X}_i^l(s)$ be the interior of $\mathscr{X}_i^l(s)$, i.e. $\text{int}\mathscr{X}_i^l(s)$ is the set of all x_i where l is the unique best reply of player i against s and let $\tilde{\pi}$ be the transformation of $\mathbb{R}^{m^*}$ which interchanges, for every $\varphi \in \Phi$, the $(\varphi\backslash k)^{\text{th}}$ and $(\varphi\backslash l)^{\text{th}}$ coordinate. Under this mapping $\mathscr{X}_i^k(s)$ is mapped into a strict subset of $\text{int}\mathscr{X}_i^l(s)$ and, therefore,

$$s_i^k \leqq \mu_i(\mathscr{X}_i^k(s)) = \mu_i(\tilde{\pi}\mathscr{X}_i^k(s)) < \mu_i(\text{int}\mathscr{X}_i^l(s)) \leqq s_i^l. \quad \square$$

5.4 Perfect Equilibria

In this section, it is investigated under which conditions on $\mathscr{P}$ only perfect equilibria are $\mathscr{P}$-firm. First it is shown that a $\mathscr{P}$-firm equilibrium may be imperfect if $\mathscr{P} = \{\mu^\varepsilon; \varepsilon > 0\}$ is such that μ^ε converges weakly to 0 as ε tends to 0. After that it is shown that a stronger mode of convergence leads to perfect equilibria.

Example 5.4.1. A $\mathscr{P}$-firm equilibrium is not necessarily perfect if $\mathscr{P}=\{\mu^\varepsilon; \varepsilon>0\}$ is such that μ^ε converges weakly to 0 as ε tends to 0.

Let Γ be the game of Fig. 4.3.1, let X be a continuous random variable with a positive density and let $\mathscr{P}=\{\mu^\varepsilon; \varepsilon>0\}$ be such that μ_i^ε is the product measure on $\mathbb{R}^4$ of the distribution of εX, for every $i\in\{1, 2\}$ and $\varepsilon>0$. Note that indeed μ^ε converges weakly to 0 as ε tends to 0. We will, however, show that the perfect equilibrium of Γ is not $\mathscr{P}$-firm if the expectation of X exists.

For $\varepsilon>0$, let $s(\varepsilon)$ be an equilibrium of $\Gamma(\mu^\varepsilon)$. Write $p(\varepsilon)=s_1^2(\varepsilon)$ and $q(\varepsilon)=s_2^2(\varepsilon)$. Since μ^ε is nonatomic, we have

$$p(\varepsilon)=\mu_1^\varepsilon[\mathscr{X}_1^2(s(\varepsilon))] \quad q(\varepsilon)=\mu_2^\varepsilon[\mathscr{X}_2^2(s(\varepsilon))]. \tag{5.4.1}$$

Let Y a random variable which is distributed as the difference of two independent random variables which both are distributed as X, and let Z be independent from Y and having the same distribution as Y. Then (5.4.1) is equivalent to

$$p(\varepsilon)=\mathbb{P}[(1-q(\varepsilon))Y+q(\varepsilon)Z\leqq -q(\varepsilon)/\varepsilon], \text{ and} \tag{5.4.2}$$

$$q(\varepsilon)=\mathbb{P}[(1-p(\varepsilon))Y+p(\varepsilon)Z\leqq -1/\varepsilon]. \tag{5.4.3}$$

Let F (resp. f) be the common distribution (resp. density) function of Y and Z. From (5.4.2), we see

$$p(\varepsilon)\geqq \mathbb{P}[Y\leqq \frac{-q(\varepsilon)}{\varepsilon} \text{ and } Z\leqq \frac{-q(\varepsilon)}{\varepsilon}]$$

$$=\mathbb{P}\left[Y\leqq \frac{-q(\varepsilon)}{\varepsilon}\right]\mathbb{P}\left[Z\leqq \frac{-q(\varepsilon)}{\varepsilon}\right]=\left(F\left(\frac{-q(\varepsilon)}{\varepsilon}\right)\right)^2,$$

which shows that $\lim_{\varepsilon\downarrow 0} p(\varepsilon)=0$ (in which case the perfect equilibrium of Γ is $\mathscr{P}$-firm) only if

$$\lim_{\varepsilon\downarrow 0}\frac{q(\varepsilon)}{\varepsilon}=\infty. \tag{5.4.4}$$

By (5.4.3), we have

$$q(\varepsilon)\leqq \mathbb{P}[Y\leqq -1/\varepsilon \quad \text{or} \quad Z\leqq -1/\varepsilon]$$

$$\leqq \mathbb{P}[Y\leqq -1/\varepsilon]+\mathbb{P}[Z\leqq -1/\varepsilon]=2F(-1/\varepsilon),$$

from which we see that (5.4.4) is fulfilled only if

$$\lim_{x\downarrow -\infty} xF(x)=-\infty. \tag{5.4.5}$$

However, if (5.4.5) is satisfied, then

$$\int_{-\infty}^{0} tf(t)dt\leqq \lim_{x\downarrow -\infty}\int_{-\infty}^{x} tf(t)dt\leqq \lim_{x\downarrow -\infty} x\int_{-\infty}^{x} f(t)dt=\lim_{x\downarrow -\infty} xF(x)=-\infty,$$

which means that the expectation of Y (and, hence, of X) does not exist. Hence, every $\mathscr{P}$-firm equilibrium of Γ is imperfect if the expectation of X exists. This leads to the conclusion that, in general, weak convergence is not sufficient to establish perfectness.

The actual reason why the perfect equilibrium is not obtained in the example above is the following. From (5.4.2), we see that player 1's aggregate strategy choice in $\Gamma(\mu^\varepsilon)$ is determined by two factors:

(i) his a priori uncertainty about his own payoffs, represented by ε, and
(ii) his uncertainty about the exact payoffs of player 2, represented by $q(\varepsilon)$.

Now, our aim in this chapter is to model the situation in which player 1's strategy choice is determined by his own payoff in Γ and by his slight uncertainty about the payoffs of player 2 (and not by his a priori uncertainty about his own payoffs). Hence, we actually want our model to be such that $q(\varepsilon)$ is much larger than ε (i.e. we want that (5.4.4) is satisfied). But, as we have seen above, this property cannot be expected if μ^ε converges only weakly to 0. Hence, a collection $\{\Gamma(\mu^\varepsilon)\}_{\varepsilon\downarrow 0}$ of disturbed games for which μ^ε converges only weakly to 0 is not a good model for the situation we want to describe.

To obtain a model which fills our needs, we have to decrease a player's a priori uncertainty about his own payoffs. This can be accomplished by looking at disturbed games in which chance chooses the original game with probability close to 1. Formally, this is represented by a sequence $\{\Gamma(\mu^\varepsilon)\}_{\varepsilon\downarrow 0}$ for which $\mu_i^\varepsilon(0)$ converges to 1 as ε tends to 0, for every $i\in N$. In this case, if ε is small, each player's strategy choice in $\Gamma(\mu^\varepsilon)$ is almost solely determined by his own payoffs in Γ and by his uncertainty about the payoffs of the other players. We will prove that, if the uncertainty is of this kind, only perfect equilibria can be obtained.

Theorem 5.4.2. *Every $\mathscr{P}$-firm equilibrium is perfect if $\mathscr{P}=\{\mu^\varepsilon;\ \varepsilon>0\}$ is such that $\lim_{\varepsilon\downarrow 0}\mu_i^\varepsilon(0)=1$ for all $i\in N$.*

Proof. Let Γ be a normal form game and let $\mathscr{P}$ satisfy the condition of the theorem. For $\varepsilon>0$, let $s(\varepsilon)\in E(\Gamma(\mu^\varepsilon))$ be such that $s(\varepsilon)$ converges to $s\in E(\Gamma)$ as ε tends to 0. By rewriting (5.2.8) we see that for every $i\in N$ and $\varepsilon>0$

$$s_i(\varepsilon)=\mu_i^\varepsilon(0)\bar{s}_i(\varepsilon)+(1-\mu_i^\varepsilon(0))\bar{\bar{s}}_i(\varepsilon)\quad\text{with } \bar{s}_i(\varepsilon)\in \operatorname{conv} B_i(s(\varepsilon)|0)\,. \tag{5.4.6}$$

Since $B_i(s(\varepsilon)|0)$ is nothing else than the set of pure best replies of player i against $s(\varepsilon)$ in Γ, this is equivalent to

$$s_i(\varepsilon)=\mu_i^\varepsilon(0)\bar{s}_i(\varepsilon)+(1-\mu_i^\varepsilon(0))\bar{\bar{s}}_i(\varepsilon)\text{with } \bar{s}_i(\varepsilon)\text{ a best reply against } s(\varepsilon)\text{ in }\Gamma. \tag{5.4.7}$$

Since $\mu_i^\varepsilon(0)$ converges to 1 as ε tends to 0, we have that $\bar{s}_i(\varepsilon)$ converges to s_i as ε tends to 0 and, therefore, every s_i is a best reply against $s(\varepsilon)$ for all sufficiently small ε. Since, every $s(\varepsilon)$ is completely mixed by Corollary 5.3.4, s is a perfect equilibrium. □

Theorem 5.4.2 shows that, in order to justify the restriction to perfect equilibria for normal form games, one does not have to rely on the assumption that the players with a small probability make mistakes. Rather one can refer to the fact that a player always has some slight uncertainty about the payoffs of the other players. However, note that perfect equilibria are obtained only if this uncertainty is of a very special kind. Note, furthermore, that the converse of Theorem 5.4.2 is not correct: there exist perfect equilibria which, for every choice of $\mathscr{P}$, fail to be $\mathscr{P}$-firm. Namely, because of the monotonicity property of equilibria of disturbed games (Theorem 5.3.6), the unreasonable perfect equilibria in which a player assigns a greater probability to a more costly mistake than to a less costly one can never be $\mathscr{P}$-firm. So, for instance, in the game of Fig. 3.1.1 with $M=1$, the perfect equilibrium in which player 1 plays his second strategy can never be $\mathscr{P}$-firm.

Theorem 5.4.3. *There exist perfect equilibria which fail to be $\mathscr{P}$-firm whatever $\mathscr{P}$ is.*

In the next section, it will be investigated whether the result of Theorem 5.4.2 can be strengthened in the sense that every $\mathscr{P}$-firm equilibrium is proper (or weakly proper) if $\mathscr{P}$ satisfies the condition of the theorem. For the sake of a clear exposition, we will restrict ourselves to disturbed games $\Gamma(\mu)$ in which every μ_i has no atoms except possibly at 0. Hence, *throughout the remainder of the chapter, we will assume*

Assumption 5.4.4. For every $i \in N$ the distribution μ_i has at most an atom at 0.

It should, however, be stressed that our results hold also in the case in which this assumption is not satisfied. If $\Gamma(\mu)$ is a disturbed game for which Assumption 5.4.4 is satisfied, then we see from (5.2.8) that s is an equilibrium of $\Gamma(\mu^\varepsilon)$ if and only if every s_i can be written as

$$s_i = \mu_i(0)\bar{s}_i + (1-\mu_i(0))b_i^{ac}(s) \quad \text{with } \bar{s}_i \text{ a best reply against } s \text{ in } \Gamma, \tag{5.4.8}$$

where $b_i^{ac}(s)$ is as in (5.2.5).

5.5 Weakly Proper Equilibria

In this section, it is investigated which conditions on $\mathscr{P}$ suffice for $\mathscr{P}$-firm equilibria to be weakly proper. We start by showing that a $\mathscr{P}$-firm equilibrium is not necessarily weakly proper if $\mathscr{P}=\{\mu^\varepsilon; \varepsilon>0\}$ is such that $\mu_i^\varepsilon(0)$ converges to 1 as ε tends to 0. The example used to demonstrate this fact (Example 5.5.1), however, suggests that some stronger mode of convergence (which we will call strong convergence) might lead to weakly proper equilibria. It will then be investigated whether this is indeed the case.

Example 5.5.1. A $\mathscr{P}$-firm equilibrium need not be weakly proper if $\mathscr{P}=\{\mu^\varepsilon; \varepsilon>0\}$ is such that $\lim_{\varepsilon \downarrow 0} \mu_i^\varepsilon(0)=1$ for all $i \in N$.

Let Γ be the game of Fig. 4.1.1 and let $\mathscr{P}=\{\mu^\varepsilon; \varepsilon>0\}$ be such that, for $i \in \{1,2\}$ and $\varepsilon>0$, the distribution μ_i^ε is given by $\mu_i^\varepsilon=(1-\varepsilon)\delta+\varepsilon\nu$, where δ is the probability measure which assigns all mass to 0 and where ν is an arbitrary nonatomic probability measure on $\mathbb{R}^6$.

We claim that, for M sufficiently large, only the non-weakly proper equilibrium in which player 1 uses his second pure strategy is $\mathscr{P}$-firm. Namely, for $\varepsilon>0$, let $s(\varepsilon)$ be an equilibrium of $\Gamma(\mu^\varepsilon)$. Since player 2 will play his first pure strategy if his exact payoff is as in Γ, we have

$$s_2^k(\varepsilon)=\varepsilon\nu[\mathscr{X}_2^k(s(\varepsilon))] \quad \text{for } k\in\{2, 3\},$$

hence, we have

$$\frac{s_2^3(\varepsilon)}{s_2^2(\varepsilon)}=\frac{\nu[\mathscr{X}_2^3(s(\varepsilon))]}{\nu[\mathscr{X}_2^2(s(\varepsilon))]} \quad \text{for all } \varepsilon>0. \tag{5.5.1}$$

Now $\cap_s\chi_2^3(s)\neq\emptyset$, so that $\inf_s\nu[\chi_2^3(s)]>0$, from which it follows that, if M is sufficiently large

$$\frac{s_2^3(\varepsilon)}{s_2^2(\varepsilon)}\geqq s_2^3(\varepsilon)=\nu[\mathscr{X}_2^3(s(\varepsilon))]>\frac{1}{M} \quad \text{for all } \varepsilon>0. \tag{5.5.2}$$

If (5.5.2) is satisfied, then player 1 will play his second pure strategy in $\Gamma(\mu^\varepsilon)$ if he gets to hear that his actual payoff is as described by Γ. This shows that, if M is sufficiently large, only the non-weakly proper equilibrium in which player 1 plays his second strategy is $\mathscr{P}$-firm.

The actual reason why the weakly proper equilibrium is not obtained is the fact that the ratio in (5.5.1) does not directly depend on ε but only indirectly via $s(\varepsilon)$. In the example above, the measure ν represents the conditional beliefs player 1 has about the payoffs of player 2 in $\Gamma(\mu^\varepsilon)$, once he knows that these payoffs are not as in Γ. More precisely if X_2^ε generates μ_2^ε, then

$$\nu(A)=\mathbb{P}(X_2^\varepsilon\in A|X_2^\varepsilon\neq 0) \quad \text{for all } A\in\mathscr{B} \text{ and } \varepsilon>0,$$

and, hence, these beliefs are independent of ε. But if a player has only slight uncertainty about the payoffs of another player, then in the case in which he knows that this player's payoffs are not as in Γ, he presumably will still think that these payoffs are close to those of Γ. This idea leads to the notion of *strong convergence*. For $\varepsilon>0$, let $\mu^\varepsilon=(\mu_1^\varepsilon,\dots,\mu_n^\varepsilon)$ be an n-tuple of probability distributions on $\mathbb{R}^{m^*}$ for which Assumption 5.4.4 is satisfied. We say that μ^ε *converges strongly* to 0 as ε tends to 0 (which is denoted by $\mu^\varepsilon\overset{s}{\to}0$ $(\varepsilon\to 0)$) if the following two conditions are satisfied:

$$\mu_i^\varepsilon(0) \text{ converges to 1 as } \varepsilon \text{ tends to 0, for all } i\in N, \text{ and} \tag{5.5.3}$$

$$\nu^\varepsilon:=\mu^{\varepsilon,ac} \text{ (the absolute continuous component of } \mu^\varepsilon) \text{ converges weakly to 0 as } \varepsilon \text{ tends to 0.} \tag{5.5.4}$$

If $\mu^\varepsilon\overset{s}{\to}0$ $(\varepsilon\to 0)$ and if X_i^ε is a random vector generating μ_i^ε, then

$$X_i^\varepsilon=\begin{cases} 0 \text{ with probability } \mu_i^\varepsilon(0) \text{ (which tends to 1 as } \varepsilon \text{ tends to 0)}, \\ Y_i^\varepsilon \text{ with probability } 1-\mu_i^\varepsilon(0), \text{ where } Y_i^\varepsilon \text{ converges in probability to 0 as } \varepsilon \text{ tends to 0.} \end{cases}$$

Hence, strong convergence expresses that disturbances occur with a small probability and that disturbances are small.

For the remainder of the section, let $\mathscr{P}=\{\mu^\varepsilon; \varepsilon>0\}$ be such that $\mu^\varepsilon \xrightarrow{s} 0(\varepsilon \to 0)$ and let us investigate whether, for every game Γ, ever $\mathscr{P}$-firm equilibrium is proper or weakly proper. Let us first consider weakly proper equilibria. Assume s is a $\mathscr{P}$-firm equilibrium of a game Γ and, for $\varepsilon>0$, let $s(\varepsilon) \in E(\Gamma(\mu^\varepsilon))$ be such that $s(\varepsilon)$ converges to s as ε tends to 0. In the proof of Theorem 5.4.2, we have seen that every $s(\varepsilon)$ is completely mixed and that s is a best reply against $s(\varepsilon)$ if ε is sufficiently small. Hence, s is weakly proper whenever $s(\varepsilon)$ satisfies condition (2.3.2). Let $i \in N$ and $k, l \in \Phi_i$ be such that $R_i(s\backslash k) < R_i(s\backslash l)$. In this case, if ε is small, $\mathscr{X}_i^k(s(\varepsilon))$ is much further away from 0 than $\mathscr{X}_i^l(s(\varepsilon))$ is and, therefore, if v_i^ε is tightly concentrated around 0, one may expect that $s_i^k(\varepsilon)$ will be of smaller order than $s_i^l(\varepsilon)$. An example of a measure which is tightly concentrated is a measure with a normal (Gaussian) density and, in fact, we have:

Theorem 5.5.2. *Let $\mathscr{P}=\{\mu^\varepsilon; \varepsilon>0\}$ be such that μ^ε converges strongly to 0 as ε tends to 0 and such that, for $i \in N$ and $\varepsilon>0$, the distribution v_i^ε* (as in (5.5.4)) *is the product measure of a measure on $\mathbb{R}$ with a normal density with parameters 0 and ε* (see Sect. 5.7). *Then every $\mathscr{P}$-firm equilibrium is weakly proper.*

The proof of this theorem is postponed till Sect. 5.7, since it is rather technical.

Another extreme case is the one in which v_i^ε is widespread (has a density with a heavy tail). In this case, if $R_i(s\backslash k) < R_i(s\backslash l)$, it might occur that $s_i^k(\varepsilon)$ is not of smaller order than $s_i^l(\varepsilon)$. Namely, if v_i^ε is widespread, then there is a relatively large probability of the actual payoffs of player i in $\Gamma(\mu^\varepsilon)$ being far away from his payoffs in Γ, which implies that his payoffs in Γ will have only a relatively small influence on his "mistake" probabilities. An example of a measure which is widespread is a measure with a Cauchy density and, in fact, we have

Theorem 5.5.3. *Let $\mathscr{P}=\{\mu^\varepsilon; \varepsilon>0\}$ be such that μ^ε converges strongly to $0(\varepsilon \to 0)$ and such that, for $i \in N$ and $\varepsilon>0$, the distribution v_i^ε* (as in (5.5.4)) *is the product measure of a measure on $\mathbb{R}$ with a Cauchy density with parameters 0 and ε* (see Sect. 5.7). *Then there exists a game for which every $\mathscr{P}$-firm equilibrium fails to be weakly proper.*

Again, we refer to Sect. 5.7 for the proof.

Next, let us turn to proper equilibria and let us investigate whether Theorem 5.5.2 can be strengthened to yield that every $\mathscr{P}$-firm equilibrium is proper if $\mathscr{P}$ is as in that theorem. This will be the case if every equilibrium $s(\varepsilon)$ of $\Gamma(\mu^\varepsilon)$ satisfies condition (2.3.1). However, this cannot be expected. Namely, if the pure strategy k of player i is only a little bit worse than the pure strategy l against $s(\varepsilon)$, then $\mathscr{X}_i^k(s(\varepsilon))$ is only a little bit smaller than $\mathscr{X}_i^l(s(\varepsilon))$, which implies, by Assumption 5.3.5, that $s_i^k(\varepsilon)$ will not be of smaller order than $s_i^l(\varepsilon)$. Hence, we have the following theorem, which is proved in Sect. 5.7.

Theorem 5.5.4. *Under the conditions of* Theorem 5.5.2, *there exists a game for which every $\mathscr{P}$-firm equilibrium fails to be proper.*

5.6 Strictly Proper Equilibria and Regular Equilibria

We have already seen that the set of equilibria which are $\mathscr{P}$-firm may depend upon the choice of $\mathscr{P}$ (cf. Theorem 5.4.2 and Example 5.4.1). This raises the question of which equilibria are $\mathscr{P}$-firm for every choice of $\mathscr{P}$. In this section, it is shown that the answer to this question is related to strictly proper equilibria and to regular equilibria.

As a first result, we have:

Theorem 5.6.1. *A strictly proper equilibrium is $\mathscr{P}$-firm for all $\mathscr{P}=\{\mu^\varepsilon;\ \varepsilon>0\}$ which are such that* $\lim_{\varepsilon\downarrow 0}\mu_i^\varepsilon(0)=1$ *for all* $i\in N$.

Proof. Let Γ be an n-person normal form game and let $\mathscr{P}=\{\mu^\varepsilon;\ \varepsilon>0\}$ satisfy the condition of the theorem. For $s\in S$ and $\varepsilon>0$, define $\eta(s,\varepsilon)$ by

$$\eta_i(s,\varepsilon):=(1-\mu_i^\varepsilon(0))\,b_i^\varepsilon(s)\quad \text{for } i\in N,$$

where $b_i^\varepsilon(s)$ is the vector with k^{th} component

$$b_i^{\varepsilon,k}(s):=v_i^\varepsilon[\mathscr{X}_i^k(s)]\quad \text{for } k\in\Phi_i.$$

From (5.4.8) we see that s is an equilibrium of $\Gamma(\mu^\varepsilon)$ if and only if every s_i can be written as

$$s_i=\mu_i^\varepsilon(0)\bar{s}_i+\eta_i(s,\varepsilon)\quad \text{with } \bar{s}_i \text{ a best reply against } s \text{ in } \Gamma,$$

from which it follows, by means of (2.2.3), that

$$s\in E(\Gamma(\mu^\varepsilon))\quad \text{if and only if}\quad s\in E(\Gamma,\eta(s,\varepsilon)),\tag{5.6.1}$$

where the perturbed game $(\Gamma,\eta(s,\varepsilon))$ is as in Def. 2.2.1. Next, assume $\hat{s}$ is a strictly proper equilibrium of Γ and let $\hat{\eta}$ be as in Definition 2.3.7. If ε is sufficiently small, then $\eta(s,\varepsilon)<\hat{\eta}$ for all $s\in S$, hence, there exists an equilibrium $\hat{s}(\eta(s,\varepsilon))$ of $(\Gamma,\eta(s,\varepsilon))$ which is close to $\hat{s}$ and which depends continuously on $\eta(s,\varepsilon)$. Consider the mapping $s\rightarrow\hat{s}(\eta(s,\varepsilon))$. Since μ_i^{ac} has a continuous density (Assumption 5.2.2), this mapping is continuous, which implies (by Brouwer's Fixed Point Theorem, Smart [1974]) that it has a fixed point $\hat{s}(\varepsilon)$. Hence, $\hat{s}(\varepsilon)$ is an equilibrium of $(\Gamma,\eta(\hat{s},\varepsilon),\varepsilon)$, and, therefore (in view of (5.6.1)) an equilibrium of $\Gamma(\mu^\varepsilon)$. This shows that $\hat{s}$ is a $\mathscr{P}$-firm equilibrium of Γ since $\hat{s}(\varepsilon)$ converges to $\hat{s}$ as ε tends to 0. □

A simple example can show that the statement in Theorem 5.6.1 is incorrect if the condition of the theorem is replaced by "μ^ε converges weakly to 0 as ε tends to 0". Namely, consider the 2-person normal form game in which both players have 2 pure strategies and in which all payoffs are 1. Then all equilibria are strictly proper, but, due to the symmetry assumption 5.3.5 only the symmetric equilibrium in which both players play $(\frac{1}{2},\frac{1}{2})$ is $\mathscr{P}$-firm.

Next, we will show that every regular equilibrium is $\mathscr{P}$-firm for all $\mathscr{P}=\{\mu^\varepsilon;\ \varepsilon>0\}$ for which $\mu^\varepsilon\xrightarrow{w}0\ (\varepsilon\rightarrow 0)$ and which are such that μ^ε depends continuously on

ε (this latter assumption is made only to keep the analysis tractable, as the reader will see from the proof of Theorem 5.6.2). The proof of Theorem 5.6.2 has the same structure as the proof of Theorem 4.6.1: it is an application of the Implicit Function Theorem in combination with Brouwer's Fixed Point Theorem. The proof is a generalization of the proof of Theorem 7 of Harsanyi [1973a]. Presenting it gives us the opportunity to correct a mathematical error in Harsanyi's proof (which occurs in his Lemma 7).

Theorem 5.6.2. *A regular equilibrium is $\mathscr{P}$-firm for all collections $\mathscr{P}=\{\mu^\varepsilon;\varepsilon>0\}$ for which there exists a random vector $X=(X_1,\dots,X_n)$ such that ν_i^ε* (as in (5.5.4)) *is the distribution of εX_i, for every $i\in N$ and $\varepsilon>0$.*

Proof. Let $\Gamma=(\Phi_1,\dots,\Phi_n,R_1,\dots,R_n)$ be a normal form game and let $\mathscr{P}$ and X be as in the theorem. We will restrict ourselves to the case in which μ_i^ε is atomless, for every i and ε. Hence, we have $\mu_i^\varepsilon=\nu_i^\varepsilon$. The reader can easily adjust the proof for the situation in which there are atoms. Assume $\hat{s}$ is a regular equilibrium of Γ and, without loss of generality, assume $(1,\dots,1)\in C(\hat{s})$. Let $\varepsilon>0$. From (5.2.8) we see that s is an equilibrium of $\Gamma(\mu^\varepsilon)$ if and only if

$$s_i^k=\mu_i^\varepsilon[\mathscr{X}_i^k(s)] \quad \text{for all } i,k, \tag{5.6.2}$$

which, by the condition of the theorem, is equivalent to

$$s_i^k=\mathbb{P}[R_i(s\backslash k)+\varepsilon X_i(s\backslash k)\geqq R_i(s\backslash l)+\varepsilon X_i(s\backslash l) \text{ for all } l\in\Phi_i] \quad \text{for all } i,k. \tag{5.6.3}$$

In order to facilitate the application of the Implicit Function Theorem, we will rewrite (5.6.3) in a way which at first sight looks rather cumbersome. Let us write

$$A_i:=\{\alpha_i,\ \alpha_i\in\mathscr{F}(C(\hat{s}_i),\mathbb{R}) \text{ with } \alpha_i^1=0\} \quad\text{and}\quad A:=\prod_{i=1}^n A_i,$$

For $s\in S$ and $\alpha\in A$, define $F(s,\alpha)$ by

$$F_i^k(s,\alpha):=\mathbb{P}[\alpha_i^k-\alpha_i^l\geqq X_i(s\backslash l)-X_i(s\backslash k) \text{ for all } l\in C(\hat{s}_i)] \quad (i\in N,\ k\in C(\hat{s}_i)).$$

Then we see from (5.6.3) that s is an equilibrium of $\Gamma(\mu^\varepsilon)$ if and only if there exist $\alpha,\beta\in A$ such that

$$s_i^1=1-\sum_{k\neq 1} s_i^k \qquad \text{for } i\in N, \tag{5.6.4}$$

$$s_i^k=F_i^k(s,\alpha)-\beta_i^k \qquad \text{for } i\in N,\ k\in C(\hat{s}_i),\ k\neq 1, \tag{5.6.5}$$

$$\alpha_i^k=[R_i(s\backslash k)-R_i(s\backslash 1)]/\varepsilon \qquad \text{for } i\in N,\ k\in C(\hat{s}_i),\ k\neq 1, \tag{5.6.6}$$

$$\beta_i^k=F_i^k(s,\alpha)-\mu_i^\varepsilon[\mathscr{X}_i^k(s)] \qquad \text{for } i\in N,\ k\in C(\hat{s}_i),\ k\neq 1, \text{ and} \tag{5.6.7}$$

$$s_i^k=\mu_i^\varepsilon[\mathscr{X}_i^k(s)] \qquad \text{for } i\in N,\ k\notin C(\hat{s}_i). \tag{5.6.8}$$

Note that, if $\hat{s}$ is completely mixed, then s is an equilibrium of $\Gamma(\mu^\varepsilon)$ if and only if there exists some $\alpha\in A$ such that (5.6.4) – (5.6.6) are satisfied with $\beta=0$. The application of the Implicit Function Theorem can already by seen by comparing (5.6.4) and (5.6.6) with (4.6.2) and (4.6.3). To apply this theorem we need some

$\hat{\alpha} \in A$ for which $F_i^k(\hat{s}, \hat{\alpha}) = \hat{s}_i^k$ for all $i \in N$ and $k \in C(\hat{s}_i)$ with $k \neq 1$. We claim that such $\hat{\alpha}$ exists. Namely, consider the mapping $E: A \to A$ defined by

$$E_i^k(\alpha) := \begin{cases} \alpha_i^k - \max(0, F_i^k(\hat{s}, \alpha) - \hat{s}_i^k) + \max(0, F_i^1(\hat{s}, \alpha) - \hat{s}_i^1) \\ \qquad (i \in N,\ k \in C(\hat{s}_i),\ k \neq 1), \\ 0 \qquad (i \in N,\ k = 1). \end{cases}$$

By using that $\alpha_i^1 = 0$ for all $\alpha_i \in A_i$, the reader can verify that for all $k \neq 1$

$$\lim_{\alpha_i^k \downarrow -\infty} F_i^k(\hat{s}, \alpha) = 0 \quad \text{and} \quad \lim_{\alpha_i^k \uparrow \infty} F_i^1(\hat{s}, \alpha) = 0,$$

which implies that there exists a nonempty, compact and convex set K in A such that $E(K) \subset K$. Since E is continuous, Brouwer's Fixed Point Theorem (Smart [1974]) guarantees that there exists a fixed point $\hat{\alpha}$ of E. For every $i \in N$ and $k \in C(\hat{s}_i)$ with $k \neq 1$, we have

$$\max(0, F_i^k(\hat{s}, \hat{\alpha}) - \hat{s}_i^k) = \max(0, F_i^1(\hat{s}, \hat{\alpha}) - \hat{s}_i^1),$$

from which it follows, by using

$$\sum_{k \in C(\hat{s}_i)} F_i^k(\hat{s}, \hat{\alpha}) = \sum_{k \in C(\hat{s}_i)} \hat{s}_i^k = 1,$$

that indeed $F_i^k(\hat{s}, \hat{\alpha}) = \hat{s}_i^k$ for all $i \in N$ and $k \in C(\hat{s}_i)$.

Next, consider the mappings G and H defined by

$$\begin{aligned} G_i^1(s, \alpha, \beta, \varepsilon) &:= \sum_k s_i^k - 1 && \text{for } i \in N, \\ G_i^k(s, \alpha, \beta, \varepsilon) &:= R_i(s \backslash k) - R_i(s \backslash 1) - \varepsilon \alpha_i^k && \text{for } i \in N,\ k \in C(\hat{s}_i),\ k \neq 1, \\ H_i^k(s, \alpha, \beta, \varepsilon) &:= s_i^k - F_i^k(s, \alpha) + \beta_i^k && \text{for } i \in N,\ k \in C(\hat{s}_i),\ k \neq 1. \end{aligned}$$

Note that the definitions of these mappings are motivated by (5.6.4) – (5.6.6). Since X has a continuous density by Assumption (5.2.2), the mappings G and H are differentiable. For $s \in S$, let us write $s = (\sigma, \tau)$, where σ is the restriction of s to $C(\hat{s})$ and where τ is the vector consisting of the remaining components of s. Since $\hat{s}$ is a regular equilibrium of Γ and since the density of X is positive everywhere (Assumption 5.3.3), we have that

$$\frac{\partial(G, H)}{\partial(\sigma, \alpha)} \Big| (\hat{s}, \hat{\alpha}, 0, 0) \text{ is nonsigular}.$$

Therefore, it follows from the Implicit Function Theorem (Dieudonné [1960], p. 268), that there exist neighborhoods U of $\tau = 0$, V of $\beta = 0$ and W of $\varepsilon = 0$ and, for every $(\tau, \beta, \varepsilon) \in U \times V \times W$, some $\sigma(\tau, \beta, \varepsilon)$ close to $\hat{s}$ and $\alpha(\tau, \beta, \varepsilon)$ close to $\hat{\alpha}$ such that

$$(\sigma(\tau, \beta, \varepsilon), \tau, \alpha(\tau, \beta, \varepsilon), \beta, \varepsilon) \text{ is a solution of } (5.6.4) - (5.6.5) \tag{5.6.9}$$

Let us write $s(\tau, \beta, \varepsilon)$ for $(\sigma(\tau, \beta, \varepsilon), \tau)$ and let $\varepsilon \in W$. Motivated by (5.6.7) – (5.6.8), define mappings M and N with domain $U \times V$ by

$$M_i^k(\tau, \beta) := \mu_i^\varepsilon[\mathscr{X}_i^k(s(\tau, \beta, \varepsilon))] \qquad \text{for } i \in N,\ k \notin \Phi_i. \tag{5.6.10}$$

and

$$N_i^k(\tau, \beta) := F_i^k(s(\tau, \beta, \varepsilon), \alpha(\tau, \beta, \varepsilon)) - \mu_i^\varepsilon[\mathscr{X}_i^k(s(\tau, \beta, \varepsilon))] \quad \text{for } i \in N, k \in \Phi_i, k \neq 1. \tag{5.6.11}$$

Since $\hat{s}$ is a quasi-strict equilibrium (Corollary 2.5.3), it follows that for sufficiently small ε

$$(M(\tau, \beta), N(\tau, \beta)) \in U \times V \quad \text{for all } (\tau, \beta) \in U \times V.$$

Since U and V can be chosen compact and convex and since M and N are continuous, we can conclude from Brouwer's Fixed Point Theorem (Smart [1974]) that there exists a fixed point $(\tau(\varepsilon), \beta(\varepsilon))$ of (M, N). Let us write $s(\varepsilon)$ for $s(\tau(\varepsilon), \beta(\varepsilon), \varepsilon)$ and $\alpha(\varepsilon)$ for $\alpha(\tau(\varepsilon), \beta(\varepsilon), \varepsilon)$. Then it follows from (5.6.9) – (5.6.11) and the fixed point property of $(\tau(\varepsilon), \beta(\varepsilon))$ that $(s(\varepsilon), \alpha(\varepsilon), \beta(\varepsilon))$ is a solution to (5.6.4) – (5.6.8), which shows that $s(\varepsilon)$ is an equilibrium of $\Gamma(\mu^\varepsilon)$. This completes the proof since $s(\varepsilon)$ converges to $\hat{s}$ as ε tends to 0. □

Next, let us return to the instability of mixed strategy equilibria which was considered in Sect. 1.6. In that section, we said that this instability is only a seeming instability. This view is justified by the Theorems 2.6.2, 5.2.4 and 5.6.2: For almost all equilibria (viz. for the regular ones) this instability disappears if the slight uncertainty which each player has about the other players' payoffs is taken into account.

5.7 Proofs of the Theorems of Sect. 5.5

In this section, we write f_α^β (resp. F_α^β) for the normal density function (resp. normal distribution function) with parameters α and β, hence

$$f_0^1(x) := \frac{1}{\sqrt{2\pi}} e^{-x^2/2}; \quad f_\alpha^\beta(x) := \frac{1}{\beta} f_0^1\left(\frac{x-\alpha}{\beta}\right); \quad F_\alpha^\beta(x) := \int_{-\infty}^{x} f_\alpha^\beta(t)\,dt. \tag{5.7.1}$$

Furthermore, we write g_α^β (resp. G_α^β) for the Cauchy density (resp. distribution) function with parameters α and β, i.e.

$$g_0^1(x) := \frac{1}{\pi(1+x^2)}; \quad g_\alpha^\beta(x) := \frac{1}{\beta} g_0^1\left(\frac{x-\alpha}{\beta}\right); \quad G_\alpha^\beta(x) := \int_{-\infty}^{x} g_\alpha^\beta(t)\,dt.$$

We use $\sim$ to denote asymptotic equivalence, i.e. for $a \in \mathbb{R} \cup \{-\infty\} \cup \{\infty\}$ and functions f and g, we write

$$f(x) \sim g(x)\ (x \to a) \quad \text{if } \lim_{x \to a} \frac{f(x)}{g(x)} = 1.$$

We need the following standard results from asymptotic analysis (cf. De Bruyn [1958]), which can be proved by elementary methods.

$$F_0^1(x) \sim \frac{-f_0^1(x)}{x} \quad (x \downarrow -\infty), \tag{5.7.3}$$

and

$$G_0^1(x) \sim \frac{-1}{\pi x} \quad (x \downarrow -\infty). \tag{5.7.4}$$

Now, we are ready to prove our theorems.

Proof of Theorem 5.5.2. Let $\Gamma = (\Phi_1, \ldots, \Phi_n, R_1, \ldots, R_n)$ be an n-person normal form game, let $\mathscr{P} = \{\mu^\varepsilon; \varepsilon > 0\}$ satisfy the condition of Theorem 5.5.2, let $\hat{s}$ be $\mathscr{P}$-firm and, for $\varepsilon > 0$, let $s(\varepsilon) \in E(\Gamma(\mu^\varepsilon))$ be such that $s(\varepsilon)$ converges to $\hat{s}$ as ε tends to 0. Since $s(\varepsilon)$ is completely mixed for every $\varepsilon > 0$ and since $\hat{s}$ is a best reply against $s(\varepsilon)$ for all sufficiently small ε, it suffices to show that

$$\lim_{\varepsilon \downarrow 0} \frac{s_i^k(\varepsilon)}{s_i^l(\varepsilon)} = 0 \quad \text{if } R_i(\hat{s}\backslash k) < R_i(\hat{s}\backslash l), \text{ for all } i, k, l. \tag{5.7.5}$$

Assume $i \in N$, $k, l \in \Phi_i$ are such that the condition of (5.7.5) is satisfied and let ν^ε be as in (5.5.4). For $\varepsilon > 0$, we have

$$s_i^k(\varepsilon) = (1 - \mu_i^\varepsilon(0))\, \nu_i^\varepsilon[\mathscr{X}_i^k(s(\varepsilon))] \qquad s_i^l(\varepsilon) \geqq (1 - \mu_i^\varepsilon(0))\, \nu_i^\varepsilon[\mathscr{X}_i^l(s(\varepsilon))],$$

hence, it suffices to show that

$$\lim_{\varepsilon \downarrow 0} \frac{\nu_i^\varepsilon[\mathscr{X}_i^k(s(\varepsilon))]}{\nu_i^\varepsilon[\mathscr{X}_i^l(s(\varepsilon))]} = 0. \tag{5.7.6}$$

For $\varepsilon > 0$, let Y_i^ε be a random vector with distribution ν_i^ε and for $s \in S$, let $Z_i^\varepsilon(s) := R_i(s) + Y_i^\varepsilon(s)$. For every $k' \in \Phi_i$, we have

$$\nu_i^\varepsilon[\mathscr{X}_i^{k'}(s(\varepsilon))] = \mathbb{P}\left[Z_i^\varepsilon(s(\varepsilon)\backslash k') = \max_{l' \in \Phi_i} Z_i^\varepsilon(s(\varepsilon)\backslash l') \right].$$

Now, $Z_i^\varepsilon(s(\varepsilon)\backslash k')$ has a normal distribution with parameters $R_i(s(\varepsilon)\backslash k)$ and $\sigma_i(\varepsilon)$, where $\sigma_i(\varepsilon)$ is given by

$$\sigma_i(\varepsilon) := \varepsilon \left(\sum_{\varphi_{-i} \in \Phi_{-i}} s_{-i}(\varepsilon, \varphi_{-i})^2 \right)^{1/2}, \tag{5.7.7}$$

in which $s_{-i}(\varepsilon, \varphi_{-i})$ denotes the probability which $s_{-i}(\varepsilon) := (s_1(\varepsilon), \ldots, s_{i-1}(\varepsilon), s_{i+1}(\varepsilon), \ldots, s_n(\varepsilon))$ assigns to $\varphi_i \in \Phi_{-i} := \prod_{j \neq i} \Phi_j$. Note that $\sigma_i(\varepsilon)$ converges to 0 as ε tends to 0 and, therefore, after suitable transformations, (5.7.6) follows the following Lemma.

Lemma 5.7.1. *For $\varepsilon > 0$, let $(Z_1^\varepsilon, \ldots, Z_m^\varepsilon)$ be an m-tuple of independent random variables, each Z_i^ε having a normal distribution with parameters α_i and ε, such that $\alpha_1 < \alpha_2$. Then*

$$\lim_{\varepsilon \uparrow 0} \mathbb{P}[Z_1^\varepsilon = \max_k Z_k^\varepsilon] / \mathbb{P}\left[Z_2^\varepsilon = \max_k Z_k^\varepsilon \right] = 0. \tag{5.7.8}$$

Proof. (5.7.8) follows easily if $\alpha_2 = \max_k \alpha_k$. Therefore assume, without loss of generality, that $\alpha_3 > \alpha_2$. Furthermore, assume, without loss of generality that

$\alpha_1=0$. Define Z^ε by

$$Z^\varepsilon := \max_{k \geqq 3} Z_k^\varepsilon$$

The reader can verify that in order to prove (5.7.8) it suffices to show that

$$\lim_{\varepsilon \downarrow 0} \frac{\mathbb{P}[Z_1^\varepsilon \geqq Z^\varepsilon]}{\mathbb{P}[Z_2^\varepsilon \geqq Z^\varepsilon]} = 0.$$

For $i \in \{1, 2\}$, we have

$$\mathbb{P}[Z_i^\varepsilon \geqq Z^\varepsilon] = \int_{-\infty}^{\infty} F^\varepsilon(x) f_{\alpha_i}^\varepsilon(x)\,dx\,, \tag{5.7.9}$$

where the distribution function F^ε of Z^ε is given by:

$$F^\varepsilon(x) = \prod_{k=3}^{m} F_{\alpha_k}^\varepsilon(x)\,.$$

For $\varepsilon > 0$, define

$$c(\varepsilon) := e^{-\alpha_2(\alpha_3-\alpha_2)/2\varepsilon^2}\,,$$

then the reader can verify that

$$F_{\alpha_3}^\varepsilon(x) \leqq c(\varepsilon) F_{\alpha_3}^\varepsilon(x+\alpha_2) \qquad \text{if } 2x \leqq \alpha_3\,, \text{ and} \tag{5.7.10}$$

$$f_{\alpha_1}^\varepsilon(x) \leqq c(\varepsilon) f_{\alpha_2}^\varepsilon(x) \qquad \text{if } 2x \geqq \alpha_3\,. \tag{5.7.11}$$

By using the monotonicity of $F_{\alpha_k}^\varepsilon$, it follows from (5.7.10) that

$$F^\varepsilon(x) \leqq c(\varepsilon) F^\varepsilon(x+\alpha_2) \quad \text{if } 2x \leqq \alpha_3\,. \tag{5.7.12}$$

By splitting the integral of (5.7.9) for $i=1$ into two parts and by using (5.7.11) – (5.7.12), if follows that

$$\mathbb{P}[Z_1^\varepsilon \geqq Z^\varepsilon] \leqq 2c(\varepsilon)\mathbb{P}[Z_2^\varepsilon \geqq Z^\varepsilon].$$

Since $c(\varepsilon)$ converges to 0 as ε tends to 0, this completes the proof of the lemma and, hence, the proof of Theorem 5.5.2. □

Proof of Theorem 5.5.3. Let Γ be the game of Fig. 4.1.1 and let $\mathscr{P}$ be as in Theorem 5.5.3. We will show that, for sufficiently large M, only the equilibrium in which player 1 plays his second strategy is $\mathscr{P}$-firm. For $\varepsilon > 0$, let $s(\varepsilon)$ be an equilibrium of $\Gamma(\mu^\varepsilon)$. Since player 2 will play his first strategy if his payoff is as in Γ, we have

$$s_2^k(\varepsilon) = (1-\mu_2^\varepsilon(0))\,v_2^\varepsilon[\mathscr{X}_2^k(s(\varepsilon))] \quad \text{for } k \in \{2, 3\}\,. \tag{5.7.13}$$

For $\varepsilon > 0$ and $s \in S$, let Y_2^ε be a random vector with distribution v_2^ε and let $Z_2^\varepsilon(s) := R_2(s) + Y_2^\varepsilon(s)$. Then (5.7.13) is equivalent to

$$s_2^k(\varepsilon) = (1-\mu_2^\varepsilon(0))\mathbb{P}\left[Z_2^\varepsilon(s(\varepsilon)\backslash k) = \max_{l \in \{1,2,3\}} Z_2^\varepsilon(s(\varepsilon)\backslash l\right] \quad \text{for } k \in \{2, 3\}\,. \tag{5.7.14}$$

Since $Z_2^\varepsilon(s\backslash l)$ has a Cauchy distribution with parameters $R_2(s\backslash l)$ and $\sigma=\sigma_1(\varepsilon)$ given by (5.7.7), it follows from (5.7.14), that

$$s_2^3(\varepsilon)=(1-\mu_2^\varepsilon(0))\int_{-\infty}^{\infty} G_2^\sigma(x)G_1^\sigma(x)g_0^\sigma(x)dx.$$

If ε is small, σ is close to 0, in which case

$$G_2^\sigma(x)G_1^\sigma(x)\geqq\tfrac{1}{2}\quad\text{for all } x\geqq 3,$$

With the consequence that

$$s_2^3(\varepsilon)\geqq(1-\mu_2^\varepsilon(0))\int_3^{\infty}\tfrac{1}{2}g_0^\sigma(x)dx=\tfrac{1}{2}(1-\mu_2^\varepsilon(0))G_0^\sigma(-3). \tag{5.7.15}$$

On the other hand we have

$$s_2^2(\varepsilon)\leqq(1-\mu_2^\varepsilon(0))\mathbb{P}[Z_2^\varepsilon(s(\varepsilon)\backslash 2)\geqq Z_2^\varepsilon(s(\varepsilon)\backslash 1)],$$

hence, since the difference of two random variables, both having a Cauchy distribution, again has a Cauchy distribution

$$s_2^2(\varepsilon)\leqq(1-\mu_2^\varepsilon(0))G_0^{\sqrt{2}\sigma}(-1). \tag{5.7.16}$$

By combining (5.7.15), (5.7.16) and (5.7.4) we see that

$$\lim_{\varepsilon\downarrow 0}\frac{s_2^3(\varepsilon)}{s_2^2(\varepsilon)}\geqq\frac{1}{12}\sqrt{2},$$

hence, if $M>6\sqrt{2}$, then only the non-weakly proper equilibrium in which player 2 plays his second strategy is $\mathscr{P}$-firm. □

Proof of Theorem 5.5.4. Let Γ be the game of Fig. 2.3.2 and let $\mathscr{P}$ be as in Theorem 5.5.4. We claim that, if $s(\varepsilon)$ is an equilibrium of $\Gamma(\mu^\varepsilon)$, then

$$s_2^3(\varepsilon)\sim s_2^2(\varepsilon)\quad(\varepsilon\downarrow 0). \tag{5.7.17}$$

Obviously, (5.7.17) has the consequence that player 1 will play his second strategy if his payoff is as in Γ and if ε is small, which implies that the unique proper equilibrium of Γ (which has player 1 playng his first strategy) is not $\mathscr{P}$-firm. Equation (5.7.17) can be proved by using the same methods as in the proof of Theorem 5.5.2. Let us give a more instructive proof of the slightly weaker result

$$\mathbb{P}[Z_2^\varepsilon(s(\varepsilon)\backslash 3)\geqq Z_2^\varepsilon(s(\varepsilon)\backslash 1)]\sim\mathbb{P}[Z_2^\varepsilon(s(\varepsilon)\backslash 2)\geqq Z_2^\varepsilon(s(\varepsilon)\backslash 1)]\quad(\varepsilon\downarrow 0), \tag{5.7.18}$$

where $Z_2^\varepsilon(s)$, for $\varepsilon>0$ and $s\in S$, is as in the proof of Theorem 5.5.2. Let us write σ for $\sigma_1(\varepsilon)$ as defined in (5.7.7). Then

$$\mathbb{P}[Z_2^\varepsilon(s(\varepsilon)\backslash 3)\geqq Z_2^\varepsilon(s(\varepsilon)\backslash 1)]=F_0^{\sqrt{2}\sigma}(-1-s_1^3(\varepsilon)),\text{ and} \tag{5.7.19}$$

$$\mathbb{P}[Z_2^\varepsilon(s(\varepsilon)\backslash 2)\geqq Z_2^\varepsilon(s(\varepsilon)\backslash 1)]=F_0^{\sqrt{2}\sigma}(-1). \tag{5.7.20}$$

Now, if we write $\bar{\sigma}$ for $\sigma_2(\varepsilon)$ as defined in (5.7.7), then

$$s_1^3(\varepsilon) \leqq \mathbb{P}[Z_1^\varepsilon(s(\varepsilon)\backslash 3) \geqq Z_1^\varepsilon(s(\varepsilon)\backslash 1)] \leqq F_0^{\sqrt{2}\bar{\sigma}}(-1),$$

which implies that $s_1^3(\varepsilon)$ is of smaller order than $\bar{\sigma}$, hence, also of smaller order than σ since $\sigma \sim \bar{\sigma} \sim \varepsilon(\varepsilon \downarrow 0)$. From this fact it follows, by combining (5.7.19) – (5.7.20) with (5.7.3), that (5.7.18) is indeed correct. □

Notes

1. One main message that Chap. 5 delivers is that for generic normal form games this particular type of payoff uncertainty does not eliminate any equilibrium. In Carlsson and Van Damme [1990] a slightly different model is studied that has drastically different consequences: For a generic 2×2 bimatrix game it only allows a unique equilibrium. Very roughly, the Carlsson/Van Damme model is a follows. A game Γ is drawn from the class of 2×2 bimatrix games according to some nonatomic distribution; the game Γ is not revealed to the players but each player i receives a noisy signal $\Gamma^i = \Gamma + \varepsilon E^i$ about (the payoffs of) the game. (The random variables E^1 and E^2 are assumed to be independent.) Hence, a player doesn't know his own payoffs exactly, he has some information about both players' payoffs, and players' informations are correlated. Carlsson and Van Damme investigate the equilibria of this incomplete information game and they show that, when the noise vanishes, each such equilibrium forces players to choose the risk dominant equilibrium (Harsanyi and Selten [1988]) of the game that corresponds to their observations. Hence, in the coordination game in which players simultaneously choose between '1' and '2', and in which each player gets 'k' if both choose 'k' and 0 if choices don't match, the Carlsson/Van Damme approximation method leads to both players choosing '2'. For an explanation for why this type of model leads to such different results, the reader is urged to consult Carlsson and Van Damme [1990]. Some of the intuition, however, can already be gleaned from Rubinstein [1989]. Other models dealing with payoff uncertainty in normal form games are Dekel and Fudenberg [1990], Fudenberg, Kreps and Levine [1988] and Monderer and Samet [1989].

6 Extensive Form Games

The comprehensive study of normal form games in Chaps. 2–5 has yielded a deeper insight into the relationships between various refinements of the Nash concept. The analysis has also shown that, for (generic) normal form games, there is actually little need to refine the Nash concept since, for almost all such games, all Nash equilibria possess all properties one might hope for.[1]

Any nontrivial extensive form game, however, gives rise to a nongeneric normal form and in this chapter it will be shown that for extensive form games the situation indeed is fundamentally different. A Nash equilibrium of an extensive game may prescribe strikingly irrational behavior off the equilibrium path and almost any nontrivial extensive form possesses such unreasonable equilibria.

This chapter provides an introduction to extensive form games. Several refined equilibrium concepts are presented of which the basic properties are derived. For specific applications of these refined concepts, however, the reader is refered to Chaps. 7, 8, and 10. The discussion of stable equilibria (in the sense of Kohlberg and Mertens [1986]) is also deferred to the concluding chapter.

The discussion will be confined to finite extensive form games with perfect recall. Section 6.1 provides the definition of such a game as well as the notation that will be used.

Section 6.2 considers equilibria and subgame perfect equilibria. It is shown that an equilibrium has to prescribe rational behavior only at those information sets which can be reached when the equilibrium is played. Furthermore, it is shown that every game possesses at least one subgame perfect equilibrium.

The concept of sequential equilibria is the subject of Sect. 6.3. The formal definition is given, the consistency of beliefs is discussed and some basic properties are derived.

Section 6.4 contains a discussion concerning perfect equilibria. It is shown that every game possesses at least one perfect equilibrium and the relation with the sequential equilibrium concept is studied. Also the difference between perfectness in the normal form and perfectness in the extensive form is stressed.

Proper equilibria are considered in Sect. 6.5. It is shown that a proper equilibrium of the normal form induces a sequential equilibrium outcome in any extensive game with this normal form, hence, a normal form analysis can already exclude some kinds of irrational behavior in the extensive form. In fact, some intuitively unreasonable sequential equilibria can be excluded by requiring normal form properness. However, not all such unreasonable equilibria can be excluded in this way.

Extensive form games with control costs are studied in Sect. 6.6. It is shown that, if control costs are present, the players will play a sequential equilibrium.

In Sect. 6.7, it is investigated which influence incomplete knowledge of the payoff functions has on the strategy choices in an extensive form game. It is demonstrated that only very specific uncertainty will force the players to play a sequential equilibrium.It should, however, be noted that many challenging problems concerning the incomplete information approach are still unsolved.

The results in this chapter are based on Kreps and Wilson [1981a], Selten [1975] and van Damme [1984].

6.1 Definitions

In this section, the formal definition of a (finite) game in extensive form is given. Furthermore, several concepts which are related to such a game are introduced. The exposition follows Selten [1975].

The extensive form representation of a game is the representation which explicitly displays the rules of the game, i.e. it specifies the following data (cf. the examples in Sects. 1.2 – 1.4):

(i) the order of the moves in the game,
(ii) for every decision point, which player has to move there,
(iii) the information a player has whenever it is his turn to move,
(iv) the choices available to a player when he has to move,
(v) the probabilities associated with the chance moves, and
(vi) the payoffs for all players.

Mathematically, this specification is provided by a sextuple (K, P, U, C, p, r) of which K specifies the order of the moves, P specifies the player who has to move, etc. Formally, a *(finite) n-person game in extensive form* is defined as a sextuple $\Gamma = (K, P, U, C, p, r)$ of which the constituents are as follows:

The Game Tree K

The game tree K is a finite tree with a distinguished node o, the *origin* (or root) of K. The unique sequence of nodes and branches connecting the root o with a node x of K is called the *path* to x, and we say that x *comes before* y (to be denoted by $x<y$) if x is on the path to y and different from y. Note that $<$ is a partial

Table 6.1.1. The terminology used with respect to the game tree

Notation	Definition	Terminology
$x<y$	x on the path to y and $x \neq y$	y comes after x, x comes before y
$P(x)$	$\max\{y; y<x\}$ $(x \neq o)$	the immediate predecessor of x
$S(x)$	$\{y; x \in P(y)\}$	the immediate successors of x
Z	$\{x; S(x) = \emptyset\}$	the endpoints of the tree
X	the complement of Z	the decision points
$Z(x)$	$\{y \in Z; x<y\}$	the terminal successors of x.

ordering. The terminology which will be used with respect to K is summarized in Table 6.1.1. The interpretation of K is as follows: the game starts at the root o and proceeds along a path from node to immediate successor until an endpoint is reached. The various paths give the "plays" that might occur.

The Player Partition P

The player partition P is a partition of X into $n+1$ sets $P_0, P_1, \ldots, P_n$. The set P_i is the set of decision points of player i. Player 0 is the *chance player* responsible for the random moves occurring in the game.

The Information Partition U

The information partition U is an n-tuple $(U_1, \ldots, U_n)$ where U_i is a partition of P_i (into so called *information sets* of player i) such that for every information set the following two conditions are satisfied

(i) every path intersects the information set at most once, and
(ii) all nodes in the information set have the same number of immediate successors.

The information set $u \in U_i$ which contains $x \in P_i$ represents the set of nodes player i cannot distinguish from x based on the information he has when he has to move at x. In our figures, we depict information sets by connecting the nodes in this set by a dotted line and we label the set with i whenever it is an information set of player i.

The Choice Partition C

The choice partition C is a collection $C = \left\{C_u; u \in \bigcup_{i=1}^{n} U_i\right\}$, where C_u is a partition of $\bigcup_{x \in u} S(x)$ into so called *choices* at u, such that every choice contains exactly one element of $S(x)$ for every $x \in u$ (note that this is possible because of condition (ii) in the definition of U).

The interpretation is that, if player i takes the choice $c \in C_u$ at the information set $u \in U_i$, then, if he is actually at $x \in u$, the next node reached by the play is that element of $S(x)$ which is contained in c. We identify the choice c at u with the collection of branches leading out of u and having an element in c as endpoint and in our graphical representations we label the choices along these branches. So, in the game of Fig. 1.4.2, the choice L_2 of player 2 consists of the two left-hand branches leading out at the information set of this player.

The Probability Assignment p

The probability assignment p specifies for every $x \in P_0$ a completely mixed probability distribution p_x on $S(x)$.

The interpretation is that, if x is reached, then $y \in S(x)$ is reached with the (positive) probability $p_x(y)$. In the figures the probabilities are depicted along the branches.

The Payoff Function r

The payoff function r is an n-tuple $(r_1, \ldots, r_n)$ where r_i is a real valued function with domain Z.

If the endpoint z is reached, then player i receives the payoff $r_i(z)$. We depict the vector $r(z)$ at the endpoint z, first the payoff of player 1 etc.

Throughout the chapter, we will restrict ourselves to extensive form games in which every player never forgets anything, i.e. at every point where a player has to make a decision, he knows what he previously has known (i.e. which of his information sets have been reached) and what he previously has done (i.e. which choices he has taken). To formalize this, let us say that a choice c comes before a node x (or $c<x$) if one of the branches in c is on the path to x. *Throughout the chapter, we will assume*:

Assumption 6.1.1. The game Γ has *perfect recall* (Kuhn [1953]), i.e. for all $i \in \{1, \ldots, n\}, u, v \in U_i, c \in C_u$ and $x, y \in v$, we have that c comes before x if and only if c comes before y.

For a discussion of Assumption 6.1.1 we refer to Selten [1975]. As a consequence of this assumption it makes sense to say that the choice c comes before the information set v $(c<v)$. We also say that the information set $u \in U_i$ comes before $v \in U_i$ (to be denoted by $u<v$) if there exists a choice c at u with $c<v$. The relation $<$ is a partial ordering on U_i, from which it follows that player i can represent his decision problem by means of a *decision tree*, once he knows the strategies chosen by his opponents (see Wilson [1972]).

Another important consequence of Assumption 6.1.1 is that a player does not have to correlate his choices at different information sets, but that he can restrict himself to behavior strategies (Kuhn [1953], see below).

In the remainder of this section, we introduce several concepts which are related to an extensive form game Γ.

Strategies

A *pure strategy* φ_i of player i is a mapping which assigns a choice $c \in C_u$ to every information set $u \in U_i$. The set of all pure strategies of player i is denoted by Φ_i. A *mixed strategy* s_i of player i is a probability distribution on Φ_i and the set of all such strategies is denoted by S_i. A *behavior strategy* of player i assigns a probability distribution on C_u to every information set $u \in U_i$. The set of all these strategies is B_i. Mixed strategies correspond to prior randomization, behavior strategies to local randomization. A *strategy combination* is an n-tuple of strategies, one for each player. B denotes the set of all behavior strategy combinations, the ones which are relevant for games with perfect recall (see (6.1.2)). If $b_i \in B_i$ and $c \in C_u$ with $u \in U_i$, then $b_i \backslash c$ denotes the strategy b_i changed so that c is taken with certainty at u. For $b \in B$ and $b'_i \in B_i$, we denote by $b \backslash b'_i$ that strategy combination in which all players play in accordance to b, except player i who plays b'_i. Finally, if c is a choice of player i, then $b \backslash c := b \backslash b'_i$ where $b'_i := b_i \backslash c$.

Realization Probabilities

If $b \in B$, then for every $z \in Z$ one can compute the probability $\mathbb{P}^b(z)$ that z will be reached when b is played. Hence, every $b \in B$ induces a probability distribution $\mathbb{P}^b$ on Z. This distribution $\mathbb{P}^b$ will also be called the *outcome* of b. If A is an arbitrary set of nodes, then we write $\mathbb{P}^b(A)$ for $\mathbb{P}^b(Z(A))$ where $Z(A)$ denotes the set of endpoints coming after A. Hence, $\mathbb{P}^b(A)$ is the probability that A will be reached when b is played.

A node x is said to be *possible* when playing $b_i \in B_i$ if there exists some $b \in B$ with i^{th} component b_i such that $\mathbb{P}^b(x) > 0$. An information set u is said to be *relevant* when playing b_i if some node in u is possible when playing b_i. $\mathrm{Poss}(b_i)$ (resp. $\mathrm{Rel}(b_i)$) denotes the set of all nodes of the tree (resp. $u \in U_i$) which are possible (resp. relevant) when playing b_i. The following lemma is an immediate corollary of Assumption 6.1.1.

Lemma 6.1.2.

(i) *If $u \in U_i$ is relevant when playing b_i, then every node is possible when playing b_i.*
(ii) *If $x \in u$ and $u \in U_i$ is relevant for both b_i' and b_i'', then $\mathbb{P}^{b \backslash b_i''}(x|u) = \mathbb{P}^{b \backslash b_i'}(x|u)$ for every $b \in B$ for which these conditional probabilities are well-defined.*

The Restriction to Behavior Strategies

If player i uses the mixed strategy s_i, then if $u \in U_i$ is reached, he will take the choice $c \in C_u$ with the probability

$$b_{iu}(c) = \sum_{\varphi_i \in \mathrm{Rel}(c)} s_i(\varphi_i) \Big/ \sum_{\varphi_i \in \mathrm{Rel}(u)} s_i(\varphi_i)\,, \tag{6.1.1}$$

where $\mathrm{Rel}(u)$ denotes the set of all pure strategies for which u is relevant and $\mathrm{Rel}(c)$ is the subset of $\mathrm{Rel}(u)$ consisting of those strategies which choose c at u. Let b_i be a behavior strategy defined by (6.1.1) whenever this quantity is well-defined, and defined arbitrary for those u which cannot be reached when s_i is played. In Kuhn [1953] it is shown that such a behavior strategy b_i is *realization equivalent* to s_i, i.e. that

$$\mathbb{P}^{\pi \backslash s_i} = \mathbb{P}^{\pi \backslash b_i} \tag{6.1.2}$$

for any (mixed or behavior) strategy combination π. This shows that whatever can be achieved by using a mixed strategy can also be achieved by using a behavior strategy. Hence, since there is no reason whatever for a player to use a strategy more general than a behavior strategy, we will restrict ourselves to this class of strategies.

Expected Payoffs and the Normal Form

If b is played, then the expected payoff $R_i(b)$ to player i is given by $R_i(b) = \sum_z \mathbb{P}^b(z) r_i(z)$. The normal form of Γ is the normal form game $\mathcal{N}(\Gamma) = (\Phi_1, \ldots, \Phi_n, R_1, \ldots, R_n)$. To compute the expected payoff of a mixed strategy combination, knowledge of the normal form is sufficient, but this is not true for behavior strategy combinations.

Conditional Realization Probabilities

Let b be a behavior strategy combination and let $x \in X$. For every $z \in Z$ one can compute the probability $\mathbb{P}_x^b(z)$ that z will be reached if b is played and if the game is started at x. If $\mathbb{P}^b(x) > 0$, then $\mathbb{P}_x^b(z)$ is just the conditional probability $\mathbb{P}^b(z|x)$, but $\mathbb{P}_x^b(z)$ is also well-defined if $\mathbb{P}^b(x) = 0$. (Note that for $\mathbb{P}_x^b$ to be well-defined it is not necessary that the subgame starting at x is well-defined.) The expectation of r_i with respect to $\mathbb{P}_x^b$ will be denoted by $R_{ix}(b)$.

Subgames

Frequently it is the case that a game naturally decomposes into smaller games. This is formalized by the notion of subgames. Let $x \in X$ and let K_x be the subtree of K rising at x. If it is the case that every information set of Γ either is completely contained in K_x or is disjoint from K_x, then the restriction of Γ to K_x constitutes a game of its own, to be called the *subgame* Γ_x *starting at* x. In this case, every behavior strategy combination b decomposes into a pair (b_{-x}, b_x) where b_x is a behavior strategy combination in Γ_x and b_{-x} is a behavior strategy combination for the remaining part of the game (the *truncated game*). If it is known that b_x will be played in Γ_x, then, in order to analyse Γ, it suffices to analyse the truncated game $\Gamma_{-x}(b_x)$ which results from Γ by replacing the subtree K_x by an endpoint with payoff $R_{ix}(b_x)$ for every player i.

6.2 Equilibria and Subgame Perfectness

Let Γ be an extensive form game and let b be a behavior strategy combination in Γ. We say that $b'_i \in B_i$ is a *best reply* (of player i) against b if

$$R_i(b \backslash b'_i) = \max_{b''_i \in B_i} R_i(b \backslash b''_i) , \tag{6.2.1}$$

where the expected payoff $R_i(\cdot)$ is as in Sect. 6.1. The strategy combination b' is said to be a *best reply* against b if every component of b' is a best reply against b and a strategy combination which is a best reply against itself is called a (*Nash*) *equilibrium* of Γ. Since the normal form of Γ possesses an equilibrium in mixed strategies and since for every mixed strategy there exists an equivalent behavior strategy (see (6.1.2)), Γ has at least one equilibrium.

If an equilibrium is played, then each player uses a strategy which maximizes his ex ante expected payoff. However, once the information set $u \in U_i$ is reached, only the payoffs after u are relevant to player i and, therefore, in the remainder of the game, player i will use a strategy which is a *best reply at* u, i.e. a strategy b'_i which satisfies

$$R_{iu}(b \backslash b'_i) = \max_{b''_i \in B_i} R_{iu}(b \backslash b''_i) , \tag{6.2.2}$$

where the conditional expected payoff $R_{iu}(b)$ is defined by

$$R_{iu}(b) = \sum_{z \in Z} \mathbb{P}^b(z|u) r_i(z) = \sum_{x \in u} \mathbb{P}^b(x|u) R_{ix}(b) \quad \text{if } \mathbb{P}^b(u) > 0 . \tag{6.2.3}$$

Note that it follows from Lemma 6.1.2 that $R_{iu}(b\backslash b_i')$ depends only on what b_i' prescribes at the information sets v with $v \geqq u$.

In the following theorem the two 'best reply' concepts which have been introduced are related to each other.

Theorem 6.2.1. *b_i' is a best reply against b if and only if b_i' is a best reply against b at all information sets $u \in U_i$ which are reached with positive probability when $b\backslash b_i'$ is played.*

Proof. Let $\{u_\alpha;\ \alpha \in A\}$ be a collection of information sets of player i such that $\{Z(u_\alpha);\ \alpha \in A\}$ is a partition of $Z(P_i)$, where P_i is the set of decision points of player i. Then, for every $b \in B$, we have

$$R_i(b) = \sum_\alpha \mathbb{P}^b(u_\alpha) R_{iu_\alpha}(b) + \sum_{z \notin Z(P_i)} \mathbb{P}^b(z) r_i(z),$$

where the summation ranges over all α for which $\mathbb{P}^b(u_\alpha) > 0$.

From this the statement of the theorem follows immediately. □

Theorem 6.2.1 shows that for a strategy combination to be an equilibrium it is only necessary that rational behavior is prescribed at every information set which might be reached when the equilibrium is played; at every other information set the behavior may be arbitrary. This is the cause for the existence of 'unreasonable' equilibria of extensive form games as we have seen in Chap. 1 and the reason why the Nash equilibrium concept has to be refined. For extensive form games, the need to refine the equilibrium concept is much more severe than for normal form games: There are 'many' extensive form games with 'unreasonable' equilibria, whereas for almost all normal form games all equilibria are nice (Theorem 2.6.2). To demonstrate the former claim, consider the game of Fig. 1.2.1. This game has 2 connected sets of Nash equilibria, viz. $\{(R_1, R_2)\}$ and $\{(L_1, pL_2 + (1-p) R_2);\ p \geqq \frac{1}{2}\}$. Hence, the game has 2 equilibrium outcomes, viz. (R_1, R_2) and L_1. (Recall from Sect. 6.1 that the outcome of a strategy combination is defined as the probability distribution that is induced on the endpoints.) As argued before, only the outcome (R_1, R_2) makes sense. Now note that the same conclusions as above still hold true for any game with payoffs close to the original game: The sets of equilibria change continuously, but the equilibrium outcomes remain unchanged and again only (R_1, R_2) is 'reasonable'. Hence, we have a class of games with positive measure, any of which admits 'unreasonable' equilibria. We immediately see that requiring essentiality (cf. Def. 2.4.1) is not effective in excluding unreasonable equilibria (therefore, this concept will not be considered in this chapter) and that Theorem 2.6.2 cannot be extended to extensive form games. It seems that there is only one general property of normal form games that is shared by games in extensive form: Generically both have finitely many equilibrium outcomes. (Note that for extensive form games, finite cannot be replaced by odd). This property was formally demonstrated in Kreps and Wilson [1982a].

Theorem 6.2.2 (Kreps and Wilson [1982a], Appendix A3). *For almost all finite extensive form games, the set of Nash equilibrium outcomes is finite (i.e. for any* *game structure, the set* $\{\mathbb{P}^b; b$ *is a Nash equilibrium*$\}$ *is finite for almost all payoff functions*).

The first concept we considered in Chapt. 1 to eliminate Nash equilibria which prescribe irrational behavior at unreached information sets was the subgame perfectness concept (Selten [1965]). An equilibrium b of Γ is said to be a *subgame perfect equilibrium* if, for every subgame Γ_x of Γ, the restriction b_x of b to Γ constitutes a Nash equilibrium of Γ_x. The following lemma is essential in establishing that every game possesses at least one subgame perfect equilibrium.

Lemma 6.2.3 (Kuhn [1953]). *If* b_x *is an equilibrium of the subgame* Γ_x *and* b_{-x} *is an equilibrium of the truncated game* $\Gamma_{-x}(b_x)$, *then* (b_{-x}, b_x) *is an equilibrium of* Γ.

The proof of Lemma 6.2.3 follows easily from the observation that for every $b \in B$

$$R_i(b) = \mathbb{P}^b(x) R_{ix}(b) + \sum_{z \notin Z(x)} \mathbb{P}^b(z) r_i(z),$$

and that, if Γ_x is well-defined, $R_{ix}(b)$ depends only on b_x.

Lemma 6.2.3 implies that a subgame perfect equilibrium of Γ can be found by dynamic programming: First one considers all minimal subgames of Γ and then one truncates Γ by assuming that in any such subgame an equilibrium will be played. This procedure is repeated until there are no subgames left. In this way one meets all subgames and lemma 6.2.3 assures that in every subgame an equilibrium results. Hence, we have [2, 3]

Theorem 6.2.4. *Every game possesses at least one subgame perfect equilibrium.*

The power (resp. weakness) of the subgame perfectness concept will be illustrated in Chaps. 7, 8. For now, it suffices to remark that this concept is not powerful enough to eliminate all 'unreasonable' equilibria (cf. the game of Fig. 1.4.1). Therefore, in the literature, more stringent refinements have been introduced. One of these is considered in the next section.[4]

6.3 Sequential Equilibria

The concept of sequential equilibria has been proposed in Kreps and Wilson [1982a] in order to exclude 'unreasonable' Nash equilibria. In this section, we give the formal definition of this concept and derive some elementary properties of it. The discussion is based on Kreps and Wilson [1982a], the elaboration on consistency follows Ramey [1985]. Throughout the section, it is assumed that an extensive form game Γ is given.

Suppose the players have agreed to play the equilibrium b of Γ. It seems reasonable to suppose that player i, upon reaching an information set u with prior probability 0, will try to reconstruct what has happened and will choose a strategy which is a best reply at u against b with respect to his beliefs about how the game has evolved thus far. The basic assumption underlying the sequential equilibrium concept is that the players indeed behave in this way (which corresponds to the notion of rationality of Savage [1954]). According to this concept, a rational solution of the game not only has to prescribe the strategies used by the players, but it also should prescribe the beliefs the players have. This leads to the following definitions. A *system of beliefs* is a mapping $\mu: X \to [0, 1]$ with $\sum_{x \in u} \mu(x) = 1$ for all information sets u. An *assessment* is a pair (b, μ) where b is a behavior strategy combination and μ is a system of beliefs.

In an assessment (b, μ), the system of beliefs μ represents the beliefs of the players when b is played, i.e. if $x \in u$ with $u \in U_i$, then $\mu(x)$ is the probability player i assigns to being at x if he gets to hear that u is reached. An assessment (b, μ) together with an information set u, determine a probability distribution $\mathbb{P}_u^{b,\mu}$ on Z by

$$\mathbb{P}_u^{b,\mu} := \sum_{x \in u} \mu(x) \mathbb{P}_x^b .$$

The expectation of r_i with respect to $\mathbb{P}_u^{b,\mu}$ will be denoted by $R_{iu}^{\mu}(b)$, hence

$$R_{iu}^{\mu}(b) = \sum_{z \in Z} \mathbb{P}_u^{b,\mu}(z) r_i(z) .$$

If player i expects b to be played and if his beliefs are given by μ, then, if $u \in U_i$ is reached, he will choose a strategy b_i' satisfying

$$R_{iu}^{\mu}(b \backslash b_i') = \max_{b_i'' \in B_i} R_{iu}^{\mu}(b \backslash b_i'') .$$

Such a strategy is called a *best reply at u against* (b, μ). Obviously, for an assessment (b, μ) to be an equilibrium, it is necessary that b is a *sequential best reply* against (b, μ), i.e. that b prescribes a best reply at every information set, but this is not sufficient since the beliefs μ also have to be *consistent* with b. In particular, the beliefs should be determined by b whenever possible, i.e.

$$\mu(x) = \mathbb{P}^b(x|u) \quad \text{if } x \in u \text{ and } \mathbb{P}^b(u) > 0 . \tag{6.3.1}$$

Note that (6.3.1) determines μ completely if b is completely mixed. But also if $\mathbb{P}^b(u) = 0$ the beliefs at u cannot be completely arbitrary since they have to respect the structure of the game. For instance, if the beliefs of player i are given by μ and if b is played, then at an information set $v \geqq u$ his beliefs should satisfy.

$$\mu(x) = \mathbb{P}_u^{b,\mu}(x|v) \quad \text{if } x \in v \text{ and } \mathbb{P}_u^{b,\mu}(v) > 0 . \tag{6.3.2}$$

To ensure that all such conditions are satisfied, Kreps and Wilson adopt the following definition of consistency which in essence means that the beliefs μ can be explained by small deviations from b, i.e. by means of mistakes.

Definition 6.3.1. An assessment (b, μ) is *consistent* if there exists a sequence $\{b^\varepsilon, \mu^\varepsilon\}_{\varepsilon \downarrow 0}$ where b^ε is a completely mixed behavior strategy combination and μ^ε is the system of beliefs generated by b^ε (i.e. is given by (6.3.1)) such that

$$\lim_{\varepsilon \downarrow 0} (b^\varepsilon, \mu^\varepsilon) = (b, \mu).$$

A *sequential equilibrium* is a consistent assessment (b, μ) for which b is a sequential best reply against (b, μ).

For an extensive discussion concerning the definition of consistency, we refer to Kreps and Wilson [1982a]. However, two points are worth noting:

(i) Kreps and Wilson admit that they themselves are not quite sure that consistency 'ought' to be defind as they have done in their paper; it might be that Def. 6.3.1 requires too much consistency in beliefs off the equilibrium path.

(ii) Kreps and Wilson claim that their consistency notion is not too weak, specifically it is claimed that, if (b, μ) is consistent, then (b, μ) is also *structurally consistent*, i.e. that for each information set u there exists a strategy combination b' such that

$$\mathbb{P}^{b'}(u) > 0 \quad \text{and} \quad \mu(x) = \mathbb{P}^{b'}(x|u) \quad \text{for all } x \in u. \tag{6.3.3}$$

(cf. (6.3.1)). Hence, structural consistency means that, if an information set is reached unexpectedly, then one can construct a 'second most likely hypothesis' (i.e. that b' is played) and one can justify one's beliefs with this alternative strategy combination. However, it has been shown in Ramey [1985] that the latter claim is unwarrantable: A consistent assessment need not be structurally consistent.[5] This is illustrated by means of the game of Fig. 6.3.1. (It should also be clear that structural consistency is not sufficient to guarantee consistency in the sense of Def. 6.3.1.)

Consider the assessment $(\bar{b}, \mu)$ defined by

$$\bar{b} = (L_1, L_2, M_3),$$

$$\mu(a) = 1, \ \mu(b) = 0$$

$$\mu(x) = \tfrac{1}{2}, \ \mu(y) = \tfrac{1}{2}, \ \mu(z) = 0.$$

This assessment is consistent since it is obtained as the limit of b^{ε} where $b_i^{\varepsilon} = (1-\varepsilon)L_i + \varepsilon R_i \ (i = 1, 2)$, however, it is not structurally consistent. Namely, if

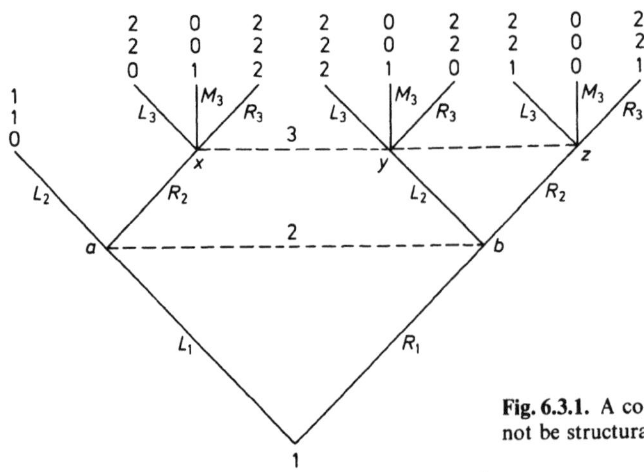

Fig. 6.3.1. A consistent assessment need not be structurally consistent

u is the information set $\{x, y, z\}$ of player 3 and b' is as in (6.3.3), then we must have

$$b'_1(R_1) > 0 \quad \text{or} \quad b'_2(R_2) > 0$$

since otherwise u is not reached. Furthermore

$$b'_1(R_1) = 0 \quad \text{or} \quad b'_2(R_2) = 0$$

since otherwise $\mu(z) > 0$. Hence, we see that for any structurally consistent assessment (b, μ)

$$\text{if } \mu(z) = 0, \text{ then } \mu(x) = 0 \text{ or } \mu(y) = 0$$

and $(\bar{b}, \mu)$ is not structurally consistent.

Note that $(\bar{b}, \mu)$ is a sequential equilibrium, hence, a sequential equilibrium need not be structurally consistent (Using the terminology of Ramey [1985]: it need not be a structural equilibrium.) In fact, the outcome (L_1, L_2) cannot be obtained by a structural equilibrium: Player i $(i = 1, 2)$ is willing to choose L_i only if M_3 is chosen with a probability of at least $\frac{1}{2}$, but M_3 is a best response of player 3 only against the structurally inconsistent belief $\mu = (\frac{1}{2}, \frac{1}{2}, 0)$.

The above example shows that the exact definition of consistency may influence the set of 'acceptable' equilibria and the set of 'acceptable' outcomes. In this book we will restrict ourselves to the original definition[6] of consistency as given by Kreps and Wilson (Definition 6.3.1) since we regard the above mentioned drawback of this concept as being only a minor one.[7] After all, what is the rationale for requiring (6.3.3)?[8] Furthermore, the original definition has the advantage that existence is almost for free: It is an easy consequence of the Theorems 6.4.3 and 6.4.4 that every game possesses at least one sequential equilibrium.[9] Sometimes we will abuse terminology a little and call a strategy combination b a sequential equilibrium if some μ can be found for which (b, μ) is a sequential equilibrium. Note that for b to be sequential it is only necessary that b is supported by *some* system of beliefs; it might very well be that another system of beliefs which is also consistent with b completely upsets the equilibrium (this is the case for the equilibrium (A, R_2) in the game of Fig. 6.5.1).

The perfect recall Assumption 6.1.7 implies that the set of information sets of player i is partially ordered and that it is possible to construct a *decision tree* for player i once the opponents of i have fixed their strategies. For a choice $c \in C_u$ with $u \in U_i$, let us denote by $S(c)$ the set of all those information sets and endpoints of player i's decision tree which come directly after c, hence

$$S(c) := \{v \in U_i \cup Z;\ c < v \text{ and } (\text{if } w \in U_i \cup Z \text{ and } c < w \leqq v, \text{ then } w = v)\}.$$

Let (b, μ) be a consistent assessment. Perfect recall implies that the beliefs of player i are not influenced by his own strategy and, therefore, for all $u \in U_i$ and $c \in C_u$, we have

$$\mathbb{P}_u^{b\backslash c, \mu} = \sum_{v \in S(c)} \mathbb{P}_u^{b\backslash c, \mu}(v)\, \mathbb{P}_v^{b, \mu},$$

where $\mathbb{P}_z^{b,\mu}$ is the degenerate distribution at z for $z \in Z$. This implies that for all $u \in U_i$ and $c \in C_u$

$$R_{iu}^{\mu}(b\backslash c) = \sum_{v \in S(c)} \mathbb{P}_u^{b\backslash c,\mu}(v) R_{iv}^{\mu}(b) \,. \tag{6.3.4}$$

Let $\tilde{R}_{iu}^{\mu}(b)$ (resp. $\tilde{R}_{ic}^{\mu}(b)$) be the maximum expected payoff player i can get against (b, μ) if u is reached (resp. u is reached and c is played at u). Then it follows from (6.3.4) that these quantities can be iteratively computed by the dynamic programming scheme

$$\tilde{R}_{ic}^{\mu}(b) = \sum_{v \in S(c)} \mathbb{P}_u^{b\backslash c,\mu}(v) \tilde{R}_{iv}^{\mu}(b) \,, \tag{6.3.5}$$

$$\tilde{R}_{iu}^{\mu}(b) = \max_{c \in C_u} \tilde{R}_{ic}^{\mu}(b) \,, \tag{6.3.6}$$

which is initialized by setting $\tilde{R}_{iz}^{\mu}(b) = r_i(z)$ for $z \in Z$. Therefore, b_i' is a sequential best reply against (b, μ) if and only if for all $u \in U_i$ and $c \in C_i$:

$$\text{if } b_{iu}'(c) > 0, \text{ then } c \text{ attains the maximum in (6.3.6)} \tag{6.3.7}$$

and so checking whether a consistent assessment is a sequential equilibrium is very easy. Since, furthermore, verifying whether an assessment is consistent can be executed by an efficient labelling procedure (Kreps and Wilson [1982a], Appendix 1), it is easy to check whether an assessment is a sequential equilibrium. This is a great advantage of this concept, when compared to Selten's perfectness concept which will be considered in the next section.

By using the methods of Theorem 6.2.1 it is not difficult to show that every sequential equilibrium is subgame perfect. The converse is not correct as we know from the game of Fig. 1.4.2.

Theorem 6.3.2. *Every sequential equilibrium is subgame perfect.*

In Chap. 1 we have already seen that not every sequential equilibrium need to be sensible. The trouble is caused by the fact that, for an equilibrium to be sequential, it is necessary only that it can be supported by *some* system of beliefs and these beliefs may be implausible. This indeed was the case in the game of Fig. 1.4.3: To support the equilibrium, player 2 had to assume that player 1 had chosen a dominated action. Hence, we see that, to get a meaningful concept, the set of admissible beliefs should be restricted. The concepts of perfect and proper equilibria that will be considered in the next two sections impose additional conditions on beliefs, but we will see that both concepts are not restrictive enough. In Chap. 10, we will return to the issue of how to define 'plausible beliefs' and there it will be shown that positive results can be obtained by invoking the stability concept of Kohlberg and Mertens [1986].

6.4 Perfect Equilibria

The perfectness concept has been proposed in Selten [1975] in order to eliminate the equilibria which prescribe irrational behavior at unreached information sets. In this section, we show that every extensive form game possesses at least one perfect equilibrium. The relation between perfect equilibria and sequential equilibria is studied and the difference between perfectness in the normal form and perfectness in the extensive form is illustrated.

An equilibrium strategy can prescribe irrational behavior only at those information sets which cannot be reached when the equilibrium is played. Selten has argued that virtually all information sets are reachable whatever equilibrium is played, as a consequence of the possibility that *mistakes* might occur. Each player should incorporate this possibility in choosing his strategy and, therefore, he should prescribe a rational choice at every information set. Mathematically, the idea of mistakes is formalized in the same way as for normal form games, i.e. by a perturbed game. In such a perturbed game, every choice at every information set has to be chosen with a strictly positive (mistake) probability.

Definition 6.4.1. Let Γ be an extensive form game. If η is a mapping which assigns to every choice in Γ a positive number η_c such that $\sum_{c \in C_u} \eta_c < 1$ for every information set u, then the *perturbed game* (Γ, η) is the (infinite) extensive form game with the same structure as Γ, but in which every player i is only allowed to use behavior strategies b_i which satisfy $b_{iu}(c) \geqq \eta_c$ for all $u \in U_i$ and $c \in C_u$.

Let (Γ, η) be a perturbed game and let $B(\eta)$ be the set of admissible strategy combinations in (Γ, η). An admissible strategy combination b is said to be an *equilibrium* of (Γ, η) if it is a best reply against itself, i.e. if it satisfies

$$R_i(b) = \max_{b'_i \in B_i(\eta)} R_i(b \backslash b'_i) \quad \text{for all } i \in \{1, \ldots, n\}. \tag{6.4.1}$$

From Theorem 6.2.1 it is seen that b is an equilibrium of (Γ, η) if and only if b prescribes a best reply at every information set, i.e.

$$R_{iu}(b) = \max_{b'_i \in B_i(\eta)} R_{iu}(b \backslash b'_i) \quad \text{for all } i, u. \tag{6.4.2}$$

An equilibrium of Γ is said to be a *perfect equilibrium* if it is still sensible to play this equilibrium if slight mistakes are taken into account. Formally

Definition 6.4.2. b is a *perfect equilibrium* of Γ if b is a limit point of a sequence $\{b(\eta)\}_{\eta \downarrow 0}$ where $b(\eta)$ is an equilibrium of (Γ, η).

Assume b is a perfect equilibrium of Γ and, for $t \in \mathbb{N}$, let $b(t)$ be an equilibrium of $(\Gamma, \eta(t))$ such that $\lim_{t \to \infty} b(t) = b$ and $\lim_{t \to \infty} \eta(t) = 0$. For $t \in \mathbb{N}$, let $\mu(t)$ be the system of beliefs generated by $b(t)$ and, without loss of generality, assume

$\mu = \lim_{t\to\infty} \mu(t)$ exists. Since for all $i \in U$, $u \in U_i$ and $b'_i \in B_i$

$$\lim_{t\to\infty} R_{iu}(b(t)\backslash b'_i) = R^{\mu}_{iu}(b\backslash b'_i) ,$$

it follows from (6.4.2) that

$$R^{\mu}_{iu}(b) = \max_{b'_i \in B_i} R^{\mu}_{iu}(b\backslash b'_i) \quad \text{for all } i, u ,$$

which shows that (b, μ) is a sequential equilibrium. Hence, every perfect equilibrium is sequential. In Sect. 1.4, we have seen that the converse of this statement is not correct. This is also clear by comparing the definitions of sequential equilibria and perfect equilibria: whereas sequential equilibria only have to be optimal with respect to mistakes made in the past, perfect equilibria also have to be optimal with respect to mistakes that might occur in the future. Hence, a perfect equilibrium, in addition to being sequential, also has to satisfy some robustness condition (cf. normal form games, in this case every equilibrium is sequential). As for normal form games, one might expect that almost all sequential equilibria possess this robustness condition. Kreps and Wilson have shown that this indeed is the case.

Theorem 6.4.3 (Kreps and Wilson [1982a]).

(i) *Every perfect equilibrium is sequential.*

(ii) *For almost all extensive form games, almost all sequential equilibria are perfect.*

(iii) *For almost all games, the set of sequential equilibrium outcomes (i.e. the set of probability distributions over the endpoints resulting from sequential equilibria) coincides with the set of perfect equilibrium outcomes.*

Note the difference between the parts (ii) and (iii) of the theorem: Part (iii) states that generically an outside observer cannot conclude whether a perfect or a sequential equilibrium is played, but even for a generic game there may exist some sequential equilibria which are not perfect (however, part (ii) guarantees that there is only a small set of such equilibria).

Theorem 6.4.3 shows that generically there is little difference between the sequential equilibrium concept and the perfectness concept. Kreps and Wilson have shown that exact coincidence can be obtained by slightly modifying Selten's perfectness definition. Namely, define a *weakly perfect equilibrium* of Γ as a strategy combination b which is a limit point of a sequence $\{b(\eta)\}_{\eta\downarrow 0}$ where $b(\eta)$ is an equilibrium of a perturbed game $(\Gamma(\eta), \eta)$ associated with an extensive form game $\Gamma(\eta)$ with payoffs close to those of Γ. Then it can be shown (Kreps and Wilson [1982a], Proposition 6) that an equilibrium is sequential if and only if it is weakly perfect.

Comparing Theorem 6.4.3 with Theorem 2.6.2, the reader might get the impression that for extensive form games it is not necessary to consider more stringent refinements than sequential equilibria, in the sense that almost always all sequential equilibria have all nice properties one wants equilibria to have. In the next section, it will be shown that this impression is wrong.

We will now show that every extensive form game possesses at least one perfect equilibrium (and, hence, also at least one sequential equilibrium). It suffices to show that every perturbed game possesses at least one equilibrium. It will be clear that in an equilibrium of a perturbed game a choice which is not optimal has to be taken with minimum probability. Hence, for an admissible strategy combination b to be an equilibrium of (Γ, η), it is necessary that

$$\text{if } b_{iu}(c) > \eta_c, \text{ then } R_{iu}(b\backslash c) = \max_{c' \in C_u} R_{iu}(b\backslash c') \text{ for all } i, u, c. \tag{6.4.3}$$

By using a dynamic programming argument, one easily sees that (6.4.3) implies (6.4.2), hence, $b \in B(\eta)$ is an equilibrium of (Γ, η) if and only if b satisfies (6.4.3). Notice that, since the choices at u do not influence the payoffs at the information sets $v \in U_i$ which do not come after u, Eq. (6.4.3) is equivalent to

$$\text{if } R_i(b\backslash c) < R_i(b\backslash c'), \text{ then } b_{iu}(c) = \eta_c \text{ for all } i, u, c, c'. \tag{6.4.4}$$

Now consider *Kuhn's interpretation* of how an extensive form game is played (Kuhn [1953]). Kuhn views a player as a collection of separate *agents*, one agent for each information set of this player, each agent having the same payoff as the player to whom he belongs. The normal form game $\mathscr{A}\mathscr{N}(\Gamma)$ corresponding to this interpretation is called the *agent normal form* of Γ (Selten [1975]). The players in $\mathscr{A}\mathscr{N}(\Gamma)$ are the agents of Γ, agent iu (i.e. the one corresponding to the information set $u \in U_i$) has C_u as pure strategy set and R_i as payoff. By comparing (6.4.4) with (2.2.3), it is seen that b is an equilibrium of (Γ, η) if and only if b is an equilibrium of $(\mathscr{A}\mathscr{N}(\Gamma), \eta)$. Therefore, we can conclude from Theorem 2.1.1 that (Γ, η) possesses at least one equilibrium, hence, that Γ has at least one perfect equilibrium.

Theorem 6.4.4 (Selten [1975]). *Every extensive form game possesses at least one perfect equilibrium. The set of perfect equilibria of Γ coincides with the set of perfect equilibria of the agent normal form of Γ.*

In general, the perfect equilibria of Γ do not coincide with the perfect equilibria of the normal form of Γ, as is illustrated by means of the games of Figs. 6.4.1 and 6.4.2.

The unique perfect equilibrium of the normal form of the game of Fig. 6.4.1 is $(L_1 l, L_2)$. In the extensive form also the equilibrium $(R_1 l, L_2)$ is perfect: if player 1 expects that player 2 makes mistakes with a smaller probability than he himself does, then it is indeed rational for him to choose R_1 at his first information set. The game of Fig. 6.4.1 shows that a perfect equilibrium of an extensive form game may involve dominated strategies.[10] It follows from (6.4.4) that a perfect equilibrium cannot involve dominated choices, but Theorem 3.2.1 cannot be generalized to extensive form games: for 2 person games, an equilibrium which is in undominated choices need not be perfect (caused by the fact that a 2-person extensive form game may have more than 2 agents).

The game of Fig. 6.4.1 might be called a degenerate game and, in fact, it can be shown that almost always a perfect equilibrium of Γ will also be perfect in the normal form of this game. The game of Fig. 6.4.2, however, shows that, for

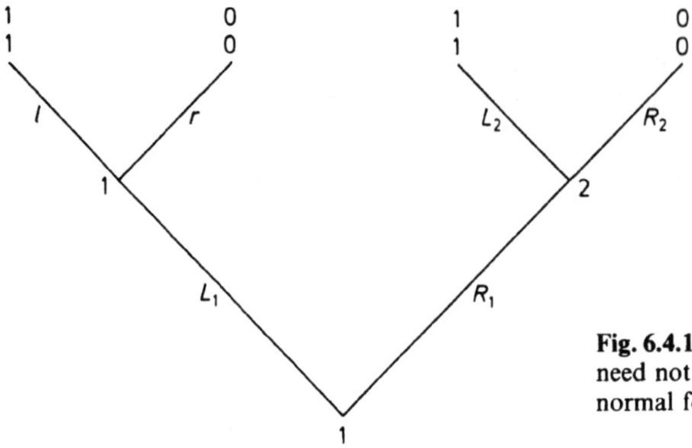

Fig. 6.4.1. A perfect equilibrium of Γ need not be a perfect equilibrium of the normal form of Γ

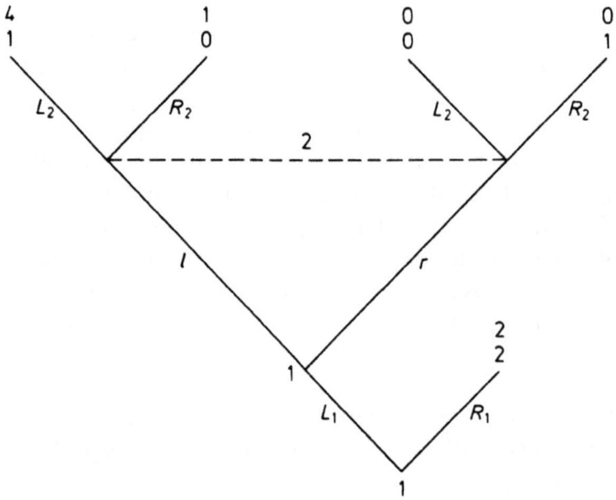

	L_2	R_2
L_1l	4, 1	1, 0
L_1r	0, 0	0, 1
R_1l	2, 2	2, 2
R_1r	2, 2	2, 2

Fig. 6.4.2. A perfect equilibrium of the normal form of Γ is not necessarily a perfect equilibrium of Γ

nondegenerate extensive form games, a perfect equilibrium of $\mathcal{N}(\Gamma)$ need not be a perfect equilibrium of Γ (also see Selten [1975], Sect. 13).

The unique subgame perfect equilibrium of Γ is $(L_1 l, L_2)$. In the normal form of Γ, also $(R_1 l, R_2)$ is perfect: if player 2 expects player 1 to make the mistake $L_1 r$ with a larger probability than the mistake $L_1 l$, then it is optimal for him to choose R_2, thereby forcing player 1 to choose $R_1 l$ or $R_1 r$. Note that the perfectness concept for normal form games does not exclude the possibility that $L_1 r$ occurs with a larger probability than $L_1 l$. In Γ, however, player 2 should not think that $L_1 r$ occurs with a larger probability than $L_1 l$: if the second information set of player 1 is reached, then l is better for player 1 than r is and, therefore, player 1 will intend to choose l, hence, l will occur with the largest probability. Consequently a perfect equilibrium of the normal form need not even induce a subgame perfect equilibrium outcome in the extensive form.

Note that the equilibrium $(R_1 l, R_2)$ is not a proper equilibrium of $\mathcal{N}(\Gamma)$. The properness concept prevents player 2 from thinking that $L_1 r$ occurs with a larger probability than $L_1 l$. The only proper equilibrium of $\mathcal{N}(\Gamma)$ is $(L_1 l, L_2)$ and so, in this example, the proper equilibrium of $\mathcal{N}(\Gamma)$ prescribes perfect behavior in Γ. In the next section, it will be shown that this is not a coincidence.

6.5 Proper Equilibria

Proper equilibria of an extensive form game Γ can be defined by means of the agent normal form $\mathcal{A}\mathcal{N}(\Gamma)$ or by means of the ordinary normal form $\mathcal{N}(\Gamma)$. In this section it is shown that for proper equilibria the situation is fundamentally different from that for perfect equilibria: In generic games, normal form properness is a more stringent requirement than properness in the agent normal form and, whether the game is generic or not, a proper equilibrium of the normal form always induces a sequential equilibrium outcome in any extensive game with that normal form. Hence, some kinds of irrational behavior in the extensive form can already be detected by means of a normal form analysis.[10] In fact, some 'unreasonable' sequential equilibria can be eliminated by requiring properness, but, as the concluding example shows, not all 'unreasonable' equilibria can be eliminated in this way. The discussion is based on van Damme [1984].

Consider the game of Fig. 6.5.1 which is a slight modification of the game of Fig. 6.4.2. (See Chap. 10 for a discussion about whether or not these games should be considered equivalent.)

A sequential (and perfect) equilibrium of this game is (A, R_2). This equilibrium is supported by the beliefs of player 2 that the mistake R_1 occurs with a larger probability than the mistake L_1. In our view these beliefs are not sensible. Since for player 1 the choice R_1 is dominated by the choice L_1, player 2 should expect L_1 to occur with the largest probability and, therefore, if his information set is reached, he should play L_2, thereby inducing player 1 to play L_1. Hence, only (L_1, L_2) is a sensible equilibrium. Notice that all games close to the game of Fig. 6.5.1 have (A, R_2) as a sequential (yet not sensible) equilibrium and so there

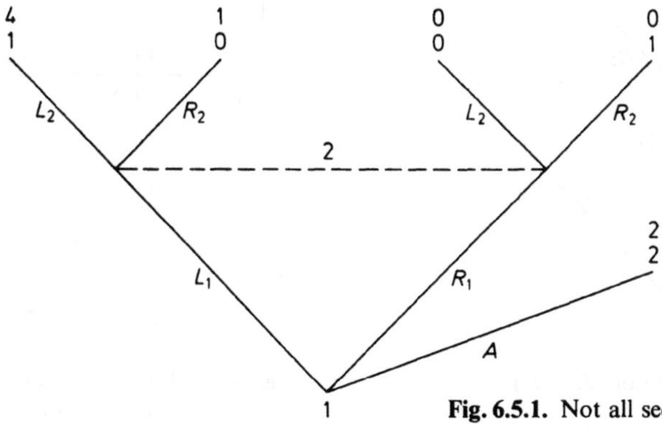

Fig. 6.5.1. Not all sequential equilibria are sensible

exists an open set of games for which not all sequential equilibria are sensible. This shows that the sequential equilibrium concept has to be refined further.

In the game Γ of Fig. 6.5.1, the equilibrium (A, R_2) can be excluded by restricting oneself to proper equilibria of the agent normal form of Γ. A *proper equilibrium* of $\mathscr{A}\mathscr{N}(\Gamma)$ is a strategy combination b which is the limit of a sequence $\{b^\varepsilon\}_{\varepsilon\downarrow 0}$ of completely mixed behavior strategy combinations which satisfy

$$\text{if } R_{iu}(b^\varepsilon\backslash c) < R_{iu}(b^\varepsilon\backslash c') \text{ then } b^\varepsilon_{iu}(c) \leqq \varepsilon b^\varepsilon_{iu}(c') \text{ for all } i, u \text{ and } c, c' \in C_u. \tag{6.5.1}$$

(cf. Def. 2.3.1). Note that every proper equilibrium of $\mathscr{A}\mathscr{N}(\Gamma)$ is a perfect equilibrium of Γ (Theorem 6.4.4). The converse is not true as we have seen above. Eq. (6.5.1) expresses the idea that if some choice c is worse than a choice c' at the *same* information set, then c is mistakenly chosen with a much smaller probability than c'. However, the fact that only choices at the same information set of a player are compared has the consequence that not all unreasonable sequential equilibria are excluded by restricting oneself to proper equilibria of $\mathscr{A}\mathscr{N}(\Gamma)$. This is demonstrated by means of the game of Fig. 6.5.2, which is a slight modification of the game of Fig. 6.5.1.

A perfect equilibrium of the game of Fig. 6.5.2 is $(R_1 r, R_2)$. Since at each information set there are just two choices, every perfect equilibrium of $\mathscr{A}\mathscr{N}(\Gamma)$ (hence, of Γ) is a proper equilibrium of $\mathscr{A}\mathscr{N}(\Gamma)$ and so $(R_1 r, R_2)$ is proper. We do not consider this equilibrium as being reasonable. Upon reaching his information set, player 2 should realize that for player 1 the choice l at v is worse than his choice L_1 at u. Therefore, he should assign the largest probability to being in the left hand node of his information set and, consequently, he should play L_2, thereby inducing player 1 to play L_1. Hence, the only sensible equilibrium is $(L_1 r, L_2)$.

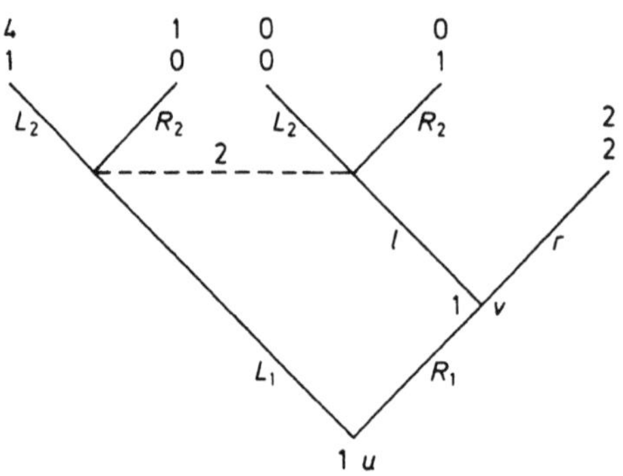

Fig. 6.5.2. A proper equilibrium of $\mathscr{A}\mathscr{N}(\Gamma)$ need not be sensible in Γ

	L_2	R_2
$L_1 l$	4, 1	1, 0
$L_1 r$	4, 1	1, 0
$R_1 l$	0, 0	0, 1
$R_1 r$	2, 2	2, 2

Fig. 6.5.3. The normal form of the game of Fig. 6.5.2

Consider the normal form $\mathcal{N}(\Gamma)$ of the game Γ of Fig. 6.5.2, which is given in Fig. 6.5.3. (Note that up to duplications, this also is the normal form of the games of the Figs. 6.4.2 and 6.5.1.)

The proper equilibria of this game are the equilibria in which player 2 plays L_2 and player 1 plays a combination of $L_1 l$ and $L_1 r$. Therefore, every proper equilibrium leads to L_1 and L_2 being played in Γ and so yields the unique sensible outcome of Γ. Note, however, that a proper equilibrium of $\mathcal{N}(\Gamma)$ may prescribe irrational behavior off the equilibrium path: $(L_1 l, L_2)$ prescribes irrational behavior at the information set v of player 1. Off the equilibrium path, rational behavior can be obtained by looking at limit points of (behavior strategy combinations induced by) ε-proper equilibria as ε tends to 0. If s^ε is an ε-proper equilibrium of $\mathcal{N}(\Gamma)$ close to $(L_1 l, L_2)$ and if b^ε is the behavior strategy combination induced by s^ε (i.e. b_i^ε is given by (6.1.1) for $i \in \{1, 2\}$), then b^ε converges to the unique sensible equilibrium $(L_1 r, L_2)$ of Γ, as follows easily from the definition of an ε-proper equilibrium of $\mathcal{N}(\Gamma)$ and Eq. (6.1.1).

Hence, in this example, proper equilibria of $\mathcal{N}(\Gamma)$ give rise to sequential equilibrium outcomes in Γ and limit points of ε-proper equilibria of $\mathcal{N}(\Gamma)$ are sequential equilibria of Γ. Next, it will be shown that this is true for any extensive form game Γ. In fact, it will be shown that a behavior strategy combination b^ε induced by an ε-proper equilibrium of $\mathcal{N}(\Gamma)$ possesses a property similar to (6.5.1) (see (6.5.5)). Let Γ be an extensive form game and let b be a completely mixed behavior strategy combination. For $c \in C_u$ with $u \in U_i$ define $\tilde{R}_{ic}(b)$ as the maximum payoff player i can get if u is reached, if he plays c at u and if his opponents play b, i.e.

$$\tilde{R}_{ic}(b) = \max_{b_i' \in B_i} R_{iu}(b \backslash b_i' \backslash c), \tag{6.5.2}$$

where $b \backslash b_i' \backslash c$ denotes the behavior strategy combination $b \backslash (b_i' \backslash c)$. Note that $\tilde{R}_{ic}(b)$ can be computed by the following dynamic programming scheme (cf. (6.3.3), (6.3.4)):

$$\tilde{R}_{ic}(b) = \sum_{v \in S(c)} \mathbb{P}_u^{b \backslash c}(v) \tilde{R}_{iv}(b) \quad \text{for } c \in C_u, \tag{6.5.3}$$

$$\tilde{R}_{iu}(b) = \max_{c \in C_u} \tilde{R}_{ic}(b), \tag{6.5.4}$$

which is initialized by setting $\tilde{R}_{iz}(b) = r_i(z)$ for $z \in Z$.

We have (cf. Eq. (6.5.1)):

Lemma 6.5.1. *If b^ε is a behavior strategy combination in Γ which is induced by an ε-proper equilibrium s^ε of $\mathcal{N}(\Gamma)$, then*

$$\text{if } \tilde{R}_{ic}(b^\varepsilon) < \tilde{R}_{i\hat{c}}(b^\varepsilon), \text{ then } b_{iu}^\varepsilon(c) \leqq \varepsilon |\text{Rel}(c)| b_{iu}^\varepsilon(\hat{c}) \text{ for all } i, u \text{ and } c, \hat{c} \in C_u, \tag{6.5.5}$$

where Rel(c) *denotes the set of all pure strategies of player i which are relevant for c.*

Proof. Assume i, u, c, $\hat{c}$ are such that the condition of (6.5.5) is satisfied and let $\tilde{\varphi}_i \in \text{Rel}(\hat{c})$ be a pure strategy which satisfies

$$\tilde{R}_{i\hat{c}}(b^\varepsilon) = R_{iu}(b^\varepsilon \backslash \tilde{\varphi}_i).$$

For any strategy $\varphi_i \in \mathrm{Rel}(c)$, we have

$$R_{iu}(b^\varepsilon \backslash \varphi_i) < R_{iu}(b^\varepsilon \backslash \tilde{\varphi}_i) . \tag{6.5.6}$$

For $\varphi_i \in \mathrm{Rel}(c)$, define $\hat{\varphi}_i \in \mathrm{Rel}(\hat{c})$ by

$$\hat{\varphi}_{iv} = \begin{cases} \tilde{\varphi}_{iv} & \text{if } v \geqq u , \\ \varphi_{iv} & \text{otherwise} . \end{cases}$$

Then it follows from (6.5.6) that

$$R_i(b^\varepsilon \backslash \varphi_i) < R_i(b^\varepsilon \backslash \hat{\varphi}_i) ,$$

which, since b^ε is induced by s^ε, is equivalent to

$$R_i(s^\varepsilon \backslash \varphi_i) < R_i(s^\varepsilon \backslash \hat{\varphi}_i) .$$

Hence, from the definition of an ε-proper equilibrium, we deduce

$$s_i^\varepsilon(\varphi_i) \leqq \varepsilon s_i^\varepsilon(\hat{\varphi}_i) ,$$

which obviously implies that

$$\sum_{\varphi_i \in \mathrm{Rel}(c)} s_i^\varepsilon(\varphi_i) \leqq \varepsilon |\mathrm{Rel}(c)| \sum_{\varphi_i \in \mathrm{Rel}(\hat{c})} s_i^\varepsilon(\varphi_i) . \tag{6.5.7}$$

Dividing both sides of (6.5.7) by $\sum_{\varphi_i \in \mathrm{Rel}(u)} s_i^\varepsilon(\varphi_i)$ yields (6.5.5). □

Let s be a proper equilibrium of $\mathcal{N}(\Gamma)$, for $\varepsilon > 0$, let s^ε be an ε-proper equilibrium of Γ converging to s as ε tends to 0 and let b^ε be induced by s^ε. Without loss of generality, assume that $(b, \mu) = \lim_{\varepsilon \downarrow 0} (b^\varepsilon, \mu^\varepsilon)$ exists, where μ^ε is the system of beliefs generated by b^ε (i.e. μ^ε is given by (6.3.1)). From (6.5.4) and (6.5.5) we see that

$$\text{if } b_{iu}(c) > 0, \text{ then } \tilde{R}_{ic}(b^\varepsilon) = \tilde{R}_{iu}(b^\varepsilon) \text{ for all sufficiently small } \varepsilon.$$

Therefore, we have

$$\text{if } b_{iu}(c) > 0, \text{ then } \tilde{R}^\mu_{ic}(b) = \tilde{R}^\mu_{iu}(b) \text{ for all } i, u, c.$$

Hence, from the characterization of sequential best replies given in (6.3.5), it follows that (b, μ) is a sequential equilibrium of Γ. Since, furthermore, the behavior strategy induced by s_i coincides with b_i at all information sets which might be reached when s_i is played, we have that b generates the same path as s does, i.e. $\mathbb{P}^s = \mathbb{P}^b$. Hence, we have proved.

Theorem 6.5.2 (Van Damme [1984], also cf. Kohlberg and Mertens [1986]).

(i) *If s is a proper equilibrium of $\mathcal{N}(\Gamma)$, then $\mathbb{P}^s$ is a sequential equilibrium outcome in Γ.*

(ii) *If b is a limit point of a sequence $\{b^\varepsilon\}_{\varepsilon \downarrow 0}$ where b^ε is induced by an ε-proper equilibrium s^ε of $\mathcal{N}(\Gamma)$, then b is a sequential equilibrium of Γ.*

In Theorem 6.5.2 'sequential' cannot be replaced by 'perfect'. Namely, consider the game Γ in which the payoffs are the same as in the game of Fig. 6.4.1, except for the

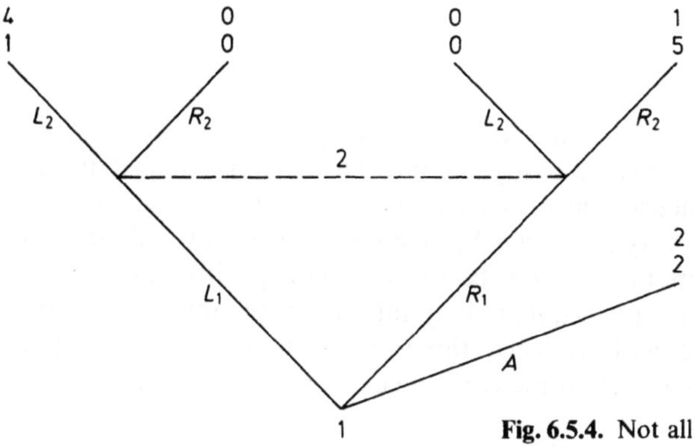

Fig. 6.5.4. Not all proper equilibria are sensible

fact that now player 1 always receives 1 if he plays R_1. The unique perfect equilibrium of this game is $(R_1 l, L_2)$. However, since $(L_1 l, L_2)$ is a proper equilibrium of the normal form of this game, also this equilibrium can be induced in the way of Theorem 6.5.2. Note, however, that the game currently under consideration is a degenerate game. By comparing Eqs. (6.5.1) and (6.5.5) one sees that, for generic games, the conclusion in Theorem 6.5.2 (ii) can be sharpened to 'then b is a proper equilibrium of $\mathscr{A}\mathscr{N}(\Gamma)$'.

Previous examples have shown that, by requiring properness, some 'unreasonable' sequential equilibria can be eliminated. By means of the game of Fig. 6.5.4, which has the same structure as Kohlberg's example in Kreps and Wilson [1982a, Fig. 14], it is shown that not all intuitively unreasonable sequential equilibria can be eliminated in this way.

A proper equilibrium of the normal form of the game of Fig. 6.5.4 is (A, R_2): if player 2 expects the mistake R_1 to occur with a larger probability than the mistake L_1, then it is optimal for him to play R_2. However, this equilibrium is not sensible. If the players have agreed to play (A, R_2) and if the information set of player 2 is nevertheless reached, then at the outset player 2 should not conclude that player 1 has made a mistake, but he should ask himself whether player 1 could have had any reason to deviate. Since for rational players only equilibria are sensible, this means that he should check whether player 1 perhaps opts for a different equilibrium. Now, since (L_1, L_2) is the only equilibrium for which the information set of player 2 is reached and since, moreover, player 1 prefers this equilibrium to (A, R_2), he should conclude that indeed player 1 has chosen L_1 and, therefore, he is forced to choose L_2. Hence, only (L_1, L_2) is sensible.

Alternatively one might argue that player 2 (whenever he is reached) should conclude that L_1 has been chosen since R_1 is dominated by A and since a rational player will not choose a dominated action. This again leads to the conclusion that only (L_1, L_2) is sensible. In this specific example, the 'unreasonable' equilibrium outcome can also be eliminated by requiring strict perfectness. In general, strict perfectness is much too stringent a requirement, however. In Chap. 10, we will show that this is caused by the single-valuedness of this concept and that its set-valued analogon (i.e. stability in the sense of Kohlberg and Mertens [1986]) yields satisfactory results.

6.6 Control Costs

A game with control costs models the idea that a player makes mistakes as a consequence of the fact that it is too costly to control his actions completely (see Chap. 4). If every player incurs control costs at any of his information sets in an extensive form game, then every choice will be taken with a small probability and, consequently, every information set will be reached with positive probability. Therefore, one expects that only sequential equilibria can be obtained as limit points of equilibria of games with control as these costs go to 0. We will show that this is indeed the case, provided that the control costs are incorporated into the model in the correct way.

A naive way to investigate the influence of control costs in an extensive form game would be to investigate the influence in the agent normal form. By means of the game of Fig. 6.6.1, it will be shown that this is not the correct way of modelling.

From the viewpoint of player 1, the game Γ is much different from $\mathscr{A}\mathscr{N}(\Gamma)$. In $\mathscr{A}\mathscr{N}(\Gamma)$ player 1 always has to move (no matter what player 2 does) and his choice R_1 is not much worse than L_1 (since player 2 will intend to play L_2). Therefore, if he incurs control costs, he will not spend much effort in trying to prevent R_1. In fact, the control costs may force him to choose R_1 with a positive probability (see Sect. 4.3). In Γ, however, player 1 has to move only if player 2 has chosen R_2 and in this case R_1 is much worse than L_1. Therefore, if his information set is reached, player 1 will try very hard to prevent R_1 and even if he incurs control costs he will choose R_1 with only a small probability.

The reason that the influence of control costs cannot be investigated by means of the agent normal form is that in the agent normal form one always has to move and, hence, always incurs control costs, whereas in the extensive form one incurs control costs only if an information set is actually reached. This motivates the following definition.

Definition 6.6.1. Let Γ be an n-person extensive form game and let $f = (f_1, \dots, f_n)$ be an n-tuple of control cost functions (see Sect. 4.1). The game Γ^f, which is called a *game with control costs*, is the extensive form game with the same structure as Γ,

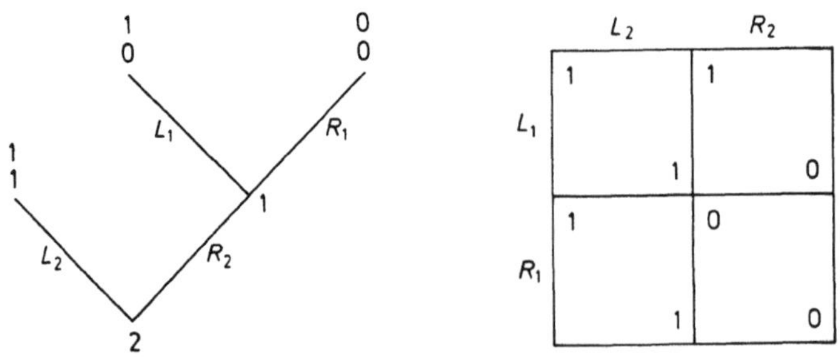

Extensive form game Γ Agent normal form $\mathscr{A}\mathscr{N}(\Gamma)$

Fig. 6.6.1. The influence of control costs cannot be investigated by means of the agent normal form

but in which the payoff to player i if b is played is given by

$$R_i^f(b):=R_i(b)-\sum_{u\in U_i}\sum_{c\in C_u}\mathbb{P}^b(u)f_i(b_{iu}(c)). \tag{6.6.1}$$

The game Γ^f models the idea that each player i incurs control costs as described by f_i at any of his information sets. One could also allow the possibility of different control costs at different information sets (in this case one just has to replace f_i by f_u in Eq. (6.6.1)), in fact, one can even allow different control costs for different choices (in this case, (6.6.1) contains a term f_c). We will not consider these extensions. Note that the term $\mathbb{P}^b(u)$ in Eq. (6.6.1) ensures that player i incurs a cost at u only if u is actually reached when b is played.

The reader should have no difficulty in proving the following generalization of Theorem 4.2.6.

Theorem 6.6.2. *The game Γ^f possesses at least one equilibrium. All equilibria of Γ^f are completely mixed.*

Similarly as in the proof of Lemma 4.2.3, it is seen that a completely mixed strategy combination b is an equilibrium of Γ^f if and only if

$$R_i(b\backslash c)-R_i(b\backslash \hat{c})=\mathbb{P}^b(u)\,[f_i'(b_{iu}(c))-f_i'(b_{iu}(\hat{c}))] \quad \text{for all } i, u, c, \hat{c}. \tag{6.6.2}$$

Since, for every $u\in U_i$ and $c\in C_u$

$$R_i(b\backslash c)=\mathbb{P}^b(u)R_{iu}(b\backslash c)+\sum_{z\notin Z(u)}\mathbb{P}^b(z)r_i(z),$$

we can conclude from (6.6.2) that b is an equilibrium of Γ^f if and only if

$$R_{iu}(b\backslash c)-R_{iu}(b\backslash \hat{c})=f_i'(b_{iu}(c))-f_i'(b_{iu}(\hat{c})) \quad \text{for all } i, u, c, \hat{c}. \tag{6.6.3}$$

From Eq. (6.6.3) we can conclude:

Theorem 6.6.3. *Let Γ be an n-person extensive form game and let f be an n-tuple of control cost functions. If b is a limit point of a sequence $\{b^\varepsilon\}_{\varepsilon\downarrow 0}$, where b^ε is an equilibrium of $\Gamma^{\varepsilon f}$, then b is a sequential equilibrium of Γ.*

Proof. Without loss of generality, assume b is the limit of $\{b^\varepsilon\}$. For $\varepsilon>0$, let μ^ε be the system of beliefs generated by b^ε and, without loss of generality, assume $\mu=\lim_{\varepsilon\downarrow 0}\mu^\varepsilon$ exists. From (6.6.3) we can deduce

$$R_{iu}(b^\varepsilon\backslash c)-R_{iu}(b^\varepsilon\backslash \hat{c})=\varepsilon[f_i'(b_{iu}^\varepsilon(c))-f_i'(b_{iu}^\varepsilon(\hat{c}))] \quad \text{for all } i, u, c, \hat{c}.$$

Since, furthermore

$$\lim_{\varepsilon\downarrow 0}R_{iu}(b^\varepsilon\backslash c)=R_{iu}^\mu(b\backslash c) \quad \text{for all } i, u, c,$$

it follows, in the same way as in the proof of Theorem 4.3.1, that

$$\text{if } R_{iu}^\mu(b\backslash c)<R_{iu}^\mu(b\backslash \hat{c}), \text{ then } b_{iu}(c)=0 \text{ for all } i, u, c, \hat{c}.$$

From this one can conclude, e.g. by using a dynamic programming argument, that b is a sequential best reply against (b,μ). □

Since for normal form games every equilibrium is sequential, it follows from our discussion in Sect 4.3 that not all sequential equilibria can be obtained as a limit in the way of Theorem 6.6.3. Furthermore, it follows from the results in Chap. 4, that to obtain perfectness instead of sequentiality, the control costs have to satisfy more stringent conditions.

6.7 Incomplete Information

In this section, it is investigated what the consequences are of inexact knowledge of the payoffs on the strategy choices in an extensive form game. To that end, we study disturbed extensive form games, i.e. games in which each player, although knowing his own payoff function exactly, knows the payoff functions of his opponents somewhat imprecisely. One might expect that, under similar conditions as in Chap. 5, only sequential equilibria can be obtained as limit points of equilibria of disturbed games in which the information about the payoffs becomes better and better. We will indicate that this is the case only if the disturbances are of a special kind, but it should be noted that the results are far from complete and that many challenging problems are still unsolved.

Let us start by giving an informal definition of a *disturbed extensive form game* (the formal definition is similar to Definition 5.2.1). Let Γ be an ordinary (complete information) n-person extensive form game. The situation we have in mind is the following: the players have to play a game of which it is common knowledge (Aumann [1976]) that it has the same structure as Γ, but of which the payoffs may be slightly different from those in Γ (due to the fact that each player's payoff is subject to small random disturbances of which the precise effects are only known to the player himself). It is assumed that the distribution μ_i of the disturbances in player i's payoff is known to all players. The game with these rules is denoted by $\Gamma(\mu)$, where μ is the n-tuple $(\mu_1, \ldots, \mu_n)$. As in Chap. 5, it is assumed that μ satisfies the Assumptions 5.2.2 and 5.3.5. Since we are interested in the case in which the players are only slightly uncertain about each other's payoffs, we will investigate what happens if the disturbances go to 0, i.e. which equilibria of Γ can be approximated by equilibria of disturbed games $\Gamma(\mu^\varepsilon)$ for which μ^ε converges weakly to 0 as ε tends to 0 (see Sect. 5.3).

The first question which has to be answered is: how should an equilibrium of $\Gamma(\mu)$ be defined? First of all, a strategy of player i in $\Gamma(\mu)$ is a rule which tells him what to do for every payoff he might have, i.e. it is a mapping σ_i: $\mathbb{R}^Z \to B_i$, where Z is the set of endpoints of K and B_i is the set of behavior strategies of player i in Γ. We will restrict ourselves to strategies which satisfy:

$$\text{if } r_i(z) > \max_{z' \neq z} r_i(z'), \text{ then } z \in \operatorname{Poss}(\sigma_i(r_i)), \tag{6.7.1}$$

where $\operatorname{Poss}(\sigma_i(r_i))$ is the set of nodes in the tree which might be reached when $\sigma_i(r_i)$ is played (see (6.1.9)). Obviously, every sensible strategy satisfies (6.7.1). Suppose the strategy combination $\sigma = (\sigma_1, \ldots, \sigma_n)$ is played in $\Gamma(\mu)$ and let u be an information set of player i. If player i gets to hear that u is reached, then he can deduce that the payoff of player j must be in the set

$$R_j(u, \sigma) := \{r_j \in \mathbb{R}^Z;\ \mathrm{Poss}(\sigma_j(r_j)) \cap u \neq \varnothing\} \tag{6.7.2}$$

and, in this case, he will think that the disturbances in the payoffs of player j have the distribution

$$\mu_j^{u,\sigma} := \mu_j(\cdot | R_j(u, \sigma)) . \tag{6.7.3}$$

Note that $\mu_j^{u,\sigma}$ is well-defined for every u and σ because of (6.7.1) and Assumption 5.3.3. Also note that this distribution depends only on the component σ_j of σ. Based on the observation that u is reached, player i will predict that player j will choose c at the information set v with the probability

$$\int \sigma_{jv}(c)\, d\mu_j^{u,\sigma}(\cdot | R_j(v, \sigma)) , \tag{6.7.4}$$

where $\sigma_{jv}(c)$ is the mapping which assigns to each payoff r_j the probability $\sigma_{jv}(c, r_j)$ that player j chooses c at v if the payoff is r_j. Note that (6.7.4) is well-defined for all v which might be reached when u is reached [i.e. $Z(u) \cap Z(v) \neq \varnothing$]. Let $b_j^{u,\sigma}$ be a behavior strategy which is defined as in (6.7.4) whenever possible and which is arbitrarily defined elsewhere. Let $b^{u,\sigma}$ be a behavior strategy combination in which every player j plays $b_j^{u,\sigma}$. If player i gets to hear that u is reached, he will predict the behavior of his opponents by $b^{u,\sigma}$. For the strategy combination σ to be an equilibrium of $\Gamma(\mu)$, we should have that at any information set u the player chooses a best reply against the strategies he expects the others to follow, based on the observation that u is reached. Therefore, we define

Definition 6.7.1. A strategy combination σ is an *equilibrium* of $\Gamma(\mu)$ if for all $i \in N$, $u \in U_i$ and $r_i \in \mathbb{R}^Z$

$$R_{iu}(b^{u,\sigma} \backslash \sigma_i(r_i)) = \max_{b_i' \in B_i} R_{iu}(b^{u,\sigma} \backslash b_i') , \tag{6.7.5}$$

where $R_{iu}(\cdot)$ denotes the expected payoff for player i after u if his actual payoff vector in $\Gamma(\mu)$ is r_i.

Note that the conditional expected payoff in (6.7.5) is well-defined since u is reached if $b^{u,\sigma}$ is played (Assumption 5.3.3 and (6.7.1)) and since $b^{u,\sigma}$ is defined by (6.7.4) in all those parts of the game tree which can effect the expected payoff $R_{iu}(\cdot)$. We conjecture that a disturbed extensive form game always (i.e. whenever Assumption 5.2.2 is satisfied) possesses an equilibrium, but a proof is not yet on paper. Instead of studying the existence problem, let us consider the question of which equilibria of Γ can be approximated by equilibria of $\Gamma(\mu^\varepsilon)$ if μ^ε converges weakly to 0 as ε tends to 0. To be more precise: if the strategy combination σ is played in $\Gamma(\mu)$, then to an outside observer it will look as if the behavior strategy combination b defined by

$$b_{iu} = b_{iu}^{u,\sigma} \quad \text{for all } i, u \tag{6.7.6}$$

is played, where $b_{iu}^{u,\sigma}$ is as in (6.7.4). b is called the behavior strategy combination *induced* by σ. We will investigate which equilibria of Γ can be obtained as limit

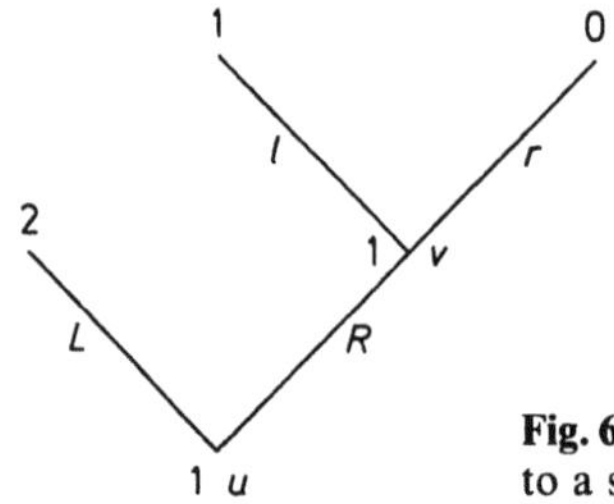

Fig. 6.7.1. Uncertainty about the payoffs does not necessarily lead to a sequential equilibrium being played

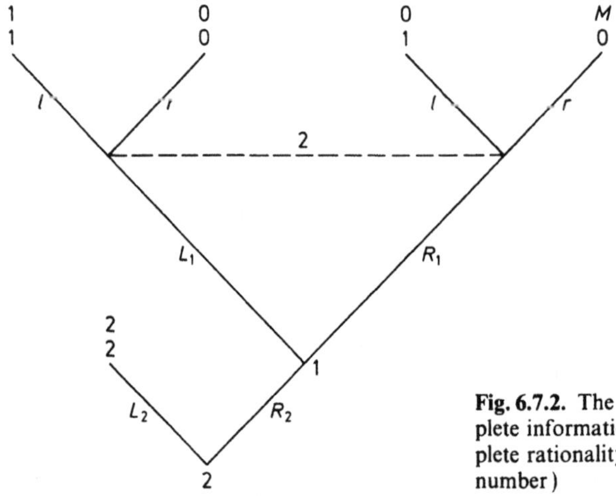

Fig. 6.7.2. The difference between the incomplete information approach and the incomplete rationality approach (M is some real number)

points of a sequence $\{b^\varepsilon\}_{\varepsilon \downarrow 0}$ where b^ε is induced by an equilibrium σ^ε of $\Gamma(\mu^\varepsilon)$. Such equilibria will be called *$\mathscr{P}$-firm* equilibria (where $\mathscr{P}=\{\mu^\varepsilon;\ \varepsilon>0\}$).

The game of Fig. 6.7.1 illustrates that in general non-sequential equilibria might be obtained as limit points.

Assume the payoffs of player 1 fluctuate around (2, 1, 0). Let the random variable X_i^ε represent the payoff at the ith endpoint of the tree (counting from left to right). If ε is close to 0, then an outside observer will see L being played with a probability close to 1, but at v he does not necessarily observe l with probability close to 1. Namely if b^ε is induced by an equilibrium of the ε-disturbed game, then

$$b_{1v}^\varepsilon(r)=\mathbb{P}[X_3^\varepsilon > X_2^\varepsilon | \max(X_2^\varepsilon, X_3^\varepsilon) > X_1^\varepsilon]$$

and although $\mathbb{P}[X_3^\varepsilon > X_2^\varepsilon]$ tends to 0 as ε tends to 0, the limit of $b_{1v}^\varepsilon(r)$ may be positive. This, for instance, will be true if X_i^ε has a Cauchy distribution (cf. the proof of Theorem 5.5.3). Hence, non-sequential equilibria may be $\mathscr{P}$-firm.

The reason why a non-sequential equilibrium may be $\mathscr{P}$-firm in the game of Fig. 6.7.1 is that, if v is actually reached, then the payoffs after v may be quite different from those displayed in Fig. 6.7.1. Hence, in the disturbed game, the subgame starting at v may be quite different from this subgame in the undisturbed game. The game of Fig. 6.7.2 is another example to demonstrate this. By means of this game we illustrate the difference between the incomplete rationality (perfectness) approach and the incomplete information approach.

Consider first the case in which there are no random disturbances in the payoffs, but in which the players might be slightly irrational. In this case, if the information set of player 1 is reached, this player concludes that player 2 has made a mistake. Yet, player 1 will think that player 2 will make mistakes only with a small probability at his second information set and, therefore, player 1 will play L_1.

Next, consider the case in which there are random disturbances in the payoffs. Ex ante, player 1 thinks that the payoffs of player 2 are as displayed in Γ and, therefore, he expects his information set not to be reached. If his information set is reached, then player 1 has to revise his priors and he might as well come to the conclusion that the payoffs of player 2 in the subgame are quite different from those displayed in Fig. 6.7.2. Therefore, he might assign a positive probability to player 2 playing r in the subgame and, consequently, it might be optimal to choose R_1 if M is large. To make the latter statement more precise, note that the normal form of the game of Fig. 6.7.2 is the game of Fig. 4.1.1 and in Theorem 5.5.3 we showed that it is optimal for player 1 to choose R_1 in the case in which the disturbances have a Cauchy distribution.

In both examples, the reason why a non-sequential equilibrium of Γ might be $\mathscr{P}$-firm is that, upon unexpectedly reaching an information set $u \in U_i$, player i might conclude that the payoffs after u are completely different from those in Γ. Hence, $\mathscr{P}$-firmness can be expected to imply sequentiality only if a player does not have to revise his beliefs about the payoffs after u once u is reached. More precisely, let σ be a strategy combination in $\Gamma(\mu^\varepsilon)$, let $\mu^{\varepsilon,u,\sigma}$ be defined as in (6.7.3) and let $\bar{\mu}_i^{\varepsilon,u,\sigma}$ be the conditional distribution of $\mu_i^{\varepsilon,u,\sigma}$ on $Z(u)$, i.e.

$$\bar{\mu}_i^{\varepsilon,u,\sigma}(B) := \mu_i^{\varepsilon,u,\sigma}(B \times \mathbb{R}^{Z \setminus Z(u)}) \quad \text{for a Borel subset } B \text{ of } \mathbb{R}^{Z(u)}. \tag{6.7.7}$$

If σ is played in $\Gamma(\mu^\varepsilon)$, then, if u is reached, player is beliefs about the payoffs after u are described by $\bar{\mu}_i^{\varepsilon,u,\sigma}$. The condition that a player, upon reaching u, should still think that the payoffs after u in $\Gamma(\mu^\varepsilon)$ are close to the payoffs in Γ amounts to requiring that

$$\bar{\mu}_i^{\varepsilon,u,\sigma} \text{ converges weakly to 0 as } \varepsilon \text{ tends to 0 for all } i, u, \sigma. \tag{6.7.8}$$

We conjecture that (6.7.8) is fulfilled if the disturbances at the different endpoints of the game are independent and have a normal distribution with parameters 0 and ε (cf. Lemma 5.7.1). Furthermore, we conjecture that only sequential equilibria are $\mathscr{P}$-firm if $\mathscr{P} = \{\mu^\varepsilon; \varepsilon > 0\}$ is such that (6.7.8) is satisfied, but, up to now, we have been unable to verify these conjectures. However, it is easy to see that not all sequential equilibria can be $\mathscr{P}$–firm in this case. Consider the game Γ of Fig. 6.5.1 and let $\mathscr{P} = \{\mu^\varepsilon; \varepsilon > 0\}$ be such that every μ_i^ε is the product distribution on $\mathbb{R}^5$ of a normal distribution on $\mathbb{R}$ with parameters 0 and ε. By (6.7.8) player 2, when he has to make a choice, believes that the payoffs of player 1 after this player's choices L_1 and R_1 are close to those of Γ (he cannot conclude anything about the payoffs of player 1 after A). Therefore, he will conclude that player 1 has chosen L_1 and, consequently, he will choose L_2. Hence, only the equilibrium (L_1, L_2) is $\mathscr{P}$-firm (cf. the discussions in Sect. 5.5 and 6.5). Note that, if (6.7.8) is satisfied, the disturbed game approach in essence means that player 2 analyses the game Γ of Fig. 6.5.1 by means of the game Γ' which is obtained from Γ by deleting the choice A of player 1: since the unique equilibrium of Γ' is (L_1, L_2), player 2 has to choose L_2.

Condition (6.7.8) definitely is a very stringent requirement, consequently, $\mathscr{P}$-firmness must be expected to be much weaker than sequentially. This is confirmed by the recent paper Fudenberg et. al. [1988] in which it is shown that for any *pure* equilibrium φ of an extensive form game Γ one can find some $\mathscr{P} = \{\mu^\varepsilon; \varepsilon > 0\}$ such that φ is $\mathscr{P}$-firm and, moreover, $\varphi = \lim_{\varepsilon \downarrow 0} \varphi^\varepsilon$ where φ^ε is a *strict* equilibrium of $\Gamma(\mu^\varepsilon)$. Hence, any pure equilibrium can be approximated by strict equilibria of slightly perturbed games.[12] Since strict equilibria possess all robustness properties one can hope for, Fudenberg et al. conclude from this result that a game theoretic analyst should have second doubts on rejecting any (pure) Nash equilibrium unless he is 100 % sure that his model is completely accurate.

Notes

1. However, see note 9 to Chap. 1.
2. It will be clear that 'generic' games of perfect information have a unique subgame perfect equilibrium. However, in 'applications' ties arise naturally and these give rise to interesting problems, see e. g. O'Neill [1986], Leininger [1986, 1988] and Ponssard [1990c].
3. Theorem 6.2.4 has been generalized to (a certain class) of perfect information finite-action, infinite-length games in Fudenberg and Levine [1983], and to perfect information games in which the action spaces are continua in Harris [1985], Hellwig and Leininger [1987] and Hellwig, Leininger, Reny and Robson [1991]. As far as this author knows, existence of SPE in (nicely behaved) continuous games of imperfect information has not yet been established, nor are counterexamples known.
4. See Note 3 to Chap. 1 for a collection of papers in which it is argued that the requirement of subgame perfection is too restrictive.
5. See also Kreps and Ramey [1987]. In that paper it is also shown that sequential equilibria entail convex structurally consistent assessments, i. e. there exist beliefs that are obtained from a convex combination of b's as in (6.3.3) that sustain the equilibrium. Furthermore, Kreps and Ramey give an example illustrating that structural consistency may conflict with sequential rationality: The concept of sequential best reply at u against (b, μ) is based on the assumption that there will be no deviations from b *after* u, while if $P^b(u) = 0$, condition (6.3.3) forces the player to assume that there has been a deviation *before* u; in some games one cannot have a deviation before u without simultaneously having a deviation after u. To put it differently: If at u one believes that b' has been played rather than b, why doesn't one optimize then against b' rather than against b?

6. Weaker concepts of consistency, leading to correspondingly weaker equilibrium notions are proposed in Fudenberg and Tirole [1989] and Weibull [1990].
7. McLennan [1989] has characterized the sequential equilibrium concept as the set of fixed points of a certain correspondence from a space of 'consistent conditional system' into itself. His framework allows the existence of a sequential equilibrium to be proved by a direct fixed point argument. Kohlberg and Reny [1991] have given a characterization of the Kreps/Wilson consistency concept without using trembles, i. e. their characterization is in terms of the basic data of the extensive form game (i. e. information sets and choices).
8. Recall from note 5 that structural consistency may conflict with sequential rationality.
9. Well-behaved imperfect information games in which the action spaces are continua need not have sequential equilibria, see Van Damme [1987, p. 29] for an example.
10. The relationship between perfectness, sequentiality and (lexicographic) domination is further investigated in Okada [1989].
11. See Mailath, Samuelson and Swinkels [1990] for a discussion of how also other extensive form equilibrium notions can already be detected in the (reduced) normal form.
12. Of course one may question whether the perturbations that are considered in Fudenberg et al. [1988] are actually slight perturbations. The results of Fudenberg et al. crucially depend on the fact that a small ex ante uncertainty can become large if an 'unexpected' action is observed.

7 Bargaining and Fair Division

In this chapter surplus sharing problems are considered, i.e. it is assumed that synergetic gains can be obtained by cooperating and the question is how these gains should be divided. Although traditionally such problems belong to the realm of cooperative game theory, we will study them by non-cooperative methods. The objectives are twofold: (1) to show how concepts from cooperative game theory can be implemented by means of noncooperative methods and (2) to illustrate the strength of the subgame perfectness concept (and the weakness of the Nash equilibrium concept) in dynamic games with perfect information.

In Sects. 7.2 and 7.3 we consider the simple situation in which a single indivisible object has to be allocated and in which side payments are possible. It is investigated what the consequences are of using the divide and choose method or some auction method to determine the allocation.

In the second half of the chapter, 2-person bargaining games without transferable utility are studied. Section 7.4 provides a general introduction in which the axioms underlying the Nash solution and the Kalai/Smorodinsky solution are presented. Sections 7.5 and 7.6 deal with noncooperative implementations of the Nash solution, specifically the demand game introduced by Nash and its infinite horizon extension as studied by Binmore and Rubinstein. Noncooperative models that generate the Kalai/Smorodinsky solution are the subject of Sect. 7.7.

Finally, in Sect. 7.8 we consider bargaining games with variable threat point and we illustrate that the solution proposed by Nash is viable only if binding commitments are possible.

Most results in this chapter are not new. The material is based on Nash [1953], Crawford [1977, 1979], Binmore [1980], Rubinstein [1982], Moulin [1984] and Güth and van Damme [1986]. No attempt has been made to achieve the most general results, rather we have made those simplifying assumptions that make the models as transparent as possible.[1]

7.1 Introduction

Thus far we have considered abstract games given either in normal form or in extensive form. This general approach has enabled us to derive results concerning the existence of various refined equilibrium concepts as well as about the relations between these refinements. However, in order to get some feeling about the actual importance of the various concepts, it is necessary to look into classes of problems that exhibit more structure. In this chapter, one such class, the class of surplus sharing games, is studied.

The situation considered is one in which a surplus becomes available when the players succeed in cooperating and the question is how this surplus should be divided. A pure exchange economy provides an example of such a situation: Each player has an initial endowment and by trading each person can possibly achieve a commodity bundle that he prefers to the status quo. Although traditionally such problems belong to the realm of cooperative game theory, our aim is to study them by means of noncooperative methods and to illustrate various refinements by means of simple specific examples. Throughout the chapter it will be assumed that the preferences of the participants are common knowledge, hence, the games to be analysed are games of complete information.[2] Furthermore, attention will be restricted to situations in which only two persons are involved. The largest part of the chapter is devoted to games with perfect information and, consequently, emphasis will be on the subgame perfectness concept. The discussion of the Nash demand game in Sect. 7.5, however, involves the concept of essential equilibria. The power (resp. weakness) of the other refined equilibrium concepts discussed in previous chapters (sequential, proper, strictly perfect equilibria) is best illustrated by means of games with incomplete information and for a discussion of these the reader is referred to Chap. 10.

In Sects. 7.2 and 7.3 of this chapter we consider a very simple surplus sharing problem: There is an indivisible object to which two players have equal rights and side payments among the players are possible. Hence, the problem is who should get the object and how much compensation this person should pay to his opponent in order to guarantee a 'fair' outcome. The obvious solution is to allocate the object to the person who values it most and to split the surplus equally, however, the question is whether this outcome will indeed result in a purely noncooperative context. Specifically, the question is whether we can design games that implement the desired solution.

Section 7.2 analyses the situation in which the players have agreed to use the divide and choose method: Both players ante a certain amount of money; one player is appointed divider; he transfers money from the pot to the object and the other player has to choose between the object plus the transfer or the remainder of the money. We compute the set of Nash equilibria of this game and show that there is a unique subgame perfect equilibrium. The analysis shows that being the divider confers an advantage (due to the complete information), that the resulting allocation is Pareto efficient, and that assuming that players are supercautious (play minimax strategies), leads to the wrong conclusions. Finally it is shown that 'equal division' results when the role of divider is allocated by means of an auction.

In Sect. 7.3, we assume that some auction method (with equal division of the revenue) is used to allocate the object. A one-parameter family of auctions is considered, among which there are the ascending (English) auction, the descending (Dutch) auction, and an auction corresponding to the 'Steinhaus fair division procedure'. We show that auctions that allow a dynamic implementation (the English and Dutch versions) admit a unique perfect equilibrium, but that otherwise there are many strict (hence, perfect) equilibria. Furthermore, the allocation resulting from the English auction is different from the one that results from the Dutch auction.

The situation discussed in Sects. 7.2 and 7.3 can be viewed as a bargaining problem with fixed threats. It is a very simple such problem since there is transferable utility. In the second half of the chapter we consider more general bargaining problems, not necessarily having transferable utility. Section 7.4 provides a general introduction, adopting the traditional axiomatic approach, with special emphasis on the solutions proposed by Nash and by Kalai and Smorodinsky. (For a survey of this approach, see Roth [1979] or Kalai [1985].)[3] The axiomatic approach has its drawbacks, however. Probably this is best illustrated by the fact that several solutions coexist, each supported by its own axiom system, without it being clear which axioms really capture the essence of rational bargaining. This has lead several authors to search for more basic models by means of which the axioms (or at least the solutions implied by the axioms) could be derived. Hence, the quest has been for noncooperative models that implement axiomatic solutions.

This line of investigation has first been persued in Nash [1953] and Nash's model is reviewed in Sect. 7.5. Nash proposes a simultaneous-move one-shot game: Each player states a demand and these demands are granted if and only if they are compatible. In this model, any feasible allocation can be obtained as the expected payoff of a Nash equilibrium, however, there is only one essential equilibrium and this results in the outcome corresponding to the axiomatic Nash solution. In other words, if the players are slightly uncertain about which demand pairs are feasible, then it is optimal to demand the payoff granted by the axiomatic Nash solution. Section 7.5 is based on Nash [1953] and Binmore [1987a, c].

In Sect. 7.6, we consider a second noncooperative implementation of the Nash solution, due to Rubinstein and Binmore. (This approach is also closely related to ideas discussed in Ståhl [1972]). The Rubinstein model allows for repeated bargaining: In turns, the players propose outcomes until an agreement is reached. (It is assumed that players discount future payoffs so that they have an incentive to make serious offers.) The main result is that there is a unique subgame perfect equilibrium and that agreement is reached immediately. Furthermore, if both players have the same discount rate and if this discount rate converges to 0, then the subgame perfect equilibrium payoff converges to the axiomatic Nash outcome. This section is based on Binmore [1987a, b] and Rubinstein [1982].

In Sect. 7.7, an auction model is considered that implements the Kalai/Smorodinsky solution. Before the actual bargaining game is played, an auction is organized to determine which player may propose an outcome in the bargaining game. The bids in the auction determine the utility level that the proposer has to guarantee his opponent in the bargaining game, and when this is done in the appropriate way, the Kalai/Smorodinsky solution results as the unique subgame perfect equilibrium outcome of the 2-stage game. This section is based on Crawford [1977] and Moulin [1984].

In the concluding Sect. 7.8, we study bargaining problems with variable threats[4] and we point out that the approach proposed by Nash to analyse such games is viable only if players can commit themselves to using their threats, hence, Nash's solution in the variable threat case involves equilibria that are not subgame perfect. This section also tries to throw light on the distinctions between cooperative games, noncooperative games with communication and noncooperative games without communication.

7.2 Divide and Choose

A fair division problem arises in situations in which a bundle of goods has to be divided among a group of persons in a fair and efficient way. In this section and the next one, we consider a very simple such situation: there is one indivisible commodity that has to be allocated. In this section, it is assumed that the players have agreed to use the divide and choose method and we investigate what the optimal behavior is in this case. For an analysis of the divide and choose method in a more general setting (with complete information) the reader is referred to Crawford [1977], Crawford and Heller [1979] or Kolm [1972]. For the consequences of incomplete information, the reader can consult van Damme [1985].

Suppose there is one indivisible commodity that has to be allocated to one of two persons who have equal rights to it. The question is who should get the object and how much should this person pay to his opponent in order to achieve a fair allocation. Suppose the players value the object differently and that player 1's value v_1 is higher than the value v_2 of player 2. Each person is completely informed about both his own value and that of his opponent. Efficiency requires that the object should be allocated to player 1, while the fact that the player have equal rights to the object implies that each player i is entitled to a payoff of at least $1/2v_i$. Hence, the problem is: On which payoff pair (r_1, r_2) with

$$r_1+r_2=v_1, \quad r_1 \geqq 1/2v_1, \quad r_2 \geqq 1/2v_2, \tag{7.2.1}$$

the players should agree? Symmetry suggests that this should be the midpoint of this interval, i.e. the point (r_1^*, r_2^*)

$$r_1^*=3/4v_1-1/4v_2, \qquad r_2^*=1/4v_1+1/4v_2, \tag{7.2.2}$$

but the question is whether this point will indeed be realized if a certain procedure is used. Note that (r_1^*, r_2^*) is the Nash (or Kalai/Smorodinsky) solution when the threat point is $(1/2v_1, 1/2v_2)$ (see Sect. 7.4). It is also the solution proposed in Steinhaus [1948]. Steinhaus calls $1/2v_i$ the fair share of player i and his solution gives each player his fair share plus half of the available surplus, the surplus being $v_1-1/2v_1-1/2v_2$.

Assume the players have agreed to use the *divide and choose* method, more specifically, the variant of this method that has been introduced in Luce and Raiffa [1957]. Both persons put an equal amount of money in the pot, the total being worth more than the object. Next, one person is appointed divider, the other is the chooser. The divider transfers a certain amount of money from the pot to the commodity and the chooser is allowed to take either the commodity plus the money that has been transfered (action C) or to take the remainder of the money (action M). The divider takes whatever is left. Since the resulting allocation does not depend on the exact ante of the players, we may assume each person antes an amount v_1. Let us first analyze the game Γ_1 in which player 1 is the divider. A (pure) strategy of player 1 then corresponds to a certain transfer, hence it is an element x of $[0, 2v_1]$. A (pure) strategy of player 2 is a rule telling this player which action to take depending on the transfer chosen by player 1, hence, it is a map $f:[0, 2v_1] \to \{C, M\}$. (Since player 2 can see what player 1 does, we have a 2-stage game with perfect information.) A strategy pair (x,f) results in a pair of net payoffs $R(x,f) = R(x,f(x))$, where

$$R(x, C) = (v_1 - x,\ v_2 + x - v_1), \quad R(x, M) = (x,\ v_1 - x). \tag{7.2.3}$$

Assume player 1 has chosen x. Then it is optimal for player 2 to choose $f^*(x)$, where

$$f^*(x) = \begin{cases} M & \text{if } x < v_1 - 1/2v_2, \\ C & \text{if } x > v_1 - 1/2v_2. \end{cases} \tag{7.2.4}$$

Both choices are optimal when $x = v_1 - 1/2v_2$. Assuming that player 2 will choose $f^*(x)$ for every value of x, player 1 can compute his payoff $R_1(x, f^*(x))$

$$R_1(x, f^*(x)) = \begin{cases} x & \text{if } x < v_1 - 1/2v_2, \\ v_1 - x & \text{if } x > v_1 - 1/2v_2. \end{cases} \tag{7.2.5}$$

His payoff is intermediate if $x = v_1 - 1/2v_2$. Hence, player 1 will choose x smaller than, but as close as possible to $v_1 - 1/2v_2$. In fact, if $(\bar{x}, \bar{f})$ is a subgame perfect equilibrium of Γ_1, then

$$\bar{x} \text{ is a best reply against } \bar{f}, \text{ and} \tag{7.2.6}$$

$$\bar{f}(x) \text{ is a best reply against } x \text{ for all } x, \tag{7.2.7}$$

which implies $\bar{f} = f^*$ and player 2 should choose M also when $x = v_1 - 1/2v_2$ since otherwise the best reply of player 1 does not exist.[5] Hence, player 1 chooses $\bar{x} = v_1 - 1/2v_2$ and there is a unique subgame perfect equilibrium. We have proved:

Theorem 7.2.1. *The divide and choose game in which player* 1 *is the divider has* $(v_1 - 1/2v_2,\ 1/2v_2)$ *as its unique subgame perfect equilibrium outcome.*

Note that this subgame perfect equilibrium outcome is that payoff pair satisfying (7.2.1) that is most favourable to player 1: The divider adds as much money as possible to the commodity subject to the constraint that player 2 still is willing to take the money, and thereby he extracts all available surplus. The game Γ_2 in which player 2 is the divider can be analysed in the same way: The optimal strategy for player 2 is to add as little money to the object as possible but such that player 1 still prefers the object. By following this optimal strategy player 2 can extract all surplus; the subgame perfect equilibrium outcome is $(1/2v_1,\ 1/2v_1)$. Comparing this result to Theorem 7.2.1, we can state

Corollary 7.2.2. *The divide and choose method confers an advantage to the divider.*

This corollary is valid for a much wider class of problems than the one considered in this section (see e.g. Crawford [1977]), however, for the result to be valid it is essential that the players are fully informed about each others values. If you have only vague information about your opponent's value, then it might be advantageous to be the chooser (see van Damme [1985]).

Because of Corollary 7.2.2, the role of divider should be allocated in a fair way. One possibility is to toss a coin, hence, each person has a 50 % chance of being the divider. In this case the expected payoff is indeed the pair (r_1^*, r_2^*) from (7.2.2). An alternative way, that has been suggested in Crawford [1979], is to auction off this role: Both players bid, the highest bidder becomes the divider and pays a certain amount to his opponent in order to obtain this privilege. If the bids coincide, then the winner is determined by tossing a fair coin. Let us assume that the price the winner has to pay in case the bids are b_1 and b_2 is equal to

$$P_\lambda(b_1, b_2) = \lambda \max(b_1, b_2) + (1-\lambda)\min(b_1, b_2), \tag{7.2.8}$$

where λ is a parameter, $\lambda \in [0, 1]$. Of special interest are the cases with $\lambda = 1$ (the highest price auction) and $\lambda = 0$ (the Vickrey or second highest price auction). From the fact that the outcome is $(v_1 - 1/2v_2, 1/2v_2)$ if player 1 wins the auction and $(1/2v_1, 1/2v_1)$ if player 2 wins, it follows that the situation can be described by the normal form game Γ^λ with payoff functions

$$R_1^\lambda(b_1, b_2) = \begin{cases} 1/2v_1 + P_\lambda(b_1, b_2) & \text{if } b_1 < b_2, \\ 3/4v_1 - 1/4v_2 & \text{if } b_1 = b_2, \\ v_1 - 1/2v_2 - P_\lambda(b_1, b_2) & \text{if } b_1 > b_2, \end{cases} \tag{7.2.9}$$

$$R_2^\lambda(b_1, b_2) = v_1 - R_1^\lambda(b_1, b_2). \tag{7.2.10}$$

It can easily be verified that (independent of λ) this game has as its unique equilibrium the pair (b_1^*, b_2^*) with $b_1^* = b_2^* = 1/4(v_1 - v_2)$, resulting in the outcome (r_1^*, r_2^*) from (7.2.2). Hence, we have shown

Corollary 7.2.3. *Whether the role of divider is allocated randomly or by using an auction, the unique subgame perfect equilibrium outcome is the payoff vector* (r_1^*, r_2^*) *from* (7.2.2).

In the next section it will be analysed what the outcome will be if an auction mechanism is used directly to allocate the commodity. We will see that in that case, the resulting allocation depends crucially on which method (i.e. which λ) is being used.

Corollary 7.2.3 shows that the desired outcome does indeed result when a subgame perfect equilibrium is played. However, one cannot draw this same conclusion for Nash equilibria. Namely, the payoff functions given in (7.2.9) and (7.2.10) presume that the subgame perfect equilibrium will be played in the divide and choose game and this need not be true in a Nash equilibrium of the auction game. Secondly, the divide and choose game itself has more equilibria than the one described in Theorem 7.2.1. In fact, the following Theorem illustrates that the Nash concept is extremely weak.

Theorem 7.2.4. *Any payoff pair* (r_1, r_2) *satisfying* (7.2.1) *is a Nash equilibrium outcome of the divide and choose game in which player* 1 *is the divider.*

Proof. The pair $(\bar{x}, \bar{f})$ is a Nash equilibrium if and only if $\bar{x}$ is a best reply against $\bar{f}$ and

$$\bar{f}(\bar{x}) \text{ is a best reply against } \bar{x}. \tag{7.2.11}$$

Note that (7.2.11) is much weaker than (7.2.7). Now, let (r_1, r_2) satisfy (7.2.1) and consider the following threat strategy $\bar{f}$ of player 2 (the chooser): Take the commodity if the transfer is more than r_1, otherwise take the money. Then player 1's best reply is to transfer r_1. Since $\bar{f}(r_1)$ is also a best reply against r_1, we have a Nash equilibrium with the desired outcome. □

Of course, a similar result to that in Theorem 7.2.4 holds in case player 2 is the divider. Hence, one cannot draw any conclusions by using the Nash concept. In the remainder of this section, it will be shown that by using the *minimax concept* one will even draw the wrong conclusions. This will once more illustrate the difference between equilibrium strategies and maximin strategies that was also considered in the discussion of the game of Fig. 1.6.2.

We have seen that in a subgame perfect equilibrium, the divider divides in such a way that the chooser is indifferent. In contrast, a supercautious (minimaxing) divider will choose that x that guarantees the highest payoff irrespective of what the chooser does. Hence, the divider's maximin strategy is to transfer such an amount that he himself is indifferent between the two choices. Therefore, if both players behave in this supercautious way, then the outcome is $(1/2v_1, 1/2v_1)$ if player 1 is the divider and $(v_1 - 1/2v_2, 1/2v_2)$ if player 2 is the divider, and one reaches conclusions that are exactly opposite to those of Theorem 7.2.1 and Corollary 7.2.2.

Of course, if one knows the value of the opponent exactly (as we have assumed throughout), then there is no reason at all to behave in this supercautious way. If player 1 is the divider, then he should exploit his knowledge about the value of his opponent. He can safely assume that player 2 will choose according to (7.2.4), hence, only the conclusions reached by the subgame perfect equilibrium concept are justified.

7.3 Auction Methods

In this section we again consider the fair division situation that was introduced in the previous section: there is a single indivisible commodity that has to be allocated to one of two players; player i assigns a value v_i to the object, $v_1 > v_2$ and both values are common knowledge. It is now assumed that an auction is used to determine the allocation: Both players bid and the object is assigned to the highest bidder for a price that is a convex combination of the two bids (as in (7.2.8)). It is shown that auctions that allow a dynamic implementation have only one subgame perfect equilibrium, whereas auctions that are essentially static have many perfect equilibria.[6]

In order to guarantee existence of equilibria, it will be assumed that there exists a smallest money unit of which any bid has to be an integer multiple. b^- will denote the highest feasible bid smaller than b and b^+ is the smallest bid larger than b. For convenience, we assume that v_i is an integer multiple of this smallest money unit.

For $\lambda \in [0, 1]$, let $P_\lambda(b_1, b_2)$ be as in (7.2.8). We will consider auctions in which the highest bidder receives the object for the price $P_\lambda(b_1, b_2)$, ties being broken by the toss of a fair coin. The revenue of the auction will be shared by both players, hence, effectively the winner only pays $1/2P_\lambda(b_1, b_2)$. Formally, we consider the normal form game $\Gamma(\lambda)$ of which the payoff function is

$$R_1^\lambda(b_1, b_2) = \begin{cases} 1/2P_\lambda(b_1, b_2) \\ 1/2v_1 \\ v_1 - 1/2P_\lambda(b_1, b_2) \end{cases} \quad R_2^\lambda(b_1, b_2) = \begin{cases} v_2 - 1/2P_\lambda(b_1, b_2) & \text{if } b_1 < b_2 \\ 1/2v_2 & \text{if } b_1 = b_2 \\ 1/2P_\lambda(b_1, b_2) & \text{if } b_1 > b_2 \end{cases} \tag{7.3.1}$$

If $\lambda = 0$, the winner pays (half of) the second highest bid. This auction is known as the *Vickrey auction* (Vickrey [1961]) and it can be viewed as the normal form of the *English or ascending auction.* This is the ordinary auction in which the price is increased until one of the player stops bidding; the player who still wanted to continue receives the object and he pays the price at which his opponent dropped out. This extensive form will be analysed later on, for the moment we confine ourselves to the normal form. Figure 7.3.1 gives a numerical example in which the values are 10 and 4 and the feasible bids are 4, 6, 8 and 10. Note that in contrast to the ordinary Vickrey auction (in which there is a seller from outside and each buyer has a payoff of zero in case he does not win the object) it is not a dominant strategy to bid your true value, since the looser's payoff depends on his own bid.

At the other extreme we have the game with $\lambda = 1$ in which the winner pays (half of) the highest bid. This is the normal form of the *Dutch or descending auction*: the auctioneer decreases the price until one of the players shouts stop; this player then receives the object for the current price. Figure 7.3.2 gives a numerical example.

In addition to the games $\Gamma(0)$ and $\Gamma(1)$, also the game $\Gamma(1/2)$ arises in a natural way. Namely, suppose there is a mediator who wants to implement the fair solution of (7.2.2) according to the *Steinhaus procedure* (see Sect. 7.2), but who does not know the players' values. This mediator will ask the players to reveal their

	4	6	8	10
4	5, 2	2, 2	2, 2	2, 2
6	8, 2 *	5, 2	3, 1	3, 1
8	8, 2	7, 3 *	5, 2	4, 0
10	8, 2	7, 3	(6, 4) *	5, 2

Fig. 7.3.1. The Vickrey auction game $\Gamma(0)$ when the values are 10 and 4 and the feasible bids are 4, 6, 8 and 10. Starred entries are equilibrium outcomes, the encircled entry is the unique perfect equilibrium payoff

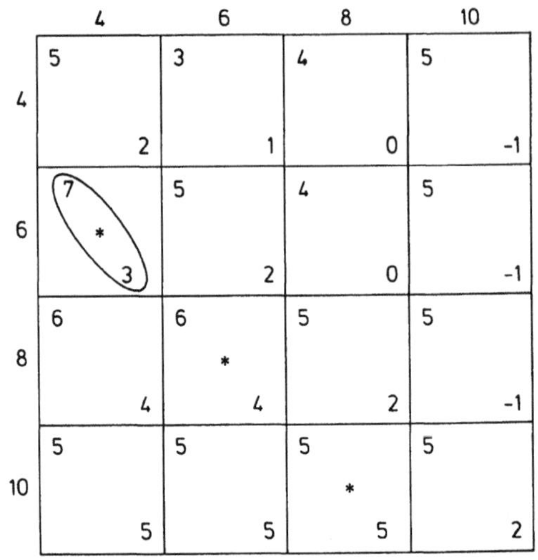

Fig. 7.3.2. The Dutch auction game $\Gamma(1)$ when the values are 10 and 4 and the feasible bids are 4, 6, 8 and 10. Starred outcomes are Nash equilibria, the encircled one is the unique perfect equilibrium

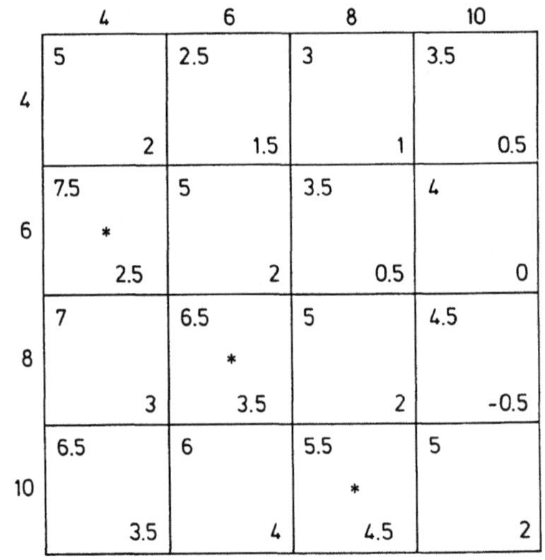

Fig. 7.3.3. The auction game $\Gamma(\frac{1}{2})$ when the values are 10 and 4 and when only the bids 4, 6, 8 and 10 are feasible. Starred outcomes are equilibria. All equilibria are strict, hence, perfect

values and he will implement the Steinhaus solution corresponding to the reported values. In such a case it is in the interest of the players to misrepresent their values, namely, player 2 has an incentive to overstate his value in order to increase his fair share and player 1 has an incentive to understate his value in order to decrease the available surplus. More specifically, if player i pretends his value to be $\bar{v}_i$, then the person pretending to have the highest value (say this is player 1) will receive the object and he pays to his opponent this player's fair share ($=1/2\bar{v}_2$) plus half of the available surplus ($=1/4(\bar{v}_1-\bar{v}_2)$). Hence, the winner pays $1/4(\bar{v}_1+\bar{v}_2)$, exactly as in the game $\Gamma(1/2)$. Figure 7.3.3 gives a numerical example of this game.

Figures 7.3.1 – 7.3.3 illustrate that the Nash equilibria of the games $\Gamma(\lambda)$ are independent of λ, but that, as far as perfect equilibria are concerned, there are sharp distinctions between $\Gamma(0)$, $\Gamma(1/2)$ and $\Gamma(1)$. It is now shown that this holds in general, i.e. for arbitrary parameter values and without imposing restrictions on the bids. We first consider Nash equilibria.

Theorem 7.3.1. *If $\lambda \in (0, 1)$, then (b_1, b_2) is a Nash equilibrium of the auction game $\Gamma(\lambda)$ if and only if*

$$b_1 \leqq v_1, \quad b_2 \geqq v_2 \quad \text{and} \quad b_1 = b_2^+ . \tag{7.3.2}$$

If $\lambda = 0$, there is one additional equilibrium to those in (7.3.2), viz.

$$b_2 = v_1 \quad \text{and} \quad b_1 = b_2^+ . \tag{7.3.3}$$

If $\lambda = 1$, there is also one additional equilibrium, viz.

$$b_1 = v_2 \quad \text{and} \quad b_2 = b_1^- . \tag{7.3.4}$$

Proof. First of all note that in equilibrium the bids will not be more than one money unit apart. This is clear if $\lambda \in (0, 1)$, since in this case the winner's payoff is decreasing in his own bid and the looser's payoff is increasing in his own bid. If $\lambda = 0$, the winner payoff is independent of his own bid, but in equilibrium his bid can only be slightly above that of the looser since otherwise the looser has an incentive to bid higher. A similar reasoning applies when $\lambda = 1$. Next, let $\lambda \in (0, 1)$. From the preceding argument it follows that the best reply b_i of player i against b_j is unique and it is easily seen that

$$\begin{aligned} &\text{if } b_j < v_i, \text{ then } b_i = b_j^+ , \\ &\text{if } b_j = v_i, \text{ then } b_i = b_j , \\ &\text{if } b_j > v_i, \text{ then } b_i = b_j^- . \end{aligned} \tag{7.3.5}$$

From these conditions it follows that the first statement of the theorem is correct. If (b_1, b_2) is an equilibrium of $\Gamma(0)$, then the first and third assertions of (7.3.5) are still correct, but if player 2 bids v_1, then player 1 is indifferent between bidding v_1 or v_1^+, hence, we have the additional equilibrium from (7.3.3). In exactly the same way one can show that (b_1, b_2) is an equilibrium of $\Gamma(1)$ if and only if either (7.3.2) or (7.3.4) is satisfied. □

Theorem 7.3.1 implies that in the limit, as the smallest money unit approaches zero, any payoff vector satisfying (7.2.1) can be obtained as a Nash equilibrium outcome of any auction game $\Gamma(\lambda)$. The proof given above also shows that any equilibrium from (7.3.2) is strict when $\lambda \in (0, 1)$, i.e. in this case, one has

$$R_1^\lambda(b_1, b_2) < R_1^\lambda(b_2^+, b_2) \quad \text{for all } b_1 \neq b_2^+ \text{, and}$$

$$R_2^\lambda(b_1, b_2) < R_2^\lambda(b_1, b_1^-) \quad \text{for all } b_2 \neq b_1^- .$$

However, this is no longer true for $\lambda \in \{0, 1\}$. If $\lambda = 0$, then player 1 can never loose by bidding v_1^+, for doing so increases his chances of being the winner in the region (of bids of his opponents) where he wants to win, and it also increases his payoff in case he does not win. Formally, if $b_1 \leqq v_1$, then

$$R_1^0(b_1, b_2) = R_1^0(v_1^+, b_2) \quad \text{if } b_2 < b_1,$$
$$R_1^0(b_1, b_2) \leqq R_1^0(v_1^+, b_2) \quad \text{if } b_1 \leqq b_2 \leqq v_1^+,$$
$$R_1^0(b_1, b_2) < R_1^0(v_1^+, b_2) \quad \text{if } b_2 > v_1^+,$$

hence, bidding v_1^+ dominates bidding any smaller amount. This shows that all equilibria from (7.3.2) are imperfect when $\lambda = 0$: the equilibrium from (7.3.3) is the unique perfect equilibrium of $\Gamma(0)$.

On the other hand, if $\lambda = 1$, then player 2 cannot loose by bidding v_2^-: doing so decreases his chances of being the winner in the region where he does not want to win and it does not influence his payoff in case he does not win. More precisely, for any bid $b_2 \geqq v_2$, we have

$$R_2^1(b_1, b_2) = R_2^1(b_1, v_2^-) \quad \text{if } b_1 > b_2$$
$$R_2^1(b_1, b_2) \leqq R_2^1(b_1, v_2^-) \quad \text{if } v_2^- \leqq b_1 \leqq b_2$$
$$R_2^1(b_1, b_2) < R_2^1(b_1, v_2^-) \quad \text{if } b_1 < v_2^-,$$

hence, for player 2 the bid v_2^- dominates any higher bid. Consequently, only the equilibrium from (7.3.4) is perfect in $\Gamma(1)$. We have shown

Theorem 7.3.2. *If $\lambda \in (0, 1)$, then any equilibrium as in (7.3.2) is a strict, hence, perfect equilibrium of $\Gamma(\lambda)$. The games $\Gamma(0)$ and $\Gamma(1)$ each have a unique perfect equilibrium which are given by (7.3.3) and (7.3.4), respectively.*

This theorem implies that, if $\lambda = 0$, then only $(1/2v_1, 1/2v_1)$ can be obtained as a perfect equilibrium payoff; if $\lambda = 1$, then the unique perfect equilibrium payoff is $(v_1 - 1/2v_2, 1/2v_2)$, and if $\lambda \in (0, 1)$ then any payoff as in (7.2.1) is a perfect equilibrium payoff. The distinction between the games $\Gamma(\lambda)$ with $\lambda \in (0, 1)$ on the one hand and those with $\lambda \in \{0, 1\}$ on the other is caused by the fact that the latter admit a dynamic implementation whereas the former do not. Let us now analyse the extensive forms of the games $\Gamma(0)$ and $\Gamma(1)$.

First consider the English auction in which the price is increased from zero until one of the players stop bidding. Since player 2 has the lower value he will stop first. However, if he plays optimally he will not stop before the price is v_1 of v_1^-. Namely, if the price is t with $t < v_1^-$, then player 1 will not want to stop since this yields only $1/2t$ while continuing to v_1 *guarantees* a payoff of $1/2v_1$. Formally, subgame perfect equilibria can be found by dynamic programming. Assume that, if some high price T is reached, the auction is stopped and the object is allocated randomly. Then at price $t = T^-$ the players face the game from Fig. 7.3.4a where S denotes 'stop' and C denotes 'continue'. For both players it is a dominant strategy to stop and this remains the case as long as $t > v_1$. When $t = v_1$, then player 1 is indifferent between C and S. Assuming that he chooses C in this case, the equilibrium outcome for $t = v_1$ is $(1/2v_1, 1/2v_1)$ and then for $t < v_1$ we get the game from Fig. 7.3.4b. This game has as its unique equilibrium (C, C). Alternatively, we have that player 1 chooses S for $t = v_1$. Then for $t = v_1^-$ the outcome will be (C, S),

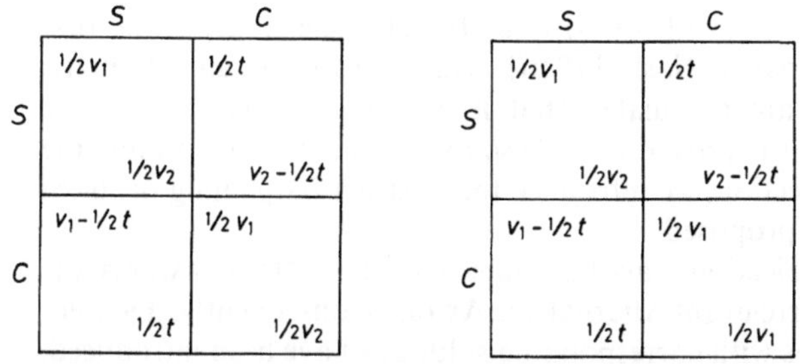

a b

Fig. 7.3.4a, b. The subgames of the English auction game, starting at a certain price t assuming that both players behave optimally for higher prices. Situation **a** corresponds to $t > v_1$ with unique equilibrium (S, S). Situation **b** arises when $t < v_1$; in this case the equilibrium is (C, C)

while (C, C) will result for any $t < v_1^-$. Hence, we see that the extensive form admits two subgame perfect equilibria, viz. the strategy pairs (f_1, f_2)

$$f_1(t) = \begin{cases} C & \text{if } t \leqq t^* \\ S & \text{if } t > t^* \end{cases} \qquad f_2(t) = \begin{cases} C & \text{if } t < t^* \\ S & \text{if } t \geqq t^* \end{cases}$$

with $t^* = v_1$ or $t^* = v_1^-$. The equilibrium with $t^* = v_1$ corresponds to the perfect equilibrium of the normal form of $\Gamma(0)$. The second subgame perfect equilibrium was not discovered by the normal form analysis because of the fact that it involves a dominated action of player 1. If indeed player 2 plays optimally (i.e. stops) at every t with $t \geqq v_1$, then player 1 is indifferent between stopping and continuing when the price is v_1, but if there is a small chance that player 2 makes a mistake and continues bidding at $t = v_1^+$, then it is better for player 1 to continue until this price is reached. Hence, in the extensive form only the equilibrium with $t^* = v_1$ is perfect and the perfect equilibria of the normal form coincide with the perfect equilibria of the extensive form.

The Dutch or descending auction can be analysed in exactly the same way. Now it is player 1 who is in the more powerful position: He can drive the price down to v_2 since player 2 is better off by buying at this price then by buying at any higher price. Straightforward dynamic programming reveals that there are two subgame perfect equilibria, viz. the strategy pairs (f_1, f_2) given by

$$f_1(t) = \begin{cases} S & \text{if } t \leqq t^* \\ C & \text{if } t > t^* \end{cases} \qquad f_2(t) = \begin{cases} S & \text{if } t < t^* \\ C & \text{if } t \geqq t^* \end{cases}$$

with $t^* = v_2$ or $t^* = v_2^+$. Only the first of these, which corresponds to the equilibrium from (7.3.4), is perfect.

7.4 Bargaining Problems and Bargaining Solutions

The fair division situation considered in the previous sections can be viewed as a bargaining problem with fixed threats. It is a very simple such problem since there is transferable utility (i.e., the Pareto efficient frontier is flat). A solution for abstract bargaining problems (with or without transferable utility) was first proposed in the seminal paper Nash [1950]. Nash proposed an axiomatic approach: several axioms are postulated that it would seem natural for the solution to have and then it is proved that these axioms actually determine the solution uniquely. In this section, we introduce the abstract bargaining problem and the axioms that Nash proposed.

In the literature many objections have been raised to one of Nash's axioms, the so called Independence of Irrelevant Alternatives Axiom. Consequently, a variety of alternative axiom systems with corresponding solutions have been introduced. Among these there is the solution of Kalai and Smorodinsky [1975] that is also discussed in this section. The lack of consensus on what the 'correct' axiom system is clearly demonstrates the weakness of the axiomatic approach and, therefore, many writers (starting with Nash) felt the need to supplement this approach with an explicit noncooperative model of the bargaining process that yields a specific axiomatic solution as its outcome. In Sects. 7.5, 7.6 and 7.7 we discuss various such models that implement respectively the Nash and the Kalai/Smorodinsky solution. In Sect. 7.8 we consider the more difficult case in which the threat point is not a priori given and we illustrate that the solution proposed by Nash is viable only if the players have the possibility to commit themselves to a course of action.

A 2-person *bargaining problem* (with fixed threats) is a pair (S, d) where S is a compact and convex subset of $\mathbb{R}^2$ and $d \in S$. The set S represents the set of payoff pairs that the players can obtain by cooperating and d, the disagreement outcome, is the payoff vector that results when they do not cooperate. The convexity of S follows from the fact that joint randomization is allowed and compactness is guaranteed if, for example, the underlying set of basic alternatives is finite. For convenience, we will assume that cooperation cannot hurt and that it can be benificial for both players simultaneously, hence

$$x \geqq d \text{ for all } x \in S \quad \text{and} \quad x > d \text{ for some } x \in S, \tag{7.4.1}$$

where $x \geqq d$ denotes $x_i \geqq d_i$ $(i=1, 2)$, while $x > d$ means $x_i > d_i$ $(i=1, 2)$. It is also assumed that S is *comprehensive*, i.e.

$$\text{if } d \leqq x \leqq y \text{ and } y \in S, \text{ then } x \in S. \tag{7.4.2}$$

We write $u(S)$ for the *utopia point* associated with S

$$u_i(S) = \max\{x_i; \exists_{x_j} (x_1, x_2) \in S\} \tag{7.4.3}$$

and ∂S denotes the strong *Pareto boundary* of S

$$\partial S = \{x \in S; \text{ if } y \in S \text{ and } y \geqq x, \text{ then } y = x\}. \tag{7.4.4}$$

Since S is comprehensive there exist (weakly) decreasing, concave functions f_1 and f_2 such that

$$\begin{aligned} S &= \{(x_1, x_2); d_1 \leqq x_1 \leqq u_1(S), d_2 \leqq x_2 \leqq f_1(x_1)\} \\ &= \{(x_1, x_2); d_2 \leqq x_2 \leqq u_2(S), d_1 \leqq x_1 \leqq f_2(x_2)\}. \end{aligned} \tag{7.4.5}$$

For convenience it is assumed that

$$f_1(u_1(S)) = d_2 \text{ and } f_2(u_2(S)) = d_1, \tag{7.4.6}$$

so that f_1 and f_2 are strictly decreasing and, hence, are each others inverses. Consequently, the weak Pareto boundary coincides with the strong Pareto boundary.

To summarize: a *bargaining problem* is a pair (S, d) where S is a compact and convex subset of $\mathbb{R}^2$ with $d \in S$ such that the conditions (7.4.1) – (7.4.6) are satisfied. We will use Σ to denote the set of all bargaining problems. It should be stressed that the regularity assumptions (7.4.1) – (7.4.6) are made only to simplify the exposition, they are not really needed to derive the results.

A *bargaining solution* is a rule F that assigns an outcome to every bargaining problem, hence

$$F: \Sigma \rightarrow \mathbb{R}^2 \quad \text{such that } F(S, d) \in S \quad \text{for all } (S, d) \in \Sigma. \tag{7.4.7}$$

We want to impose conditions that a reasonable solution should satisfy. First of all, in this complete information context there is nothing that prevents the players from obtaining an efficient outcome, hence, we require

$$F(S, d) \in \partial S \quad \text{for every } (S, d) \in \Sigma, \tag{7.4.8}$$

i.e., the solution should be *Pareto optimal*. Next, it is required that the solution be *symmetric*, i.e. that it gives both players the same payoff in case the situation is symmetric

$$F_1(S, d) = F_2(S, d) \quad \text{for every symmetric problem.} \tag{7.4.9}$$

(A problem is symmetric if $d_1 = d_2$, and $(x_1, x_2) \in S$ if and only if $(x_2, x_1) \in S$.) Finally we note that players' payoffs are von Neumann-Morgenstern utilities and that such utilities are determined only up to an additive and a positive multiplicative constant. Hence, if A is a positive affine transformation of $\mathbb{R}^2$

$$A(x_1, x_2) = (a_1 x_1 + b_1, a_2 x_2 + b_2) \quad \text{with } a_i, b_i \in \mathbb{R}, a_i > 0, \tag{7.4.10}$$

then a bargaining situation that is represented by (S, d) can equally well be represented by (AS, Ad) and the solution should not depend on the representation that is chosen:

$$F(AS, Ad) = A(F(S, d)) \quad \text{for all } (S, d) \in \Sigma \text{ and } A \text{ as in (7.4.10).} \tag{7.4.11}$$

Hence, the solution should be *scale invariant*.

To summarize: a *bargaining solution* is a map F for which the conditions (7.4.7) – (7.4.9) and (7.4.11) are satisfied.

The conditions (7.4.7) – (7.4.11) are minimal conditions that a reasonable solution should satisfy. However, there exist many solutions satisfying these axioms and, even worse, there exist counterintuitive solutions that satisfy them. In order to exclude such perverse solutions one needs an additional axiom. We will now consider two such axioms that each single out a unique solution. The first of these was introduced by Nash and is called the *Independence of Irrelevant Alternatives* Axiom. Formally stated it requires that for all (S, d) and (T, d) in Σ

$$\text{if } S \subset T \text{ and } F(T, d) \in S, \text{ then } F(S, d) = F(T, d). \tag{7.4.12}$$

This condition can be interpreted as follows. Suppose the bargaining situation is perceived to be (T, d) and the players agree on $F(T, d)$ as its solution.

However, now it turns out that some alternatives are not available and that the situation really is (S, d). Then, as long as the solution that was previously agreed upon is still available, it should also be accepted as the outcome in the new situation. Condition (7.4.12) implies that the solution depends only on the local properties of the Pareto boundary close to the solution point itself, and this may explain why the Rubinstein model of Sect. 7.6 yields this solution.

Nash has proved the following theorem:

Theorem 7.4.1 (Nash [1950]). *There is exactly one bargaining solution that satisfies (7.4.12). This Nash solution F^N prescribes as the outcome of (S, d) that point in S for which the product $(x_1 - d_1)(x_2 - d_2)$ is maximal, hence*

$$F^N(S, d) \in \operatorname*{argmax}_{x \in S} (x_1 - d_1)(x_2 - d_2) . \tag{7.4.13}$$

Many objections have been raised against Nash's independence of irrelevant alternatives axiom (see for example Roth [1979] or Luce and Raiffa [1957]).

A particularly annoying property of the Nash solution has been discovered in Kalai and Smorodinsky [1975]. Loosely speaking, they noted that when the set of feasible payoffs is expanded such that player i gains relatively more than player j, then the Nash solution might prescribe that player i's profit should fall as a consequence of this expansion. More precisely, Kalai and Smorodinsky showed that the Nash solution does *not* satisfy the following *individual monoticity* property

$$\text{if } S \subset T \text{ and } u_i(S) = u_i(T), \text{ then } F_j(S, d) \leqq F_j(T, d) \quad (i \neq j \in \{1, 2\}) . \tag{7.4.14}$$

Note that the bargaining pairs (S, d) and (T, d) satisfy the condition in (7.4.14) only if $f_j^S \leqq f_j^T$ where f_j^S and f_j^T describe, as in (7.4.5), the Pareto boundaries of S and T, respectively. Hence, (7.4.14) states that if for every utility level that player i can demand the maximum feasible level that player j can obtain simultaneously is increased, then the utility level that the solution assigns to player j should also be increased.

Kalai and Smorodinsky were able to prove the following result:

Theorem 7.4.2 (Kalai and Smorodinsky [1975]). *There exists exactly one bargaining solution F^K that satisfies (7.4.14). This solution prescribes as the outcome of (S, d) the intersection of the Pareto boundary of S with the line through d and $u(S)$.*

Clearly the *Kalai/Smorodinsky solution* F^K depends crucially on the utopia point, hence, on the maximum payoffs the players can get. However, except for this special point, the solution depends only on local properties of the Pareto boundary. Namely, it immediately follows from (7.4.14) that F^K satisfies the following weaker version of (7.4.12)

$$\text{if } S \subset T, u(S) = u(T) \text{ and } F(T, d) \in S, \text{ then } F(S, d) = F(T, d) . \tag{7.4.15}$$

There exist bargaining problems (S, d) for which $F^N(S, d) \neq F^K(S, d)$, however, when there is transferable utility (i.e. the functions f_i are affine), then the solutions coincide. In particular, for the situation considered in the previous sections, both concepts prescribe the Steinhaus outcome (7.2.2) as the solution when the threat point is $(1/2v_1, 1/2v_2)$, i.e. when the object is allocated randomly in case of disagreement.

In the literature, one can find several other solution concepts besides the ones introduced above, however, these will not be discussed. Furthermore, for convenience we will confine ourselves in Sects. 7.5–7.7 to *normalized bargaining problems*, i.e. problems (S, d) for which

$$d = (0, 0) \quad \text{and} \quad u(S) = (1, 1). \tag{7.4.16}$$

In view of (7.4.11) this entails no loss of generality since every bargaining problem is 'equivalent' to exactly one normalized problem. Σ^1 is used to denote the set of normalized bargaining problems and we write $S \in \Sigma^1$ rather than $(S, 0) \in \Sigma^1$.

7.5 The Nash Negotiation Game

We already remarked in the previous section that Nash himself felt the need to supplement his axiomatic analysis by means of an explicit (noncooperative) model of the negotiation process for which the 'reasonable' equilibria actually generate the outcome derived by the axiomatic approach. In this section, we describe and analyse the noncooperative model introduced in Nash [1953]. Corollary 7.5.6 states the main result: the Nash outcome is the unique essential equilibrium of a certain negotiation game.

Throughout the section we will assume that a fixed normalized bargaining game S is given. Nash associates with S the negotiation game Γ described by the following rules:

simultaneously and independently the players state their (non-negative) demands x_1 and x_2, and (7.5.1)

if the demands are feasible, i.e. $(x_1, x_2) \in S$, then each player receives his demand, otherwise each player receives his disagreement outcome of zero. (7.5.2)

Hence, Nash's negotiation game is the normal form game $\Gamma = (\mathbb{R}_+, \mathbb{R}_+, R_1, R_2)$ of which the payoff function R_i is given by

$$R_i(x_1, x_2) = x_i \chi_S(x_1, x_2),$$

in which χ_S denotes the characteristic function of the set S

$$\chi_S(x) = \begin{cases} 1 & \text{if } x \in S, \\ 0 & \text{otherwise.} \end{cases}$$

The idea underlying this game is that a player's demand is the minimal amount that he is willing to accept: player i will not cooperate unless the mode of cooperation yields a utility of at least x_i. Nash remarks that the choice of payoffs in the case of compatible demands may seem unreasonable since the players might end up in a Pareto inferior outcome, but he argues that this choice does not contribute a bias to the final solution and that it gives the players a strong incentive to increase their demands as much as possible.

The next theorem characterizes the (pure) equilibrium payoffs of the negotiation game Γ

Theorem 7.5.1. (Nash [1953]). *The set of (pure) equilibrium payoffs of Γ is the union of the Pareto boundary of S and the threat point* $(0, 0)$.

Proof. If players i's demand is at least 1, then player j $(j \neq i)$ has the payoff of 0 no matter what he does which shows that $(1, 0)$ and $(0, 1)$, and each pair (x_1, x_2) with $x_1 \geqq 1$ and $x_2 \geqq 1$ are equilibria. Hence, disagreement is an equilibrium outcome. On the other hand, if $x_i < 1$, then the unique best reply is $x_j = f_i(x_i)$, which shows that every Pareto optimal point is an equilibrium outcome. the argument also shows that there are no other equilibrium outcomes. □

Theorem 7.5.1 shows that the Nash equilibrium concept is too weak to draw definite conclusions: any efficient outcome can be 'rationalized' by means of this concept. The next theorem (the only instance in this chapter at which we explicitly consider mixed strategies) shows that the situation is even worse when mixed strategies are allowed: in this case any outcome in S can be rationalized.

Theorem 7.5.2. *Every point in S can be obtained as the expected payoff of some Nash equilibrium (in mixed strategies) of Γ.*

Proof. Let $(x_1, x_2) \in S$ with $x_1 < 1$ and $x_2 < 1$. We will construct an equilibrium of Γ with expected payoff (x_1, x_2). This is sufficient in view of Theorem 7.5.1, since the only points that are excluded by the condition are $(1, 0)$ and $(0, 1)$. Consider the points $(x_1, f_1(x_1))$ and $(f_2(x_2), x_2))$ on the Pareto boundary of S. Clearly

$$x_1 \leqq f_2(x_2) \quad \text{and} \quad x_2 \leqq f_1(x_1)$$

with both inequalities being strict if $(x_1, x_2) \notin \partial S$. Define $p_i \in [0, 1]$ by

$$p_i = x_j / f_i(x_i) \quad (i \neq j \in \{1, 2\})$$

and consider the following mixed strategy s_i of player i:

s_i: demand x_i with prob. p_i and demand $f_j(x_j)$ with prob. $1 - p_i$.

Then it is easily verified that for all $x_1', x_2' \in \mathbb{R}_+$

$$R_1(x_1', s_2) \leqq R_1(x_1, s_2) = R_1(f_2(x_2), s_2) = x_1,$$

and

$$R_2(s_1, x_2') \leqq R_2(s_1, x_2) = R_2(s_1, f_1(x_1)) = x_2,$$

which shows that (s_1, s_2) is a Nash equilibrium of Γ with expected payoff (x_1, x_2). □

From now on attention will again be restricted to pure strategies. The nonuniqueness of Nash equilibria derived in Theorem 7.5.2 is troublesome and the reader might even have the impression that this renders a unique solution impossible. Namely, all equilibria that result in an outcome on the Pareto boundary of S are strict (in the sense of p. 23) and, at least for mixed extensions of discrete normal form games, strict equilibria have all stability properties that one could want. However, Nash has shown that there is a crucial distinction between that class of games and the class of games with a continuum of strategies since in the latter case there is a greater freedom in the perturbations that are allowed. As a result it turns out that it is indeed possible to discriminate equilibria by studying their relative stabilities.

Specifically, Nash proposes "to smooth the game Γ to obtain a continuous payoff function and then to study the equilibrium outcomes of the smoothed game as the amount of smoothing approaches zero" (Nash [1953], p. 132). Hence, Nash advocates an approach that is in the same spirit as the one underlying the essential equilibrium concept. To smooth Γ, the function χ_S will be replaced by a continuous funtion h that has values close to χ_S except at points near the boundary of S where χ_S is discontinuous. One can think of $h(x_1, x_2)$ as the probability that the demands x_1 and x_2 are compatible. The function h then represents some (slight) uncertainty that the players have about parameters of the bargaining problem.[7]

For convenience, attention will be restricted to perturbations in the class $H = \bigcup_{\varepsilon>0} H^\varepsilon$ where H^ε is the set of functions that satisfy

$$h: \mathbb{R}^2_+ \to (0, 1], \quad h \text{ continuous}, \ h(x) = 1 \text{ for } x \in S, \text{ and} \tag{7.5.4}$$

$$\max\{h(x), x_1 x_2 h(x)\} < \varepsilon \quad \text{if } \varrho(x, S) > \varepsilon, \tag{7.5.5}$$

where $\varrho(x, S)$ denotes the (Euclidian) distance from x to S. The last condition expresses that h decreases to zero sufficiently fast when x moves away from S, although h never actually reaches zero. Let us write $\Gamma(h) = (\mathbb{R}_+, \mathbb{R}_+, R_1^h, R_2^h)$ for the game that results when the uncertainty is described by h

$$R_i^h(x_1, x_2) = x_i h(x_1, x_2). \tag{7.5.6}$$

Equilibria which are still viable when there is some slight uncertainty about the underlying bargaining problem will be called *H-essential equilibria*. Formally:

Definition 7.5.3. An equilibrium x of Γ is an *H-essential equilibrium* if associated with every sequence $\{h^\varepsilon\}_{\varepsilon\downarrow 0}$ with $h^\varepsilon \in H^\varepsilon$ there is a sequence $\{x^\varepsilon\}_{\varepsilon\downarrow 0}$ such that x^ε is an equilibrium of $\Gamma(h^\varepsilon)$ and such that x^ε converges to x as ε approaches zero.

The similarity with Def. 2.4.1 justifies the use of the word essential, the prefix H denotes that only perturbations in H are allowed. It will be shown that $F^N(S)$, the Nash bargaining solution of S, is the unique H-essential equilibrium of Γ. First we show

Theorem 7.5.4 (Nash [1953]). *The Nash solution $F^N(S)$ is an H-essential equilibrium of Γ.*

Proof. Write $x^*=(x_1^*, x_2^*)$ for the Nash bargaining solution of S. Note that $x_1^*x_2^*>0$ and that, if $0<\varepsilon<x_1^*x_2^*$ and $h\in H^\varepsilon$, then there exists a point where the function $x_1x_2h(x_1, x_2)$ attains its maximum. If $\bar{x}=(\bar{x}_1, \bar{x}_2)$ is any such point then

$$\varrho(\bar{x}, S)\leqq\varepsilon, \tag{7.5.7}$$

since at all other points the function has a smaller value than at (x_1^*, x_2^*). Furthermore $\bar{x}_1>0$ and $\bar{x}_2>0$, hence

$$x_1h(x_1, \bar{x}_2)\leqq\bar{x}_1h(\bar{x}_1, \bar{x}_2) \quad \text{for all } x_1\in\mathbb{R}_+,$$

and

$$x_2h(\bar{x}_1, x_2)\leqq\bar{x}_2h(\bar{x}_1, \bar{x}_2) \quad \text{for all } x_2\in\mathbb{R}_+,$$

so that $\bar{x}$ is a Nash equilibrium of $\Gamma(h)$. Since h is equal to 1 on S and at most equal to 1 outside of S, we also have

$$x_1^*x_2^*=x_1^*x_2^*h(x^*)\leqq\bar{x}_1\bar{x}_2h(\bar{x})\leqq\bar{x}_1\bar{x}_2. \tag{7.5.8}$$

Since the Nash solution is the unique point of S where the product x_1x_2 is maximized it follows from (7.5.7) and (7.5.8) that $\bar{x}$ converges to x^* as ε approaches zero. □

Theorem 7.5.4 shows that all points where $x_1x_2h(x)$ is maximal are equilibria of $\Gamma(h)$ and that all these equilibria converge to the Nash solution of S as h converges to χ_S. Now if all equilibria of $\Gamma(h)$ would indeed be points where $x_1x_2h(x)$ is maximal, then it would follow that the Nash solution is the only H-essential equilibrium, but, a priori it is not clear that this is the case. Nash remarks that, if h varies regularly, then the game $\Gamma(h)$ has a unique equilibrium which corresponds to the unique maximum of $x_1x_2h(x)$, so that one can indeed draw the desired conclusion, but he does not provide an example of such a regular function h. We will now provide such an example (also see Binmore [1987a, c] for a more general class of examples). For convenience, attention will be restricted to the case in which S has a smooth boundary, i.e. the functions f_i are differentiable. The more general case can then be established by using an approximation argument since the Nash solution varies continuously with S.

To establish our goal, consider the *gauge function* γ

$$\gamma(x)=\min\{\gamma;\ x\in\gamma S\} \quad (x\in\mathbb{R}_+^2). \tag{7.5.9}$$

measuring the distance of x to S when $x\notin S$. This function is convex and differentiable (see Rockefellar [1979]) and for every $x\neq0$ we have that $\gamma(x)^{-1}x\in\partial S$, hence

$$f_1(\gamma(x)^{-1}x_1)=\gamma(x)^{-1}x_2. \tag{7.5.10}$$

For convenience, let us write γ instead of $\gamma(x)$, hence, (7.5.10) simplifies to

$$f_1(\gamma^{-1}x_1)=\gamma^{-1}x_2. \tag{7.5.11}$$

Since (7.5.11) holds for all $x\neq0$ we can differentiate both sides and this yields

$$f_1'(\gamma^{-1}x_1)(\gamma-x_1\gamma_1)=-x_2\gamma_1, \tag{7.5.12}$$

and

$$-f_1'(\gamma^{-1}x_1)x_1\gamma_2=\gamma-x_2\gamma_2, \tag{7.5.13}$$

in which γ_i denotes the partial derivative of γ with respect to the ith coordinate. From (7.5.12) and (7.5.13) we see that

$$(\gamma-x_1\gamma_1)(\gamma-x_2\gamma_2)=x_1x_2\gamma_1\gamma_2,$$

or equivalently

$$\gamma=x_1\gamma_1+x_2\gamma_2. \tag{7.5.14}$$

Now, for $\varepsilon>0$ consider the function h^ε defined by

$$h^\varepsilon(x)=\begin{cases}1 & \text{if } x\in S\\ e^{-(\gamma-1)^2/\varepsilon} & \text{if } x\notin S,\end{cases}$$

with associated game $\Gamma(h^\varepsilon)$. In this game, the probability that a pair of demands x is feasible depends only on the distance $\gamma(x)$ and the probability decreases to zero in an exponential way as with the normal distribution. Hence, h^ε satisfies (7.5.4) and, if $\delta>0$, then $h^\varepsilon\in H(\delta)$ for every ε that is sufficiently small. We now show that every game $\Gamma(h^\varepsilon)$ has a unique equilibrium and that this converges to the Nash solution as ε goes to zero.

Theorem 7.5.5. *For $\varepsilon>0$ the game $\Gamma(h^\varepsilon)$ has a unique equilibrium x^ε, and this converges to the Nash solution of S as ε tends to zero.*

Proof. Note that the payoff functions of $\Gamma(h^\varepsilon)$ are differentiable except on the boundary of S and that outside of S we have

$$\frac{\partial}{\partial x_i}R_i^\varepsilon(x)=h^\varepsilon(x)\{1-2x_i\gamma_i(\gamma-1)/\varepsilon\}\quad(x\notin S). \tag{7.5.15}$$

Now Eq. (7.5.15) is also valid for points on the Pareto boundary of S if the derivative is understood to be the right derivative (i.e. approaching from outside of S), and for such a point x^* we have

$$\frac{\partial}{\partial x_i}R_i^\varepsilon(x^*)=1,$$

since $\gamma=1$ at the Pareto boundary. This shows that points on the Pareto boundary of S cannot be equilibria of $\Gamma(h^\varepsilon)$: if player i chooses $x_i\in[0,1]$ then player j will take $x_j>f_i(x_i)$. Obviously, also points in the interior of S cannot be equilibria, hence, any equilibrium lies outside of S. Therefore, if x is an equilibrium, we have

$$\frac{\partial}{\partial x_1}R_1^\varepsilon(x)=0\quad\text{and}\quad\frac{\partial}{\partial x_2}R_2^\varepsilon(x)=0,$$

which, in view of (7.5.15), is equivalent to

$$\varepsilon=2x_1\gamma_1(\gamma-1)=2x_2\gamma_2(\gamma-1), \tag{7.5.16}$$

and we see that $x_1\gamma_1 = x_2\gamma_2$. Substituting this into (7.5.14) leads to

$$x_1\gamma_1 = x_2\gamma_2 = \gamma/2. \tag{7.5.17}$$

Substituting (7.5.17) into (7.5.16) shows that

$$\gamma = (1+\sqrt{1+4\varepsilon})/2, \tag{7.5.18}$$

illustrating once more that γ converges to 1 as ε approaches zero. Substitution of (7.5.17) into (7.5.12) shows that

$$f'_1(\gamma^{-1}x_1) = -\frac{x_2}{x_1} = -\frac{\gamma^{-1}x_2}{\gamma^{-1}x_1},$$

and by comparing this to (7.4.13) we see that $\gamma^{-1}x$ is the Nash solution of S. Hence, if x is an equilibrium of $\Gamma(h^\varepsilon)$, then $x = \gamma F^N(S)$ where γ is as in (7.5.18). This shows that $\Gamma(h^\varepsilon)$ has a unique equilibrium and that this equilibrium converges to $F^N(S)$ as ε approaches zero. □

By combining the Theorems 7.5.4 and 7.5.5 we obtain the main result of this section:

Corollary 7.5.6. *The Nash bargaining solution is the unique H-essential equilibrium of the negotiation game* Γ.

7.6 The Rubinstein/Binmore Model

In this section, we consider a second noncooperative implementation of the Nash bargaining solution. The model to be discussed has been introduced in Rubinstein [1982] in which it was used to analyse the simpler situation of dividing a cake. The model and the results were extended to abstract 2-person bargaining problems in Binmore [1980] and the n-person case was considered in Moulin [1984]. Throughout the section it will be assumed that a normalized bargaining game S is given.

Consider again the negotiation game Γ from the previous section. We have already remarked that because of the static character of this game the negotiations may result in a Pareto inferior outcome. This is undesirable since in actual life nothing prevents the players from continuing the bargaining process in such a case. Hence, it is more realistic to model the negotiations by means of a multistage game. One such dynamic model that suggests itself is the supergame of Γ: the players play Γ (finitely or infinitely) many times until an agreement is reached. However, in this supergame the players have no incentive to make serious offers in all but the last round since the costs of bargaining have not been taken into account. Hence, in a more realistic model one would have a discount factor and the players maximize discounted payoffs. Equivalently[8], at least from the mathematical point of view, one assumes that there is a fixed positive probability that negotiations end after each round. The variant of the latter model in which the moves are alternating rather than simultaneous will be considered in this section.[9]

Formally, for every $\delta \in (0,1)$, we consider the games $\Gamma^1(\delta)$ and $\Gamma^2(\delta)$ which only differ in the fact that in $\Gamma^1(\delta)$ it is player 1 who starts the bargaining whereas player 2 has this privilege in $\Gamma^2(\delta)$. The game $\Gamma^1(\delta)$ is the multi-stage game described by the following rules:

Round t (t *odd*, $t \geqq 1$): Player 1 proposes an outcome $x \in S$ which player 2 can either accept (action Y) or reject (action N). If player 2 accepts x, the game ends and player i receives the payoff x_i. If x is rejected, then a random move is performed: with probability $1-\delta$ the game ends and each player receives his disagreement outcome of zero; with probability δ the game moves to round $t+1$.
Round t (t *even*, $t \geqq 2$): The procedure is the same as in any odd round except that now player 2 proposes an outcome to player 1.

The game $\Gamma^2(\delta)$ is described by the same rules as $\Gamma^1(\delta)$ except that the roles of the players are reversed, hence, player 2 proposes an outcome in rounds with an odd number while player 1 does so in any round with an even number.

We will be interested in the equilibria of the games $\Gamma^i(\delta)$ for which the continuation probability δ is close to one. Note that, although being the initiator provides a big advantage when δ is small, this is less the case when δ is close to 1. Also note that $\Gamma^i(\delta)$ is a game with perfect information.

Finally, observe that the fact that there is a positive probability that the game ends after each round gives the players a strong incentive to reach an agreement quickly. In fact, in all equilibria that we consider, agreement is reached in round 1. There are also equilibria in which agreement is reached only at a later stage, but these are imperfect: in a subgame perfect equilibrium, agreement is always reached immediately.

Before we can begin analysing the games $\Gamma^i(\delta)$ we have to introduce some notations and terminology. This will be done with respect to the game $\Gamma^1(\delta)$ only. A (pure) *strategy* for player 1 in this game is a rule telling this player what to do in every possible situation, hence, it is a sequence $\sigma_1 = \{\sigma_1^t\}_{t \in \mathbf{N}}$ with

$$\sigma_1^t: S^{t-1} \to S \quad (t \text{ odd}) \quad \text{and} \quad \sigma_1^t: S^t \to \{Y, N\} \quad (t \text{ even}),$$

where S^t is the set of all sequences of length t with elements in S. Similarly, a strategy of player 2 is a sequence $\sigma_2 = \{\sigma_2^t\}_{t \in \mathbf{N}}$

$$\sigma_2^t: S^{t-1} \to S \quad (t \text{ even}) \quad \text{and} \quad \sigma_2^t: S^t \to \{Y, N\} \quad (t \text{ odd}).$$

A pair of strategies σ uniquely determines a time of agreement (possibly ∞) as well as a pair of expected payoffs $R^1(\sigma|\delta)$. Hence, we have a well-defined game and Nash equilibria are defined in the standard way. Let $x^{t-1} \in S^{t-1}$ denote a sequence of (rejected) proposals. Then $\Gamma_t^1(x^{t-1})$ denotes the *subgame* starting with a player making a proposal in round t after the history x^{t-1}. If x is the proposal in round t and $x^t = (x^{t-1}, x)$, then $\Gamma_t^1(x^t)$ denotes the subgame starting in round t after the proposal x has been done. Any subgame of $\Gamma^1(\delta)$ is of one of these two types. Note that, if t is odd, then $\Gamma_t^1(x^{t-1})$ is a copy of $\Gamma^1(\delta)$, whereas if t is even, $\Gamma_t^1(x^{t-1})$ is a copy of $\Gamma^2(\delta)$. The fact that subgames have this special structure makes it easy to analyse the game $\Gamma^i(\delta)$. Subgame perfect equilibria are defined in the standard way. $\mathrm{PEP}^i(\delta)$ (resp. $\mathrm{EP}^i(\delta)$ will be used to denote the set of subgame perfect equilibrium payoffs (resp. equilibrium payoffs) in the game $\Gamma^i(\delta)$ and $\mathrm{PEP}^i_j(\delta)$ is the projection of $\mathrm{PEP}^i(\delta)$ on the jth coordinate, hence for example

$$\mathrm{PEP}^1_2(\delta)=\{x_2;\ (x_1, x_2)\in \mathrm{PEP}^1(\delta) \text{ for some } x_1\}.$$

The first theorem illustrates the weakness of the Nash equilibrium concept.

Theorem 7.6.1. *Any outcome in S is a Nash equilibrium outcome, hence,* $\mathrm{EP}^i(\delta)=S$ *for $i\in\{1, 2\}$ and $\delta\in(0, 1)$.*

Proof. Take $i=1$ and let $x\in S$. Then the following pair of strategies constitute a Nash equilibrium that results in the outcome x. Each player j proposes x in any round $2t+j\,(t\geqq 0)$, and in any round $2t+j+1$ he accepts the outcome x but he rejects any outcome different from x. □

The equilibrium described above clearly is not subgame perfect. For example, let S be the triangle with corners $(0, 0)$, $(1, 0)$ and $(0, 1)$, let $i=1$, $\delta=1/2$ and $x=(1/4, 3/4)$ and suppose player 1 deviates and proposes $(1/3, 2/3)$ in round 1. The equilibrium requires player 2 to reject this proposal, but by doing so he bounds his payoff by $1/2\,(=\delta)$, hence, he would be better off by accepting 2/3.

From now on, attention will be restricted to subgame perfect equilibria. It will be shown that $\Gamma^i(\delta)\,(i=1, 2)$ has a unique subgame perfect equilibrium payoff and that this payoff converges to the Nash solution of S and δ goes to 1. First it is shown that $\Gamma^i(\delta)$ admits at least one subgame perfect equilibrium. Note that we cannot apply Theorem 6.2.4 since this is an infinite game. (For general existence results on subgame perfect equilibria in infinite horizon games, see Fudenberg and Levine [1983, 1986], Harris [1985], and Hellwig and Leininger [1987]).

Suppose player 1 has the opinion that he is entitled to the payoff x_1 in the game $\Gamma^1(\delta)$, i.e. in this game player 1 will propose and accept only those outcomes that yield him a utility of at least x_1. If this is known to player 2 and if player 2 considers this claim to be reasonable, then in $\Gamma^2(\delta)$ player 2 will propose the outcome $(\delta x_1, f_1(\delta x_1)$, for this is the best outcome for player 2 that is acceptable to player 1 (player 1 cannot reject the offer of δx_1 because of the possibility that the game might terminate immediately after round 1). Similarly, if player 2 has a *'justified claim'* of x_2 in $\Gamma^2(\delta)$, then player 1 will propose $(f_2(\delta x_2), \delta x_2)$ in the game $\Gamma^1(\delta)$. Now suppose we have such a pair of justified claims (x_1^δ, x_2^δ) that is consistent, i.e.

$$x_1^\delta=f_2(\delta x_2^\delta) \quad\text{and}\quad x_2^\delta=f_1(\delta x_1^\delta), \tag{7.6.1}$$

or equivalently

$$f_1(x_1^\delta)=\delta x_2^\delta \quad\text{and}\quad f_2(x_2^\delta)=\delta x_1^\delta, \tag{7.6.2}$$

then we are in a stable situation and the players will come to an agreement in round 1. The next lemma shows that such a pair of consistent justified claims can be supported by a subgame perfect equilibrium.

Lemma 7.6.2. *If* (x_1^δ, x_2^δ) *is a solution to* (7.6.1), *then* $(x_1^\delta, f_1(x_1^\delta)) \in \mathrm{PEP}^1(\delta)$ *and* $(f_2(x_2^\delta), x_2^\delta) \in \mathrm{PEP}^2(\delta)$.

Proof. Consider the following pair of (stationary) strategies for $\Gamma^1(\delta)$:

player 1: in odd rounds propose $(x_1^\delta, f_1(x_1^\delta))$ and in even rounds accept any outcome yielding a utility of at least δx_1^δ,
player 2: in even rounds propose $(f_2(x_2^\delta), x_2^\delta)$ and in odd rounds accept any outcome yielding a payoff of at least δx_2^δ.

To verify that this pair constitutes a subgame perfect equilibrium in $\Gamma^1(\delta)$ one has to show that it is not profitable for a player to deviate in any round, assuming that later on in the game the players play in accordance to the strategies prescribed. Because of the stationarity this actually has to be done only for the rounds 1 and 2, and it is easily verified that this follows from (7.6.1). For example, if player 1 deviates in round 1 he will be strictly worse off whereas player 2 will not be better off if he deviates in this round. This shows that $(x_1^\delta, f_1(x_1^\delta)) \in \mathrm{PEP}^1(\delta)$. The second assertion of the lemma can be proved similarly. □

It will now be shown that there is a solution to (7.6.1), in fact, that there is a unique solution. Condition (7.6.1) is equivalent to

$$f_1(x_1^\delta) = \delta f_1(\delta x_1^\delta) \quad \text{and} \quad x_2^\delta = f_1(\delta x_1^\delta), \tag{7.6.3}$$

hence, x_1^δ is a zero point of the map g defined by

$$g(x_1) = f_1(x_1) - \delta f_1(\delta x_1). \tag{7.6.4}$$

Now, $g(0) > 0$, $g(1) < 0$ and g is continuous to that such a zero point does indeed exist. Furthermore, since f_1 is concave and decreasing

$$g'(x_1) = f_1'(x_1) - \delta^2 f_1'(\delta x_1) < f_1'(x_1) - f_1'(\delta x_1) \leq 0,$$

hence, g is strictly decreasing and there exists a unique zero. Consequently, the solution to (7.6.1) is unique. We have shown

Lemma 7.6.3. *There exists a unique solution to* (7.6.1).

By combining Lemmas 7.6.2 and 7.6.3 we obtain

Corollary 7.6.4. *The game* $\Gamma^i(\delta)$ *has at least one subgame perfect equilibrium.*

Next it will be shown that the equilibrium payoff identified above is in fact the unique subgame perfect equilibrium payoff. (The following argument is taken from Shaked and Sutton [1984].) Write M_j^i (resp. m_j^i) for the maximal (minimal) payoff that player j can get in any subgame perfect equilibrium of $\Gamma^i(\delta)$, hence

$$M_j^i = \sup \mathrm{PEP}_j^i(\delta), \quad m_j^i = \inf \mathrm{PEP}_j^i(\delta). \tag{7.6.5}$$

To make the argument more transparent, let us assume that there exist equilibria attaining the values M_j^i and m_j^i, hence, sup (resp. inf) can be replaced by max (min) in (7.6.5). Since player j is guaranteed a payoff m_j^j in $\Gamma^j(\delta)$, he will reject any payoff less than δm_j^j in $\Gamma^i(\delta)$. On the other hand, if in $\Gamma^j(\delta)$ the players will always play the equilibrium leading to m_j^j, then player j should accept any payoff slightly above δm_j^j in $\Gamma^i(\delta)$. These observations lead to the conclusions that

$$m_j^i = \delta m_j^j \qquad (i \neq j \in \{1, 2\}), \tag{7.6.6}$$

and

$$M_i^i = f_j(\delta m_j^j) \quad (i \neq j \in \{1, 2\}) \tag{7.6.7}$$

By invoking a similar argument, one can show that

$$M_j^i = \delta M_j^j \qquad (i \neq j \in \{1, 2\}) \tag{7.6.8}$$

$$m_i^i = f_j(\delta M_j^j) \quad (i \neq j \in \{1, 2\}). \tag{7.6.9}$$

By combining (7.6.7) with (7.6.9), we obtain

$$M_i^i = f_j(\delta f_i(\delta M_i^i)) \quad \text{and} \quad m_i^i = f_j(\delta f_i(\delta m_i^i)) \quad (i \neq j)$$

or equivalently

$$f_i(M_i^i) = \delta f_i(\delta M_i^i) \quad \text{and} \quad f_i(m_i^i) = \delta f_i(\delta m_i^i) \quad (i = 1, 2).$$

In particular, both M_1^1 and m_1^1 are zero points of the map g defined in (7.6.4), hence, $M_1^1 = m_1^1$ in view of Lemma 7.6.3. Similarly, it follows that $M_2^2 = m_2^2$ and if we substitute these identities in (7.6.6) – (7.6.9), we see that each game indeed has a unique subgame perfect equilibrium payoff. We have derived:

Theorem 7.6.5. *The game $\Gamma^i(\delta)$ (with $i \in \{1, 2\}$, $\delta \in (0, 1)$) has exactly one subgame perfect equilibrium payoff, specifically*

$$\mathrm{PEP}^1(\delta) = \{(x_1^\delta, f_1(x_1^\delta))\} \quad \textit{and} \quad \mathrm{PEP}^2(\delta) = \{(f_2(x_2^\delta), x_2^\delta)\},$$

where (x_1^δ, x_2^δ) is the solution to (7.6.1).

Theorem 7.6.5 implies that the final outcome is a Pareto efficient point of S when a subgame perfect equilibrium is played in $\Gamma^i(\delta)$. Since such a Pareto optimal point can arise only if the players reach an agreement immediately, we have

Corollary 7.6.6. *In the unique subgame perfect equilibrium of $\Gamma^i(\delta)$ the players reach an immediate agreement on a Pareto optimal point of S.*

Next, let us investigate what happens when δ, the continuation probability, converges to 1. Note that, if $\delta < 1$, then $(x_1^\delta, f_1(x_1^\delta))$ and $(f_2(x_2^\delta), x_2^\delta)$ are distinct points. However, since f_1 and f_2 are continuous, it follows from (7.6.1) that

$$\lim_{\delta \uparrow 1} (x_1^\delta, f_1(x_1^\delta)) = \lim_{\delta \uparrow 1} (f_2(x_2^\delta), x_2^\delta)). \tag{7.6.10}$$

(The limits exist since x_i^δ decreases as δ increases.) Now note that (7.6.1) implies that

$$x_1^\delta f_1(x_1^\delta) = x_2^\delta f_2(x_2^\delta), \tag{7.6.11}$$

the points $(x_1^\delta, f_1(x_1^\delta))$ and $(f_2(x_2^\delta), x_2^\delta)$ have the same Nash product. Since these points are different Pareto efficient points for every δ, it follows that their common limit must be that point on the Pareto boundary where the product $x_1 x_2$ is maximal. Hence, the limit in (7.6.10) is the Nash solution of S. We have arrived at the main result of this section:

Theorem 7.6.7. *The subgame perfect equilibrium outcome of the game $\Gamma^i(\delta)$ converges to the Nash solution of S as δ approaches* 1.

Next, let us briefly investigate what happens when the players have different perceptions about the probability that the game terminates after each round, or equivalently, when the players have different discount rates. Assume player i believes that the continuation probability is δ_i and write $\delta = (\delta_1, \delta_2)$. Define the game $\Gamma^i(\delta)$ as before, but with the understanding that in the expected payoff of player i the probability δ_i occurs. Similarly as above one can then show that the games $\Gamma^1(\delta)$ and $\Gamma^2(\delta)$ each have a unique subgame perfect equilibrium payoff, given by $(x_1^\delta, f_1(x_1^\delta))$ and $(f_2(x_2^\delta), x_2^\delta)$, respectively, where

$$f_1(x_1^\delta) = \delta_2 x_2^\delta \quad \text{and} \quad f_2(x_2^\delta) = \delta_1 x_1^\delta. \tag{7.6.12}$$

Now define λ by means of

$$\lambda = \lambda(\delta) = \frac{\ln \delta_2}{\ln \delta_1 \delta_2}. \tag{7.6.13}$$

Note that $\lambda = 1/2$ in case $\delta_1 = \delta_2$. It follows from (7.6.12) that

$$(x_1^\delta)^\lambda (f_1(x_1^\delta))^{1-\lambda} = (f_2(x_2^\delta))^\lambda (x_2^\delta)^{1-\lambda}, \tag{7.6.14}$$

hence, if $(\bar{x}_1, \bar{x}_2)$ is the common limit of $(x_1^\delta, f_1(x_1^\delta))$ and $(f_2(x_2^\delta), x_2^\delta)$, then

$$\bar{x}_1^\lambda \bar{x}_2^{1-\lambda} = \max_{x \in S} x_1^\lambda x_2^{1-\lambda}. \tag{7.6.15}$$

If $\lambda = 1/2$, then (7.6.15) determines the Nash solution. If $\lambda \neq 1/2$, then the solution concept that assigns to each bargaining problem S the point $\bar{x}$ from (7.6.15) is known as a nonsymmetric Nash solution (see Harsanyi and Selten [1972] and Kalai [1977]). This concept satisfies all axioms that the ordinary Nash solution does except (7.4.9). It is easily verified that

$$\frac{\partial \bar{x}_1^\lambda}{\partial \lambda} > 0, \quad \frac{\partial \lambda(\delta)}{\partial \delta_1} > 0 \quad \text{and} \quad \frac{\partial \lambda(\delta)}{\partial \delta_2} < 0, \tag{7.6.16}$$

which implies that a player who perceives the higher probability that the game continues will generally be better off. This is intuitively clear: a player who believes there is a high probability that the game ends will make more generous offers and will accept smaller payoffs. Alternatively, one might say that the player i with the higher value of δ_i is less risk averse and it is well known that the Nash solution treats such players more generously than more risk averse ones (see Kannai [1977], Kihlstrom, Roth and Schmeidler [1981], de Koster, Peters Tijs and Wakker [1983] and Roth [1985]).

The most attractive (and most surprising) feature of the Binmore/Rubinstein model is that the requirement of subgame perfectness suffices to ensure uniqueness of the outcome. However, it should be noted that uniqueness can be established only in the 2-person case[10] (the argument preceding Theorem 7.6.5 hinges on the fact that the game is strictly competitive, i.e. that it has a zero-sum character). It can be shown (see Moulin [1984]) that the n-person version of the Rubinstein/Binmore model admits a unique stationary subgame perfect equilibrium (which has a payoff that converges to the Nash solution), but in general there may be many nonstationary perfect equilibria as well: Avner Shaked has shown that in the 3-person case every efficient outcome can be obtained by means of a subgame perfect equilibrium (see Herrero [1985] or Sutton [1986]). In this case, the stationary equilibrium is the unique one in which strategies vary continuously with the history (Binmore [1985]) and it is also the unique one that can be approximated by means of subgame perfect equilibria of finite period truncations, hence, one might question the viability of the non-stationary equilibria.[11] Much interesting work remains to be done in this area.[12]

Finally, it should be noted that uniqueness is also lost if offers are simultaneous rather than sequential (Chatterjee and Samuelson [1990]) or as soon as there is (even a little bit of) incomplete information, for example concerning the discount rates (see Rubinstein [1985], or the survey Rubinstein [1986]).

7.7 The Crawford/Moulin Model

In this section we consider two types of games that implement the Kalai/Smorodinsky solution concept. The first type has the somewhat undesirable property that the mediator has to know the Pareto efficient frontier of the underlying bargaining problem to implement the outcome. The rules of the second class of games, however, are completely independent of the underlying situation, hence, these procedures completely decentralize the decision making. The models described in this section are based on the papers Crawford [1977], Demange [1982] and Moulin [1982, 1984].

For the remainder of the section, suppose a normalized bargaining game S is given. Note that for such a game, the Kalai/Smorodinsky solution is the point (x_1^*, x_2^*) with

$$x_1^* = x_2^* = f_1(x_1^*) = f_2(x_2^*)\,. \tag{7.7.1}$$

Now, consider the game in which each player states a demand (between 0 and 1) and in which the mediator implements that Pareto optimal outcome that yields the player with the most modest demand exactly the amount that he asks, ties being broken by the toss of a fair coin. Hence, we have the *demand game* Γ^d in which the outcome $R^d(x_1, x_2)$ associated with demands x_1 and x_2 is given by

$$R^d(x_1, x_2)=\begin{cases}(x_1, f_1(x_1)) & \text{if } x_1<x_2, \\ (x_1, f_1(x_1))/2+(f_2(x_2), x_2)/2 & \text{if } x_1=x_2, \text{ and} \\ (f_2(x_2), x_2) & \text{if } x_1>x_2.\end{cases} \qquad (7.7.2)$$

Small demands have a high probability of being accepted, but they are not very attractive. On the other hand, stating a high demand is not very likely to be effective either. Hence, the optimal demand should be somewhere in the middle. In fact, we have

Theorem 7.7.1. *The demand game Γ^d described by* (7.7.2) *has a unique Nash equilibrium. In this equilibrium, player i demands x_i^* as in* (7.7.1), *hence, the game results in the Kalai/Smorodinsky outcome.*

Proof. Suppose player i demands x_i. Then if x_j is a best reply of player j against x_i, we have

if $x_i<f_i(x_i)$, then $x_j>x_i$,

if $x_i=f_i(x_i)$, then $x_j \geqq x_i$, and

if $x_i>f_i(x_i)$, then $x_j=x_i^-$,

in which x_i^- denotes the highest feasible demand smaller than x_i. Hence, if there is no smallest money unit and all real numbers are allowed, then player j actually does not have a best reply in case $x_i>f_i(x_i)$. In an equilibrium we also cannot have that $x_i<f_i(x_i)$, for, if we would, then $x_j>x_i$ and player i could improve by demanding x_i' with $x_i<x_i'<x_j$. Hence, in equilibrium we must have $x_i=f_i(x_i)$. This proves the theorem since both players demanding their Kalai/Smorodinsky payoffs is indeed an equilibrium. □

Dual to the game described by (7.7.2), one can consider the *offer game* in which the mediator enforces that outcome that is most preferred by the player who is most generous to his opponent. Formally, this offer game Γ^o is described by the following payoffs in which x_i denotes the amount that player i offers his opponent

$$R^o(x_1, x_2)=\begin{cases}(x_2, f_1(x_2)) & \text{if } x_1<x_2, \\ (x_2, f_1(x_2))/2+(f_2(x_1), x_1)/2 & \text{if } x_1=x_2, \text{ and} \\ (f_2(x_1), x_1) & \text{if } x_1>x_2.\end{cases} \qquad (7.7.3)$$

Analogous to Theorem 7.7.1 we have

Theorem 7.7.2. *The unique Nash equilibrium outcome of the offer game Γ^o is the Kalai/Smorodinsky solution of S.*

The games Γ^d and Γ^o discussed above suffer from the fact that the mediator has to know the Pareto efficient frontier of S and this might be private information of the players. We will now introduce two games that also implement the Kalai/Smorodinsky solution, but that do not require the mediator to know the actual bargaining situation. In the analysis of these games we will use the Theorems 7.7.1 and 7.7.2, in fact the games Γ^d and Γ^o arise as subgames.

Consider again the games $\Gamma^i(\delta)$ from the previous section, but now with $\delta=0$ in which a proposal can be made only once. In this static context, the role of proposer confers an advantage, since in a take it or leave it situation his opponent will accept any small (but positive) payoff. Hence, this role has to be allocated in a fair way. Now, it is not attractive to allocate this role randomly, since, if the players are risk averse (i.e. f_i is strictly concave), then the resulting outcome is Pareto inefficient. Below it is shown that it is better to auction off the role of proposer since in that case Pareto efficiency is guaranteed. Two variants will be discussed: the highest price auction and the second highest price auction and it will be shown that they result in the same outcome.

For $\lambda \in [0, 1]$, let us denote by $D_i(\lambda)$ the game in which the mediator enforces the disagreement outcome with probability $1-\lambda$ and in which player i is appointed dictator with the complementary probability λ. A dictator can enforce any outcome that he likes, he does not need the consensus of his opponent. Now, consider the following 2-stage game Γ:

simultaneously and independently the players bid amounts x_1 and x_2 between 0 and 1, (7.7.4)

if $x_i > x_j$, then player i may propose an element y of S; if player j accepts y, then y is enforced; if player j rejects, then the game $D_j(x_j)$ is played, hence, player j becomes dictator with a probability equal to his own bid. (If the bids coincide, the proposer is appointed randomly.) (7.7.5)

In this game, if you bid high, your opponent will attempt to slightly underbid and you will have to propose an outcome that is attractive to him, hence, unattractive to you. On the other hand, if you bid low, then your opponent has an incentive to bid more, since he only has to offer you some small amount. Hence, the optimal bid is somewhere in the middle. This argument, however, is not supported by the Nash equilibrium concept since it assumes optimal behavior in every contingency. Because of the dynamic character of the game Γ, the Nash concept is extremely weak as the following theorem illustrates:

Theorem 7.7.3. *If $x \in S$, then there exists a Nash equilibrium of the game (7.7.4) – (7.7.5) that results in the outcome x.*

Proof. Let $x \in S$. If both players adopt the strategy

in round 1: bid 0,
in round 2: propose x; if you are not the proposer, accept x but reject any other outcome,
if you are the dictator: enforce the disagreement outcome,

then an equilibrium results which gives the outcome x. □

The equilibrium described above clearly is not subgame perfect: if you are the dictator then you should enforce your most preferred outcome rather than the conflict outcome. For subgame perfect equilibria, the intuitive argument given in the paragraph preceding Theorem 7.7.3 applies and we have

Theorem 7.7.4. *The game described by* (7.7.4) – (7.7.5) *has exactly one subgame perfect equilibrium outcome. This is the outcome proposed by the Kalai/ Smorodinsky solution.*

Proof. Assume the bids are x_1 and x_2 with $x_1 > x_2$. Then player 1 may make a proposal. If player 2 rejects, he becomes dictator with probability x_2, hence, rejection yields player 2 an expected payoff of x_2. Consequently, in a subgame perfect equilibrium, player 2 will accept any outcome that yields him a utility of at least x_2, hence, the optimal proposal of player 1 is $(f_2(x_2), x_2)$. Similarly, if $x_2 > x_1$, then player 2 should propose $(x_1, f_1(x_1))$ and we see that, if the players behave optimally in round 2, then the game from round 1 reduces to the demand game Γ^d from (7.7.2). But we have already shown that this game results in the Kalai/Smorodinsky outcome. □

The game described by (7.7.4) – (7.7.5) might be called a second price auction game since the looser becomes the dictator with a probability equal to the losing bid. Alternatively, one might consider the first price auction game in which the looser becomes dictator with a probability equal to the winning bid, i.e. the game described by (7.7.4) and

exactly as (7.7.5) but with $D_j(x_j)$ replaced by $D_j(x_i)$. (7.7.6)

In this first price auction game, if player 1 is the winner, then his optimal proposal is $(f_2(x_1), x_1)$, hence, in a subgame perfect equilibrium, the first round reduces to the offer game Γ^o. From Theorem 7.7.2 it follows that

Corollary 7.7.5. *The unique subgame perfect equilibrium payoff of the game* (7.7.4), (7.7.6) *is the Kalai/Smorodinsky bargaining solution.*

7.8 Bargaining Games with Variable Threat Point

In this section we consider bargaining problems (S, d) in which the threat point d is not a priori given but rather depends on the actions taken by the players. Attention is focussed on the model proposed in Nash [1953] of which the underlying assumptions are extensively discussed. It is illustrated that Nash's approach is viable only if the players can irrevocably commit themselves to carry out their threats, and it is argued that, therefore, Nash's model cannot be used for selection among correlated equilibria. The discussion should also throw light on the distinction between cooperative games, noncooperative games with communication and noncooperative games without communication.

Suppose a bimatrix game $\Gamma = (\Phi_1, \Phi_2, R_1, R_2)$ is given. However, in contrast to the other chapters, now assume that we are in a *cooperative context* in which the players can sign any contract they wish and in which all such contracts are binding.

Then any *correlated strategy* $\mu \in C$

$$C = \left\{ \mu\colon \Phi \to \mathbb{R};\ \textstyle\sum_{\varphi} \mu(\varphi) = 1,\ \mu(\varphi) \geqq 0 \text{ all } \varphi \right\} \tag{7.8.1}$$

can be agreed upon and any payoff pair in the convex hull S of pure payoffs

$$S = \mathrm{co}\{(R_1(\varphi), R_2(\varphi));\ \varphi \in \Phi\} \tag{7.8.2}$$

can be obtained. Hence, once we know which outcome d will result in case the players do not cooperate, we have a well-defined bargaining problem (S, d) and we could use a bargaining solution F to determine the outcome (S need not satisfy all regularity assumptions that were imposed in Sect. 7.4, but this is not important). Suppose the players have already agreed to a specific bargaining solution F. Then the outcome is completely determined by the threat point d. In Nash [1953] it is proposed that, at the beginning of the game, each player chooses a mixed strategy t_i (his so called *threat strategy*) which he will be *forced* to use if the players cannot come to an agreement. Hence, if the threats t_1 and t_2 are chosen, then the disagreement outcome is

$$d = d(t) = (R_1(t), R_2(t)), \tag{7.8.3}$$

where $t = (t_1, t_2)$, and the final outcome is $F(S, d(t))$. Nash remarks that it is not necessary that these threats be chosen simultaneously, but that the following two aspects are extremely important:

Both players are committed to carry out their threats, and (7.8.4)

Commitments are simultaneous and no player can commit himself to more than a threat at the beginning of the game. (7.8.5)

Nash is very explicit about the first issue. He writes

"If one considers the process of making a threat, one sees that its elements are as follows: A threatens B by convincing B that if B does not act in compliance with A's demands, then A will follow a certain policy T. Supposing A and B to be rational beings, it is essential for the success of the threat that A be *compelled* to carry out his threat T if B fails to comply. Otherwise it will have little meaning. For, in general, to execute the threat will not be something that A would want to do, just for itself." (Nash [1953], p. 130, emphasis in original.)

Hence, Nash considers a threat that hurts the threatener himself to be credible only if he can commit himself to carry it out. In noncooperative games such commitments are impossible (by definition), hence, the above argument gives exactly the reason why in such games attention has to be restricted to subgame perfect (or sequential) equilibria. Therefore, Nash is forced to assume a *cooperative framework* in which *binding commitments* can be made and *binding contracts* can be signed. He writes:

"The point of this discussion is that we must assume there is an adequate mechanism for forcing the players to stick to their threats and demands once made; and one to enforce the bargain, once agreed. Thus, we need a sort of umpire, who will enforce contracts or commitments." (Nash [1953], p. 130.)

5, 1	0, 0
0, 0	1, 5

Fig. 7.8.1. The outcome will be (5, 1) if player 1 can commit himself to a demand first. According to the Nash model the outcome is (3, 3)

Nash is much less explicit about (7.8.5) although this condition is equally necessary for his model to be valid. Namely, consider the game of Fig. 7.8.1 and suppose player 1 commits himself to the following strategy: "I will not cooperate unless the mode of cooperation yields a payoff of at least 5; if player 2 does not accept this demand, then *I* will choose my first strategy". If player 1 can communicate this commitment before player 2 can commit himself, then player 2 faces a take it or leave it situation and the best that he can do is indeed agree on the payoff (5, 1). Condition (7.8.5) rules out such strategies, and in the Nash model, the outcome will be (3, 3) because of symmetry. The consequences of asymmetric commitment possibilities have been investigated in Rosenthal [1976]. For more extensive discussions on the role of commitment, the reader is referred to Schelling [1956], Schelling [1960] or Crawford [1982].

Let us assume that the conditions (7.8.4) and (7.8.5) hold, so that the Nash model is indeed valid. Then the question of which threat strategy to choose boils down to analysing the *threat game* $\Gamma^* = (T_1, T_2, R_1^*, R_2^*)$, where

$$R_i^*(t_1, t_2) = F_i(S, d(t_1, t_2)) . \tag{7.8.6}$$

(To avoid confusion we use T_i to denote the set of mixed strategies in the underlying game Γ.) We will show that this game has a definite solution in case F is well-behaved. In particular, we require that for fixed S

$$F(S, d) \text{ is continuous in } d, \text{ and} \tag{7.8.7}$$

$$F^{-1}(x) = \{d \in S;\ F(S, d) = x\} \text{ is convex, for every } x \in S . \tag{7.8.8}$$

In the survey paper Jansen and Tijs [1981], a bargaining solution F is called *regular* if condition (7.8.8) is satisfied. The Kalai/Smorodinsky solution is continuous, but it is not regular (see e.g. Peters and van Damme [1987]). The Nash solution clearly is continuous and it is also regular since (7.4.13) implies that for every $x \in \partial S$

$$F^{-1}(x) = \{d \in S;\ d_2 + \lambda d_1 = x_2 + \lambda x_1 \text{ for some } \lambda \in [f_1'^{+}(x_1), f_1'^{-}(x_1)]\} , \tag{7.8.9}$$

where $f_1'^{+}(x_1)$ denotes the right derivative of f_1 evaluated at x_1. Note that $F^{-1}(x)$ is a line segment when the boundary is smooth at x.

The following theorem shows that any 2-person cooperative game is resolved completely in case the Nash model is appropriate and the players have agreed on a regular and continuous bargaining solution. The theorem was first proved in Nash [1953] for the special case of the Nash solution. Nash's proof was generalized in Owen [1971], Kalai and Rosenthal [1978] and Tijs and Jansen [1982].

Theorem 7.8.1. *If the bargaining solution F satisfies* (7.8.7) *and* (7.8.8), *then*

(i) *the threat game Γ^* has at least one Nash equilibrium, and*
(ii) *all equilibria of Γ^* are interchangeable and result in the same outcome.*

Proof. The proof of (i) is a standard application of the Kakutani Fixed Point Theorem. If $B_i(t_j)$ denotes the set of best replies of player i against the threat t_j of player j in Γ^*, then $B_i(t_j)$ is nonempty and compact because of (7.8.7), and it is a convex set in view of (7.8.8). The set depends in an upper semi-continuous way on t_j because of (7.8.7). Hence, Kakutani's theorem guarantees the existence of an equilibrium.

The proof of (ii) follows from the fact that $(R_1^*(t), R_2^*(t))$ is a point on the efficient frontier of S for every combination of threats t. If $(\bar{t}_1, \bar{t}_2)$ and (t_1^*, t_2^*) are equilibria of Γ^*, then

$$R_1^*(\bar{t}_1, t_2^*) \leqq R_1^*(t_1^*, t_2^*) \quad \text{and} \quad R_2^*(\bar{t}_1, t_2^*) \leqq R_2^*(\bar{t}_1, \bar{t}_2).$$

However, because of Pareto optimality this implies that

$$R_2^*(\bar{t}_1, t_2^*) \geqq R_2^*(t_1^*, t_2^*) \quad \text{and} \quad R_1^*(\bar{t}_1, t_2^*) \geqq R_1^*(\bar{t}_1, \bar{t}_2)$$

and by combining these inequalities we get $R_1^*(\bar{t}_1, \bar{t}_2) \leqq R_1^*(t_1^*, t_2^*)$. Because of symmetry we must therefore have $R^*(\bar{t}) = R^*(t^*)$, hence, all equilibria result in the same payoffs. Furthermore, one can conclude from the inequalities above that also $(t_1^*, \bar{t}_2)$ and $(\bar{t}_1, t_2^*)$ are equilibria, hence, all equilibria are interchangeable. □

Let us consider an example to illustrate why condition (7.8.4) is that important. Assume the underlying bimatrix game is such that the Pareto efficient frontier is contained in a hyperplane with slope -1 and suppose the players have agreed on the Nash bargaining solution. Eq. (7.8.9) shows that for $x \in \partial S$

$$F^{-1}(x) = \{d \in S;\ d_1 - d_2 = x_1 - x_2\}. \tag{7.8.10}$$

Now player 1 wants to have the final outcome as far as possible to the right while player 2 wants to have it as far as possible to the left. Hence, player 1 wants to maximize $x_1 - x_2$ (or equivalently $d_1 - d_2$), while player 2 wants to minimize this quantity. Hence, the threat game Γ^* reduces to the zero-sum game with payoff matrix $R_1 - R_2$. Figure 7.8.2 gives a numerical example.

The Pareto efficient frontier associated with the game of Fig. 7.8.2a is contained in the hyperplane $x_1 + x_2 = 6$ (note that $\alpha < 5$). Figure 7.8.2b displays the threat game $R_1 - R_2$. The starred entry is the unique (pure) equilibrium, hence, the optimal threat strategies are (3, 2) and the disagreement outcome is $(-1, -1)$. (Actually, the optimal threat of player 2 is not unique since this player might choose his first strategy with a small probability, however, the optimal threat of player 1 and the disagreement outcome are unique). This symmetric disagreement outcome results in a final payoff of 3 for each player, hence, the final

5 / 1	0 / α
0 / 0	1 / 5
-1 / -1	-1 / -1

a

4	$-\alpha$
0	-4
0	0*

b

Fig. 7.8.2a, b. A bimatrix game to illustrate the need for binding commitments in the Nash model. α is a parameter, $\alpha \in (0, 5)$. The threat game Γ^* reduces to the game in **b**, the optimal threats are $(3, 2)$; the threat of player 1 is credible only if he can commit himself to it

solution is independent of α. This is indeed reasonable if the conditions (7.8.4) and (7.8.5) are satisfied since in this case the solution should depend only on the Pareto efficient frontier and the threat point. Both of these are symmetric since both players have the same commitment possibilities.

However, now consider the (hypothetical) situation in which it is possible to sign binding contracts (bilateral agreements), but in which it is impossible to make unilateral binding commitments. Then the threat of player 1 to use his third strategy is not credible since it is a dominated strategy: player 1 will never want to execute it. Player 2's threat is still credible, hence, this player is in a more powerful position and it seems that the higher α, the larger the share this player can demand.

The reader might think that it is still possible to use Nash's model after player 1's third strategy has been deleted, but this is not the case. Namely, in the reduced game one finds the optimal threats are $(1, 2)$ as long as $\alpha \leqq 4$, hence, each player sticks to his most preferred oucome. However, the pair $(1, 2)$ is not an equilibrium of the game of Fig. 7.8.2a and so, if it comes to executing the threats, at least one player has an incentive to deviate. This is particularly clear when $\alpha > 1$. In this case, player 2's second strategy dominates his first and $(2, 2)$ is the unique equilibrium. If $\alpha > 1$, then player 2 will get a payoff 5 if the players don't cooperate, hence, individual rationality implies that the outcome will be $(1, 5)$ also in case cooperation is allowed. When $\alpha < 1$ then it is not immediately clear what the final outcome will be, but it is clear that the answer cannot be found by Nash's model.

The preceding discussion presumes that binding contracts are possible, but that binding commitments are not, and it may be difficult to imagine a context in which this is a realistic assumption. However, when communication is possible, there may exist contracts which are viable even in a noncooperative context since one does not need an external agency to enforce then. Following Aumann [1974] such self-enforcing contracts are called *correlated equilibria* and, in fact, in the game of Fig. 7.8.2a any Pareto efficient outcome can be obtained by means of such a correlated equilibrium if $\alpha \leqq 1$. Namely, suppose the players decide to toss a fair coin and to play (5, 1) if heads comes up and to play (1, 5) otherwise. This correlated strategy is self-binding: once it is agreed upon, no player has an incentive to deviate as long as the other abides the contract, hence, it is a correlated equilibrium resulting in an expected payoff of 3 for each player. In a similar way all other Pareto optimal points can be obtained. Now, if the players can communicate, but cannot sign binding contracts (both unilateral and bilateral) in the game from Fig. 7.8.2a, then we exactly face the situation discussed in the preceding paragraph: in order for threats to be credible, they should constitute an equilibrium and, consequently, Nash's model cannot be used to determine optimal threats.

Correlated equilibrium is the relevant solution concept for games in which communication is possible, but in which no contracts are binding. For a formal definition of this concept the reader is referred to Aumann [1974], (also see Aumann [1987], Moulin and Vial [1978], Forges [1984] and Myerson [1985, 1986]. From the discussion above it will be clear that any convex combination of Nash equilibria is a correlated equilibrium, however, there also exist correlated equilibria that are not convex combinations of Nash equilibria. To implement these, one needs a mediator or noisy channel. Anyhow, the set of correlated equilibrium payoffs is convex and the selection among these gives rise to a bargaining game with variable threat point. This threat point should represent the solution of the game when communication has broken down, i.e. the pure noncooperative solution. In particular, subgame perfectness requires that the threat point be a Nash equilibrium payoff of the game and, consequently Nash's model is not appropriate. Hence, it seems that to solve the selection problem for correlated equilibria, one first has to solve the selection problem for Nash equilibria . (For advances in the latter direction, see Harsanyi and Selten [1988] or Guth and Kalkofen [1989].)

Notes

1. Recent surveys of noncooperative models of bargaining and their applications are Osborne and Rubinstein [1990] and Binmore, Osborne and Rubinstein [1991]. For surveys of experimental tests of bargaining models, see Roth [1988] and Güth and Tietz [1990].
2. A recent survey on bargaining under incomplete information is Kennan and Wilson [1990].
3. Recent surveys on the axiomatic approach to bargaining are Thomson [1990, 1991].
4. Okada [1988] investigates bargaining in the context of a supergame in which the disagreement point is endogenously determined.
5. The property that later moving players have to break indifference in a certain way to keep 'earlier' players in equilibrium is typical for games in which the strategy spaces are continuous. It is this property which makes it hard to establish existence of SPE for these games, see Hellwig, Leininger, Reny and Robson [1991].

6. Recent surveys of auction theory are McAfee and McMillan [1987], Milgrom [1987] and Wilson [1990a].
7. See Carlsson [1987] for a model in which the noise arises because of the fact that each player slightly trembles when stating his demand.
8. However, see Binmore, Rubinstein and Wolinsky [1986).
9. Chatterjee and Samuelson [1990] study the repeated game with simultaneous offers.
10. The uniqueness result also crucially depends on the fact that the space of alternatives is a continuum, see Van Damme, Selten and Winter [1990]. Muthoo [1990] shows that uniqueness is also lost if a proposer can withdraw the offer if his opponent accepts it.
11. However, see Rubinstein [1988] for a critique on the stationarity assumption. Baron and Kalai [1990] justify stationarity on the basis of simplicity.
12. Noncooperative models of bargaining have also been applied to issues of coalition formation in general n-person cooperative games, see Binmore [1985], Chatterjee et al. [1990], and Bennett and Van Damme [1990] and the references therein.

8 Repeated Games

In this chapter we study whether repetition can lead to cooperation. Specifically, it is investigated which outcomes can be sustained by means of subgame perfect (or Nash) equilibria when a game is repeated finitely or infinitely many times. The main result is the Perfect Folk Theorem, which states that, for almost all games, every outcome that is feasible and individually rational in the one-shot game can be approximated by subgame perfect equilibrium outcomes of the discounted supergame as the discount rate tends to zero, and that, for almost all games with more than one Nash equilibrium, any such outcome can be even approximated by a subgame perfect equilibrium payoff of the finitely repeated game as the number of repetitions tends to infinity.[1]

These results imply that, in this context, the subgame perfectness concept is not very powerful. This contrasts sharply with the results obtained in the previous chapter. The difference can partly be traced to the fact that most games in the previous chapter (especially those in Sect. 8.6) were of perfect information, whereas in repeated games there are simultaneous moves.

In Sect. 8.1, we provide a brief historical sketch, mention some alternative models that rationalize cooperation in repeated games, and discuss the informational assumption that is used to derive the Folk Theorem.

Section 8.2 introduces the relevant notation and the different models studied.

The perfect folk theorem for infinitely repeated games without discounting is derived in Sect. 8.3. We concentrate on the model of Rubinstein [1979], hence, it is assumed that players use the overtaking evaluation relation.

Sections 8.4 and 8.5 deal with infinitely repeated games with discounting. Section 8.4 deals with Nash equilibria, while Sect. 8.5 studies subgame perfectness. Characterizations of the sets of Nash and subgame perfect equilibrium paths are derived and the folk theorem is proved for this class of games. The discussion of prisoners' dilemma in Sect. 8.4 was motivated by Sorin [1986]. Section 8.5 is based on Abreu, Pearce and Stachetti [1986] and Fudenberg and Maskin [1986].

Sections 8.6 and 8.7 discuss Nash, resp. subgame perfect equilibria in finitely repeated games. These sections are structured similarly as Sects. 8.4 and 8.5 and the results are similar. These sections are based on Benoit and Krishna [1984, 1985]. Finally, in Sect. 8.8, we introduce the concept of renegotiation proofness according to which a threat is credible only if it does not hurt the threatener himself, and we show that, in repeated prisoners' dilemma, the Pareto efficient points can be obtained by means of such credible threats provided that the discount rate is sufficiently small.

8.1 Introduction

It is a familiar idea that there are significant differences between long term strategic competition and interactions that occur only once. In repeated games, players can build reputations and they can react to the decisions taken by their opponents in the past. In such games, the fear of retaliation when one is perceived as being too greedy, combined with the anticipation of future gains resulting from current altruistic behavior, might make players to behave in a way that is not in their short term interests. Therefore, lucrative outcomes that in the one-shot game can be obtained only when binding contracts are possible, can frequently be sustained by means of informal, self-binding contracts (gentlemen's agreements) when the game is repeated sufficiently often.

However, it is not always true that repetition allows cooperation. The classic example is prisoners' dilemma (Figs. 1.1.1 and 8.4.1). No matter how often this game is repeated (as long as this is a finite number of times, the number being known in advance), it only allows one equilibrium: play the one-shot equilibrium at every stage. The reason is that the last stage becomes of predominant importance: The players know that, no matter what has happened before, they will always play the one-shot equilibrium in the last round. Therefore, the second to last round will be treated as a single game in which they will always play the one-shot equilibrium, etc., the game unravels completely to the beginning (see Luce and Raiffa [1957]).[2]

It has been argued that in most actual situations one cannot exclude the possibility of meeting the opponents once more, hence, that a model in which the exact number of repetitions is known in advance is unrealistic, and that infinitely repeated games offer a better approximation to model long term competition. Furthermore, it has been shown that infinitely repeated games correspond more closely to intuition: as long as the players discount the future sufficiently little, any feasible and strictly individually rational payoff vector can be obtained by some Nash equilibrium of the infinitely repeated game. This result is known as the Folk Theorem and it was probably first formally proved by Aumann (Aumann [1960, 1961]) for the special case in which there is no discounting at all.

After the concept of perfect equilibria had been introduced, it was shown that also a (subgame) perfect folk theorem holds for undiscounted games, i.e. if the players do not discount the future, any feasible and strictly individually rational outcome can be obtained as the average payoff of some subgame perfect equilibrium of the infinitely repeated game (Aumann and Shapley [1976], Rubinstein [1977, 1979]). It was also realized that any outcome that yields each player more than some one-shot equilibrium can be sustained by means of a subgame perfect equilibrium of the discounted game, provided that the discount rate is sufficiently small (Friedman [1971, 1977]). Actually Friedman only considered Nash equilibria, but the equilibria that he constructed are subgame perfect.[3]

For several years, this described the state of the art. However, this picture was known to be both incomplete and misleading:

(i) It was known that the finitely repeated prisoners' dilemma gives a misleading impression. A distinctive feature of prisoners' dilemma is that it has a unique equilibrium and that this results in minmax values. Examples were known of games not having this property which, when repeated finitely many times, allow subgame perfect equilibria that do not consist of always playing a one-shot equilibrium (e.g. van Damme [1981]), but (until recently) no general theory was available.

(ii) It was known that in infinitely repeated games with discounting, a player can be punished more severely than by reverting to the one-shot equilibrium that is worst for this player (see Abreu [1982]), but it was not known to which level a player could be held down to.

It is only very recent that these issues have been investigated thoroughly and have been resolved satisfactorily (notably in Fudenberg and Maskin [1986] and Benoit and Krishna [1985] so that we now have a fairly complete picture of what is going on in repeated games. Our aim in this chapter is to survey these recent developments. In particular, we concentrate on the results obtained since the publication of the survey paper Aumann [1981]. We investigate both infinitely repeated games with and without discounting as well as finitely repeated games in order to see whether it makes a difference of how one models long term competition. Furthermore, we consider both Nash equilibria and subgame perfect equilibria to bring out the differences as well as the common elements in these concepts and to show that, in this context, subgame perfectness usually is not a significant restriction. Finally, our main goal is to prove the following[4]

Perfect Folk Theorem

For every normal form game Γ, the set of feasible and individually rational outcomes of the one-shot game coincides with the set of subgame perfect equilibrium outcomes (and, hence, the set of Nash equilibrium outcomes) of the undiscounted supergame of Γ in which the players use the limiting average evaluation relation (Theorem 8.3.6). (8.1.1)

For almost all normal form games Γ, the set of feasible and individually rational outcomes of Γ coincides with the set of subgame perfect (and Nash) equilibrium outcomes of the undiscounted supergame of Γ in which the players use the overtaking evaluation relation (Theorem 8.3.5). (8.1.2)

For almost all normal form games Γ, the set of subgame perfect (resp. Nash) equilibrium outcomes of the discounted supergame of Γ converges to the set of feasible and individually rational outcomes of Γ as the discount factor tends to 1 (Theorem 8.5.11). (8.1.3)

For almost all normal form games Γ with more than one Nash equilibrium, the set of subgame perfect (resp. Nash) equilibrium outcomes of the T-fold repetition of Γ converges to the set of feasible and individually rational outcomes of Γ as T tends to infinity (Theorem 8.7.7). (8.1.4)

This theorem shows that, generically, it does not make much difference of how one models long term competition and that, generically, the subgame perfectness concept imposes no real restriction in repeated games. However, it should be noted that (8.1.3) and (8.1.4) are limit results and for a fixed discount factor or a fixed horizon, the set of subgame perfect equilibrium outcomes will usually be smaller than the set of Nash outcomes. Furthermore, (8.1.2) – (8.1.4) are 'almost all theorems' and the set of games for which the limit result of (8.1.3) holds is a strict subset of the set of games satisfying (8.1.2). Furthermore, for any type of repeated game one can find examples for which the limit set of the subgame perfect equilibrium outcomes is a strict subset of the limit set of the Nash equilibrium outcomes. All such examples can be found in the text. Hence, we have a complete picture except for the fact that we only have sufficient conditions (and not necessary and sufficient conditions) under which the results (8.1.2) – (8.1.4) hold. However, for the class of 2-person games, necessary and sufficient conditions for the limit results to be valid are derived.

Up to now, we have not discussed the information that is revealed during the repeated game. Clearly, if no information would be revealed at all, the Perfect Folk Theorem would not hold: In this case the players can do nothing but play a Nash equilibrium of the one-shot game at every stage.[5] The standard assumption in repeated games is that all players get to hear the action that each player has chosen after every round. This assumption clearly is unrealistic in some situations: it might be that players just get to hear their payoffs after every round and this may not be sufficient to deduce which actions were taken (see Kaneko [1982] and Lehrer [1990] for how relaxations of this assumption affect the Folk Theorem).[6] To make a unified treatment of the different types of repeated games as simple as possible, we will make an even more unrealistic assumption, viz. that not only the actual actions but also the mixed strategies that produce these actions can be observed. Formally,

Assumption 8.1.1. Mixed actions can be observed, hence, when a player has to decide which action to take at time t, then he knows which action combinations have been played previously.

To defend this assumption, let us remark that

(i) This assumption simplifies the analysis considerably since deviations from an agreed upon path can be determined unambiguously; under the standard assumption one has the additional statistical problem of testing whether a deviation has indeed occurred. Furthermore, Assumption 8.1.1 allows the identification of expected payoffs and actual payoffs so that there are no expected payoffs that are not approachable (Aumann [1959]).

(ii) As far as Nash Folk Theorems are concerned it does not make any difference of whether one makes the standard assumption or this one: one only has to look along the equilibrium path and this can be chosen to consist of pure action combinations only. For subgame perfect equilibria it might make a difference since mixed action combinations usually allow more severe punishments. However, this difference can occur only for finitely repeated games: In Fudenberg and Maskin [1986] it is shown that (8.1.3) remains valid when the standard assumption is made. (However, cf. the discussion of Theorem 8.4.8 for the case of a fixed discount factor). To the author's knowledge, up to now no example has been found for which (8.1.4) is false when only actual outcomes can be observed.

(iii) The analysis of this chapter applies without modification to the case in which the standard assumption is made but in which the underlying game Γ is as in Theorem 2.1.1, i.e. the sets of pure strategies are continua (an example is the Cournot oligopoly game).

It has already been remarked that most of the results in this chapter are not new. Sect. 8.3 is based on Rubinstein [1979], Sect. 8.4 is inspired by Sorin [1986], Sect. 8.5 builds on Abreu, Pearce and Stachetti [1986] and Fudenberg and Maskin [1986] and Sects. 8.6 and 8.7 are based on Benoit and Krishna [1985]. Only the material in Sect. 8.8 is entirely new. Sometimes our results are somewhat more general than those of the original papers, however, and some proofs are slightly different. Our main aim for including this chapter is to present these results in a general and unified way. In particular, we emphasize the fact that Nash equilibria and subgame perfect equilibria are structured similarly. Specifically, a path can be supported as a subgame perfect (resp. Nash) equilibrium path if and only if the threat to switch to the deviator's worst equilibrium path (resp. worst path) is a sufficient deterrent, provided of course that such a path exists. For Nash equilibria this existence is guaranteed (minmax the deviator forever), but for subgame perfect equilibria this existence can be readily shown only for discounted games and for finitely repeated games (this is the reason why Sect. 8.3 is structured differently). Note that this in particular implies that there is no need to consider 'history-dependent' punishments; there is no need to let the punishment fit the crime, one can always use the most severe punishment (again the situation is slightly different in Sect. 8.3).

Note that for the Perfect Folk Theorem for finitely repeated games to be valid it is necessary that the basic game has at least 2 equilibria. Hence, our approach does not justify cooperation in the finitely repeated prisoners' dilemma. There are at least 3 models that attempt to rationalize the cooperation[7] that is experimentally[8] observed in finitely repeated prisoner's dilemma:

(i) Models that assume that players do not optimize but satisfice, hence, one investigates ε-equilibria rather than exact equilibria (Radner [1978, 1980]). This model is closely related to the infinitely repeated game with discounting (see Fudenberg and Levine [1986]).

(ii) Models of bounded rationality and/or bounded complexity, hence, players can only retain aggregate information about the past and/or they are finite automata (Smale [1980], Rubinstein [1986a], Neyman [1985]).

(iii) Models of incomplete information[9], hence, there is some small probability that the actual game is different (cf. Chap. 5 and Sect. 10.7, see Kreps and Wilson [1982b], Kreps et al [1982], Milgrom and Roberts [1982b], Fudenberg and Maskin [1986] and Aumann and Sorin [1989]).

In this chapter, we will not consider these models, hence, our players do optimize, they are fully rational and completely informed.

8.2 Preliminaries

Throughout this chapter it is assumed that a fixed normal form game $\Gamma = (\Phi_1, \dots, \Phi_n, R_1, \dots, R_n)$ is given. As usual, S_i denotes the set of mixed strategies of player i in Γ. Furthermore, $S = X_i S_i$, while $\Phi = X_i \Phi_i$ and $R = (R_1, \dots, R_n)$. In order to avoid confusion between strategies in supergames (repetitions of Γ) and strategies in Γ itself, the latter will be referred to as actions. Three different payoff sets will be relevant

$$R(\Phi) := \{R(\varphi); \varphi \in \Phi\}, \quad R(S) := \{R(s); s \in S\}, \quad R(C) := \mathrm{co}R(\Phi). \tag{8.2.1}$$

which are the sets of payoff vectors obtainable by pure, mixed and correlated action combinations, respectively (co denotes the convex hull). The *minmax payoff* of player i in Γ is

$$v_i := \min_{s \in S} \max_{s_i' \in S_i} R_i(s/s_i'), \tag{8.2.2}$$

It is the smallest payoff level to which the opponents of player i can hold this player down to. For 2-person games, this minmax level coincides with a player's maxmin level, which is the highest payoff that a player can guarantee himself (the formal definition of this quantity is obtained by interchanging the max and the min in (8.2.2)). However, if there are more than 2 players, then the maxmin payoff is usually lower since the opponents of player i cannot concert their actions when punishing this player, that is, they cannot use a correlated strategy. An example is given in Fig. 8.2.1, which displays the payoffs to player 1 (the row chooser) in a 3-person game. Player 1 can guarantee himself a payoff 0 by choosing both rows with probability 1/2 and this is the highest payoff that he can guarantee himself. On the other hand, if the players 2 and 3 both play (1/2, 1/2) then they hold player 1 down to 1 and this is the best that they can do if they want to punish player 1 since player 1's best reply against any action combination of 2 and 3 yields him at least 1.

We will use m^i to denote some action combination that is an optimal punishment of player i in Γ, hence

$$R_i(m^i) = \max_{s_i \in S_i} R_i(m^i/s_i) = \min_{s \in S} \max_{s_i' \in S_i} R_i(s/s_i') = v_i. \tag{8.2.3}$$

-2	2
2	2

2	2
2	-2

Fig. 8.2.1. The payoff that the row player can guarantee himself is strictly smaller than the payoff to which this player can be held down to

For convenience, we will write

$$v_j^i := R_j(m^i) \quad i, j \in N, \tag{8.2.4}$$

hence $v_i^i = v_i$. If s is any action combination, then by switching to his best reply, player i obtains at least v_i:

$$\max_{s_i' \in S_i} R_i(s/s_i') \geqq \min_{s \in S} \max_{s_i' \in S_i} R_i(s/s_i') = v_i \quad \text{for all } s \in S, \tag{8.2.5}$$

from which it follows that any equilibrium of Γ yields player i a payoff of at least v_i. We write

$$F := \{x \in R(C);\ x_i \geqq v_i \text{ for all } i \in N\} \tag{8.2.6}$$

and we refer to F as the set of *feasible and individually rational payoff vectors* of Γ. If Γ would be played cooperatively, then the resulting outcome would be an element of this set. In the proofs it will be convenient to use the following upper bound on the payoffs in Γ

$$M_i := \max_{\varphi} |R_i(\varphi)|, \quad M := \max_i M_i. \tag{8.2.7}$$

In this chapter, our main interest is not in the one-shot version of Γ, rather it is to investigate which payoffs can be obtained when Γ is repeated frequently. First, consider the case in which Γ is repeated infinitely often, the play starting at time 0. Assumption 8.1.1 implies that one can identify a subgame at time t, as well as an information set of player $i (i \in N)$ at time t with an element of S^t, the t-fold Cartesian product of S (for convenience set $S^0 = \{0\}$). An element of S^t is called a history at time t and is denoted by h^t

$$h^t = (h^t(0), \ldots, h^t(t-1)) \quad \text{with } h^t(\tau) \in S \text{ all } \tau < t. \tag{8.2.8}$$

The set of all histories is denoted by H

$$H := \bigcup_{t=0}^{\infty} S^t, \tag{8.2.9}$$

with typical element h. A *behavior strategy* of player i is a mapping $\sigma_i: H \to S_i$. The set of all such strategies is Σ_i and $\Sigma := X_i \Sigma_i$. Assumption 8.1.1 implies that we have a game with almost perfect information so that the restriction to behavior strategies is indeed justified (Aumann [1964]). Each strategy combination $\sigma \in \Sigma$ induces an outcome path (or shortly path) $\pi(\sigma) = \{\pi^t(\sigma)\}_t$ given by

$$\pi^0(\sigma) := \sigma(0), \quad \pi^t(\sigma) := \sigma(\pi^0(\sigma), \ldots, \pi^{t-1}(\sigma)) \quad (t \geqq 1). \tag{8.2.10}$$

With some mild abuse of notation we write

$$R(\sigma^t) := R(\pi^t(\sigma)), \quad R(\sigma) := \{R(\sigma^t)\}_t, \tag{8.2.11}$$

hence, $R(\sigma)$ is the stream of payoff vectors generated by σ. Furthermore, $R_i^\infty(\sigma)$ denotes player i's average payoff resulting from σ

$$R_i^\infty(\sigma) := \varliminf_{T \to \infty} \frac{1}{T} \sum_{t=0}^{T-1} R_i(\sigma^t). \tag{8.2.12}$$

Most papers dealing with undiscounted supergames assume that players try to maximize their average payoffs, more precisely, if $x=\{x^t\}_t$ and $y=\{y^t\}_t$ are streams of payoffs, then player i's strict preference ordering $\succ_i^a$ is assumed to be

$$x \succ_i^a y \text{ if and only if } \varliminf_{T\to\infty} \frac{1}{T} \sum_{t=0}^{T-1} (x^t - y^t) > 0. \tag{8.2.13}$$

This evaluation relation has the drawback that anything that happens in a finite time interval does not matter at all, which makes the game somewhat pathological. Following Rubinstein [1979], we will therefore assume that players use the Ramsey/Weiszacker overtaking criterion to evaluate streams of payoffs. In this case player i's strict preference ordering $\succ_i$ is given by

$$x \succ_i y \text{ if and only if } \varliminf_{T\to\infty} \sum_{t=0}^{T} (x^t - y^t) > 0. \tag{8.2.14}$$

This preference ordering is stronger, i.e. separates more sequences, than the one from (8.2.13). Therefore, all results from Sect. 8.3 remain correct when $\succ_i$ is replaced by $\succ_i^a$. In this case, some results might be sharpened, however, and it will be indicated when this is possible.

The *supergame* $\Gamma(\infty)$ is the infinitely repeated game in which the preferences of each player i are described by (8.2.14). A strategy combination σ is a *Nash equilibrium* of this game if

$$R_i(\sigma) \succsim_i R_i(\sigma/\sigma_i') \quad \text{for all } i\in N \text{ and } \sigma_i' \in \Sigma_i, \tag{8.2.15}$$

where $\succsim_i$ is the weak ordering associated with $\succ_i$. We write

$$E(\infty) := \{\sigma\in\Sigma;\ \sigma \text{ is a Nash equilibrium of } \Gamma(\infty)\}, \tag{8.2.16}$$

$$E\Pi(\infty) := \{\pi(\sigma);\ \sigma\in E(\infty)\}, \tag{8.2.17}$$

and

$$EP(\infty) := \{R^\infty(\sigma);\ \sigma\in E(\infty)\}. \tag{8.2.18}$$

If $\sigma_i \in \Sigma_i$ and $h\in H$, then σ_{ih}, the continuation of σ_i after h is the strategy defined by

$$\sigma_{ih}(h') := \sigma_i(h, h'), \tag{8.2.19}$$

where (h, h') is the history h followed by the history h' (with appropriate conventions for histories of length 0). If $\sigma\in\Sigma$ and $h\in H$, then $\sigma_h := (\sigma_{1h}, \ldots, \sigma_{nh})$ and σ is said to be a *subgame perfect equilibrium* of $\Gamma(\infty)$ if σ_h is a Nash equilibrium of $\Gamma(\infty)$ for all $h\in H$. The sets $PE(\infty)$, $PE\Pi(\infty)$ and $PEP(\infty)$ are defined similarly as in (8.2.16) – (8.2.18), but now with σ being a subgame perfect equilibrium.

In the second type of supergame that will be considered, the *infinitely repeated game with discounting*, $\Gamma(\delta)$, each player i is assumed to have the utility function R_i^δ given by

$$R_i^\delta(\sigma) := (1-\delta) \sum_{t=0}^{\infty} \delta^t R_i(\sigma^t), \tag{8.2.20}$$

where $\delta \in [0, 1)$ is a common discount factor. The assumption that all players have the same discount factor is made to ensure that $R^\delta(\sigma)$ belongs to $R(C)$ for all $\sigma \in \Sigma$ and this enables an easy comparison of the one-shot game and the supergame. Nash equilibria and subgame perfect equilibria can be defined for $\Gamma(\delta)$ as was done for $\Gamma(\infty)$ and the definitions of $E(\delta)$, $E\Pi(\delta)$ and $EP(\delta)$ are obtained by replacing ∞ by δ in (6.2.16) – (8.2.18). Again the prefix P denotes subgame perfectness, hence, for example, $PEP(\delta)$ is the set of normalized discounted payoffs that can be generated by some subgame perfect equilibrium of $\Gamma(\delta)$, hence $PEP(\delta) = \{R^\delta(\sigma);\, \sigma \in PE(\delta)\}$.

In addition to the two types of infinitely repeated games discussed above, we will also consider the *T-fold repetition* $\Gamma(T)$ *of* Γ for any finite T. In this game, player i is assumed to maximize his T-stage average payoff $R_i^T(\sigma)$

$$R_i^T(\sigma) := \frac{1}{T} \sum_{t=0}^{T-1} R_i(\sigma^t)\,. \tag{8.2.21}$$

Nash equilibria of $\Gamma(T)$ are defined in the standard way and σ is a subgame perfect equilibrium of $\Gamma(T)$ if σ_h is a Nash equilibrium of $\Gamma(T-t)$ for all $h \in S^t$ and all $t \in \{0, \ldots, T-1\}$. Again we define the sets $E(T)$, $PE(T)$, etc., similarly as in (8.2.16) – (8.2.18).

We have already seen that, if Γ is played cooperatively, the outcome must be in the set F of (8.2.6). Eq. (8.2.5) shows that, if Γ is played repeatedly, player i can obtain at least v_i in every round by always choosing a one-shot best reply against the action combination of his opponents. Therefore, for all δ and T.

$$PEP(\infty) \subset EP(\infty) \subset F, \quad PEP(\delta) \subset EP(\delta) \subset F, \quad PEP(T) \subset EP(T) \subset F \tag{8.2.22}$$

The (Perfect) Folk Theorem states that, under mild conditions, the first two inclusions are in fact identities and that the second and third tuple become identities in the limit. We will show that the sets involved are all compact, so that one can indeed speak of convergence and limit, namely in the sense of the Hausdorff distance for compact sets. For example, the statement that $PEP(T)$ *converges* to F as T tends to infinity means that

$$\forall_{\varepsilon > 0}\, \exists_{T^*} \forall_{T \geq T^*} \left[F \subset \bigcup_{x \in PEP(T)} B_\varepsilon(x) \right], \tag{8.2.23}$$

where $B_\varepsilon(x)$ is the ball with radius ε centered around x. In words: for any $\varepsilon > 0$ there exists some T^* such that any feasible and individually rational payoff vector can be approximated to within ε by some subgame perfect equilibrium of any repetition of Γ that lasts at least T^* stages. Because of compactness, condition (8.2.23) is equivalent to

$$\forall_{\varepsilon > 0} \forall_{y \in F}\, \exists_{T^*} \forall_{T \geq T^*} \exists_{x \in PEP(T)} [y \in B_\varepsilon(x)]\,, \tag{8.2.24}$$

hence, it is suffices to show that each point separately can be approximated.

8.3 Infinitely Repeated Games Without Discounting

In this section, we consider the infinitely repeated game $\Gamma(\infty)$ in which the players use the overtaking evaluation relation. The results are based on Rubinstein [1979], a paper that in turn is based on Roth [1976]. Our results are slightly more general than Rubinstein's, however, since Rubinstein restricts himself to stationary paths and such paths may not be sufficient to sustain all equilibrium payoffs (cf. Fig. 8.3.1). The main result is that $F = PEP(\infty)$ in case F contains an outcome that yields each player strictly more than his minmax value. Examples demonstrate that the result need longer be true when the condition on F is dropped.

We first consider Nash equilibria since these are especially easy to deal with. Let $\pi = \{s^t\}_t$ be a path. π is a Nash equilibrium path if and only if along this path deviations are not profitable. Now, Eq. (8.2.5) shows that, after a deviation, player i gets at least v_i per round in an equilibrium. Furthermore, this player can be held down to this level. Hence, π is an equilibrium path if and only if the threat to minmax a player forever after a deviation is a sufficient deterrent. Formally, π is an equilibrium path if and only if the strategy combination σ described by

$$\sigma(0) = s^0 \tag{8.3.1}$$

$$\sigma(h^t) = \begin{cases} m^i & \text{if } h^t(t-1) = m^i \text{ or } h^t(t-1) = s^{t-1}/s_i \text{ with } s_i \neq s_i^{t-1} \\ s^t & \text{otherwise,} \end{cases} \tag{8.3.2}$$

is a Nash equilibrium of $\Gamma(\infty)$. This latter condition is easily seen to be equivalent with

$$\varliminf_{T\to\infty} \left[\sum_{\tau=t}^{T} R_i(s^\tau) - \max_{s_i} R_i(s^t/s_i) - (T-t)v_i \right] \geqq 0 \quad \text{all } i, t \tag{8.3.3}$$

so that we have proved

Theorem 8.3.1. *$\pi = \{s^t\}_t$ is a Nash equilibrium path of $\Gamma(\infty)$ if and only if the conditions (8.3.3) are satisfied.*

If π is a path that results for each player in an average payoff that is higher than this player's minmax payoff, then (8.3.3) is certainly satisfied. Hence, we have

Corollary 8.3.2. *If $x \in F$ and $x_i > v_i$ for all $i \in N$, then $x \in EP(\infty)$.*

The corollary implies that the supergame may have equilibria of which the payoffs are smaller than those of any equilibrium of Γ. This happens, for example, in the modified prisoners' dilemma game of Fig. 1.3.2: that game has a unique equilibrium and this yields each player of payoff of 3. The supergame $\Gamma(\infty)$, however, has an equilibrium that yields each player the payoff 2 (since each player's minmax value is 1).

If the players would use the limiting average criterion, then a similar characterization of the set of equilibrium paths can be derived and one can show

	L_2	R_2
L_1	1, 1	0, 0
R_1	2, 1	1, 0

Fig. 8.3.1. The payoff vector $(1, 1)$ cannot be obtained by means of a stationary equilibrium path

that any payoff vector in F can be obtained by means of some stationary path. However, if the players use the overtaking relation, then stationary paths may not be sufficient to obtain payoffs that are only weakly individually rational. This is illustrated by the game of Fig.8.3.1.

In the game of Fig. 8.3.1 we have $v_1 = v_2 = 1$. The payoff vector $(1, 1)$ can be obtained by the stationary path in which the players always play (L_1, L_2). However, this path is not an equilibrium path: If player 1 deviates to R_1 then he receives a payoff of 2 once and a payoff of at least 1 forever thereafter, and this is strictly preferred to always receiving 1. There exists an equilibrium path that does result in the average payoff vector $(1, 1)$, however. The only modification necessary is that player 2 should allow player 1 to play R_1 once in a while, although not too often. Formally, let $Z \subset \mathbb{N}$ be an infinite set with zero density

$$|Z| = \infty \quad \text{and} \quad \lim_{T \to \infty} |Z \cap \{0, \ldots, T\}|/T = 0, \tag{8.3.4}$$

and let π be the path in which the players play (L_1, L_2) when $t \notin Z$ and (R_1, L_2) when $t \in Z$. Then π is a path with average payoff vector $(1, 1)$ and π satisfies condition (8.3.3).

Using the insight obtained above it is not difficult to show that, for any 2-person game, any payoff vector in F can be obtained by means of some Nash equilibrium of $\Gamma(\infty)$.

Theorem 8.3.3. *If Γ is a 2-person game, then $EP(\infty) = F$.*

Proof. Because of (8.2.22), it suffices to show that $F \subset EP(\infty)$. This inclusion is trivially true if $F = \{(v_1, v_2)\}$: In this case Γ has an equilibrium with payoff vector (v_1, v_2) and playing this at every stage results in an equilibrium of $\Gamma(\infty)$. If, on the other hand, F contains points that are strictly individually rational for both players, then for any $x \in F$ one can find a path $\pi = \{s^t\}_t$ with average payoff x and such that $R_i(s^t) > v_i$ for infinitely many t, hence, $F \subset EP(\infty)$ in view of Theorem 8.3.1. Finally, if, say $x_1 = v_1$ for all $x \in F$ but $x_2 > v_2$ for some $x \in F$, then v_1 is the maximal payoff that player 1 can get in Γ so that this player never has an incentive to deviate from a path $\pi = \{s^t\}_t$ with $R_1(s^t) = v_1$ for all t. By choosing π such that $R_2(s^t) > v_2$ for infinitely many t's one also ensures that player 2 cannot profitably deviate. □

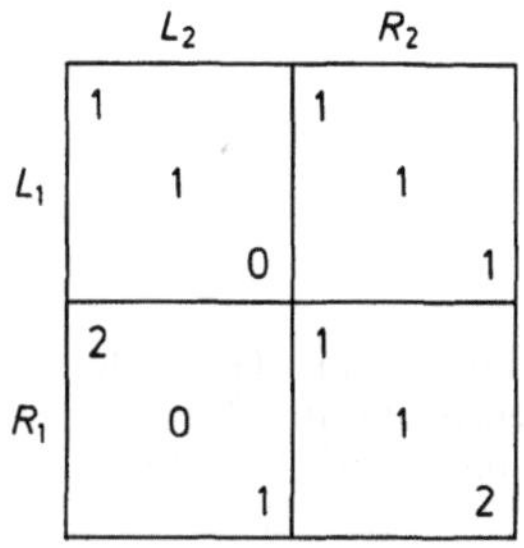

Fig. 8.3.2. A 3-person game (in which player 3 is a dummy) for which $EP(\infty)$ is a strict subset of F

We have already seen that, if the players use the limiting average criterion, then Theorem 8.3.3 can be generalized to n-person games. The game of Fig. 8.3.2 shows that this is no longer true in case the overtaking criterion is used.

In the game of Fig. 8.3.2 we have $v_1=1$, $v_2=1$ and $v_3=0$. Actually, this is a constant sum game between the players 1 and 2 with a dummy player 3. The equilibria of Γ are the action combinations in which player 2 chooses R_2 and the equilibrium payoffs are $EP(1)=\{(1, 1, \alpha);\ 1\leqq\alpha\leqq 2\}$. The set F is the line segment from $(1, 1, 0)$ to $(1, 1, 2)$, $F=\{(1, 1, \alpha);\ 0\leqq\alpha\leqq 2\}$. We claim that $EP(\infty)=EP(1)$. Namely, the maximum payoff that player 2 can get is 1 and player 2 can guarantee this payoff in every round of $\Gamma(\infty)$ by playing always R_2. Therefore, in any equilibrium of $\Gamma(\infty)$, player 2 must get the payoff 1 in every round. Since the game is constant sum among the players 1 and 2, this implies that also player 1 has a payoff of 1 in each round. Now, assume that player 2 would play L_2 with positive probability in some round of $\Gamma(\infty)$. Then, by switching to R_1 in this round, player 1 would obtain a stream of payoffs that he strictly prefers to the constant stream of 1, hence, this cannot be an equilibrium. Therefore, in any equilibrium of $\Gamma(\infty)$, player 2 chooses R_2 in every round and player 3's payoff is at least 1. Hence $EP(\infty)=EP(1)\neq F$.

The reason that $EP(\infty)\neq F$ for the game of Fig. 8.3.2 is that the two non-dummy players in that game cannot be rewarded above their minmax levels. It is easy to show that $EP(\infty)=F$ in case all players can be rewarded.

Theorem 8.3.4. *If Γ is an n-person game for which there exists some $x\in F$ such that $x_i>v_i$ for all $i\in N$, then $EP(\infty)=F$.*

Proof. For $x\in F$, let $\pi=\{s^t\}_t$ be a path with average payoff x and such that each player i gets more than v_i infinitely often. Then π is an equilibrium path of $\Gamma(\infty)$. □

Next, let us turn to subgame perfect equilibria. Note that the strategy combination σ as in (8.3.1) – (8.3.2) need not be a subgame perfect equilibrium since it may not be in the interest of one player to minmax another forever. The game of Fig. 8.3.3 provides an example where this is the case.

In the game of Fig. 8.3.3, we have $v_1=1$ and $v_2=0$. Consider the path in which (L_1, L_2) is played at every stage and let σ be the equilibrium that supports this

	L_2	R_2
L_1	2, 1	1, 2
R_1	0, 0	0, 0

Fig. 8.3.3. The equilibrium described in (8.3.1) – (8.3.2) to support the stationary path with average payoffs (2, 1) is not subgame perfect: it is not in the interest of player 1 to minmax player 2 forever

path as in (8.3.1) – (8.3.2). This equilibrium is not subgame perfect since player 1's threat to switch to R_1 forever after a deviation of player 2 is not credible: If this player would execute the threat, he would have a payoff of 0, while he would have at least 1 if he would continue to play L_1. Consequently, if one wants to support the payoff vector (2, 1) by means of a subgame perfect equilibrium, then one can only use punishments that have finite duration. In this specific example, playing R_1 once is sufficient to deter deviation. However, player 2 will be deterred only if he is sure that player 1 will indeed play R_1 at least once. Now, if player 2's actions would be independent of what player 1 does (or has done), then player 1 does not have an incentive to play R_1. Therefore, to obtain (2, 1) as a subgame perfect equilibrium payoff, player 2 must give player 1 an incentive to play R_1, hence, *player* 2 *must reward player* 1 *for having punished him.* Incorporating these two elements it is not difficult to construct a subgame perfect equilibrium that leads to (L_1, L_2) being played at every stage: play (L_1, L_2) until a player deviates; if one player deviates, the players punish each other by playing (R_1, R_2) *once*; if a player does not cooperate in punishing, then punish again. Formally, if the strategy combination σ is defined by

$$\sigma(0) = (L_1, L_2),$$

$$\sigma(h^t) = \begin{cases} (L_1, L_2) & \text{if } h^t(t-1) = (L_1, L_2) \text{ or if } h^t(t-1) = (R_1, R_2), \\ (R_1, R_2) & \text{otherwise,} \end{cases}$$

then σ is a subgame perfect equilibrium of $\Gamma(\infty)$ that results in the average payoff (2, 1).

In the next theorem we show that $PEP(\infty) = F$ for any game in which each player can be rewarded above his minmax level. The equilibria that are constructed in the proof have the two properties discussed above: (1) punishments are of finite duration and (2) if a player does not cooperate in punishing another, he is punished himself.

Theorem 8.3.5 (Rubinstein [1979]). *If Γ is an n-person game for which there exists some $x \in F$ such that $x_i > v_i$ for all $i \in N$, then $PEP(\infty) = F$.*

Proof. Without loss of generality, assume $v_i > 0$ for all i and first consider $x \in F$ with $x_i > v_i$ for all $i \in N$. For notational convenience, assume $x = R(s)$ for some $s \in S$. We will construct a subgame perfect equilibrium by first classifying the subgames into several states. If a subgame belongs to state s, then the action combination s has to be played; if it belongs to state $m_i(t)$, then first player i has to be minmaxed for t periods before s is again reached. The law of motion is as follows: If nobody deviates (i.e. s is played in state s and m^i is played in state $m_i(t)$) or if at least two players deviate simultaneously, we move from s to s and from $m_i(t)$ to $m_i(t-1)$. If player i is the unique deviator in state s, we move to $m_i(T_i)$, if he deviates in $m_i(t)$ we just continue in $m_i(t-1)$ and if he is the unique deviator in state $m_j(t)$ $(j \neq i)$ then we move to state $m_i(T_i(t))$. We claim that the parameters T_i and $T_i(t)$ can be chosen such that a subgame perfect equilibrium results. Namely, deviating in state s is not profitable for player i if

$$\max_{s_i'} R_i(s/s_i') + T_i v_i \leqq (1+T_i)x_i,$$

and choosing T_i such that

$$T_i \geqq \frac{M_i}{x_i - v_i} \tag{8.3.5}$$

guarantees that this inequality is satisfied. Similarly, deviating in state $m_j(t)$ is not profitable for player i $(j \neq i)$ if

$$\max_{s_i'} R_i(m^j/s_i') + T_i(t)v_i \leqq tv_i^j + (T_i(t) - t + 1)x_i,$$

i.e. player i prefers minmaxing player j for t periods to being minmaxed himself for $T_i(t)$ periods. This condition is satisfied if

$$T_i(t) \geqq \frac{(2t+1)M_i}{x_i - v_i}. \tag{8.3.6}$$

Hence, choosing T_i as in (8.3.5) and $T_i(t)$ as in (8.3.6) for all t guarantees that a subgame perfect equilibrium results. (Note that also it is not profitable for i to deviate in $m_i(t)$). The case in which $x_i = v_i$ for some i can be dealt with by taking a path π with average payoff x but in which each player receives more than his minmax level infinitely many times. The only difference is that the formulae for T_i and $T_i(t)$ become more complicated since these numbers will now also depend on the time at which the deviation occurs and the set of times at wich player i receives more than his minmax level. □

Let us briefly consider the case in which the players use the limiting average evaluation relation (8.2.13) rather than the overtaking relation (8.2.14). For $x \in F$, let $\pi = \{s^t\}_t$ be a path with average payoff x, such that $R_i(s^t) > v_i$ for infinitely many t for each player i who can be rewarded above his minmax level (i.e. for any player i with $v_i < M_i$). In this case, if a player deviates from π, he can be punished until his total payoff is smaller than the payoff that he would have had if he had not deviated. Therefore, any payoff vector in F can be obtained by means of some subgame perfect equilibrium and we have the following analogue to Theorem 8.3.4, due to Aumann and Shapley [1976].

Corollary 8.3.6. *If the players use the limiting average evaluation relation (8.2.13), then, for any game Γ, any payoff vector in F can be obtained by means of some subgame perfect equilibrium of the undiscounted supergame.*

	L_2	R_2
L_1	1, 1	1, 0
R_1	0, 0	0, 0

Fig. 8.3.4. A 2-person game for which $PEP(\infty) \neq F$

We already know from the game of Fig. 8.3.2 that Corollary 8.3.6 is no longer correct when the players use the overtaking evaluation relation, i.e. the condition in Theorem 8.3.5 that there exist strictly individually rational payoff vectors cannot be dropped. In view of Theorem 8.3.3 one could still think that $PEP(\infty) = F$ for any 2-person game. However, even this need not be true as the game of Fig. 8.3.4 illustrates.

In the game of Fig. 8.3.4 we have $v_1 = 1$ and $v_2 = 0$, hence F is the line segment from $(1, 0)$ to $(1, 1)$. We claim that $PEP(\infty) = \{(1, 1)\}$. Namely, since player 1 can guarantee a payoff of 1 per stage by playing L_1 and since this is the maximal payoff that this player can get, he always has to play L_1 in any subgame perfect equilibrium of $\Gamma(\infty)$. Hence, the set of subgame perfect equilibria remains unchanged when R_1 is deleted from the game. But in this reduced game, player 2 can guarantee 1, hence, indeed $PEP(\infty) = \{(1, 1)\}$. (Note that $(1, 0)$ belongs to $EP(\infty)$ and that $(1, 0)$ is a perfect equilibrium payoff if the players use the limiting average criterion.)

The game of Fig. 8.3.4 is degenerate in the sense that $M_1 = v_1$, player 1 cannot get more than his minmax value. If Γ is such a game, then in a subgame perfect equilibrium of $\Gamma(\infty)$, at every stage player 1 has to choose an action in $S_1(v_1)$.

$$S_1(v_1) := \{s_1 \in S_1;\ R_1(s_1, s_2) = v_1 \text{ for some } s_2 \in S_2\}\ . \tag{8.3.7}$$

Hence, in a subgame perfect equilibrium of $\Gamma(\infty)$, player 2's average payoff cannot be lower than

$$\tilde{v}_2 := \min_{s_1 \in S_1(v_1)} \max_{s_2} R_2(s_1, s_2)\ . \tag{8.3.8}$$

(The reader can easily provide an example in which $\tilde{v}_2$ is lower than player 2's smallest equilibrium payoff of Γ.) Therefore, we have that $PEP(\infty) \subset \tilde{F}_2$ where

$$\tilde{F}_2 := \{x \in F;\ x_2 \geqq \tilde{v}_2\} \tag{8.3.9}$$

($\tilde{F}_1$ can be defined similarly). By reviewing the proof of Theorem 8.3.5, it is easily seen that in fact $PEP(\infty) = \tilde{F}_2$ for any game with $x_1 = v_1 = M_1$ for all $x \in F$ and $x_2 > v_2$ for some $x \in F$. Since $PEP(\infty) = F$ in case $F = \{(v_1, v_2)\}$ we have a complete description of $PEP(\infty)$ for 2-person games.

Theorem 8.3.7. *If Γ is a 2-person game, then $PEP(\infty) = F$ if $F = \{(v_1, v_2)\}$ or if there exists some $x \in F$ with $x_1 > v_1$ and $x_2 > v_2$. If only player i can get more than his minmax value at some point in F, then $PEP(\infty) = \tilde{F}_i$.*

8.4 Infinitely Repeated Games with Discounting: Nash Equilibria

In this section, we study Nash equilibria of the discounted game $\Gamma(\delta)$. Characterizations are derived of the sets of equilibrium paths and equilibrium payoffs and it is shown that $EP(\delta)$ converges to F if F contains strictly individually rational points. The theory is illustrated by means of Prisoners' Dilemma for which $EP(\delta)$ is computed explicitly for several values of δ.

Let $\delta \in [0, 1)$ and let $\pi = \{s^t\}_t$ be a path in $\Gamma(\delta)$. Similarly as in Sect. 8.3, it is seen that π is an equilibrium path if and only if the threat to minmax a player forever after a deviation from π is sufficient to deter deviations. Hence, π is a Nash equilibrium path of $\Gamma(\delta)$ if and only if

$$R_{it}^{\delta}(\pi) \geqq (1-\delta) R_i(s^t/s_i) + \delta v_i \quad \text{all } i, \text{ all } s_i \neq s_i^t, \text{ all } t, \tag{8.4.1}$$

where $R_{it}^{\delta}(\pi)$ denotes player i's conditional payoff at time t along the path π

$$R_{it}^{\delta}(\pi) := (1-\delta) \sum_{\tau=t}^{\infty} \delta^{\tau-t} R_i(s^{\tau}) . \tag{8.4.2}$$

Since R_i is continuous, (8.4.1) holds if and only if

$$R_{it}^{\delta}(\pi) \geqq (1-\delta) \max_{s_i} R_i(s^t/s_i) + \delta v_i \quad \text{all } i, \text{ all } t \tag{8.4.3}$$

and we have

Theorem 8.4.1 $\pi = \{s^t\}_t$ *is a Nash equilibrium path of* $\Gamma(\delta)$ *if and only if the conditions* (8.4.3) *are satisfied.*

We will now derive a characterization of $EP(\delta)$, the set of equilibrium payoffs of $\Gamma(\delta)$. At first reading, the tools to be developed now might seem too heavy for the results derived later on. However, these tools will also be of use in the next section and they will facilitate the comparison of Nash equilibria to subgame perfect equilibria.

Let $x^0 \in EP(\delta)$ and let $\pi = \{s^t\}_t$ be an equilibrium path of $\Gamma(\delta)$ with payoff x^0. Then the subpath $\pi^1 = \{s^t\}_{t=1}^{\infty}$ is also an equilibrium path of $\Gamma(\delta)$, hence, if $x^1 = R_1^{\delta}(\pi) = (R_{11}^{\delta}(\pi), \ldots, R_{n1}^{\delta}(\pi))$, then $x^1 \in EP(\delta)$. Consequently, we have

$$x^0 = (1-\delta) R(s^0) + \delta x^1 \quad \text{for some } x^1 \in EP(\delta), \tag{8.4.4}$$

and

$$x_i^0 \geqq (1-\delta) \max_{s_i} R_i(s^0/s_i) + \delta v_i \quad \text{for all } i \in N. \tag{8.4.5}$$

Conversely, if x^0 satisfies (8.4.4) – (8.4.5), then there exists an equilibrium path π with payoff x^0 that satisfies (8.4.3). Hence, we have

Lemma 8.4.2. $x^0 \in EP(\delta)$ *if and only if* x^0 *is a solution to* (8.4.4) – (8.4.5) *for some* $s^0 \in S$.

Let V be a nonempty subset of F. If, at the beginning of round 1, the players can get a total discounted payoff vector in V, then at $t=0$ they can get any payoff vector x^0 satisfying (8.4.5) and

$$x^0 = (1-\delta)R(s^0) + \delta x^1 \quad \text{for some } s^0 \in S \text{ and } x^1 \in V. \tag{8.4.6}$$

Let $E^\delta(V)$ be the set of payoff vectors satisfying (8.4.5) and (8.4.6). Then $E^\delta(V) \neq \emptyset$, since the set $(1-\delta)EP(1) + \delta V$ is contained in $E^\delta(V)$. Eq. (8.2.22) implies $E^\delta(V) \subset F$, hence E^δ is a map that associates to each nonempty subset of F another nonempty subset of F. Using this notation, Lemma 8.4.2 can be paraphrased as

Corollary 8.4.3. *$EP(\delta)$ is a fixed point of E^δ, i.e., if $V = EP(\delta)$, then $E^\delta(V) = V$.*

The following properties easily follow from the definition of E^δ.

Corollary 8.4.4.
(i) *If V is compact, then $E^\delta(V)$ is compact,*
(ii) *If $V \subset W$, then $E^\delta(V) \subset E^\delta(W)$.*

Furthermore, from Theorem 8.4.1 and the definition of E^δ, it follows that

Corollary 8.4.5. *If $V \subset E^\delta(V)$, then $E^\delta(V) \subset EP(\delta)$.*

By combining Corollaries 8.4.3 and 8.4.5 we obtain the following characterization of $EP(\delta)$.

Corollary 8.4.6. *$EP(\delta)$ is the largest fixed point of E^δ.*

Finally, the results obtained thus far allow us to conclude that $EP(\delta)$ is compact.

Theorem 8.4.7. *$EP(\delta)$ is a nonempty, compact set.*

Proof. $EP(\delta)$ is nonempty since repeatedly playing an equilibrium of Γ yields an equilibrium in $\Gamma(\delta)$. Eq. (8.2.22) shows that $EP(\delta) \subset F$. Write $V = EP(\delta)$ and let $\mathrm{cl}V$ be the closure. From Corollaries 8.4.3 and 8.4.4 it follows that $V = E^\delta(V) \subset E^\delta(\mathrm{cl}V)$. Since the largest set is compact $\mathrm{cl}V \subset E^\delta(\mathrm{cl}V)$. But Corollary 8.4.5 shows that $\mathrm{cl}V \subset E^\delta(\mathrm{cl}V) \subset V$, hence V is compact. □

Next, we will illustrate the theory developed above (especially Theorem 8.4.1) by means of the prisoners' dilemma game of Fig. 8.4.1. The example will show that $EP(\delta)$ need not be the ony fixed point of E^δ. Namely, the one-shot prisoners' dilemma has a unique equilibrium with payoff vector $(1, 1)$ so that $\{(1, 1)\}$ is a fixed point of E^δ for every δ. On the other hand, Theorem 8.4.8 shows that $EP(1/4) \neq \{(1, 1)\}$. The discussion that follows is based on Sorin [1986].

In this prisoners' dilemma game, R_i is a strictly dominant strategy for player i, hence, (R_1, R_2) is the unique Nash equilibrium of Γ. Furthermore, R_i is the

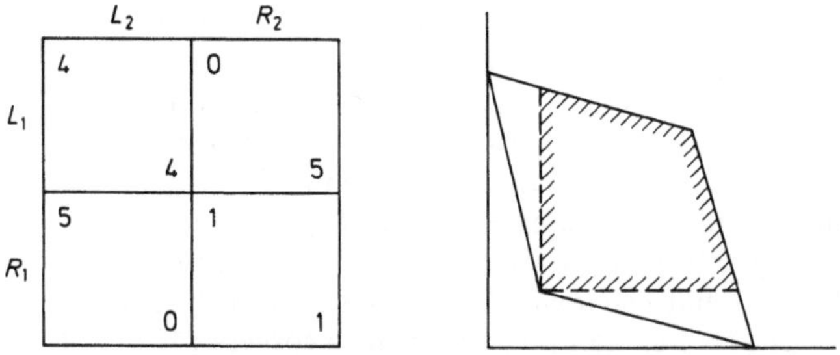

Fig. 8.4.1. Prisoner's Dilemma (*left*) and the set of feasible and individually rational payoff vectors of this game (*right*)

strategy that hurts player i's opponent most, hence $R_i = m_i^j \,(j \neq i)$ and $v_i = 1 \,(i = 1, 2)$. If $s = (s_1, s_2)$ is the mixed strategy combination with $s_i = (p_i, 1 - p_i)$, then

$$R(s) = (1 - p_1 + 4p_2, 1 - p_2 + 4p_1), \tag{8.4.7}$$

which shows that $R(C) = R(S)$: any feasible payoff vector can be obtained by means of a mixed strategy combination. The set of feasible and individually rational payoff vectors is given by

$$F = \text{co}\{(1, 1), (4, 4), (1, 19/4), (19/4, 1)\}. \tag{8.4.8}$$

Let $\delta \in [0, 1)$ and consider the repeated game $\Gamma(\delta)$. If δ is close to 0, then the first stage is of predominant importance and one can expect that $EP(\delta) = EP(0)$. On the other hand, if δ is close to 1, the first stage is only of minor importance and intuition suggests that implicit collusion can be obtained as an equilibrium outcome. We will now show that this intuition is correct.

Let $\pi = \{s^t\}_t$ be a path with $s^t = (s_1^t, s_2^t)$ and $s_i^t = (p_i^t, 1 - p_i^t)$. Since R_i dominates L_i, the equilibrium condition (8.4.3) reduces to

$$R_{it}^{\delta}(\pi) \geqq (1 - \delta)(1 + 4p_j^t) + \delta \qquad i \neq j \in \{1, 2\},\ t \in \mathbb{N},$$

which is equivalent to

$$p_i^t \leqq \sum_{\tau = t+1}^{\infty} \delta^{\tau - t}(4p_j^{\tau} - p_i^{\tau}) \qquad i \neq j \in \{1, 2\},\ t \in \mathbb{N}. \tag{8.4.9}$$

Writing $p^t = p_1^t + p_2^t$ and adding the two inequaltities in (8.4.9) for a fixed value of t, one obtains

$$p^t \leqq \sum_{\tau = t+1}^{\infty} 3\delta^{\tau - t} p^{\tau} \quad (t \in \mathbb{N}).$$

Setting $p = \sup\{p^t,\ t \in \mathbb{N}\}$ yields

$$p \leqq 3\delta(1 - \delta)^{-1} p,$$

hence $p = 0$ or $\delta \geqq 1/4$. Therefore, if $\delta < 1/4$, then $p^t = 0$ for all t, hence, $\Gamma(\delta)$ has a unique Nash equilibrium path and this path consists of playing the one-shot Nash equilibrium at every stage. (Note that this does not imply that $\Gamma(\delta)$ has a unique Nash equilibrium.) In particular $EP(\delta) = \{(1, 1)\}$ for all $\delta \in [0, 1/4)$.

We will now show that $EP(1/4)=F$. First, it is easily checked that the stationary path in which the players play $s=(s_1, s_2)$ at every stage is an equilibrium path of $\Gamma(1/4)$ if and only if $s_1=s_2$, hence, $EP(1/4)$ contains the line segment from $(1,1)$ to $(4,4)$. (Note that, in particular, the cooperative outcome in which the players always play (L_1, L_2) can be obtained in $\Gamma(1/4)$). Next, consider the path in which the players alternate between (s_1, s_2) and (s_2, s_1) where $s_i=(p_i, 1-p_i)$, starting with (s_1, s_2). This path results in the discounted payoff vector $(1+3p_2, 1+3p_1)$ and Theorem 8.4.1 shows that all such paths are equilibrium paths. Hence, $EP(1/4)$ contains the square with corners $(1,1)$, $(4,1)$, $(1,4)$ and $(4,4)$. Finally, suppose the players play $s=(s_1, s_2)$ at stage 0 and that they switch to the equilibrium with payoff x (x in the square described above) at $t=1$. The resulting payoff is $3/4R(s)+1/4x$ and this path is an equilibrium path if and only if deviating at time 0 is not profitable, i.e. if and only if $1+3p_i\leqq x_i (i=1, 2)$. Hence, $EP(1/4)$ contains the set

$$\{(y_1, y_2);\ y_i=3p_j+1/4(x_i+3-3p_i);\ 0\leqq p_i\leqq 1,\ 1+3p_i\leqq x_i\leqq 4\}\,.$$

Taking $p_2=1$ yields $x_2=4$ and we get the set

$$\{(y_1, y_2);\ y_1=4+1/4(x_1-1-3p_1),\ y_2=1+3p_1, 0\leqq p_1\leqq 1, 1+3p_1\leqq x_1\leqq 4\}\,,$$

which is the triangle with corners $(4, 1)$, $(4, 4)$ and $(19/4, 1)$. By symmetry we can also get the triangle with corners $(1, 4)$, $(4, 4)$ and $(1, 19/4)$, hence, $EP(1/4)=F$. We have shown

Theorem 8.4.8. *If Γ is the prisoners' dilemma game from* Fig. 8.4.1 *then $EP(\delta)=\{(1, 1)\}$ for all $\delta\in[0, 1/4)$ and $EP(1/4)=F$, where F is as in* (8.4.5).

In Sorin [1986], the standard assumption is made that the players can only observe the outcomes of the lotteries, i.e. they cannot observe the lotteries themselves. Sorin shows that in this case $EP(1/4)$ is a proper subset of F (it is the union of the square with corners $(1,1)$, $(4,1)$, $(1,4)$ and $(4,4)$, the line segment from $(4,1)$ to $(19/4,1)$ and the line segment from $(1,4)$ to $(1,19/4)$. Hence, in this example, there are more equilibrium payoffs if the players can observe mixed strategies. It is not clear, however, that this is a general property.

Similarly as above it can be shown that, for prisoners' dilemma, $EP(\delta)=F$ for any $\delta\geqq 1/4$. This may not be directly concluded from Theorem 8.4.8, however. Namely, although $E\Pi(\delta)$ is monotonic in δ (as follows easily from Theorem 8.4.1.), $EP(\delta)$ need not be monotonic in δ. The game of Fig. 8.4.2 (taken from Sorin [1986]) illustrates this property.

	L_2	R_2
L_1	1, 0	0, 0
R_1	0, 0	0, 1

Fig. 8.4.2. $EP(\delta)$ need not be monotonic in δ

If Γ is the game of Fig. 8.4.2, then playing (L_1, L_2) at stage 0 and playing (R_1, R_2) forever thereafter is an equilibrium path of $\Gamma(\delta)$ for every value of δ. If $\delta = 1/8$, this path results in the payoff vector $(7/8, 1/8)$. It is clear that this is not an equilibrium payoff of $\Gamma(0)$ and we claim that this vector does not belong to $EP(1/4)$ either. In fact, it is not even a feasible payoff vector in $\Gamma(1/4)$. Namely, since $(7/8, 1/8)$ is Pareto optimal in F, the players have to play either (L_1, L_2) or (R_1, R_2) at every stage of $\Gamma(1/4)$ in order to get this payoff vector. At $t=0$, they should play (L_1, L_2) since otherwise player 1's payoff is at most 1/4. This implies that the conditional expected payoff vector starting from stage 1 should be $(1/2, 1/2)$. However, this is not feasible in $\Gamma(1/4)$: If $x=(x_1, x_2)$ is Pareto optimal in F and feasible in $\Gamma(1/4)$, then $x_1 \geqq 3/4$ or $x_2 \geqq 3/4$.

One of the purposes of the discussion above has been to illustrate that, although Theorem 8.4.1 gives a complete characterization of the set of equilibrium paths, the determination of the set of equilibrium payoffs may not be a trivial task. We now turn to a much easier task: the determination of the limit of $EP(\delta)$ as δ tends to 1.

Theorem 8.4.9. *If Γ is an n-person game for which there exists some $x \in F$ such that $x_i > v_i$ for all $i \in N$, then $EP(\delta)$ converges to F as δ tends to 1.*

Proof. In view of the equivalence of (8.2.23) and (8.2.24) and because of Theorem 8.4.6 it is sufficient to show that for any $\varepsilon > 0$ and $x \in F$, we have that $B_\varepsilon(x) \cap EP(\delta) \neq \emptyset$ for all δ close to 1. Let x and ε be given and let $\{s^t\}_{t=0}^{T-1}$ be a finite cycle of strategy combinations such that the average (undiscounted) payoff during this cycle, $\bar{x}$, is strictly individually rational and within ε distance of x,

$$\bar{x} := \frac{1}{T} \sum_{t=0}^{T-1} R(s^t), \qquad \|\bar{x} - x\| < \varepsilon, \qquad \bar{x}_i > v_i \text{ for all } i.$$

Let $x(\delta)$ be the payoff that results in $\Gamma(\delta)$ when the cycle $\{s^t\}_{t=0}^{T-1}$ is played repeatedly. Theorem 8.4.1 shows that $x(\delta) \in EP(\delta)$ if for all $i \in N$ and $t < T$

$$(1-\delta) \max_{s_i} R_i(s^t/s_i) + \delta v_i \leqq (1-\delta) \sum_{\tau=t}^{T-1} \delta^{\tau-t} R_i(s^\tau) + \delta^{T-t} x_i(\delta). \qquad (8.4.10)$$

As δ tends to 1, the left hand side of (8.4.10) converges to v_i while the right hand side converges to $\bar{x}_i$. Since $\bar{x}$ is strictly individually rational, $x(\delta) \in EP(\delta)$ for all δ sufficiently close to 1. This completes the proof since $x(\delta) \in B_\varepsilon(x)$ for δ close to 1. □

The game of Fig. 8.3.2 shows that the condition in Theorem 8.4.9 cannot be dispensed with completely. For that game, $EP(\delta) = EP(1)$ for all $\delta \in [0, 1)$, while $F = \{(1, 1, \alpha); 0 \leqq \alpha \leqq 2\} \neq EP(1)$. On the other hand, the condition is not a necessary one. We will not derive necessary and sufficient conditions for $EP(\delta)$ to converge to F, rather, we confine ourselves to showing that for 2-person games no additional condition is necessary.

Theorem 8.4.10. *If Γ is a 2-person game, then $EP(\delta)$ converges to F as δ tends to 1.*

Proof. If Γ is a 2-person game for which $F = \{(v_1, v_2)\}$, then $EP(\delta) = F$ for all $\delta \in [0, 1)$. Hence, we only have to consider the case in which, say $x_1 = v_1$ for all $x \in F$ and $x_2 > v_2$ for some $x \in F$. In this case $v_1 = \max_s R_1(s)$, so that player 1 never has an incentive to deviate from a path that yields him v_1. On the other hand, if $x_2 > v_2$, then player 2 can be punished and the argument of the proof of Theorem 8.4.9 shows that, in this case, $x \in EP(\delta)$ for all δ sufficiently close to 1. □

It will be clear to the reader that the equilibria discussed in this section need not be subgame perfect: the threat to minmax forever generally is not credible. As for undiscounted games, one could try to remedy this by considering punishments of finite duration and by letting a player be punished if he does not cooperate in punishing another. Typically, in this case, succesive punishments will have to be of longer and longer duration. In undiscounted games this causes no problems since later period punishments hurt as much as earlier punishments and since future rewards are as attractive as current ones. With discounting, however, if a reward is postponed for too long, it may not be sufficiently attractive to induce a player to punish. Therefore, there may not be a subgame perfect equilibrium within the class of strategies used in the proof of Theorem 8.3.5. The game of Fig. 8.4.3 (taken from Fudenberg and Maskin [1986]) provides an example of this situation.

The game of Fig. 8.4.3 has (L_1, L_2) as its unique equilibrium. This equilibrium has a payoff 1, but each player's minmax value is only 0.

Note that, if player 1 minmaxes player 2 and if player 2 chooses his best response, then player 1 himself is hurt most. Let $\delta \in [0, 1)$ and consider the discounted game $\Gamma(\delta)$. Since a player can be rewarded with a payoff of at most 1 per period and since a player's total discounted payoff cannot fall below 0 in a subgame perfect equilibrium, player i cannot minmax player j for more than T periods where

$$-2(1-\delta^T) + 1\delta^T \geqq 0,$$

hence

$$T \leqq \frac{\ln 2/3}{\ln \delta}. \tag{8.4.11}$$

Consider the stationary path in which each player i chooses L_i with probability $1/2\sqrt{3}$ at each round, yielding the payoff vector $(1/2, 1/2)$, and let strategies as in the proof of Theorem 8.3.5 sustain this path. Since $(1/2, 1/2)$ is not an equilibrium

	L_2	R_2
L_1	1, 1	0, -2
R_1	-2, 0	-1, -1

Fig. 8.4.3. If the future is discounted, there is no subgame perfect equilibrium within the class of strategies described in the proof of Theorem 8.3.5

payoff vector of Γ, these strategies must prescribe at least a 1-period punishment. We claim that they prescribe punishments of unbounded duration. Namely, in order to induce player i to punish for t periods, player j must threaten to minmax for $T_i(t)$ periods where

$$(1-\delta)+1/2\delta^{T_i(t)+1} \leqq -2(1-\delta^t)+1/2\delta^t,$$

and it is easily checked that this implies $T_i(t)>t$. Hence, unbounded punishment sequences are necessary. However, (8.4.11) shows that such strategies cannot constitute a subgame perfect equilibrium.

The discussion above shows that subgame perfect equilibria of discounted games have a different structure than the equilibria that we constructed for undiscounted games in Sect. 8.3. Of course, the example does not warrant the conclusion that 'Perfect Folk Theorems' for discounted games are impossible. In fact, in the next section, it will be shown that such results are possible provided that the set F of feasible and individually rational payoffs in Γ is full dimensional. In that section, we will also investigate the structure of subgame perfect equilibria of discounted games in somewhat greater detail.

8.5 Infinitely Repeated Games with Discounting: Subgame Perfect Equilibria

This section is devoted to the study of subgame perfect equilibria in the discounted game $\Gamma(\delta)$. In the first part, characterizations are derived of the set of subgame perfect equilibrium paths $PE\Pi(\delta)$ and the set of subgame perfect equilibrium payoffs $PEP(\delta)$. This part builds on Abreu, Pearce and Stachetti [1986], but our proofs are more direct than those in that paper. In the second part it is shown that $PEP(\delta)$ converges to F as δ tends to 1 for any game for which F is full dimensional. This part is based on Fudenberg and Maskin [1986].

The equilibria discussed in the previous section need not be subgame perfect since the threat to minmax a player forever need not be credible. The subgame perfectness concept requires that an equilibrium be played in every contingency, hence, at every stage the play has to continue with a subgame perfect equilibrium and for every possible deviation there should be a subgame perfect equilibrium which renders the deviation unprofitable. Now, clearly a player can be prevented from deviating if and only if the threat to switch to this player's worst subgame perfect equilibrium is a sufficient deterrent, provided of course that this worst subgame perfect equilibrium exists. Our first goal is to show that this is indeed the case.

Let $\delta\in[0,1)$ and $x\in PEP(\delta)$. Let σ be a subgame perfect equilibrium of $\Gamma(\delta)$ with payoff vector x, let s^0 be the action combination prescribed by σ at $t=0$ and, for $s\in S$, let $w(s)$ be the expected payoff vector generated by σ, conditional on having reached the subgame s at $t=1$. Then $w(s)\in PEP(\delta)$ for all $s\in S$ and

$$x=(1-\delta)R(s^0)+\delta w(s^0), \quad \text{and}$$

$$x_i \geqq (1-\delta)R_i(s^0/s_i)+\delta w(s^0/s_i) \quad \text{all } i,\ s_i.$$

These conditions are equivalent to saying that x is a Nash equilibrium payoff vector of the game $\Gamma(\delta, w)$ with payoff function

$$R^{\delta,w}(s) = (1-\delta)R(s) + \delta w(s) \quad (s \in S). \tag{8.5.1}$$

Hence, if $x \in PEP(\delta)$, then there exists a function $w: S \to PEP(\delta)$ such that x is an equilibrium payoff of $\Gamma(\delta, w)$. The converse is also true; if x is an equilibrium payoff vector of $\Gamma(\delta, w)$ for some function $w: S \to PEP(\delta)$, then the equilibria of the subgames and the equilibrium of $\Gamma(\delta, w)$ can be combined to give a subgame perfect equilibrium of $\Gamma(\delta)$ with payoff x. Hence, we have the following analogue to Lemma 8.4.2.

Lemma 8.5.1. *$x \in PEP(\delta)$ if and only if there exists a function $w: S \to PEP(\delta)$ such that x is an equilibrium payoff of the game $\Gamma(\delta, w)$.*

Let W be a nonempty subset of F and let $w: S \to W$. Then the game $\Gamma(\delta, w)$ as in (8.5.1) is well-defined and has a clear interpretation: given that playing s will be rewarded with $w(s)$ at $t=1$, which action should I choose at $t=0$? We will say that x is admissible with respect to W in $\Gamma(\delta)$ if x is an equilibrium payoff vector of some game $\Gamma(\delta, w)$ with $w: S \to W$, and the set of all such admissible payoff vectors will be denoted by $P^\delta(W)$. Note that the set $(1-\delta)EP(1) + \delta W$ is contained in $P^\delta(W)$ for all δ and W, hence $P^\delta(W) \neq \varnothing$. Furthermore, $P^\delta(W) \subset F$ as a consequence of (8.2.22). Hence, P^δ is a mapping that assigns to each nonempty subset of F another such subset. Using this terminology, Lemma 8.5.1 can be paraphrased as

Corollary 8.5.2. *$PEP(\delta)$ is a fixed point of P^δ, hence, if $V = PEP(\delta)$, then $P^\delta(V) = V$.*

Note that Corollary 8.4.3 states a similar property for the Nash equilibrium payoffs of $\Gamma(\delta)$. We will now show that the operators P^δ and E^δ have several properties in common. First of all, the definition of admissibility implies that P^δ is monotonic.

Corollary 8.5.3. *If $V \subset W$, then $P^\delta(V) \subset P^\delta(W)$.*

Next, let W be a compact subset of F and let $w_i^* = \min\{w_i, w \in W\}$ be the worst payoff in W from player i's point of view. In this case, $x \in P^\delta(W)$ if and only if there exist $s^* \in S$ and $w(s^*) \in W$ such that

$$x = (1-\delta)R(s^*) + \delta w(s^*), \text{ and} \tag{8.5.2}$$

$$x_i \geqq (1-\delta)\max_{s_i} R_i(s^*/s_i) + \delta w_i^* \quad \text{for all } i \in N, \tag{8.5.3}$$

and one easily checks that the set of all such x is again compact. Hence

Corollary 8.5.4. *If W is compact, then $P^\delta(W)$ is compact.*

Next, we derive the analogue of Corollary 8.4.5. For the following lemma to be valid, it is not necessary that W is compact, however, imposing this condition allows a simpler proof and yields more insight.

Lemma 8.5.5. *If W is compact and $W \subset P^\delta(W)$, then $P^\delta(W) \subset PEP(\delta)$.*

Proof. Let W satisfy the condition, let $w_i^* := \min\{w_i,\ w \in W\}$ and let $x^0 \in P^\delta(W)$. There exist $s^{00} \in S$ and $x^1 \in W$ such that $x^0 = (1-\delta)R(s^{00}) + \delta x^1$ and such that (8.5.3) is satisfied with $s^* = s^{00}$. Since $W \subset P^\delta(W)$, we have $x^1 \in P^\delta(W)$ and the process can be continued. Continuing inductively, we get

$$x^t = (1-\delta)R(s^{0t}) + \delta x^{t+1} \quad t \in \mathbb{N}, \text{ and}$$

$$x_i^t \geqq (1-\delta)\max_{s_i} R_i(s^{0t}/s_i) + \delta w_i^* \quad t \in N,\ i \in \mathbb{N}.$$

Let $\pi^0 = \{s^{0t}\}_t$ be the path that is constructed in this way. This procedure can also be followed for any $x \in W$ with $x_i = w_i^*\ (i \in N)$. Let $\pi^i = \{s^{it}\}_t$ be such a 'worst path' for player i. Using these notations, the above equations yield

$$R_{it}^\delta(\pi^0) = x_i^t \quad i \in N,\ t \in \mathbb{N}, \tag{8.5.4}$$

$$R_{i0}^\delta(\pi^i) = w_i^* \quad i \in N, \text{ and} \tag{8.5.5}$$

$$R_{it}^\delta(\pi^k) \geqq (1-\delta)\max_{s_i} R_i(s^{kt}/s_i) + \delta w_i^* \quad i \in N,\ t \in \mathbb{N},\ k = 0, 1, \ldots, n. \tag{8.5.6}$$

The desired equilibrium σ with payoff vector x^0 is completely determined by the $(n+1)$ – tuple of paths $\Pi = (\pi^0, \pi^1, \ldots, \pi^n)$ constructed above. In words, σ is described as follows: the players start with π^0; as long as no deviations from a path $\pi^k\ (k=0, \ldots, n)$ have occurred, the players continue with π^k; if player i is the unique deviator, the players switch to the path π^i, i.e. to the path that is worst for player i; if a situation cannot be classified as one of the above, the players restart π^0. Formally, σ is described by

$$\sigma(0) = s^{00} \tag{8.5.7}$$

and, for each $t \geqq 1$ and $h^t \in S^t$ a history at time t

$$\sigma(h^t) = \begin{cases} s^{k\tau} & \text{if } h^t(t-1) = s^{k\tau-1} \text{ for some } k,\ \tau \\ s^{i0} & \text{if } h^t(t-1) = s^{k\tau-1}/s_i \text{ with } s_i \neq s_i^{k\tau-1}, \text{ for some } k,\ \tau, \text{ and} \\ s^{00} & \text{otherwise,} \end{cases} \tag{8.5.8}$$

where $h^t(t-1)$ is the outcome at time $t-1$ associated to the history h^t. If σ is played, the path π^0 results, hence $R^\delta(\sigma) = x^0$. We claim that σ is a subgame perfect equilibrium of $\Gamma(\delta)$. Because of the simple structure of σ, to verify this claim, it is sufficient to check that all paths $\pi^k\ (k=0, 1, \ldots, n)$ are equilibrium paths. Comparing Eq. (8.5.6) to Eq. (8.4.3) we see that this is true provided that $w_i^* \geqq v_i$. However, this follows from the fact that $W \subset F$. □

By combining the results obtained thus far we can show

Theorem 8.5.6. *$PEP(\delta)$ is a non-empty, compact set.*

Proof. Repeatedly playing an equilibrium of Γ yields a subgame perfect equilibrium in $\Gamma(\delta)$, hence $PEP(\delta) \neq \emptyset$. The compactness can be shown as was done for $EP(\delta)$ in Theorem 8.4.7. □

Let V be a fixed point of P^δ. Then $V \subset \mathrm{cl}V \subset P^\delta(\mathrm{cl}V)$ (cf. the proof of Theorem 8.4.7), hence $V \subset \mathrm{cl}V \subset PEP(\delta)$. Since $PEP(\delta)$ is a fixed point of P^δ, we have the following characterization of the set of subgame perfect equilibrium payoffs (cf. the characterization of $EP(\delta)$ in Corollary 8.4.6).

Corollary 8.5.7. *$PEP(\delta)$ is the largest fixed point of P^δ.*

(From Corollary 8.5.9, Theorem 8.4.8 and the fact that $\{(1, 1)\}$ is a fixed point of P^δ for the prisoners' dilemma game, it follows that $PEP(\delta)$ need not be the unique fixed point of P^δ.)

Because of Theorem 8.5.6, each player i's worst subgame perfect equilibrium payoff, $v_i(\delta)$, is well defined

$$v_i(\delta) := \min\{x_i;\ x \in PEP(\delta)\}, \tag{8.5.9}$$

and we have the following analogue to Theorem 8.4.1.

Theorem 8.5.8. *$\pi = \{s^t\}_t$ is a subgame perfect equilibrium path in $\Gamma(\delta)$ if and only if*

$$R^\delta_{it}(\pi) \geqq (1-\delta)\max_{s_i} R_i(s^t/s_i) + \delta v_i(\delta) \quad \textit{for all } i \textit{ and } t. \tag{8.5.10}$$

In $\Gamma(\delta)$, a possibility of punishing player i is to revert to that equilibrium of Γ that yields player i the smallest payoff, hence

$$v_i \leqq v_i(\delta) \leqq v_i(0) \quad \text{for all } i \text{ and } \delta. \tag{8.5.11}$$

Therefore, by comparing Theorems 8.4.1 and 8.5.8, we see

Corollary 8.5.9. *If Γ is a game for which $v_i = v_i(0)$ for all $i \in N$, then $PE\Pi(\delta) = E\Pi(\delta)$ and $PEP(\delta) = EP(\delta)$.*

The condition of this Corollary is satisfied for the prisoners' dilemma game of Fig. 8.4.1, hence, Theorem 8.4.3 describes the sets $PEP(\delta)$ for $\delta \in [0, 1/4]$. The equilibria that we constructed for this prisoner's dilemma game are all subgame perfect[10], but note that one cannot draw the conclusion that $E(\delta) = PE(\delta)$ for a game that satisfies the condition of Corollary 8.5.9.

Clearly, for Theorem 8.5.8 to be most useful, one has to know how to compute $v_i(\delta)$. Generaly, this computation is not straightforward, especially since for most games the inequalities in (8.5.11) are strict when $\delta \in (0, 1)$, hence, reverting to the worst one-period equilibrium need not be an optimal punishment (cf. Theorem 8.5.11). We will not provide a solution here, rather we confine ourselves to showing that optimal punishments have a stick and carrot character. (This has first been shown in Abreu [1982]. Also see Abreu [1988]).

Let σ be a subgame perfect equilibrium of $\Gamma(\delta)$ that is worst from player i's point of view: $R^\delta_i(\sigma) = v_i(\delta)$. Let $\pi(\sigma) = \pi = \{s^t\}_t$ be the path induced by σ.

Then, for any $T \in \mathbb{N}$, the subpath $\pi^T = \{s^t\}_T^\infty$ is also a subgame perfect equilibrium path of $\Gamma(\delta)$. Therefore $R_{iT}^\delta(\pi) \geqq v_i(\delta)$, or equivalently

$$(1-\delta^T)^{-1} \sum_{t=0}^{T-1} \delta^t R_i(s^t) \leqq (1-\delta)^{-1} \sum_{t=T}^{\infty} \delta^{t-T} R_i(s^t), \tag{8.5.12}$$

which shows that player i's average payoff during the first T stages is not higher than his average payoff during the later stages. In general, this inequality will be strict, hence, the initial stages of the punishment will be more severe than later stages. In this case, the optimal punishment indeed has a *stick and carrot* flavor: a player is willing to 'cooperate' in his punishment since, if he would defect, he would have to go through the initial punishment stages again.

Since optimal punishments may be difficult to construct explicitly, we follow an alternative approach to prove "Perfect Folk Theorems" for discounted games. The key to this approach is to restrict oneself to the simple strategy combinations used in the proof of Lemma 8.5.5. Recall that the strategy combination σ defined by (8.5.7), (8.5.8) is completely specified by $n+1$ paths $\pi^0, \pi^1, \ldots, \pi^n$: the players start with π^0; if player i deviates they restart π^i, etc. In the proof of Lemma 8.5.5, the paths π^i with $i \in N$ had special properties, but the strategy combination σ is well-defined for any $(n+1)$-tuple of paths, although, of course it need not be a subgame perfect equilibrium in this case. If $\Pi = (\pi^0, \pi^1, ., \pi^n)$ is an arbitrary $(n+1)$-tuple of paths with $\pi^k = \{s^{kt}\}_t$ and $\sigma = \sigma(\Pi)$ is the *simple strategy combination* induced by Π (i.e. σ is given by (8.5.7) – (8.5.8)), then σ is a subgame perfect equilibrium if and only if

$$R_{it}^\delta(\pi^k) \geqq (1-\delta) \max_{s_i} R_i(s^{kt}/s_i) + \delta R_{i0}^\delta(\pi^i) \quad \text{all } i, k, t. \tag{8.5.13}$$

In the remainder, attention will be restricted to such simple strategy combinations. The proof of Lemma 8.5.5 (with $W = PEP(\delta)$) shows that this restriction is justified since any subgame perfect equilibrium path can be sustained by such a simple strategy combination. This result was first derived in Abreu [1983] and Harris [1983]. (Also see Abreu [1988]). Hence,

Corollary 8.5.10. *Any $x \in PEP(\delta)$ can be obtained by means of a simple strategy combination that is a subgame perfect equilibrium.*

From the results obtained in the previous section it follows that $PEP(\delta)$ can be expected to converge to F only if F contains strictly individually rational elements. The game from Fig. 8.5.1 (taken from Fudenberg and Maskin [1986]) shows that this condition is not sufficient for $PEP(\delta)$ to converge to F.

The game of Fig. 8.5.1 is degenerate in the sense that all players always have the same payoffs (consequently, we will drop the index i and write $v(\delta)$ for $v_i(\delta)$). The game has 3 equilibria, with respective payoffs 1, 1 and 1/4 (in the third equilibrium, each player mixes (1/2, 1/2)). Each player's minmax value is 0 and F is the line segment from (0, 0, 0) to (1, 1, 1). Hence, Theorem 8.4.4 shows that $EP(\delta)$ converges to F as δ tends to 1. We claim that $v(\delta) = v(0) = 1/4$ for all $\delta \in [0, 1)$ so that $PEP(\delta)$ does not converge to F. Namely, let $\pi = \{s^t\}_t$ with $s_i^0 = (p_i, 1-p_i)$ be a subgame perfect equilibrium path of $\Gamma(\delta)$ with payoff $v(\delta)$.

1, 1, 1	0, 0, 0
0, 0, 0	0, 0, 0

0, 0, 0	0, 0, 0
0, 0, 0	1, 1, 1

Fig. 8.5.1. A 3-person game for which $EP(\delta)$ converges to F as δ tends to 1, but for which $PEP(\delta)$ converges to a proper subset of F. Furthermore $PEP(\infty)=F$

By defecting at time 0, player i can obtain

$$\gamma_i=\max\{p_jp_k, (1-p_j)(1-p_k)\} \quad (i\neq j\neq k).$$

Hence, the most fortunate defector can obtain

$$\gamma=\max\{\gamma_1, \gamma_2, \gamma_3\} \geqq 1/4.$$

Because of Theorem 8.5.8 (with $t=0$) we must have that

$$v(\delta) \geqq (1-\delta)\gamma+\delta v(\delta),$$

hence, $v(\delta) \geqq \gamma$. By (8.5.11), $v(\delta) \leqq v(0)$, hence

$$1/4=v(0) \geqq v(\delta) \geqq \gamma \geqq 1/4,$$

and $v(\delta)=1/4$ for all $\delta \in [0, 1)$. Therefore $PEP(\delta)$ does not converge to F as δ tends to 1. (It will be clear that $PEP(\delta)$ converges to the line segment joining $1/4(1, 1, 1)$ and $(1, 1, 1)$). Note that Theorem 8.3.4 implies that $PEP(\infty)=F$, hence, for this game it makes a big difference whether the players discount only a little or whether there is no discounting at all.

The fact that the 3 players always have the same payoffs in the game of Fig. 8.5.1 implies that it is impossible to reward one player while simultaneously punishing another and it is this property that causes the discontinuity of $PEP(\delta)$ at $\delta=1$. However, if Γ is a game for which F is full dimensional, then it is possible to reward one player and to punish another simultaneously. We will show that in this case indeed $PEP(\delta)$ converges to F as δ tends to 1. Hence, to obtain $\lim_{\delta \uparrow 1} PEP(\delta)=F$, a condition that is only slightly more restrictive than the one of Theorem 8.4.9 is sufficient.

Theorem 8.5.11 (Fudenberg and Maskin [1986]). *If Γ is an n-person game for which* $\dim F=n$ *then $PEP(\delta)$ converges to F as δ tends to* 1.

Proof. Since F is full dimensional, it is sufficient to show that for any $x \in \operatorname{int} F$ and $\varepsilon>0$ we have that $B_\varepsilon(x) \cap PEP(\delta) \neq \varnothing$ for all δ sufficiently close to 1 ($\operatorname{int} F$ is the interior of F). Let $x \in \operatorname{int} F$ and let $\gamma>0$ be such that $x^i(\gamma) \in F$ for all $i \in N$, where

$x^i(\gamma)$ is defined by

$$x_j^i(\gamma) = \begin{cases} x_j + \gamma & \text{if } j \neq i, \\ x_i & \text{if } j = i. \end{cases} \tag{8.5.14}$$

For notational convenience, let us assume that there exist action combinations s^0 and $s^i (i \in N)$ such that $R(s^0) = x$ and $R(s^i) = x^i(\gamma)$ for $i \in N$. (In the general case, one has to approximate by means of finite cycles, but this does not change the main argument.) For $k = 0, 1, \ldots, n$, let $\pi^k = \{s^{kt}\}_t$ be the path given by

$$s^{0t} := s^0,$$

$$s^{kt} := \begin{cases} m^k & \text{if } t < T \\ s^k & \text{if } t \geqq T \end{cases} \quad k \in N,$$

let Π be the $(n+1)$-tuple $\Pi = (\pi^0, \pi^1, \ldots, \pi^n)$ and let σ be the simple strategy combination induced by Π. Hence, the players play s until a deviation occurs; if player i deviates he is minmaxed for T periods, whereafter player i's opponent will be rewarded for having punished this player; if a player does not cooperate in punishing another, he is punished himself. We claim that there exists some $T \in \mathbb{N}$ and some $\delta^* \in [0, 1)$ such that $\sigma \in PE(\delta)$ for all $\delta > \delta^*$. We have

$$R_{it}^{\delta}(\pi^0) = x_i$$

$$R_{it}^{\delta}(\pi^k) = \begin{cases} (1 - \delta^{T-t}) v_i^k + \delta^{T-t} x_i^k(\gamma) & t < T \\ x_i^k(\gamma) & t \geqq T \end{cases} \quad k \in N$$

and we must check that the conditions (8.5.13) are satisfied. For convenience, let us assume that $v_i = 0$ for all $i \in N$ (rescale payoffs if necessary). Condition (8.5.13) will certainly be satisfied for $k = 0$ if

$$(1 - \delta) M_i + \delta^{T+1} x_i \leqq x_i \quad \text{for all } i \in N$$

where M_i is as in (8.2.7). This inequality holds if

$$\frac{1 - \delta^{T+1}}{1 - \delta} \geqq \frac{M_i}{x_i} \quad \text{for all } i, \tag{8.5.15}$$

The right hand side of (8.5.15) is a finite number since $x_i > v_i = 0$. As δ tends to 1, the left hand side converges to $T + 1$, hence, if we choose T such that

$$T \geqq \max_i \frac{M_i}{x_i}, \tag{8.5.16}$$

then (8.5.15) is satisfied for all δ sufficiently close to 1. Next, consider a path π^k with $k \in N$. Choosing T as in (8.5.16) guarantees that player k cannot profitably deviate during the reward phase of π^k. Since this player is at a one-shot best reply during each of the first T stages of π^k and since by deviating he only postpones his reward x_k, he also cannot profitably deviate during one of the first T stages. Finally, let $i \neq k$. Choosing T as in (8.5.16) guarantees that player i has no incentive to deviate during the reward phase of π^k. Player i is willing to minmax player k for t times $(t \leqq T)$ if

$$(1-\delta)M_i+\delta^{T+1}x_i \leqq (1-\delta^t)v_i^k+\delta^t x_i(\gamma) \quad (i \in N).$$

Substituting $v_i^k \geqq -M_i$ and $t \leqq T$, we see that this inequality holds if

$$\frac{\delta^T}{1-\delta^T} \geqq 2\frac{M_i}{\gamma} \quad \text{for all } i.$$

For a fixed T as in (8.5.16), this condition is satisfied for all δ that are sufficiently close to 1. Hence, for such T we have that σ is a subgame perfect equilibrium of $\Gamma(\delta)$ for all δ sufficiently close to 1. This completes the proof since $R^\delta(\sigma)=x$. □

To conclude this section, let us consider 2-person games. The game from Fig. 8.3.4 shows that $PEP(\delta)$ need not converge to F if only one player can get more than his minmax value in F. The same argument as in Sect. 8.3 shows that, for this game, $PEP(\delta)=\{(1, 1)\} \neq F$. The following theorem shows that for 2-person games this is the only instance in which $PEP(\delta)$ does not converge to F.

Theorem 8.5.12 (Fudenberg and Maskin [1986]). *If Γ is a 2-person game, then* $\lim_{\delta \uparrow 1} PEP(\delta)=F$ *if* $F=\{(v_1, v_2)\}$ *or if there exists* $x \in F$ *with* $x_1>v_1$ *and* $x_2>v_2$.

Proof. If $F=\{(v_1, v_2)\}$, then Γ has an equilibrium with value (v_1, v_2) and $PEP(\delta)=F$ for all $\delta \in [0, 1)$. Hence, assume there exists $x \in F$ that Pareto dominates (v_1, v_2). If $\dim F=2$, the result follows from the previous theorem, so that it suffices to consider the case in which F is a line segment with positive slope. By rescaling the payoffs, we can obtain that $R_1(s)=R_2(s)$ for all $s \in S$. Furthermore, we may assume $v_2 \leqq v_1=0$. Let $s \in S$ be such that $x_1=R_1(s)>0$ and consider the paths $\pi^0=\{s^{0t}\}_t$ and $\pi=\{s^t\}_t$ defined by

$$s^{0t}=s$$

$$s^t=\begin{cases}(m_1^2, m_2^1) & \text{if } t<T \\ s & \text{if } t \geqq T\end{cases}$$

Hence, along the path π, the players minmax each other for T periods whereafter they switch to s. Let y_1 be the payoff when the players minmax each other. Then $y_1 \leqq v_2 \leqq 0$ and

$$R_{1t}^\delta(\pi^0)=x_1$$

$$R_{1t}^\delta(\pi)=\begin{cases}(1-\delta^{T-t})y_1+\delta^{T-t}x_1 & t<T \\ x_1 & t \geqq T\end{cases}$$

We claim that, if T is as in (8.5.16), then the simple strategy combination σ that is induced by $\Pi=(\pi^0, \pi, \pi)$ is a subgame perfect equilibrium of $\Gamma(\delta)$ for all δ sufficiently close to 1. Namely, since $y_1 \leqq 0$, deviating from π^0 is not profitable if (8.5.15) is satisfied. In this case, also deviating during the reward phase of π is not profitable. Since a player gets at most 0 in the period that he deviates from (m_1^2, m_2^1), he will be willing to punish for T periods as long as $R_{1T}^{\delta}(\pi) \geqq 0$, i.e. as long as

$$(1-\delta^T)y_1 + \delta^T x_1 \geqq 0.$$

For T as in (8.5.16) this condition is satisfied for all δ sufficiently close to 1. □

Finally, let Γ be a 2-person game in which only one player can get more than his minmax value at some point in F, say $x_1 = v_1$ for all $x \in F$ and $x_2 > v_2$ for some $x \in F$. Let $\tilde{F}_2$ be defined as in (8.3.9). By replacing the minmax pair (m_1^2, m_2^1) by an action combination s with $R_1(s) = v_1$ and $R_2(s) \leqq \tilde{v}_2$ and by taking $x_2 \geqq \tilde{v}_2$ in the proof above, one sees that in this case $PEP(\delta)$ converges to $\tilde{F}_2$ as δ tends to 1. Hence, the following Corollary completes the description of $\lim_{\delta \uparrow 1} PEP(\delta)$ for 2-person games.

Corollary 8.5.13. *If Γ is a 2-person game for which $x_i = v_i$ for all $x \in F$ and $x_j > v_j$ for some $x \in F$ $(i \neq j)$, then* PEP(δ) *converges to $\tilde{F}_j$ as δ tends to* 1.

By comparing Theorem 8.5.12 and Corollary 8.5.13 to Theorem 8.3.7 we see that, for 2-person games, $PEP(\infty) = \lim_{\delta \uparrow 1} PEP(\delta)$, hence, there is no discontinuity in the set of perfect equilibrium payoffs if $\delta = 1$. If there are more than 2 players, however, there may be a discontinuity as we know from Fig. 8.5.1.

8.6 Finitely Repeated Games: Nash Equilibria

In the next two sections it will be investigated what happens when Γ is repeated finitely many times. In this section, Nash equilibria are considered and it is shown that the set of equilibrium payoffs of the T-stage game converges to the set of feasible and individually rational payoffs of Γ as T tends to infinity for any game for which each player can get more than his minmax value at some equilibrium. The presentation is based on Benoit and Krishna [1987], in which this theorem has first been proved. (Also see Friedman [1985]).

Consider, $\Gamma(T)$ the T-fold repetition of Γ $(T \in \mathbb{N}, T \geqq 1)$. It will be convenient to let time run from 1 to T and to *index time backwards*. Hence, when $\pi = \{s^t\}_{t=1}^T$ is a path in $\Gamma(T)$, then s^t is the action combination taken when there are still t stages to go, s^1 is the action combination that is played in the final stage. A player can be deterred from deviating from π if and only if the threat to minmax this player till the end of the game is a sufficient deterrent. Hence, we have the following analogue to Theorems 8.3.1 and 8.4.1.

Theorem 8.6.1. $\pi = \{s^t\}_{t=1}^T$ *is a Nash equilibrium path of $\Gamma(T)$ if and only if*

$$\sum_{\tau=1}^{t} R_i(s^\tau) \geqq \max_{s_i} R_i(s^t/s_i) + (t-1)v_i \quad \textit{all } i, \textit{ all } t \leqq T. \tag{8.6.1}$$

Since repeatedly playing an equilibrium of Γ yields an equilibrium in $\Gamma(T)$, this repeated game has at least one equilibrium. Theorem 8.6.1 shows that the set of equilibrium paths of $\Gamma(T)$ is closed, hence, that $EP(T)$ is closed. Since $EP(T) \subset F$, we have

Corollary 8.6.2. *$EP(T)$ is a nonempty, compact set for every T.*

Let Γ be a game for which $EP(1) = \{(v_1, v_2, \ldots, v_n)\}$ and let (s^2, s^1) be an equilibrium path of $G(2)$. Since endpaths of equilibrium paths are again equilibrium paths, $s^1 \in E(1)$ and $R_i(s^1) = v_i$ for all i. Consequently, for $t=2$, condition (8.6.1) reduces to $R_i(s^2) \geqq \max R_i(s^2/s_i)$ for all i, hence, s^2 must be an equilibrium of Γ. Proceeding inductively we see that, if $\pi = \{s^t\}_{t=1}^{T}$ is an equilibrium path of $\Gamma(T)$, then s^t is an equilibrium of Γ for all t, hence, $E\Pi(T) = E(1)^T$ and $EP(T) = EP(1)$.

Corollary 8.6.3. *If Γ is an n-person game for which $EP(1) = \{(v_1, \ldots, v_n)\}$, then $E\Pi(T) = E(1)^T$ and $EP(T) = EP(1)$ for all T.*

In particular, if Γ has a unique equilibrium and if this equilibrium yields each player his minmax value (as is, for example, the case with prisoners' dilemma), then $\Gamma(T)$ has a unique equilibrium path for every T. One cannot conclude, however, that $\Gamma(T)$ has a unique Nash equilibrium in this case since in a Nash equilibrium a player's action may be completely arbitrary in any subgame that cannot be reached as long as this player adheres to his equilibrium strategy.

Note that for Corollary 8.6.3 to be valid, it is essential that the unique equilibrium payoff is the minmax payoff. This is illustrated by means of the game of Fig. 8.6.1, which is a slight modification of the prisoners' dilemma game of Fig. 8.4.1 (also cf. the game of Fig. 1.3.2).

In the game of Fig. 8.6.1, R_i is a dominant strategy, hence, $R = (R_1, R_2)$ is the unique Nash equilibrium. Although adding the action A_i has not changed the set of equilibria of Γ, it has lowered each player's minmax value: in the enlarged game, a player can only guarantee the value 0. This has important consequences for the set of equilibria of $\Gamma(T)$ where $T \geqq 2$. Namely, if x is a feasible and individually

	L_2	R_2	A_2
L_1	4, 4	0, 5	0, 0
R_1	5, 0	1, 1	0, 0
A_1	0, 0	0, 0	-1, -1

Fig. 8.6.1. Prisoners' dilemma with an additional strictly dominated strategy

rational payoff in $\Gamma(x \in F = \mathrm{co}\{0,0), (5,0), (0,5), (4,4)\})$, then $x = R(s)$ for some $s \in S$ for which

$$\max_{s_i'} R_i(s/s_i') \leqq R_i(s) + 1 \quad (i = 1, 2), \tag{8.6.2}$$

hence, each player can gain at most 1 by deviating. Now, this gain can be completely annihilated by punishing the deviator in the last round, so that (s, R) is an equilibrium path of $\Gamma(2)$ for every s as in (8.6.2). More generally, if $T \geqq 2$ and s is as in (8.6.2), then the path in which s is played in all but the last round and in which R is played in this last round is an equilibrium path of $\Gamma(T)$. Hence, for this modified prisoners' dilemma, $EP(T)$ converges to F as T tends to infinity, which is in sharp contrast to the stationary of $EP(T)$ for the original version of the game.

In finitely repeated games, it makes a crucial difference whether or not the basic game Γ admits equilibria with payoffs larger than the minmax values. In Corollary 8.6.3, one extreme case was considered. In the next theorem, we consider the other extreme case in which each player can get more than this minmax value at some equilibrium of Γ.

Theorem 8.6.4 (Benoit and Krishna [1987]). *If Γ is an n-person game for which* $\max EP_i(1) > v_i$ *for all $i \in N$, then $EP(T)$ converges to F as T tends to infinity.*

Proof. For $i \in N$, let s^i be an equilibrium of Γ with $R_i(s^i) > v_i$. Let c be the cycle $c = (s^1, \ldots, s^n)$ and let y_i be player i's average payoff during this cycle. Equation (8.2.5) implies that $y_i > v_i$ for all $i \in N$. Let $x \in F$ and, for convenience, assume $x = R(s)$ for some $s \in S$. (If such s cannot be found one has to approximate x by means of a finite cycle, but this does not change the main argument). If a player who cooperates in playing s is rewarded by going through the cycle c a large number of times, while a deviating player is punished by v_i, it is possible to get the outcome s in equilibrium. Specifically, the path $(s, c^{(T)})$ in which s is played followed by T times c is an equilibrium path of $\Gamma(nT+1)$ if

$$M_i + nTv_i \leqq x_i + nTy_i \quad \text{for all } i \in N,$$

hence

$$T \geqq \frac{M_i - x_i}{n(y_i - v_i)} \quad \text{for all } i \in N \tag{8.6.3}$$

Since x is individually rational, choosing T as in (8.6.3) also guarantees that the path $(s^{(t)}, c^{(T)})$ is an equilibrium path of $\Gamma(nT+t)$ for all $t \geqq 1$. Keeping T fixed and letting t go to infinity, the average payoff resulting from this path approaches x. $\square$

For 2-person games, the condition of Theorem 8.6.4 can be relaxed as we show in the following theorem.

Theorem 8.6.5. *If Γ is a 2-person game for which* $\max EP_i(1) > v_i$ *for some i, then $EP(T)$ converges to F as T tends to infinity.*

Proof. Assume Γ is a 2-person game with $EP_2(1) > v_2$ and let $\bar{s}$ be an equilibrium of Γ with $R_2(\bar{s}) > v_2$. First consider the case in which $R_1(s) \leqq v_1$ for all $s \in S$, hence, F is a line segment parallel to the x_2-axis. In this case, player 1 cannot profitably deviate from a path that yields v_1, while player 2 can be punished in case he deviates. Therefore, if $R(s) \in F$, then there exists some T such that the path $(s^{(t)}, \bar{s}^{(T)})$ is an equilibrium path of $\Gamma(T+t)$ for all $t \geqq 1$ and the result holds. Next, suppose $R_1(s) > v_1$ for some $s \in S$. Let s be such that s_1 is a best reply against s_2 and $R_1(s) > v_1$. Then there exists some T such that $(s, \bar{s}^{(T)})$ is an equilibrium path of $\Gamma(T+1)$. By replacing c with $(s, \bar{s}^{(T)})$ in the proof of Theorem 8.6.4, one sees that the result also holds in this case. □

The result of Theorem 8.6.5 no longer holds if there are more than 2 players as is illustrated by means of the game of Fig. 8.3.2: the dummy player in this game has an equilibrium payoff that is strictly higher than his minmax value, but this player is strategically irrelevant and Corollary 8.6.3 implies that $EP(T) = EP(1)$ for all T, hence $EP(T)$ does not converge to F.

By combining Theorem 8.6.5 with Corollary 8.6.3, we obtain

Corollary 8.6.6. *If Γ is a 2-person game, then $EP(T) = EP(1)$ for all T or $\lim_{T\to\infty} EP(T) = F$.*

It is unknown to the author whether this Corollary remains correct if there are more than 2 players.

8.7 Finitely Repeated Games: Subgame Perfect Equilibria

This section deals with subgame perfect equilibria in the finitely repeated game $\Gamma(T)$. In the first part, some general properties of such equilibria are derived. In the second part it is shown that every feasible and individually rational payoff vector can be approximated by subgame perfect equilibrium payoffs of $\Gamma(T)$ if T is sufficiently large, provided that (1) each player has at least 2 equilibrium payoffs in Γ and (2) F is full dimensional. The discussion is based on Benoit and Krishna [1985].

The characterization of subgame perfect equilibrium paths derived in Theorem 8.7.1 will be used repeatedly:

Theorem 8.7.1. *For any $T \in \mathbb{N}$, $T \geqq 1$ we have*

(i) *$PE\Pi(T)$ and $PEP(T)$ are nonempty, compact sets,*

(ii) *if $\pi = \{s^t\}_{t=1}^{T}$, then $\pi \in PE\Pi(T)$ if and only if*

$$\sum_{\tau=1}^{t} R_i(s^\tau) \geqq \max_{s_i} R_i(s^t/s_i) + (t-1)v_i(t-1) \quad \textit{all } i, \textit{ all } t = 1, \ldots, T, \tag{8.7.1}$$

where $v_i(t)$ is player i's worst subgame perfect equilibrium payoff in $\Gamma(t)$:

$$v_i(t) := \begin{cases} \min PEP_i(t) & t \geqq 1 \\ 0 & t = 0. \end{cases} \tag{8.7.2}$$

Proof. The proof is by induction with respect to T. If $T=1$, then (8.7.1) are the conditions for a Nash equilibrium of Γ and Nash's theorem guarantees the existence of at least one equilibrium. Clearly, the set of solutions to (8.7.1) is closed, hence, also the set of equilibrium payoffs is closed. Since $E\Pi(1)=PE\Pi(1)$ and $EP(1)=PEP(1)$, the assertion is correct for $T=1$. Assume the assertions (i) and (ii) have been proved for $T\geqq 1$. The path $\pi=\{s^t\}_{t=1}^{T+1}$ is a subgame perfect equilibrium path in $\Gamma(T+1)$ if and only if (1) the endpath $\pi^T=\{s^t\}_{t=1}^{T}$ is an element of $\Gamma(T)$ and (2) for every deviation s_i of player i at time $T+1$ $(s_i \neq s_i^{T+1})$ there exists a subgame perfect equilibrium continuation at time T that renders the deviation unprofitable. The induction hypothesis implies that (1) is equivalent to (8.7.1) and that $v_i(t)$ is well defined for $t\leqq T$. Therefore, (2) is equivalent to the inequality in (8.7.1) with $t=T+1$. Hence, assertion (ii) is correct for $T+1$. Since the set of solutions to (8.7.1) is closed and since repeatedly playing an equilibrium of Γ yields an equilibrium in $\Gamma(T)$ for any T, also assertion (i) is correct. □

We will now derive some consequences of Theorem 8.7.1. The theorem implies that subpaths (or more precisely endpaths) of subgame perfect equilibrium paths are again subgame perfect equilibrium paths. The same need not be true for the initial part of a path (see the example below), but, if both the initial part and the final part are subgame perfect, then their conjunction is also guaranteed to have this property.

Corollary 8.7.2. *If $T=T_1+T_2$ and $\pi=(\pi_1,\pi_2)$ where π_i is a path of length T_i, then*

(i) *If $\pi\in PE\Pi(T)$, then $\pi_2\in PE\Pi(T_2)$, and*
(ii) *If $\pi_1\in PE\Pi(T_1)$ and $\pi_2\in PE\Pi(T_2)$, then $\pi\in PE\Pi(T)$.*

By patching together the worst subgame perfect equilibrium of the T_1 period game with the worst subgame perfect equilibrium of $\Gamma(T_2)$, one obtains a subgame perfect equilibrium of $\Gamma(T_1+T_2)$, therefore, the sequence $\{tv_i(t)\}_t$ must be subadditive for all $i\in N$:

$$\text{if } T=T_1+T_2, \text{ then } Tv_i(T)\leqq T_1 v_i(T_1)+T_2 v_i(T_2) \quad (i\in N). \tag{8.7.3}$$

In particular, $v_i(T)\leqq v_i(1)$ since one can always punish player i by playing the worst one-period Nash equilibrium at every stage. Let $T\in\mathbb{N}$, $T\geqq 1$ and let $\pi=\{s^t\}_{t=1}^{2T}$ be that element of $PE\Pi(2T)$ that is worst for player i. Since the endpath $\pi^T=\{s^t\}_{t=1}^{T}$ is a subgame perfect equilibrium path in $\Gamma(T)$, we have

$$Tv_i(T)\leqq \sum_{t=1}^{T} R_i(s^t).$$

On the other hand, (8.7.3) implies that

$$\sum_{t=1}^{2T} R_i(s^t)\leqq 2Tv_i(T)$$

so that we find that

$$\sum_{t=T+1}^{2T} R_i(s^t)\leqq \sum_{t=1}^{T} R_i(s^t). \tag{8.7.4}$$

Hence, player i's payoff during the first half of the optimal punishment is lower (or at least not higher) than the payoff during the second half: As in the discounted case, we have that optimal punishments have a *stick and carrot* character. This is illustrated by means of the game of Fig. 8.7.1 which shows that the inequalities (8.7.3) may be strict.

The game of Fig. 8.7.1 has 3 equilibria with respective payoffs 6, 3 and 2, hence $v(1)=2$ (since the players have identical payoffs, we drop the index i). It is easily checked that the path (R, M) (where $R=(R_1, R_2)$, etc) satisfies condition (8.7.1), hence, $(R, M)\in PE\Pi(2)$ and $v(2)\leqq 3/2<v(1)$. We claim that $v(2)=3/2$. Namely, assume $(s^2, s^1)\in PE\Pi(2)$. Then $s^1\in E(1)$ and, if $s^1\in\{L, M\}$, we have $R^2(\pi)\geqq 3/2$. If s^1 is the mixed equilibrium, then Theorem 8.7.1 implies that $s^2\in E(1)$ and in this case $R^2(\pi)\geqq 2$. Hence, indeed $v(2)=3/2$ and (R, M) is the worst 2-period subgame perfect equilibrium. Note that this equilibrium has the stick and carrot character mentioned above and that the inequalities (8.7.3) and (8.7.4) are strict.

In the example above we have seen that a player need not be at a one-shot best reply at every stage along the equilibrium path that is worst for him. The example also shows that an endpath of the worst subgame perfect equilibrium path need not be the worst path possible. This latter property can be contrasted with the fact that any endpath of the best subgame perfect equilibrium path also has to be the most attractive one possible. (This latter property follows easily from Corollary 8.7.2.). Perhaps the most important lesson that can be learnt from the game of Fig. 8.7.1 is that the threat to play the worst one-shot equilibrium at every stage need not be the most effective threat possible. In many situations, this implies that not all modes of beneficial cooperation can be sustained by means of such threats. The game of Fig. 8.7.2 provides an example.

The game Γ of Fig. 8.7.2 results from that of Fig. 8.7.1 by adding for each player i a strictly dominated action A_i. Therefore, the equilibrium payoffs of Γ are 6, 3 and 2. Consider the 3-fold repetition $\Gamma(3)$ of Γ. By threatening to switch to the worst one-period Nash equilibrium one cannot sustain the path (A, L, L): this path yields a payoff of 22 and if the threat is executed after a deviation to L_i in the first round, player i has a payoff of 23. However, it is possible to sustain this path by using the more sophisticated threat discussed in the game of Fig. 8.7.1: If a

	L_2	M_2	R_2
L_1	6, 6	0, 0	0, 0
M_1	0, 0	3, 3	1, 1
R_1	0, 0	1, 1	0, 0

Fig. 8.7.1. The worst subgame perfect equilibrium of the 2-period game results in a smaller average payoff than the worst equilibrium of the one-shot game

	L_2	M_2	R_2	A_2
L_1	6, 6	0, 0	0, 0	19, -19
M_1	0, 0	3, 3	1, 1	19, -19
R_1	0, 0	1, 1	0, 0	19, -19
A_1	-19, 19	-19, 19	-19, 19	10, 10

Fig. 8.7.2. The threat to revert to the worst one-period equilibrium is not sufficient to sustain all modes of implicit collusion

	L_2	R_2
L_1	2, 2	1, 2
R_1	-1, 0	0, 0

Fig. 8.7.3. A game for which $v_2(1) > v_2(3) > v_2(2)$, hence, the sequence $\{v_i(t)\}$ need not be monotonic

player is punished optimally after a deviation in the first round, his total payoff is at most $19+3=22$, hence, no player has an incentive to deviate.

The reader might have the intuition that games of larger duration admit a greater flexibility in punishment and, consequently, that such games admit more severe punishments. Hence, one might think that the sequence $\{v_i(t)\}_t$ is monotonically decreasing. The game from Fig. 8.7.3 shows that this intuition is incorrect.

In the game of Fig. 8.7.3 we have $EP(1) = \{(\alpha, 2); \alpha \in [1,2]\}$, hence $v_1(1) = 1$ and $v_2(1) = 2$. Since $v_1(1) = v_1$, we have $v_1(t) = 1$ for all t. We claim $v_2(2) = 1$. Namely, $v_2(2) \geqq 1$ since player 2 has a payoff of 2 in the last round and $v_2(2) \leqq 1$ since (R, L) is a subgame perfect equilibrium path that yields player 2 an average payoff of 1. Next, consider $\Gamma(3)$ and let $\pi = (s^3, s^2, s^1)$ with $s^t = (s_1^t, s_2^t)$ and $s_i^t = (p_i^t, 1-p_i^t)$ be a subgame perfect equilibrium path in $\Gamma(3)$ that is worst for player 2. Then $s^1 = (L_1, L_2)$ as a consequence of the fact that player 1 is also hurt when he punishes player 2 so that he has to be rewarded for having done so. Theorem 8.7.1 shows that this path π can be found by solving the nonlinear programming problem

$$\text{Minimize} \quad \frac{1}{3}\{R_2(s^3)+R_2(s^2)+2\}\,,$$

$$\text{subject to} \quad R_1(s^2)+2 \geqq \max_{s_1} R_1(s^2/s_1)+1\,,$$

$$R_1(s^3)+R_1(s^2)+2 \geqq \max_{s_1} R_1(s^3/s_1)+2\,,$$

$$s^2,\ s^3 \in S\,.$$

Since player 2 is indifferent between L_2 and R_2, the constraints for this player are redundant. This problem reduces to

$$\text{Minimize} \quad \frac{2}{3}\{p_1^3+p_1^2+1\},$$

$$\text{subject to} \quad p_1^2(1+2p_2^2) \geqq p_2^2,$$

$$p_1^3(1+2p_2^3)+p_1^2(1+2p_2^2) \geqq 1+2p_2^3+p_2^2,$$

$$p_i^t \in [0,\ 1],$$

and it is easily seen that the optimal solution is given by $p_1^3=0$, $p_2^3=0$, $p_2^2=1$ and $p_1^2=2/3$. Hence, $v_2(3)$, the value of this program, is 10/9. Therefore $v_2(1)>v_2(3)>v_2(2)$.

Although the sequence $\{v_i(t)\}_t$ need not be monotonic, this sequence definitely is well behaved, as the following theorem demonstrates.

Theorem 8.7.3. *For $i \in N$, let $v_i^* := \inf_t v_i(t)$. Then $v_i^* = \lim_{t\to\infty} v_i(t)$.*

Proof. v_i^* is well-defined since $v_i \leqq v_i(t) \leqq v_i(1)$ for all t. Let $\varepsilon>0$ and let T be such that $v_i(T)<v_i^*+\varepsilon$. For $t \geqq T$ write $t=kT+l$ with $l<T$. From (8.7.3) it follows that

$$tv_i(t) \leqq kTv_i(T)+lv_i(1)\,,$$

hence,

$$v_i(t) \leqq v_i(T)+\frac{T}{t}v_i(1)\,,$$

and $v_i(t)<v_i^*+\varepsilon$ for all t sufficiently large. □

Next, let us investigate the behavior of $PEP(T)$ as T tends to infinity. First of all, it immediately follows from Theorem 8.7.1 that, if Γ has a unique equilibrium payoff, then in any subgame perfect equilibrium of $\Gamma(T)$, an equilibrium of Γ has to be played at every stage. Hence

Corollary 8.7.4. *If Γ is an n-person game for which $EP(1)=\{(v_1(1),\ldots,v_n(1))\}$ then $PEP(T)=EP(1)$ for all T.*

Note the difference between the Corollaries 8.6.3 and 8.7.4: the latter requires only that Γ has a unique equilibrium payoff vector, this need not be the vector of minmax values. Hence, for the game of Fig. 8.4.1 as well as for the game of Fig. 8.6.1 we have that $PEP(T) = EP(1) = \{(1, 1)\}$. Also note that, if the basic game Γ has a unique equilibrium, then $\Gamma(T)$ has a unique subgame perfect equilibrium for any T and recall that this need not be true for Nash equilibria, not even in the case in which the unique equilibrium of Γ results in minmax values.

Next, consider the case opposite to the one treated in Corollary 8.7.4, viz. each player has at least 2 equilibrium payoffs in Γ. In this case, by switching to a better or a worse equilibrium, it is possible to reward or to punish a player. Theorem 8.7.5 shows that in this case, implicit collusion can be sustained by means of subgame perfect equilibria. More specifically, when the game is repeated sufficiently many times, any payoff vector above v^* (defined as in Theorem 8.7.3) can be sustained by a subgame perfect equilibrium.

Theorem 8.7.5. *If Γ is an n-person game for which* $\max EP_i(1) > \min EP_i(1)$ *for all $i \in N$, then, as T tends to infinity, $PEP(T)$ converges to F^**

$$F^* := \{x \in F;\ x_i \geqq v_i^* \quad \text{for all } i\}\,. \tag{8.7.5}$$

Proof. Let Γ satisfy the condition. Theorem 8.7.3 and Eq. (8.2.23) imply that $PEP(T) \subset F^*$ for all T. If suffices to show that any element in the relative interior of F^* (i.e. $x_i > v_i^*$ for all i) can be approximated by subgame perfect equilibrium payoffs of games of sufficiently long duration. Hence, let such x be given and choose $\varepsilon > 0$ such that

$$\varepsilon < x_i - v_i^* \quad \text{all } i \in N, \text{ and}$$

$$\varepsilon < \max EP_i(1) - \min EP_i(1) \quad \text{all } i \in N\,.$$

For convenience, assume $x = R(s)$ for some $s \in S$. There exists some T and a path $\pi \in PE\Pi(T)$ with average payoff vector y such that

$$M_i - x_i \leqq T(y_i - v_i^* - \varepsilon) \quad \text{all } i \in N\,.$$

Furthermore, T can be chosen such that

$$v_i(t) \leqq v_i^* + \varepsilon \quad \text{for all } t \geqq T,\ i \in N\,.$$

For $t \in \mathbb{N}$, consider the path $(s^{(t)}, \pi)$ in $\Gamma(T+t)$ in which first s is played t times, whereafter the players switch to π. Since for $t \geqq 1$ and $i \in N$

$$\begin{aligned}\max_{s_i'} R_i(s/s_i') + (t-1+T)\,v_i(t-1+T) &\leqq M_i + (t-1+T)\,(v_i^* + \varepsilon)\\ &\leqq M_i + (t-1)\,x_i + T(v_i^* + \varepsilon)\\ &\leqq t x_i + T y_i\,,\end{aligned}$$

we have that any such path is a subgame perfect equilibrium path. As t tends to infinity, the average payoff along this path approaches x. □

In view of the previous theorem, to determine $\lim_{T \to \infty} PEP(T)$, it is sufficient to compute v_i^*. The game of Fig. 8.5.1 shows that, without imposing additional conditions on F, one cannot guarantee that $v_i^* = v_i$: by using backwards induction, it is easily seen that for this game $v_i(T)=v_i(1)=1/4$ for all T, while $v_i=0$. Hence, some kind of dimensionality condition on F is needed: it should be possible to reward player j for having punished player i without simultaneously rewarding player i. Below we will show that, if Γ is a game for which each player has at least two equilibrium payoffs and for which F is full dimensional, then indeed $v_i^*=v_i$ and $PEP(T)$ converges to F as T tends to infinity. We first introduce some notation.

Let Γ be an n-person game. For $x\in F$ and $\gamma\geqq 0$ define $x^i(\gamma)$ as in (8.5.14) and let $\gamma(x)$ be defined by

$$\gamma(x):=\max\{\gamma\colon x^i(\gamma)\in F \text{ for all } i\in N\}\,. \tag{8.7.6}$$

γ is a continuous function on F and $\gamma(x)>0$ for $x\in\operatorname{int}(F)$. Let $v=(v_1,\ldots,v_n)$ be the vector of minmax values of Γ and, for $x\in F$, let $H(x)$ be the convex combination of x and v defined by

$$H(x):=\frac{2Mx+\gamma(x)v}{2M+\gamma(x)}\,, \tag{8.7.7}$$

where M is as in (8.2.7). Then $H\colon F\to R^n$ is a continuous function and $H_i(x)<x_i$ for all $i\in N$, if $x\in\operatorname{int} F$. We first show that, if $x\in F^*$, then $v_i^*\leqq H_i(x)\leqq x_i$, hence, the function H can be used to construct more severe punishments. The 'Perfect Folk Theorem' for finitely repeated games (Theorem 8.7.7) is an easy consequence of this result.

Lemma 8.7.6. *If Γ is an n-person game for which* $\max EP_i(1)>\min EP_i(1)$ *for all $i\in N$, then $v_i^*\leqq H_i(x)$ for all $x\in F^*$ and $i\in N$.*

Proof. Let $x\in F^*$. If $\gamma(x)=0$, then $H_i(x)=x_i$ and the result follows from (8.7.5). Hence, assume $\gamma(x)>0$. Write $\gamma=\gamma(x)$ and $x^i(\gamma)=x^i(\gamma(x))$ $(i\in N)$. The previous theorem shows that, for all $\varepsilon>0$, one can find some T^* such that $B_\varepsilon(x^i(\gamma))\cap PEP(T)\neq\emptyset$ for all $i\in N$ and all $T\geqq T^*$.

For convenience, assume $T^*=1$ (which amounts to considering $\Gamma(T^*)$ as the basic game that is repeated) and that there exist equilibria s^i of Γ such that $R(s^i)=x^i(\gamma)$ for $i\in N$. (The general case can be proved by the same methods, using the continuity of H.)

Let $i\in N$ and for T, $t\in\mathbb{N}$ consider the path $\pi(T,t)$ in $\Gamma(T+t)$ in which first m^i is played for t periods, whereafter s^i is played T times. No player has an incentive to deviate from this path during one of the last T stages. Suppose that, if player i deviates during one of the first t stages, the players just continue with this path, but that, if some player j $(j\neq i)$ deviates, the players switch to the one-shot equilibrium s^j for the remainder of the game. In this case player i has no incentive to deviate, since he is at a one-shot best reply during any of the first t stages. Player j will have no incentive to deviate provided that

$$\max_{s_j} R_j(s^i/s_j)+(\tau-1+T)x_j\leqq \tau v_j^i+Tx_j^i(\gamma)\qquad 1\leqq\tau\leqq t\,.$$

This condition is satisfied when

$$M_j-x_j+\tau(x_j-v_j^i)\leqq T\gamma\qquad 1\leqq\tau\leqq t$$

and this inequality certainly holds if

$$(1+t)2M \leqq T\gamma. \tag{8.7.8}$$

For $T \in \mathbb{N}$, let $t(T)$ be the largest t for which (8.7.8) is satisfied and let $\pi(T)$ be the path $\pi(T, t(T))$. The path $\pi(T)$ is a subgame perfect equilibrium path and player i's average payoff along this path is equal to

$$\frac{t(T)v_i + Tx_i}{t(T)+T}.$$

As T tends to infinity, this average payoff converges to $H_i(x)$, hence, $v_i^* \leqq H_i(x)$. □

If Γ is a game for which the payoff set F is full dimensional, then there exists $x \in F$ with $\gamma(x) > 0$. It immediately follows from the lemma that $v_i^* = v_i$ in this case, hence $PEP(T)$ converges to F:

Theorem 8.7.7 (Benoit and Krishna [1985]). *If Γ is an n-person game for which* $\max EP_i(1) > \min EP_i(1)$ *for all $i \in N$ and for which* $\dim F = n$, *then $PEP(T)$ converges to F as T tends to infinity.*

Proof. It suffices to show that $v_i^* = v_i$ for $i \in N$. Let $i \in N$ and let $\bar{v}_i$ be such that

$$v_i = \min\{x_i;\ x \in F\} < \bar{v}_i < \max\{x_i;\ x \in F\}.$$

Then there exists some $x \in \operatorname{int}F$ such that $x_i = \bar{v}_i$. We have $H_i(x) < x_i$ and the previous lemma shows that $v_i^* \leqq H_i(x) < x_i = \bar{v}_i$. Hence, $v_i^* = v_i$. □

To conclude this section, 2-person games are investigated in somewhat greater detail. We first show that for such games the dimensionality condition is not needed.

Theorem 8.7.8 (Benoit and Krishna [1985]). *If Γ is a 2-person game for which* $\max EP_i(1) > \min EP_i(1)$ *for $i = 1, 2$, then $PEP(T)$ converges to F as T tends to infinity.*

Proof. If Γ is a game that satisfies this condition, but that does not satisfy the condition of Theorem 8.7.7, then F is a line segment with positive slope. Let Γ be such a game and, without loss of generality, assume $R_1(s) = R_2(s)$ for all $s \in S$ and $v_2 \leqq v_1 = 0$. Let $x = \max EP_1(1)$ and $y = \min EP_1(1)$. If $y = 0$, the result follows from Theorem 8.7.5, so assume $y > 0$. Let $z = R_1(m_1^2, m_2^1)$ be the payoff when the players minmax each other: Then $z < 0$ since otherwise (m_1^2, m_2^1) would be an equilibrium with value 0. Choose T_1 such that

$$M + T_1 z \leqq 0, \tag{8.7.9}$$

i.e. each player prefers being minmaxed $T_1 + 1$ periods above getting his maximal payoff once and having to play (m_1^2, m_2^1) for T_1 times thereafter. For T_1 as in

(8.7.9) choose T_2 such that

$$T_2 \geqq \frac{T_1(2y-z)+M}{x-y}. \qquad (8.7.10)$$

For T_1 and T_2 fixed and $t \in \mathbb{N}$ consider the path $\pi(t)$ in which first the players minmax each other (i.e. play (m_1^2, m_2^1)) for T_1 times, then m^1 is played for t times and finally the equilibrium with payoff x is played T_2 times. Condition (8.7.10) guarantees that the threat to switch to the worst one-period Nash equilibrium is sufficient to sustain $\pi(t)$ as a subgame perfect equilibrium path for all $t \leqq T_1$. We claim that $\pi(t)$ is a subgame perfect equilibrium path for all t. The proof is by induction with respect to t. Assume the claim has already been proved for $t \leqq T$ (with $T \geqq T_1$) and consider $\pi(T+1)$. The induction hypothesis implies that deviating from this path at one of the last $T+T_2$ periods is not profitable. Assume that, if a deviation occurs at time t with $t \geqq T+T_2+1$, the players switch to the subgame perfect equilibrium path $\pi(t-1-T_1)$. Hence, a deviation will be punished by having to play (m_1^2, m_2^1) for another T_1 periods. Condition (8.7.9) guarantees that deviating is not profitable when $t=T+T_2+1$: in the period in which one deviates the payoff is at most M, but one will receive z for T_1 periods. It is also not profitable to deviate if $t>T+T_2+1$ since one can get at most 0 by deviating at such t and since one will have to go through the most severe phase of the punishment again. This establishes the claim. Since the average payoff along $\pi(t)$ approaches $0(=v_1)$ as t tends to infinity, we have $v_1^*=v_1$ and the result follows from Theorem 8.7.5. □

Finally, consider a 2-person game in which only player 2 has more than 1 equilibrium value. First suppose that there exists $x \in F$ such that $x_1 > v_1$. This implies that there exists $s \in S$ such that s_1 is a best reply against s_2 and $R_1(s) \neq v_1(1)$. Let $\bar{s}$ be the most attractive equilibrium for player 2 in Γ. Then there exists some T such that $(s, \bar{s}^{(T)})$ is an element of $PE\Pi(T+1)$: player 1 cannot profitable deviate since he is always at a one-shot best reply, and player 2 can be punished if he deviates in period $T+1$. In this case, $\Gamma(T+1)$ satisfies the condition of Theorem 8.7.8 and we have

Corollary 8.7.9. *If Γ is a 2-person game with* $\max EP_i(1) > \min EP_j(1)$ *for some i and $x_i > v_i$ for all i, then* PEP(T) *converges to F as T tends to infinity.*

Next, let Γ be a game for which player 2 has at least 2 equilibrium values and for which $x_1 = v_1$ for all $x \in F$, hence $v_1 = M_1$. In this case, $PEP(T) \subset \tilde{F}_2$ where $\tilde{F}_2$ is as in (8.3.9). By replacing, in the proof of Theorem 8.7.8, (m_1^2, m_2^1) by an action combination $\tilde{m}$ for which $R_1(\tilde{m}) = v_1$ and $R_2(\tilde{m}) \leqq \tilde{v}_2$, one can show that

Corollary 8.7.10. *If Γ is a 2-person game with $\max EP_2(1) > \min EP_2(1)$ and $x_1 = v_1$ for all $x \in F$, then $PEP(T)$ converges to $\tilde{F}_2$ as T tends to infinity.*

Hence, for 2-person games we have a complete description of $\lim_{T \to \infty} PEP(T)$. Note that the description closely parallels that of $\lim_{\delta \uparrow 1} PEP(\delta)$. The only difference is that $PEP(T) = EP(1)$ whenever $EP(1)$ in a singleton and this need not be true for $PEP(\delta)$.

8.8 Renegotiation-Proof Equilibria

In this chapter it has been shown that repetition *can* lead to cooperation, i.e. mutually beneficial outcomes may be reached when a game is repeated sufficiently often. However, in general, by repeating a game not only better outcomes can be obtained, one can also get equilibria that yield each player less than any one-shot equilibrium does. Hence, we have not succeeded in showing that repetition *must* lead to cooperation, and this is the result one ideally would want to obtain. Of course, it is impossible to obtain this stronger result with the current approach because of the fact that playing a one-shot equilibrium at every stage always generates an equilibrium of the supergame. In this section, we briefly investigate whether a refinement of the subgame perfectness concept, obtained by requiring some kind of local optimality, does force the players to cooperate.[11] This refinement is introduced to eliminate a certain drawback of the subgame perfectness concept. This drawback is discussed first.

Consider the game of Fig. 8.8.1. $L = (L_1, L_2)$ is the unique Nash equilibrium of this game, and this equilibrium is Pareto optimal. When this game is repeated finitely many times, there is a unique subgame perfect equilibrium and this requires playing L at every stage. However, in the infinitely repeated game, new subgame perfect equilibria emerge if the discount factor δ is sufficiently high. The path π in which the players first play T times R and then switch to playing L forever is subgame perfect in $\Gamma(\delta)$ as long as $\delta^T \geqq 1/2$ (cf. Theorem 8.5.12). However, this is not a very sensible equilibrium : by carrying out the threat to restart π, a player also hurts himself, so why would he execute it? We consider this threat not to be credible and, hence, the equilibrium not to be viable: At every stage in the game, no matter what has happened before, the players will realize that only one attractive subgame perfect equilibrium is available (viz. always play L) and, in our opinion, playing this equilibrium is the only sensible thing to do.

Above we have argued that threats that hurt the threatener himself are not credible and that subgame perfect equilibria supported by such threats are not viable. By reviewing the proofs of the previous sections it is seen that many of the subgame perfect equilibria described in this chapter do use such self-lacerating threats and, hence, are not credible in this sense. More precisely, many subgame perfect equilibria, although possibly being globally optimal, need not be locally optimal, i.e. in certain subgames they may prescribe to continue with a specific equilibrium although a uniformly better equilibrium is available.

	L_2	R_2
L_1	1, 1	0, 0
R_1	0, 0	-1, -1

Fig. 8.8.1. The unique equilibrium of this game is Pareto optimal. The infinitely repeated game admits non Pareto optimal equilibria, provided that the discount rate is sufficiently small

Imposing local optimality excludes the 'bad' equilibria in Fig. 8.8.1., however, in many games, excluding 'bad' equilibria will also exclude some 'good' ones. An example is the game of Fig. 8.7.2. We have shown that (A, L, L) is a subgame perfect equilibrium path in the 3-fold repetition of this game. (In fact, it is the best symmetric one). This path is supported by the threat to switch to the equilibrium (R, M) after a deviation in round 1. Now, note that (R, M) is not the best equilibrium of $\Gamma(2)$: the equilibrium (L, L) is better for both players. Therefore, we find it hard to accept that the players would switch to (R, M) after a deviation: the dupe will make the best of a bad bargain and the deviator will have no difficulty in convincing him that they should continue with (L, L) rather than with (R, M). Hence, in our view, the subgame perfect equilibrium (A, L, L) is not viable.

The general principle invoked in the above examples is that threats involving subgame perfect equilibria that are Pareto inferior are not credible. Following Farrell [1983] we will say that an equilibrium is *renegotiation-proof* if it does not involve such threats. Before we define this concept formally, it is convenient to derive an alternative characterization of subgame perfect equilibria in finitely repeated games.

Let Γ be an n-person game and let V be a compact set of payoff vectors, $V \subset \mathbb{R}^n$. For $t \in \mathbb{N}$, write $Q^t(V)$ for the set of pairs (x, s) satisfying (8.8.1) and (8.8.2)

$$x = \frac{1}{t}R(s) + \frac{t-1}{t}v \qquad \text{for some } v \in V, \tag{8.8.1}$$

$$x_i \geqq \frac{1}{t}\max_{s'_i \in S_i} R_i(s \backslash s'_i) + \frac{t-1}{t}\min_{v \in V} v_i \qquad \text{for all } i \in N. \tag{8.8.2}$$

Furthermore, let $P^t(V)$ (resp. $S^t(V)$) be the projection of $Q^t(V)$ on the set of payoff vectors x (resp. the set of strategy combinations s). Then it is clear that the subgame perfect equilibrium payoffs of $\Gamma(T)$ can be found by the recursive scheme

$$PEP(t) := P^t(PEP(t-1)) \quad 1 \leqq t \leqq T, \tag{8.8.3}$$

($PEP(0)$ may be defined arbitrarily), and that σ is a subgame perfect equilibrium of $\Gamma(T)$ if and only if for every possible history h at time t, the action combination $s^t = \sigma(h)$ that is prescribed by σ satisfies

$$s^t \in S^t(PEP(t-1)). \tag{8.8.4}$$

Next, let us return to renegotiation-proof equilibria. The basic idea is that a Pareto efficient equilibrium should be played in every subgame. Hence, a set V qualifies as a set of renegotiation-proof equilibrium payoffs only if no element in V Pareto dominates another element of V, i.e. if V coincides with its Pareto boundary V^+

$$V^+ := \{x \in V;\ \text{if } y \in V, \text{ then } y_i \leqq x_i \text{ for some } i\}. \tag{8.8.5}$$

Furthermore, given that continuation payoffs at time $t-1$ belong to V, we have that the strategy combination s is admissible in period t if and only if $s \in S^t(V)^+$, where

$$S^t(V)^+ := \{s \in S;\ (x, s) \in Q^t(V) \text{ for some } x \in P^t(V)^+\}. \tag{8.8.6}$$

Formally then, the set of renegotiation-proof equilibrium payoffs is defined by the recursive scheme

$$RPEP(t) := P^t(RPEP(t-1))^+ \quad (1 \leqq t \leqq T) \tag{8.8.7}$$

(cf. (8.8.3)) and σ is a *renegotiation-proof equilibrium* (briefly *RPE*) if and only if for every possible history h at time t, the action combination $s^t = \sigma(h)$ satisfies

$$s^t \in S^t(RPEP(t-1))^+ \tag{8.8.8}$$

(cf. (8.8.4)).

The game of Fig. 8.7.2. has $L = (L_1, L_2)$ as its unique Pareto optimal equilibrium. From the definition of renegotiation-proofness it easily follows that, in this case, $\Gamma(T)$ has a unique RPE for every T: the players should play L in every round. Hence, we have the following analogue to Corollary 8.7.4.

Theorem 8.8.1. *If Γ has a unique Pareto optimal equilibrium, then $\Gamma(T)$ has a unique renegotiation-proof equilibrium for every T.*

Next, we illustrate by means of an example that the requirement of renegotiation-proofness can force the players to cooperate. More precisely, playing a one-shot Pareto optimal equilibrium in every round need not yield a renegotiation-proof equilibrium in the repeated game.[12]

In the game of Fig. 8.8.2, the action M_i is strictly dominated. The game has 3 equilibria, L, R and a mixed one, A, in which player i chooses $A_i = 1/2L_i + 1/2R_i$. The respective equilibrium payoffs are (10, 4), (4, 10) and (5, 5), hence, all 3 equilibria are Pareto efficient. Consider the game $\Gamma(2)$. The path (M, A) is a subgame perfect equilibrium path which results in a payoff of 25 for each player and it is easily checked that it is renegotiation-proof (it suffices to show that there does not exist an equilibrium that yields both players more than 25). This, however, implies that playing an equilibrium of Γ twice is not renegotiation-proof in $\Gamma(2)$ since any such path results in a payoff less than 25 for both players. (Also

	L_2	M_2	R_2
L_1	10, 4	21, 0	0, 0
M_1	0, 21	20, 20	0, 21
R_1	6, 6	21, 0	4, 10

Fig. 8.8.2. Playing a Pareto optimal equilibrium of Γ twice is not renegotiation proof in $\Gamma(2)$

note that, in any renegotiation-proof path of $\Gamma(2)$, one must play A in the last stage; If L would be played, then player 2 must play a one-shot best response in the first stage, but one easily verifies that in this case both players always get less than 25.)

The above discussion shows that one can hope that the renegotiation-proofness concept might force the players to cooperate in games of sufficiently long duration, provided that the original game at least 2 Pareto optimal equilibria. We hope to deal with this issue in more detail in a forthcoming paper.[13]

In the remainder of this section, we consider infinitely repeated games with discounting and we show that, in prisoners' dilemma, the requirement of renegotiation-proofness indeed forces the players to cooperate if the discount rate is sufficiently small.

Assume a game Γ and a discount factor δ are given. The problem is how to translate (8.8.7) – (8.8.8) to the infinite horizon context. Note that associated with every strategy combination σ in $\Gamma(\delta)$ there is a set of payoff vectors, viz. one normalized discounted continuation payoff $R^{\delta}(\sigma|h)$ for every history h. It will be more convenient to define directly sets of renegotiation-proof equilibrium payoffs rather than renegotiation-proof equilibria. Let V be a compact set of payoffs. For V to be renegotiation-proof, it is definitely necessary that no payoff in V dominates another payoff in V, hence $V = V^{+}$. Next, let P^{δ} be the map defined in Sect. 8.5 (immediately below Lemma 8.5.1). From the arguments in Sect. 8.5, it follows that also $P^{\delta}(V)^{+}$ should be considered renegotiation-proof, when V satisfies this condition. As in Sect. 8.5, we are interested in maximal attainable sets, hence, if V is renegotiation-proof, then V should satisfy

$$V = P^{\delta}(V)^{+}. \tag{8.8.9}$$

(Note that (8.8.9) implies that $V = V^{+}$.) Hence, V should be a fixed point of a certain mapping. However, not all fixed points should necessarily be considered as renegotiation-proof. Namely, if V and W are fixed points and any payoff in V is dominated by a payoff in W, then V is not eligible as a solution since it is not viable when the renegotiation-proofness idea is applied a second time. Hence, V should satisfy

$$\text{if } W \neq V \text{ and } W = P^{\delta}(W)^{+}, \text{ then it is not true that for any } v \in V \text{ there exists } w \in W \text{ with } v \leqq w. \tag{8.8.10}$$

Formally, let us call a set of payoffs V in $\Gamma(\delta)$ *renegotiation-proof* if it satisfies (8.8.9) – (8.8.10). Corollary 8.5.7 shows that, if V is renegotiation-proof, then $V \subset PEP(\delta)$. We now come to the main result of this section.

Theorem 8.8.2. *If Γ is the prisoners' dilemma game from* Fig. 8.4.1, *then the Pareto efficient frontier of F is the unique renegotiation-proof set of payoffs of $\Gamma(\delta)$ for any $\delta \geqq 1/4$.*

Proof. It suffices to show that F^{+}, the Pareto efficient frontier of F, is renegotiation-proof for any $\delta \geqq 1/4$. Let $\delta \geqq 1/4$ and let $x = (x_1, x_2)$ be a convex combination of (5, 0) and (4, 4) with $x_2 \geqq 1$. If $1 \leqq x_2 \leqq 4\delta$, then x results when

(R_1, L_2) is played at $t=0$ and when at $t=1$ the players switch to $x' \in F^+$ with $\delta x'_2 = x_2$. Furthermore, the threat to switch to the worst element of F^+ at $t=1$ is sufficient to deter deviations at $t=0$. If $4\delta \leqq x_2 \leqq 4$, then x results when, at $t=0$, player 2 chooses L_2 and player 1 chooses $pL_1 + (1-p)R_1$ where

$$p = \frac{x_2 - 4\delta}{4(1-\delta)},$$

and, when the players switch to playing (L_1, L_2) forever at $t=1$. Again, the threat to revert to the deviator's worst point in F^+ is a sufficient deterrent. Hence $(P^\delta(F^+))^+$ contains all convex combinations of $(5, 0)$ and $(4, 4)$ with $x_2 \geqq 1$. By symmetry, therefore, $F^+ \subset P^\delta(F^+)^+$ hence $F^+ = P^\delta(F^+)^+$. □

This theorem shows that the analogue of Theorem 8.8.1 is not valid for infinitely repeated games with discounting. Furthermore, it shows that the concept of renegotiation-proofness might be fruitful in investigating cooperation in repeated games. However, a careful analysis of for which games the set of Pareto admissible payoffs of $\Gamma(\delta)$ converges to the efficient frontier of F, must be postponed to a forthcoming paper.[14]

Notes

1. The theory on repeated games with complete information is also surveyed in Sorin [1988]. At the advanced level, a comprehensive overview is provided in Mertens, Sorin and Zamir [1990].
2. For the unraveling to occur it is not sufficient that both players *know* what the last stage of the game is, there has to be a finite upper bound on the game that is *common knowledge*, see Neyman [1987].
3. The equilibria constructed in Friedman [1971] are not subgame perfect. In that paper, Friedman considers equilibria in reaction functions, i. e. at each time t each player conditions his actions only on the opponents' actions at time $t-1$. In repeated duopoly games with discounting, such equilibria can be subgame perfect only if they are trivial, i. e. if they prescribe to play the static Cournot equilibrium in every stage, see Robson [1986] and Stanford [1986a]. In Stanford [1986b] it is shown that in undiscounted supergames such strategies can give rise to nontrivial SPEa. Friedman and Samuelson [1988] obtained SPEa in continuous reaction functions that specify the action of a player in each period as a function of the actions of *all* players in the previous period (i. e. a player also looks at what he himself did in the previous period).
4. Güth, Leininger and Stephan [1988] criticize folk theorems. They argue for the imposition of subgame consistency (Harsanyi and Selten [1988]), i. e. the requirement that two isomorphic games should be played in the same way. Notice that all subgames of the discounted supergame $\Gamma(\delta)$ are isomorphic, hence, subgame consistency implies stationarity. For a critique on stationarity see Rubinstein [1988]. Also see Mertens [1989b]. For existence and

robustness results concerning Markov equilibria, see Maskin and Tirole [1989].

5. See Kaneko [1982]. If there are infinitely many players and there is no discounting this result only holds under additional regularity assumptions, see Masso and Rosenthal [1989].
6. Some recent papers dealing with the issue of how much information the per period outcomes must convey to obtain a folk theorem are Fudenberg and Levine [1989b, c], Fudenberg, Levine and Maskin [1989] and Abreu, Milgrom and Pearce [1990].
7. Fershtman, Judd and Kalai [1987] show that cooperation can result in static games if players can delegate their decisions to agents (and if contracts between players and agents are binding and observable).
8. For a bchavioral theory of the finitely repeated Cournot duopoly game, see Selten, Mitzkewitz and Ulrich [1988] and Mitzkewitz [1988] as well as the references therein.
9. Neyman [1987] shows that cooperation may result in the finitely repeated PD if the length of the game is not common knowledge. Neyman's model also falls under the heading of incomplete information.
10. In Stahl [1989] it has been argued that the prisoners' dilemma from Fig. 8.4.1 is a knife-edge case since, irrespective of what the other player does, defecting yields one unit of utility more than cooperating. Stahl describes the set of subgame perfect *correlated* equilibrium payoffs for each discount factor for general prisoners' dilemma games. He shows that the set of payoffs is monotonically nondecreasing and upper hemicontinuous in δ.
11. Other models that force the players to cooperate in some classes of games are described in Aumann and Sorin [1989], Matsui [1989], Fudenberg and Maskin [1990], Binmore and Samuelson [1990] and Robson [1989]. In Aumann/Sorin the driving force is that each player thinks that his opponent may be an automaton playing a perfect recall strategy. In Matsui cooperation is caused by the fact that each player with a small probability may find out the opponent's supergame strategy. The other papers invoke arguments of evolutionary stability (see Chap. 9).
12. For another example, see Bernheim and Whinston [1987].
13. The problem has been solved in a satisfactory way in Benoit and Krishna [1989]. In that paper it is shown that, if $R=\lim_{t\to\infty} RPEP(T)$ exists, then either R is a singleton or R is a subset of the weak Pareto frontier of the feasible set of payoffs. ($RPEP(t)$ is the set of average payoffs obtainable by renegotiation-proof equilibria in $\Gamma(t)$ as defined in (8.8.7).) Benoit and Krishna also give examples to illustrate that, if R is not a singleton it need not be a subset of the strong Pareto frontier, and that even if Γ has multiple equilibria, R may be a single inefficient point.
14. The issue of renegotiation in repeated games has recently received considerable attention. Papers that employ a definition of renegotiation-proofness that is closely related to the one discussed in this section are Bernheim and Ray [1989], Farrell and Maskin [1989], Evans and Maskin [1989], Asheim [1990], Ray [1989] and Van Damme [1989]. Most of these papers make a distinction between internal consistency (i. e. there should be

no Pareto dominance relations between two payoff vectors in the same renegotiation proof set (cf. (8.8.9)) and external consistency (i.e. a renegotiation proof set should not be dominated by another one, cf. (8.8.10)). Farrell/Maskin define a set V to be weakly renegotiation-proof if (i) $V=V^+$ and (ii) $V \subset P^\delta(V)$. (The notation is as in the main text. Note that this condition is weaker than Eq. (8.8.9), instead of equality, set inclusion is required.) They characterize the payoffs that belong to some weakly renegotiation-proof set and then they continue by proposing several concepts of external stability. The necessity for such external conditions was already established in Van Damme [1989]: If $\delta \geq \frac{1}{4}$, then in the prisoners' dilemma from Fig. 8.4.1 any payoff belongs to a weakly renegotiation-proof set. Evans/Maskin show that for generic 2-person games there exists a weakly renegotiation-proof payoff that is Pareto-efficient provided that the discount factor δ is near enough to 1. Bernheim and Ray [1989] adopt the same definition of weak renegotiation-proofness under the name internal consistency. Also these authors propose various notions of external consistency. Asheim [1990] expounds on the theory of social situations as developed in Greenberg [1990] (also see Greenberg [1989a, b]) and defines external stability by using an idea similar to that of Von Neumann and Morgenstern's stable sets: It is required that a 'nonviable' equilibrium is "defeated" by a 'viable' one. Asheim also does not insist on the assumption of stationarity that is implicit in the definition of weak renegotiation proofness. Ray [1989] argues that the notion of weak renegotiation-proofness does not fully capture the idea of internal stability and he argues in favor of condition (8.8.9) instead, i.e. Ray insists on the equality of the sets rather than the inclusion of V in $P^\delta(V)$. Ray shows that the limits (as δ tends to 1) of sets satisfying (8.8.9) are either singletons or are contained in the weak efficient frontier of the feasible set (cf. the similar result obtained in Benoit and Krishna [1989] for finitely repeated games). The relation between finitely and infinitely repeated games is also investigated in Blume [1987]. An entirely different concept of renegotiation-proofness has been proposed in Pearce [1987]. The intuition underlying this concept is also discussed in Abreu and Pearce [1989]. Bergin and MacLeod [1989] present a unifying framework in which the different concepts can be discussed.

9 Evolutionary Game Theory

Game Theory has been developed as a theory of rational behavior in interpersonal conflict situations, with economics and the other social sciences being the intended fields of application. Since the theory is based on an idealized picture of human rationality, it is by no means obvious that it can be applied to situations in which the players cannot be attributed any intellectual capabilities. However, in their seminal paper 'The logic of animal conflict', Maynard Smith and Price showed that animal contests can be modeled as games and that game theory can be applied successfully in biology. The objective of this chapter is to review some of the developments in this biological branch of Game Theory and to point out the distinctions and similarities with the classical branch (also see Parker and Hammerstein [1985]). The main emphasis will be on the mathematics involved, lack of space prevents an extensive discussion of the underlying biological assumptions as well as an analysis of specific examples. For these, the reader is referred to the very stimulating book 'Evolution and the Theory of Games' by John Maynard Smith.

In Sect. 9.1, the solution concept of evolutionarily stable strategies (ESS) is introduced. An ESS is a Nash equilibrium strategy of a symmetric bimatrix game which satisfies the additional stability requirement that it cannot be beaten by any rare, alternative strategy.

In Sect. 9.2, some elementary properties of ESS are derived. Regular ESS are characterized and it is shown that every game has a finite number of ESS (possibly zero).

Section 9.3 relates the ESS concept to the other equilibrium concepts discussed in this book. It is shown that an ESS induces a proper equilibrium and that a regular ESS induces a regular equilibrium. An ESS need not be strictly perfect, but it is stable against symmetric perturbations in payoffs and strategies. This section is based on Bomze [1986].

In Sect. 9.4, a dynamical system modelling the evolution of a polymorphic population is introduced and the relationships between the fixed points of this dynamic and various equilibrium concepts are investigated. The results are taken from Taylor and Jonker [1978], Zeeman [1980] and Bomze [1986].

Sections 9.5 and 9.6 extend the definitions of an ESS to asymmetric games. It is argued that, due to the spurious duplication inherent in mixed strategies, evolutionary stability cannot be sensibly defined by means of the normal form and that a definition using the agent normal form is more appropriate. Furthermore, it is shown that an ESS must be a strict equilibrium if role identification cannot occur. These sections are based on Selten [1980].

Extensive form games are the subject of the final two sections. It is demonstrated that a direct generalization of the ESS concept is too restrictive and that a more liberal definition of evolutionary stability, based on perturbed games, is more appropriate. It is also shown that every ESS can be found by dynamic programming, but that, even for games with a very special structure, not every strategy found in that way has to be an ESS. This material is based on Selten [1983].[1]

9.1 Introduction

Conflicts occur in nature whenever animals compete for limited resources such as territories, food or mates. Although such contests can be modeled as games, it is by no means obvious that Game Theory can be used to analyse them. After all, the central assumption of Game Theory is completely out of place in the biological context: Animals do not behave consciously rational and they do not consciously maximize some utility function.

Nevertheless, the forces of evolution are at work and these create a tendency towards optimization: An individual not having maximal fitness (reproductive success, expected number of offspring) will be selected against and, in an equilibrium state of evolution, all individuals must have equal and maximal fitness. Hence, in biological game theory, Darwinian fitness replaces the utility concept from classical game theory. It is not assumed that animals are rational (maximize fitness), but in an evolutionary stable state it just appears as if they are doing so.

In the biological context, a strategy is defined as a complete specification of what an animal will do, however, animals do not consciously choose strategies. A strategy is viewed as a preprogrammed behavioral policy which is not under the control of the animal itself: by some unspecified mechanism, the animal behaves as if it is following these instructions. In other words, biological game theory allows one to analyse evolution at the phenotypic level without having to model in detail the underlying genotypic level. It is assumed that strategies are heritable traits and that individuals breed true, i.e. if an individual plays a certain strategy, then so do his offspring. Note that this implies asexual reproduction. If randomizing can be benificial, there is no reason why evolution could not produce it, hence, it is assumed that genotypes specifying to play mixed strategies can exist.

In biological game theory, the emphasis is on strategies rather than players. A player can be thought of as a randomly selected animal, but the exact interpretation is not very important.

The solution concept of biological game theory is that of *evolutionarily stable strategies* (ESS). An ESS is a stable state of the process of evolution: if all individuals of a population adopt this strategy, then no mutant can invade. Note that this concept is relevant only for monomorphic populations. In Sect. 4 it will be investigated to what extend an ESS corresponds to a stable state of a polymorphic population in which only pure strategies can be played, but in which different pure strategies can co-exist.

Let us now illustrate the above ideas by means of a specific example, the Hawk-Dove game of Maynard Smith and Price [1973] (the names refer to the character of the strategies, they have no connections to the birds from which they are

	H	D
H	$\frac{1}{2}(V-C)$, $\frac{1}{2}(V-C)$	V, 0
D	0, V	$\frac{1}{2}V$, $\frac{1}{2}V$

Fig. 9.1.1. The Hawk-Dove game

derived). Imagine 2 animals are contesting a resource (such as a territory in a favourable habitat) of value V, i.e. by obtaining the resource, an animal increases Darwinian fitness (the expected number of offspring) by V. For simplicity, assume that only 2 pure strategies, hawk and dove, are possible. An animal adopting the hawk strategy always fights as hard and unrestrained as it can, retreating only when seriously injured. A dove merely threatens in a conventional way and quickly retreats when seriously challenged, without ever being wounded. Two doves can share the resource peacefully, but two hawks go on fighting until one is wounded and forced to retreat. It is assumed that a wound reduces fitness by an amount C. If we furthermore assume that there are no differences in size or age that influence the probability of winning or on which behavior can be conditioned, then the contest can be represented by the symmetric bimatrix game shown in Fig. 9.1.1.

If $V>C$, the Hawk-Dove game has a unique Nash equilibrium, (H, H), hence, it always pays to fight. In this case, H strictly dominates D, the equilibrium (H, H) is strict, and hence, is stable in every conceivable way. In a population of hawks and doves, the hawks have greater reproductive success, the doves will gradually die out and in the long run only hawks will exist.

If $V<C$, then (H, H) is not an equilibrium. Consequently, a monomorphic population of hawks is not stable: In such a population, a mutant dove has greater reproductive success and, therefore, doves will spread through the population. Similarly, a population consisting only of doves is not stable since it can be invaded by hawks ((D, D) is not a Nash equilibrium). If $V<C$, the game has a unique symmetric Nash equilibrium, viz. both players adopt the mixed strategy

$$p=p_H H+(1-p_H)D \qquad \text{with } p_H=V/C. \tag{9.1.1}$$

(The game also has 2 pure asymmetric equilibria, viz. (H, D) and (D, H) but these are not relevant, an animal cannot condition its behavior on whether it is called player 1 or player 2.)

It will be clear that, if $V<C$, a monomorphic population can be stable only if all animals adopt the mixed strategy p from (9.1.1). To see whether such a population is indeed stable, one has to investigate whether mutants that (spontaneously) arise in the population are selected against. Now, assume that a mutant playing $q\,(\neq p)$ arises. Then, since p is a mixed equilibrium, q fares equally good against p as p does, hence, q will be selected against if and only if p fares better

against q than q does. It can be shown (Theorem 9.2.3) that this is indeed the case. Therefore, p is stable: p is the unique ESS of the game of Fig. 9.1.1 if $V<C$.

It can also be shown that a random mixing polymorphic population of pure hawks and pure doves is stable if and only if the proportion of hawks is p_H as in (9.1.1), hence, in the game of Fig. 9.1.1, the ESS coincides with the unique stable state of the polymorphic population. In general, however, this need not be true (see Sect. 9.4).

In order to formally define the concept of evolutionarily stable strategies, we now introduce some notation.

Let $\Gamma = (\Phi_1, \Phi_2, A, A')$ be a *bimatrix game*. Γ is said to be *symmetric* if

$$\Phi_1 = \Phi_2 \quad \text{and} \quad A(\varphi_1, \varphi_2) = A'(\varphi_2, \varphi_1) \quad (\text{all } \varphi_1, \varphi_2)\,. \tag{9.1.2}$$

Note that this definition of symmetry is not invariant with respect to permutations of one of the strategy sets. This is natural, since, in the biological context, strategies have special meanings such as attack or flee. In the first four sections of this chapter attention will be restricted to such symmetric games and it will be convenient to have a simplified notation. If $|\Phi_1| = m$, we identify Φ_1 with $\{1, \ldots, m\}$, pure strategies are denoted by i or j, and we write a_{ij} instead of $A(i,j)$. The matrix A of player 1's payoffs is also called the *fitness matrix* of the game, and since A determines the game completely, we sometimes identify Γ and A and speak of the game A. Mixed strategies in A are elements of the simplex S^m

$$S^m := \left\{ x \in \mathbb{R}^m;\ \sum_i x_i = 1,\ x_i \geqq 0 \text{ all } i \right\} \tag{9.1.3}$$

and the expected payoff to player 1 when the mixed strategies x and y are played is equal to $x^T A y$

$$x^T A y := \sum_{i,j} x_i a_{ij} y_j\,. \tag{9.1.4}$$

If $p \in S^m$, we write $C(p)$ for the Carrier of p and $B(p)$ for the set of pure best replies against p in A

$$C(p) := \{i;\ p_i > 0\} \quad B(p) := \left\{ i;\ e_i^T A p = \max_j e_j^T A p \right\}\,. \tag{9.1.5}$$

Assume a monomorphic population is playing the mixed strategy p in the game with fitness matrix A and suppose that a mutant playing q arises. Then the population is in a perturbed state in which a small fraction ε of the individuals is playing q. The population will return to its original state if the mutant is selected against, i.e. if the fitness of a q-individual is smaller than that of an individual playing p. If opponents are paired at random, this is the case if

$$q^T A((1-\varepsilon)p + \varepsilon q) < p^T A((1-\varepsilon)p + \varepsilon q) \quad (\varepsilon > 0,\ \varepsilon \text{ small})\,. \tag{9.1.6}$$

The strategy p is stable against invasion if condition (9.1.6) is satisfied for all $q \neq p$. Any strategy that is stable in this sense is called an *evolutionarily stable strategy* or ESS. Clearly, p is an ESS if and only if for all $q \neq p$

$$\text{either } q^T A p < p^T A p \text{ or } (q^T A p = p^T A p \text{ and } q^T A q < p^T A q)\,. \tag{9.1.7}$$

Hence, if p is an ESS, then (p, p) is a symmetric Nash equilibrium of A that satisfies an additional stability condition. Formally

Definition 9.1.1. A mixed strategy p is an ESS of the symmetric bimatrix game with fitness matrix A if it satisfies the *equilibrium condition*

$$\text{if } q \in S^m, \text{ then } q^T Ap \leqq p^T Ap\,, \tag{9.1.8}$$

as well as the *stability condition*

$$\text{if } q \neq p \text{ and } q^T Ap = p^T Ap, \text{ then } q^T Aq < p^T Aq\,. \tag{9.1.9}$$

In Maynard Smith [1982] also the concept of a *neutrally stable strategy* is introduced. p is said to be neutrally stable if it satisfies (9.1.8) as well as the weaker version of (9.1.9) in which the strong inequality is replaced by a weak inequality. We will encounter this latter concept only in Sect. 9.8, hence, emphasis will be on ESS.

Recall that the following implicit assumptions underlie Def. 9.1.1:

(i) there is a large, random-mixing population,[2]
(ii) this population is monomorphic,
(iii) there is asexual reproduction,[3]
(iv) mixed strategies can exist and do breed true,
(v) the individuals involve in pairwise contests only, and[4]
(vi) contests are symmetric and static.

These assumptions will be maintained throughout the first three sections. In Sect. 9.4, we investigate what happens when (ii) and (iv) are dropped. Sects. 9.5 and 9.6 are devoted to asymmetric contests and the final two sections deal with contests in extensive form.

9.2 Evolutionarily Stable Strategies

In this section, it is investigated under which conditions a symmetric equilibrium strategy is an ESS and some elementary properties of ESS are derived. It is shown that every game has a finite number of ESS, but that there might be none. Throughout a fixed symmetric bimatrix game with fitness matrix A is assumed to be given.

First, let us study existence of ESS. Consider the correspondence

$$p \to \{q \in S^m;\ C(q) \subset B(p)\}\,.$$

This correspondence satisfies all conditions of Kakutani's Fixed Point Theorem and, hence, it has a fixed point. If p is a fixed point, then (p, p) is a symmetric Nash equilibrium of A, so that we have proved

Lemma 9.2.1 (Nash [1951]). *Every symmetric bimatrix game has a symmetric Nash equilibrium.*

The lemma shows that the equilibrium condition (9.1.8) is not too restrictive. If p is the unique best reply against p, then condition (9.1.9) is trivially satisfied, hence

Lemma 9.2.2. *If (p, p) is a strict equilibrium, then p is an ESS.*

Not every game possesses strict equilibria, however, and for equilibria in mixed strategies, condition (9.1.9) is very restrictive and can easily lead to nonexistence of ESS. For example, if A is the 2×2 matrix in which all entries are equal to 1, then $p^T Aq = 1$ for all p, q, and it is impossible to satisfy (9.1.9). However, this payoff matrix is degenerate and so one might still hope for generic existence of ESS. For 2×2 matrices this can indeed be established.

Theorem 9.2.3. *If A is a 2×2 fitness matrix with $a_{11} \neq a_{21}$ and $a_{12} \neq a_{22}$, then A has an ESS.*

Proof. If $a_{11} > a_{21}$ or $a_{22} > a_{12}$, the result follows from Lemma 9.2.2, hence, assume $a_{11} < a_{21}$ and $a_{22} < a_{12}$. In this case, the game has a unique symmetric equilibrium (p, p) and this is completely mixed $(C(p) = B(p) = \{1, 2\})$. If $q \neq p$, then

$$\begin{aligned} q^T Aq - p^T Aq &= (q-p)^T Aq = (q-p)^T A(q-p) \\ &= (q_1 - p_1)^2 (a_{11} - a_{21} + a_{22} - a_{12}) < 0, \end{aligned}$$

which shows that (9.1.9) is satisfied for p, hence p is an ESS. □

The game of Fig. 9.2.1* shows that Theorem 9.2.3 cannot be generalized to games with more than 2 pure strategies. If $\varepsilon \in (0, 1)$, the game is nondegenerate (in every column, all entries are different) and there is a unique symmetric equilibrium, viz. $p = (1/3, 1/3, 1/3)$. However, every strategy is a best reply against p and we have $e_1^T Ae_1 = \varepsilon > 1/3\varepsilon = p^T Ae_1$, so that condition (9.1.9) is violated for this unique equilibrium strategy.

ε	1	-1
-1	ε	1
1	-1	ε

Fig. 9.2.1. If $\varepsilon \in [0, 1]$, the game does not have an ESS

* Recall our convention that we identify a symmetric bimatrix game with its associated fitness matrix of player 1. Hence, in Fig. 9.2.1 only player 1's payoffs are displayed.

The above example shows that there are nontrivial games without an ESS. This should not be viewed as a drawback of the ESS concept, however. If a game does not have an ESS, no stable state exists and a population will cycle indefinitely.

Next, it will be shown that there is always a finite number of ESS. First of all, note that, if (p, p) and (q, q) are Nash equilibria of A with $q \neq p$ and $C(q) \subset B(p)$, then p cannot be an ESS, since q is a best reply against both p and q. Hence

Lemma 9.2.4. *If p is an ESS of A and (q, q) is a symmetric Nash equilibrium of A with $C(q) \subset B(p)$, then $p = q$.*

For $n \in \mathbb{N}$, let (p^n, p^n) be a symmetric Nash equilibrium of A such that $p^n \to p \, (n \to \infty)$. Then, if n sufficiently large

$$C(p) \subset C(p^n) \subset B(p^n) \subset B(p),$$

hence, if p is an ESS, then Lemma 9.2.4 implies that $p^n = p$ for all sufficiently large n. We have shown

Corollary 9.2.5. *If p is an ESS, then p is isolated within the set of symmetric equilibrium strategies.*

If there would be infinitely many ESS, there would be a cluster point and Corollary 9.2.5 shows that this is impossible, hence

Corollary 9.2.6 (Haigh [1975]). *The number of ESS is finite (but possibly zero).*

If an ESS exists, then it need not be unique, for example, there might exist several strict symmetric equilibria. However, Lemma 9.2.4 implies that, if there exists an ESS that is completely mixed, then this is the unique ESS.

Next, let us investigate under which conditions a symmetric equilibrium strategy is an ESS. For a nonempty subset V of $\{1, \ldots, m\}$, let us say that *A is negative definite with respect to V* if

$$x^T A x < 0 \quad \textit{for all } x \in \mathbb{R}^m \textit{ with } x \neq 0,\ \Sigma x_i = 0, \textit{ and } x_i = 0 \textit{ if } i \notin V. \tag{9.2.1}$$

(By definition, A is negative definite with respect to any set that is a singleton.) Note that, if $V = \{1, \ldots, k\}$ and $k \geqq 2$, then A is negative definite w.r.t. V if and only if the matrix $D^T A D$ is negative definite, where D is the $m \times (k-1)$ matrix defined by

$$d_{ij} = \begin{cases} \delta_{ij} & i<k, j<k \\ -1 & i=k, j<k \\ 0 & i>k, j<k \end{cases} \tag{9.2.2}$$

$$\delta_{ij} = \begin{cases} 1 & i=j, \\ 0 & i \neq j. \end{cases} \tag{9.2.3}$$

Let (p, p) be a symmetric Nash equilibrium of A. Then p is an ESS if and only if

$$q^T A q < p^T A q \quad \text{for all } q \neq p \text{ with } C(q) \subset B(p).$$

0	1	1
0	0	0
0	0	0

a

0	2	0
2	0	0
1	1	0

b

Fig. 9.2.2. The converses of the statements (i) and (ii) in Theorem 9.2.7 are not correct

Since $q^T Ap = p^T Ap$ for all such q, this condition is equivalent to

$$(q-p)^T A(q-p) < 0 \quad \text{for all } q \neq p \text{ with } C(p) \subset B(p)$$

and we see that p is an ESS if A is negative definite with respect to $B(p)$. However, p can be an ESS without A being negative definite with respect to $B(p)$ as the game of Fig. 9.2.2a illustrates. In that game, the first strategy is an ESS, but A is not negative definite with respect to $B(p)$ ($=\{1, 2, 3\}$). On the other hand, if $p \in S^m$ and $x \in \mathbb{R}^m$ with $x_i = 0$ for $i \notin C(p)$ and $\Sigma x_i = 0$, then there exists $q \in S^m$ with $C(q) = C(p)$ and $\varepsilon \neq 0$ such that $\varepsilon x = q - p$. Hence, if p is an ESS and $x \neq 0$, then

$$x^T Ax = \varepsilon^{-2}(q-p)^T A(q-p) < 0,$$

and we see that A is negative definite with respect to $C(p)$. Again the converse is not correct. In the game of Fig. 9.2.2b, A is negative definite with respect to $\{1, 2\}$, but the equilibrium strategy $(\frac{1}{2}, \frac{1}{2}, 0)$ is not an ESS.

Note that both counterexamples in Fig. 9.2.2 involve equilibria that are not quasi-strict. If (p, p) is quasi-strict, then $C(p) = B(p)$ and p is an ESS if and only if A is negative definite with respect to $C(p)$. We have proved

Theorem 9.2.7 (Haigh [1975], Abakuks [1980]). *Let (p, p) be a symmetric Nash equilibrium of A. Then*

(i) *If p is an ESS, then A is negative definite with respect to $C(p)$,*
(ii) *If A is negative definite with respect to $B(p)$, then p is an ESS,*
(iii) *If (p, p) is quasi-strict, then p is an ESS if and only if A is negative definite with respect to $C(p)$.*

To conclude this section, we derive a characterization of ESS that will be of use in Sect. 9.4.

Theorem 9.2.8. *p is an ESS of A if and only if there exists a neighbourhood U of p in S^m such that $p^T Ax > x^T Ax$ for all $x \in U$ with $x \neq p$.*

Proof. For p, $q \in S^m$ and $\varepsilon > 0$ write $p_\varepsilon(q) = (1-\varepsilon)p + \varepsilon q$. Note that condition (9.1.6) is equivalent to

$$p^T A p_\varepsilon(q) > p_\varepsilon^T(q) A p_\varepsilon(q) \quad (\varepsilon > 0,\ \varepsilon \text{ small}). \tag{9.2.4}$$

This proves the 'if' part of the theorem since $p_\varepsilon(q) \in U$ for every $q \in S^m$ if ε is sufficiently small.

Next, assume that p is an ESS. If p is completely mixed, one can take $U = S^m$, so assume that p is not completely mixed. Let S be the boundary of S^m, minus the interior of that facet of S^m that contains p, hence

$$S := \{q \in S^m;\ q_i = 0 \text{ for some } i,\ C(q) \neq C(p)\}.$$

Let the map ε and the number ε^* be defined by

$$\varepsilon(q) = \sup\{\varepsilon > 0;\ p^T A p_\varepsilon(q) > q^T A p_\varepsilon(q)\}$$

$$\varepsilon^* = \inf\{\varepsilon(q);\ q \in S\}.$$

If $q \neq p$, then $\varepsilon(q) > 0$ since p is an ESS. Furthermore, $\varepsilon^* > 0$ since ε is a continuous function and S is a compact set. Let U be the set

$$U := \{p_\varepsilon(q);\ q \in S,\ \varepsilon \in [0, \varepsilon^*)\}.$$

Then U is a neighbourhood of p and U satisfies the condition of the theorem. □

9.3 Strategic Stability of ESS

In this section, we relate the concept of evolutionarily stable strategies to the refinements of the Nash equilibrium concept discussed in earlier chapters. The main conclusion that can be drawn is that the stability requirement (9.1.9) is very restrictive. Most of the results in this section are taken from Bomze [1986].[5]

For 2-person games, the weakest refinement of the Nash equilibrium concept discussed before is the perfectness concept. We first show that every ESS induces a perfect equilibrium.

Theorem 9.3.1. *If p is an ESS, then (p, p) is a perfect equilibrium.*

Proof. Assume p is an ESS in the game with fitness matrix A. Then (p, p) is a Nash equilibrium and in view of Theorem 3.2.2 it suffices to show that p is undominated. If p would be dominated by q, then $q \neq p$ and $q^T A r \geqq p^T A r$ for all $r \in S^m$. In particular

$$q \neq p, \quad q^T A p \geqq p^T A p \quad \text{and} \quad q^T A q \geqq p^T A q,$$

but this contradicts (9.1.9). □

Our goal in this section is to investigate to what extend this theorem can be generalized. The most stringent refinement of the Nash concept considered thus far is that of strict equilibria and the game of Fig. 9.2.2a shows that an ESS need not be strict (however, cf. Theorem 9.6.2). The ESS of that game is not even quasi-strict, hence, it does not induce a regular equilibrium. In fact, the equilibrium (1, 1) is not even essential: one can perturb the payoffs of the game such that the unique equilibrium is (2, 1): just add ε to player 2's payoffs in the first column and to player 1's payoffs in the second row (leaving the other payoffs unchanged). However, this perturbance leads to an asymmetric game and this does not make sense in the biological context under consideration. Hence, one might ask whether any symmetric game has an equilibrium close to the given equilibrium and it is easily checked that in the game of Fig. 9.2.2a this is indeed the case. More generally, we say that a symmetric equilibrium (p, p) of a symmetric game with fitness matrix A is *symmetrically essential* if any symmetric game with fitness matrix close to A has a symmetric equilibrium close to (p, p). We have

Theorem 9.3.2. *If p is an ESS, then (p, p) is a symmetrically essential equilibrium.*

Proof. Let p be an ESS of the game A. Since pure strategies that are not best replies are irrelevant, both for evolutionary stability and for essentially, we may assume that all pure strategies are best replies against p. Let A^ε be a matrix with $A^\varepsilon \to A$ as $\varepsilon \to 0$ and let $(p^\varepsilon, p^\varepsilon)$ be a symmetric equilibrium of the game with fitness matrix A^ε (cf. Lemma 9.2.1). If q is a limit point of p^ε (as ε tends to 0), then (q, q) is a Nash equilibrium of A and $C(q) \subset B(p)$. Lemma 9.2.4 implies that $q = p$, hence (p, p) is symmetrically essential. □

Let Γ be a symmetric bimatrix game and let (Γ, η) be a symmetric perturbed game of Γ, i.e. $\eta = (\eta_1, \eta_2)$ is as in (2.2.1) with $\eta_1 = \eta_2$. An equilibrium (p, p) is said to be *symmetrically strictly perfect* if any such perturbed game with sufficiently small η has a symmetric equilibrium close to (p, p). Since the game Γ^η that occurs in the proof of Theorem 2.4.3 is symmetric whenever Γ and η are symmetric, it follows from that theorem and the above result that every ESS induces a symmetrically strictly perfect equilibrium.

Corollary 9.3.3. *If p is an ESS, then (p, p) is a symmetrically strictly perfect equilibrium.*

2, 2	1, 2	0, 0
2, 1	0, 0	1, 0
0, 0	0, 1	0, 0

Fig. 9.3.1. If p is an ESS, then (p, p) need not be strictly perfect

If p is an ESS, then (p, p) need not be strictly perfect, however. The equilibrium (p, p) need not be stable with respect to asymmetric perturbances as is illustrated by the game of Fig. 9.3.1, where for purposes of clarity, we have listed both the payoff matrix of player 1 and that of player 2.

In the game of Fig. 9.3.1, the first pure strategy is an ESS, however, the strategy pair $(1, 1)$ is not a strictly perfect equilibrium. If player 1 expects that player 2 will (mistakenly) choose his third strategy with a larger probability than his second one, then he should play his second strategy. In fact, in this game, there is no strictly perfect equilibrium. Note that the strategy pair $(1, 1)$ is a proper equilibrium. The following theorem shows that this is not a coincidence.

Theorem 9.3.4. *If p is an ESS, then (p, p) is a proper equilibrium.*

Proof. Assume p is an ESS of the game with fitness matrix A. For $\varepsilon > 0$, let p^ε be a completely mixed strategy that satisfies

$$\text{if } \begin{cases} i, j \in B(p) \text{ and } e_i^T A p^\varepsilon < e_j^T A p^\varepsilon, \text{ or} \\ i \notin B(p) \text{ and } j \in B(p), \text{ or} \\ i, j \notin B(p) \text{ and } e_i^T A p^\varepsilon < e_j^T A p^\varepsilon \end{cases}, \text{ then } p_i^\varepsilon \leqq \varepsilon p_j^\varepsilon. \tag{9.3.1}$$

By using an argument as in the proof of Theorem 2.3.3 one can show that indeed such p^ε exists for every $\varepsilon > 0$. Let q be a limit point of $\{p^\varepsilon\}_{\varepsilon \downarrow 0}$. Since $p_i^\varepsilon \leqq \varepsilon$ for all $i \notin B(p)$, we have $C(q) \subset B(p)$. Eq. (9.3.1) also implies that

$$\text{if } i \in C(q), \text{ then } e_i^T A q = \max_{j \in B(p)} e_j^T A q .$$

Hence, we have

$$q^T A p = p^T A p \quad \text{and} \quad q^T A q \geqq p^T A q ,$$

and, therefore, (9.1.9) implies $q = p$. Consequently, if ε is sufficiently small, we have

$$\text{if } i \notin B(p) \text{ and } j \in B(p), \text{ then } e_i^T A p^\varepsilon < e_j^T A p^\varepsilon$$

and the definition of p^ε implies

$$\text{if } e_i^T A p^\varepsilon < e_j^T A p^\varepsilon, \text{ then } p_i^\varepsilon \leqq \varepsilon p_j^\varepsilon \quad \text{all } i, j .$$

Hence, $(p^\varepsilon, p^\varepsilon)$ is an ε-proper equilibrium of A if ε is sufficiently small and, therefore, (p, p) is a proper equilibrium. □

Next, let us turn to the concept of regular equilibria. We already know that if p is an ESS, then (p, p) need not be quasi-strict and, hence, it need not be regular (Fig. 9.2.2a). In the next theorem, we show that, for ESS, quasi-strictness implies regularity.

Theorem 9.3.5. *If p is an ESS, then the following statements are equivalent:*

(i) *(p, p) is a regular equilibrium,*
(ii) *(p, p) is a quasi-strict equilibrium, and*
(iii) *(p, p) is an isolated equilibrium.*

1	1	0
$1+\varepsilon$	0	0
$1+\varepsilon$	0	0

Fig. 9.3.2. The first pure strategy is an inessential ESS of the game with $\varepsilon=0$

Proof. Theorem 3.4.4 shows that (ii) is implied by (i). Assume (p,p) is a quasi-strict equilibrium and, for $\varepsilon>0$, let $(p_1^\varepsilon, p_2^\varepsilon)$ be a Nash equilibrium such that $p_i^\varepsilon \to p$. Then, if ε is sufficiently small

$$C(p) \subset C(p_i^\varepsilon) \subset B(p_j^\varepsilon) \subset B(p) \quad i \neq j \in \{1, 2\}.$$

Since $C(p)=B(p)$, this implies that also $(p_1^\varepsilon, p_1^\varepsilon)$ and $(p_2^\varepsilon, p_2^\varepsilon)$ are equilibria, hence, from Corollary 9.2.5 we can conclude that $p_1^\varepsilon = p_2^\varepsilon = p$ for all ε sufficiently small. Therefore, (p, p) is an isolated equilibrium and (ii) implies (iii).

Finally, since (p, p) is a proper equilibrium if p is an ESS (Theorem 9.3.4), one can conclude from Theorem 2.3.4 and Theorem 3.4.4 that (iii) implies (i). □

In Taylor and Jonker [1978] an ESS p is called a *regular ESS* if $C(p)=B(p)$, i.e. if (p, p) is a quasi-strict equilibrium. Theorem 9.3.5 shows that this terminology cannot lead to confusion: p is a regular ESS if and only if p is an ESS and (p, p) is a regular equilibrium.

The reader might consider Theorem 9.3.2 to be unsatisfactory and incomplete: If the concept of ESS is the relevant one and if stability with respect to perturbations in payoffs is desirable, one should not only establish that every perturbed game has an equilibrium close to the given ESS, but one should establish that every such game has an ESS close to the one given. Let us call an ESS p of a fitness matrix A an *essential ESS* if every matrix A' with payoffs close to those of A has an ESS p' close to p. The game of Fig. 9.3.2 shows that not every ESS is essential. If $\varepsilon=0$, the first pure strategy is the unique ESS. If $\varepsilon>0$, however, then $p \in S^3$ is a symmetric Nash equilibrium strategy if and only if

$$p_2+p_3=\frac{\varepsilon}{1+\varepsilon}.$$

Hence, if $\varepsilon>0$, there are no isolated symmetric equilibrium strategies, and the game does not have an ESS (Corollary 9.2.5).

Note that the above example involves an ESS that is not regular. To conclude this section we show that every regular ESS is an essential ESS.

Theorem 9.3.6 (Selten [1983]). *If p is a regular ESS, then p is an essential ESS.*

Proof. If (p,p) is a strict equilibrium of game A, then (p,p) is a strict equilibrium of any game with payoffs close to A, hence, p is an ESS of any such game.

Next, assume p is a regular ESS of the game A with $|C(p)|>1$ and let $A^\varepsilon \to A$ as $\varepsilon \to 0$. Theorem 9.3.2 shows that A^ε admits a symmetric equilibrium $(p^\varepsilon, p^\varepsilon)$ with $p^\varepsilon \to p\,(\varepsilon \to 0)$. It is easily verified that, since p is regular, $(p^\varepsilon, p^\varepsilon)$ must be a quasi-strict equilibrium of A^ε and that, furthermore, $C(p^\varepsilon)=C(p)$ when ε is small. Now, A is negative definite w.r.t. $C(p)$, hence A^ε is negative definite w.r.t. $C(p)$ if ε is small. Theorem 9.2.7 shows that p^ε is an ESS of A^ε for ε small. □

One can also define the concept of (strictly) perfect ESS, but this concept will not be considered in this chapter. However, let us remark that perturbed games play an important role for defining a satisfactory concept of evolutionary stability in extensive form games (cf. Def. 9.7.2).

9.4 Population Dynamics

The concept of ESS is based on the assumption that genotypes specifying to play mixed strategies can exist. In this section, we turn to the question of what happens when only pure strategies can be played, but when the population itself might be polymorphic. It will be shown that an ESS gives rise to a stable state of such a polymorphic population, but that the converse need not be true.[6] Along the way, the dynamic stability of various other refined equilibrium concepts will be investigated.[7] This section is based on Taylor and Jonker [1978], Zeeman [1980] and Bomze [1986].

Suppose we have a (large) population of which each individual follows some strategy $i \in \{1,\ldots,m\}$. Let x_i be the proportion of i-strategists in the population, hence $x=(x_1,\ldots,x_m)\in S^m$. Assume that individuals are paired at random, that each individual engages in exactly one contest and that the payoff (fitness, expected number of offspring) to an i-strategist as a result of a contest with a j-strategist is a_{ij} where A is some $m \times m$ matrix. Then the expected payoff to an i-strategist is $e_i^T Ax$ and the average fitness of the population is $x^T Ax$. If individuals breed true (i.e. the offspring of an i-strategist are again i-strategists), then in the next generation, the percentage of i-strategists is equal to

$$x_i' = x_i \frac{e_i^T Ax}{x^T Ax} .$$

Hence, the population evolves according to the difference equation

$$x_i' - x_i = x_i \frac{e_i^T Ax - x^T Ax}{x^T Ax} \quad (i=1,\ldots,m) . \tag{9.4.1}$$

Let us assume that changes between generations are not too great, so that it is appropriate to replace (9.4.1) by the differential equation

$$\dot{x}_i = x_i \frac{e_i^T Ax - x^T Ax}{x^T Ax} \quad (i=1,\ldots,m) . \tag{9.4.2}$$

The trajectories of this differential equation are the same as those of the simpler (cubic) equation

$$\dot{x}_i = x_i(e_i^T Ax - x^T Ax) \quad (i=1,\dots,m), \tag{9.4.3}$$

and it is the latter that will be studied in this section. Eqs. (9.4.3) have been suggested in Taylor and Jonker [1978] and Zeeman [1980] as the appropriate ones for the situation under consideration. Clearly, for (9.4.3) to model the evolution of a population it is necessary that any trajectory starting in S^m remains in S^m. The following lemma shows that this indeed holds.

Lemma 9.4.1. *S^m as well as all its faces are invariant under* (9.4.3).

Proof. The plane $\Sigma x_i = 1$ is invariant since

$$\Sigma \dot{x}_i = \Sigma x_i(e_i^T Ax - x^T Ax) = x^T Ax - (\Sigma x_i) x^T Ax = 0.$$

Furthermore, if $x_i = 0$, then $\dot{x}_i = 0$ so that indeed S^m and all its faces are invariant. □

$p \in S^m$ is said to be a *dynamic equilibrium* of A if it is a fixed point of the dynamic (9.4.3). A dynamic equilibrium p is said to be *stable* if for any neighbourhood U of p in S^m there exists a neighbourhood V of p in S^m such that any trajectory of (9.4.3) that originates in V remains in U. The equilibrium p is said to be *asymptotically stable* if there exists a neighbourhood U of p in S^m such that any trajectory of (9.4.3) that originates in U converges to p. Clearly, asymptotic stability is a more stringent requirement than stability. Later on, we will see examples of dynamic equilibria that are not stable and of stable equilibria that are not asymptotically stable.

Let $F: \mathbb{R}^m \to \mathbb{R}^m$ be the map defined by the right hand side of (9.4.3), hence, $F = (F_1, \dots, F_m)$ and

$$F_i(x) = x_i(e_i^T Ax - x^T Ax). \tag{9.4.4}$$

$p \in S^m$ is a dynamic equilibrium of A if and only if $F(p) = 0$, or, equivalently

$$\text{if } p_i > 0, \text{ then } e_i^T Ap = p^T Ap. \tag{9.4.5}$$

Hence, if (p, p) is a symmetric Nash equilibrium of A, then p is a dynamic equilibrium of A. Since any pure strategy satisfies (9.4.5), the converse of this statement is false.

Corollary 9.4.2. *If (p, p) is a Nash equilibrium, then p is a dynamic equilibrium, but the converse need not be true.*

Recall that, in Sect. 2.5, regular equilibria have been defined with the aid of a function of which every equilibrium as well as every pure strategy combination is a zero. In the context under consideration, the map F from (9.4.4) has similar properties and in the next theorem we show that regularity can also be defined by means of this map. The assumption in the theorem that $p^T Ap > 0$ is immaterial since one can always rescale payoffs such that this is satisfied. Namely, if the $m \times m$

payoff matrices A and A' are related according to

$$a'_{ij} = a_{ij} + b_j \quad (b_j \in \mathbb{R} \text{ all } j), \tag{9.4.6}$$

then A and A' induce the same function F (as in (9.4.4)) and also the set of regular equilibria is invariant with respect to this transformation (note that also the set of ESS is invariant).

Theorem 9.4.3. *A Nash equilibrium (p, p) with $p^T Ap > 0$ is regular if and only if the matrix $DF(p)$ of partial derivatives of F evaluated at p is nonsingular.*

Proof. It is easily checked that for $x \in \mathbb{R}^m$

$$DF_{ij}(x) = \delta_{ij}(e_i^T Ax - x^T Ax) + x_i(a_{ij} - e_j^T Ax - x^T Ae_j),$$

where δ_{ij} is as in (9.2.3). Hence, if p is a dynamic equilibrium

$$DF_{ij}(p) = \begin{cases} p_i(a_{ij} - e_j^T Ap - p^T Ae_j) & i \in C(p), \\ \delta_{ij}(e_i^T Ap - p^T Ap) & i \notin C(p). \end{cases} \tag{9.4.7}$$

Now, let (p, p) be a Nash equilibrium of A. Then p is a dynamic equilibrium and (9.4.7) shows that (p, p) is quasi-strict if $DF(p)$ is nonsingular. Therefore, in view of (9.4.7) and Lemma 3.3.1, it suffices to show that the following two assertions are equivalent.

(I) the matrix A^* with $a^*_{ij} = a_{ij} - e_j^T Ap - p^T Ae_j$ is singular ($i, j \in C(p)$), and
(II) $e_i^T Ax = p^T Ax$ for all $i \in C(p)$ and some $x \neq 0$ with $C(x) = C(p)$ and $\Sigma x_i = 0$.

Using that $e_i^T Ap = p^T Ap > 0$ for all $i \in C(p)$, it is easily seen that these assertions are indeed equivalent. □

Let p be a dynamic equilibrium of A. (Asymptotic) stability of p depends on the eigenvalues of $DF(p)$. From (9.4.7) it follows that

$$DF(p)p = -(p^T Ap)p,$$

hence p is an eigenvector associated with the eigenvalue $-p^T Ap$. Since p points outside of the simplex S^m, this eigenvalue is not relevant for studying the stability of p with respect to S^m at least, as long as it has multplicity 1. Let us call $\lambda \in \mathbb{C}$ an *eigenvalue of p* if (i) λ is an eigenvalue of $DF(p)$ with $\lambda \neq -p^T Ap$ or (ii) $\lambda = -p^T Ap$ and $-p^T Ap$ is eigenvalue of $DF(p)$ with multiplicity more than 1. Then we have the following standard result (see e.g. Hirsch and Smale [1974]).

Theorem 9.4.4. *Let p be a dynamic equilibrium of A.*

(i) *If all eigenvalues of p have negative real parts, then p is asymptotically stable.*
(ii) *If p is stable, then all eigenvalues of p have nonpositive real parts.*

From (9.4.7) it follows that, if p is a dynamic equilibrium, then $e_i^T Ap - p^T Ap$ is an eigenvalue of p for every $i \notin C(p)$. Hence, if p is stable, then $e_i^T Ap - p^T Ap \leqq 0$ for all $i \notin C(p)$. Combining this with (9.4.5), we see that we have proved the first part of the following corollary.

Corollary 9.4.5. *If p is a stable dynamic equilibrium, then (p, p) is a Nash equilibrium, but the converse need not be true.*

To establish the second part of the Corollary, consider the game of Fig. 9.4.1. Both players playing their second strategies is a Nash equilibrium, but this is not stable. Any trajectory of (9.4.3) starting at a completely mixed strategy converges to the first strategy.

The above example involves a Nash equilibrium that is not perfect and the reader might conjecture that every perfect equilibrium is dynamically stable. The game from Fig. 9.4.2 shows that this conjecture is false.

The game from Fig. 9.4.2 results from that in Fig. 9.4.1 by adding a dominated third strategy, which converts (2, 2) into a perfect equilibrium. However, since every trajectory of (9.4.3) starting at $(x_1, x_2, 0)$ with $x_1 > 0$ is attracted to the first strategy, the second strategy is not dynamically stable.

Next, one might consider the question of whether a stable dynamic equilibrium is perfect. Again the answer is negative as is demonstrated by the game of Fig. 9.4.3.

1	0
0	0

Fig. 9.4.1. A Nash equilibrium need not be dynamically stable

1	0	0
0	0	1
0	-1	0

Fig. 9.4.2. A perfect equilibrium need not be dynamically stable

0	0	1
0	0	0
-1	0	0

a

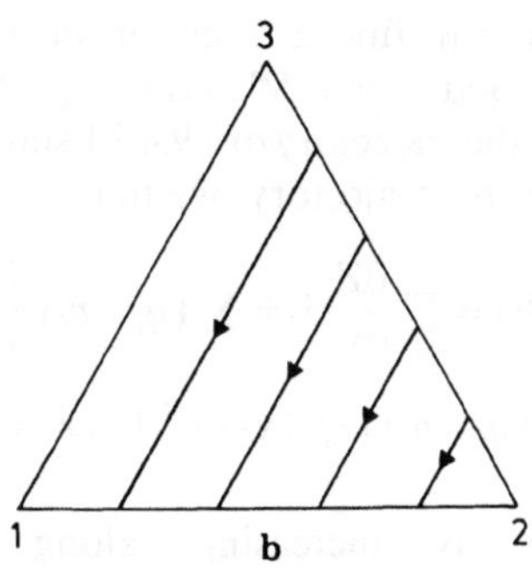

b

Fig. 9.4.3a, b. A stable dynamic equilibrium need not be perfect. Figure **a** is the game, **b** gives the phase portrait of the dynamic (9.4.3)

In the game of Fig. 9.4.3, the second pure strategy is dominated, hence, (2, 2) is not a perfect equilibrium. Since A is skew-symmetric, $x^TAx=0$ for all x and the dynamic (9.4.3) reduces to

$$\dot{x}_i=x_1x_3, \quad \dot{x}_2=0, \quad \dot{x}_3=-x_1x_3.$$

This shows that the phase portrait is indeed as in Fig. 9.4.3b and that the second strategy is a stable dynamic equilibrium.

Note that, in the example above, the second strategy is not asymptotically stable. In the next theorem we show that every asymptotically stable equilibrium is perfect and isolated. Note that the game of Fig. 1.4.2 shows that the converse is not correct: The second strategy is isolated and perfect, but not even dynamically stable.

Theorem 9.4.6 (Bomze [1986]). *If p is an asymptotically stable dynamic equilibrium, then (p, p) is a perfect and isolated equilibrium, but the converse need not be true.*

Proof. Corollary 9.4.2 shows that an asymptotically stable equilibrium must be isolated. Assume (p, p) is an equilibrium of the game with fitness matrix A that is not perfect. Then there exists q such that q dominates p and for any $x \in S^m$ we have $q^TAx \geqq p^TAx$. Define Z: $S^m \to \mathbb{R}$ by means of

$$Z(x) := \prod_i x_i^{q_i-p_i},$$

then $Z(x)>0$ if x is completely mixed, and

$$\frac{\partial Z}{\partial x_i} = \frac{q_i-p_i}{x_i} Z(x) \quad \text{if } x_i>0.$$

We claim that p is not a local maximum of Z. This is clear if $C(q) \not\subset C(p)$ since in that case $Z(p)=0$. If $C(q) \subset C(p)$, then $Z(p)>0$ and

$$\frac{\partial Z}{\partial x_i}(p)>0 \quad \text{if } q_i>p_i$$

so that one can find a direction in which Z increases. Let U be an arbitrary neighbourhood of p in S^m and let $x_0 \in U \cap \operatorname{int} S^m$ be such that $Z(x_0)>Z(p)$. Let $\{x(t)\}_t$ be the trajectory of (9.4.3) starting in x_0. Then, taking the time derivative of Z along this trajectory, we find

$$\dot{Z} = \nabla Z \cdot \dot{x} = \sum_i \frac{\partial Z}{\partial x_i}\dot{x}_i = \sum_i (q_i-p_i)\frac{\dot{x}_i}{x_i} Z$$
$$= \sum_i (q_i-p_i)(e_i^TAx-x^TAx)Z = (q^TAx-p^TAx)Z \geqq 0,$$

hence, Z is increasing along trajectories of (9.4.3). Therefore, $Z(x(t)) \geqq Z(x_0) > Z(p)$ and the trajectory cannot converge to p. Consequently, p is not asymptotically stable. □

A dynamic equilibrium is said to be *hyperbolic* if all its eigenvalues have negative real parts (hence, any hyperbolic equilibrium is asymptotically stable). Let p be such a hyperbolic equilibrium.Since (9.4.3) is invariant with respect to transformations as in (9.4.6), we may assume that $p^T Ap>0$, hence, from Corollary 9.4.5 and Theorem 9.4.3 we can conclude that (p, p) is a regular equilibrium. The converse is not true: if A is the 2×2 identity matrix and $p=(\frac{1}{2},\frac{1}{2})$ is the completely mixed equilibrium strategy, then (p,p) is a regular equilibrium, but p is not even stable: any trajectory starting at a point different from p converges to a pure strategy. We have shown

Theorem 9.4.7. *If p is a hyperbolic equilibrium, then (p,p) is a regular equilibrium, but the converse need not be true.*

Next, we relate evolutionary stability to dynamic stability. We show that an ESS (which, by definition, is a stable state of a monomorphic population) is a stable state of a polymorphic population in which only pure strategies can exist. However, there can exist stable states that are not ESS.

Theorem 9.4.8 (Taylor and Jonker [1978], Zeeman [1980]). *Every ESS is an asymptotically stable dynamic equilibrium, but not conversely.*

Proof. Assume p is an ESS of the game with fitness matrix A and let the neighbourhood U of p be as in Theorem 9.2.8. Without loss of generality, assume $x_i>0$ for all $x\in U$ and $i\in C(p)$. Define $Z\colon U\to\mathbb{R}$ by

$$Z(x)=\prod_i x_i^{p_i}. \tag{9.4.8}$$

Then $Z(x)>0$ for all $x\in U$ and p is the unique global maximum of Z. We will show that Z is a Lyapunov function for (9.4.3), i.e.

$$\nabla Z(x)\cdot(p-x)>0 \quad\text{and}\quad \dot Z(x)>0 \quad \text{for all } x\in U,\ x\neq p. \tag{9.4.9}$$

These conditions guarantee that p is the unique local maximum of Z on U, that Z increases radially towards p and that Z increases along orbits of (9.4.3), hence, (9.4.9) implies that orbits originating in U converge to p and that p is asymptotically stable. Note that, if $x\in U$, then

$$\frac{\partial Z}{\partial x_i}=\begin{cases} Z(x)\dfrac{p_i}{x_i} & \text{if } i\in C(p).\\ 0 & \text{if } i\notin C(p).\end{cases}$$

Consequently, if $x\neq p$, then

$$\begin{aligned}\nabla Z\cdot(p-x) &=\sum_i \frac{\partial Z}{\partial x_i}(p_i-x_i)\\ &=Z\sum_{i\in C(p)}\frac{p_i}{x_i}(p_i-x_i)\\ &=Z\sum_{i\in C(p)}\frac{(p_i-x_i)^2}{x_i}+Z\Big(1-\sum_{i\in C(p)}x_i\Big)>0,\end{aligned}$$

0	1	1
-2	0	4
1	1	0

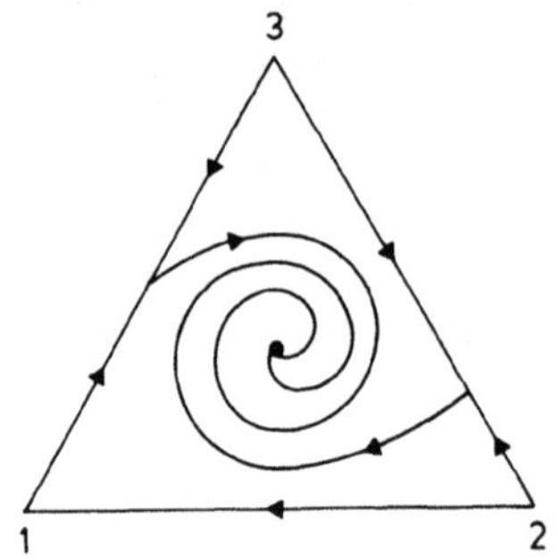

Fig. 9.4.4. $p=(1/3, 1/3, 1/3)$ is an asymptotically stable equilibrium that is not an ESS

so that the first condition in (9.4.9) is indeed satisfied. To verify the second condition, note that, if $x \neq p$

$$\begin{aligned}\dot{Z} &= \nabla Z \cdot \dot{x} = Z \sum_{i \in C(p)} p_i \frac{\dot{x}_i}{x_i} \\ &= Z \sum_{i \in C(p)} p_i (e_i^T Ax - x^T Ax) \\ &= Z(p^T Ax - x^T Ax) > 0,\end{aligned}$$

where the inequality follows from Theorem 9.2.8. This shows that every ESS is asymptotically stable.

To prove that not every asymptotically stable equilibrium is an ESS, consider the game of Fig. 9.4.4. The unique symmetric equilibrium strategy of this game is $p=(1/3, 1/3, 1/3)$, however, this strategy is not an ESS since it can be invaded by $x=(0, \frac{1}{2}, \frac{1}{2})$ ($x^T Ap = p^T Ap$ and $x^T Ax > p^T Ax$). On the other hand, $DF(p)$ is the matrix

$$\frac{1}{9}\begin{pmatrix} -1 & -1 & -4 \\ -7 & -4 & 5 \\ 2 & -1 & -7 \end{pmatrix}$$

with eigenvalues $-1/3$, $-1/3$, $-2/3$. Hence, the eigenvalues of p are $-1/3$ and $-2/3$ so that p is asymptotically stable. (p is even globally stable as the phase portrait in Fig. 9.4.4b) shows). □

Note, that, if p is a completely mixed ESS, then the above proof shows that p is even globally stable. It can be shown that p is a hyperbolic equilibrium if p is a regular ESS (Taylor and Jonker [1978]).

In 2×2 games, there is no discrepancy between evolutionary stability and asymptotic stability, as we prove in the following theorem.

Theorem 9.4.9. *If A is a 2×2 matrix, then p is an ESS of A if and only if p is an asymptotically stable equilibrium of A.*

Proof. Without loss of generality, we can assume that the diagonal elements of A are zero. One of the following must be the case:

(i) $a_{12}<0$ and $a_{21}<0$: there are 2 strict, symmetric equilibria,
(ii) $A\neq 0$ and $a_{12}a_{21}\leqq 0$: one pure strategy dominates the other,
(iii) $A=0$: both pure strategies are equivalent, or
(iv) $a_{12}>0$ and $a_{21}>0$: there is a unique completely mixed equilibrium.

It is easily checked that the assertion of the theorem is true in any of the first three cases. It follows from Theorem 9.2.3 and Theorem 9.4.8 that the statement is also correct in case (iv). □

It should not be too surprising that, for games with more than 2 pure strategies, evolutionary stability is a stronger concept than asymptotic stability. After all, the latter concept is based on the assumption that only pure strategies can be played. Therefore, it is actually more appropriate to consider the question of whether an ESS corresponds to a stable state of a polymorphic population in which different mixed strategies can co-exist.[8] Such a population can be represented by a probability density f on S^m and making the same assumption as before, viz. that growth rate equals difference in fitness, the evolution of the population is modeled by the equation

$$\dot{f}(x)=f(x)[x^T A\mu_f-\mu_f^T A\mu_f] \quad (x\in S^m), \tag{9.4.10}$$

where μ_f denotes the mean of population, $\mu_f=\int f(x)dx$. This dynamic has been investigated to some extent in Zeeman [1981], but, unfortunately, Eq. (9.4.10) is much less tractable than Eq. (9.4.3) and the results are far from complete. However, it can be seen that, in the game of Fig. 9.4.4, the Dirac distribution at $p=(1/3, 1/3, 1/3)$ is not a stable fixed point of (9.4.9), hence, $(1/3, 1/3, 1/3)$ is not a stable state of a polymorphic population of individuals playing mixed strategies. It is easy to show that, if (p, p) is a strict equilibrium of A, the Dirac distribution at p is a stable fixed point of (9.4.10), however, not much more is known about the general properties of (9.4.10).

To conclude this section, we return to the dynamic (9.4.3) and we show that certain equilibria admit an interpretation in terms of time averages of periodic orbits.

Theorem 9.4.10 (Schuster et al. [1981]). *If* $\{x(t)\}_{t=0}^T$ *is a periodic orbit of* (9.4.3), *i.e.* $x(0)=x(T)$, *then the time average* p *of this orbit,*

$$p=\frac{1}{T}\int_0^T x(t)dt$$

is a dynamic equilibrium of (9.4.3).

Proof. If $i\notin C(p)$, then $x_i(t)=0$ for all t. If $i\in C(p)$, then $x_i(t)>0$ for all t and

$$\frac{d}{dt}\log x_i(t)=\frac{\dot{x}_i}{x_i}=e_i^T Ax-x^T Ax.$$

If we integrate this identity from 0 to T, the left hand side vanishes, and by dividing through T we obtain

$$e_i^T Ap = \frac{1}{T}\int_0^T x(t)^T Ax(t)\,dt \qquad (i \in C(p)),$$

hence, p satisfies (9.4.5). □

As an illustration of this theorem, consider the game of Fig. 9.2.1 with $\varepsilon=0$ (the ordinary rock-paper-scissors game). Let $p=(1/3, 1/3, 1/3)$ be the unique equilibrium strategy and let Z be as in (9.4.7). Then

$$\dot{Z} = Z\cdot(p-x)^T Ax = -Z(p-x)^T A(p-x) = 0,$$

since A is skew-symmetric. Hence, Z is constant along orbits of (9.4.3). Consequently, the orbits of (9.4.3) are the level curves of Z, they are closed and symmetric around p. Theorem 9.4.10 shows that p is the time average of every orbit.

Note that the time average p in Theorem 9.4.10 need not be a stable equilibrium, hence, it need not be Nash. For example, if one appends the game of Fig. 9.2.1 (with $\varepsilon=0$) with a row and a column of which all entries are equal to 2, then $p=(1/3, 1/3, 1/3, 0)$ can be obtained as the time average of closed orbits, but p is unstable.

Theorem 9.4.10 can be generalized considerably. Namely, if $\{x(t)\}_t$ is a trajectory in a face of S^m that remains bounded away from the boundary of this face, then every accumulation point of $\{x(t)\}_t$ is a dynamic equilibrium of (9.4.3). The proof parallels that of the theorem, the only difference is that the left hand side does not vanish but remains bounded and, hence, approximately vanishes after dividing through T.

9.5 Asymmetric Contests: Examples and the Model

Thus far, only symmetric games have been considered, i.e. contestants start in identical situations, they have the same strategies and the same payoffs. In fact, the concept of ESS has been defined for symmetric games only. However, most actual contests are asymmetric: usually there are differences in size, sex, age or status that can be perceived by both contestants and upon which behavior can be conditioned. In the biological game theory literature, such situations are usually symmetrized, i.e. a symmetric normal form is constructed of which the ESS are analysed. Two examples in this section illustrate this procedure, and the second one clearly demonstrates its drawback: due to spurious duplication of mixed strategies in the normal form, the normal form might not have an ESS although the original situation has a stable outcome.[9] The examples lead to the conclusion that, in asymmetric contests, evolutionary stability is more sensibly defined by means of behavior strategies. The general model and the formal definitions are given at the end of this section. The discussion is based on Selten [1980, 1983]. For

		Intruder	
		H	D
Owner	H	$\frac{1}{2}(V-C)$; $\frac{1}{2}(v-C)$	V ; 0
	D	0 ; v	$\frac{1}{2}V$; $\frac{1}{2}v$

Fig. 9.5.1. The Hawk-Dove Game with ownership. $(0<v<V<C)$

further examples concerning the role of asymmetries, the reader is referred to Maynard Smith and Parker [1976] and Hammerstein [1981].

Consider the Hawk-Dove game of Fig. 9.1.1, but now assume that one of the contestants is the owner of the territory and that the other is an intruder. Also suppose that, since the owner knows local conditions better, the value V of the territory to the owner is larger than the intruder's value v, but both values are smaller than the cost C of an escalated conflict. The situation can then be represented by the (asymmetric) bimatrix game of Fig. 9.5.1.

In this situation, an animal can condition its action on the role it is in, hence, the two asymmetric equilibria (H, D) and (D, H) do make sense. Note that both of these are strict and, hence, are stable in every respect. To see whether there are other stable outcomes, let us convert the game into a symmetric one. Assume that the random elements determining ownership are independent of the genes determining fighting behavior. Then an individual can find itself both in the position of owner and in the position of intruder; both possibilities are equally likely and independent of the strategy adopted. Consequently, the situation can be represented by the symmetric game of Fig. 9.5.2 in which $\alpha\beta$ denotes 'choose α when owner, β when intruder'.

It is easily checked that the strict equilibria (H, D) and (D, H) of the game of Fig. 9.5.1, induce ESS in the symmetric normal form of Fig. 9.5.2 (in fact (HD, HD) and (DH, DH) are again strict equilibria). From Lemma 9.2.4 it follows that there is only one remaining candidate for an ESS, viz.

$$M = pHH + (1-p)DD \quad \text{with } p = \frac{V+v}{2C}\,. \tag{9.5.1}$$

This strategy corresponds to the ESS of the game without ownership. However, since HD fares better against M than M does, (M, M) is not even a symmetric equilibrium. Hence, there are only 2 ESS, viz. the *bourgeois strategy HD* (choose hawk when owner, dove when intruder) and the *paradoxical strategy DH* (dove when owner, hawk when intruder). (These names have been introduced by John Maynard Smith.) Note that, when an ESS is played, the conflict is resolved peacefully, there are no escalated conflicts.

Although, in the game of Fig. 9.5.2, both HD and DH are strict equilibrium strategies one usually observes the bourgois convention in nature (see, however, Burgess [1976] for an example of the paradoxical convention being used).

	HH	HD	DH	DD
HH	$V+v-2C$	$2V+v-C$	$V+2v-C$	$2V+2v$
HD	$V-C$	$2V$	$V+v-C$	$2V+v$
DH	$v-C$	$V+v-C$	$2v$	$V+2v$
DD	0	V	v	$V+v$

Fig. 9.5.2. The symmetric normal form of the Hawk-Dove game with ownership. To obtain actual fitness, all entries should be divided by 4. As in the previous sections, only the payoff matrix of player 1 is listed

Actually, there are at least two game theoretic reasons why it is justified to call *DH* paradoxical:

(i) This convention leads to infinite regress: As soon as the owner looses a contest, he becomes the intruder and by challenging the current owner, he again becomes the owner, etc. If one incorporates this aspect and models the situation as a dynamic game, *DH* disappears.

(ii) The bourgeois convention *HD* can evolve from a population that ignores the asymmetry (and, hence, plays the strategy *M* from (9.5.1)), but the paradoxical convention *DH* cannot. Specifically, *HD* fares better against *M* than *M* does, but *DH* fares worse, hence, the trajectory of (9.4.3) starting in *M* converges to *HD*. (Note that also *HD* is Pareto superior to *DH*, hence, this is better for the species, but it is not clear that this aspect is relevant.)

In the example above, the normal form analysis is satisfactory: the ESS of the normal form correspond to the outcomes dictated by common sense. We will now present an example in which this is not the case.

Again, consider the Hawk-Dove game of Fig. 9.1.1. Now, suppose that there are no differences between the contestants, but that behaviour can be conditioned on atmospheric conditions. Specifically, it may rain or shine and both possibilities are equally likely. Weather conditions influence the value of the territory: if it is sunny, this value is V, while if it rains the value is v. Again we assume $0<v<V<C$. Hence, payoffs depend on atmospheric conditions as in Fig. 9.5.3.

It is clear what the solution of this game should be: if it is sunny, the animals should play the ESS of the game of the left hand side of Fig. 9.5.3, while if it is raining they should play the ESS of the right hand side game. Hence, the solution should be the behavior strategy defined by

$$b_s(H) = \frac{V}{C} \quad \text{and} \quad b_r(H) = \frac{v}{C}, \tag{9.5.2}$$

Shine

	$\frac{1}{2}(V-C)$ / $\frac{1}{2}(V-C)$	V / 0
	0 / V	$\frac{1}{2}V$ / $\frac{1}{2}V$

Rain

	$\frac{1}{2}(v-C)$ / $\frac{1}{2}(v-C)$	v / 0
	0 / v	$\frac{1}{2}v$ / $\frac{1}{2}v$

Fig. 9.5.3. The Hawk-Dove Game in which the value depends on weather conditions

	HH	HD	DH	DD
HH	$V+v-2C$	$V+2v-C$	$2V+v-C$	$2V+2v$
HD	$V-C$	$V+v-C$	$2V$	$2V+v$
DH	$v-C$	$2v$	$V+v-C$	$V+2v$
DD	0	v	V	$V+v$

Fig. 9.5.4. The normal form game of the contest described by Fig. 9.5.3. The pure strategy $\alpha\beta$ prescribes to choose α when it is sunny and β when it is rainy. To obtain actual fitnesses all entries should be divided by 4

($b_s(H)$ denotes the probability of choosing H when the sun is shining). However, this solution is not obtained from a normal form analysis. The normal form of this contest is represented by Fig. 9.5.4, in which $\alpha\beta$ denotes the strategy: 'choose α when sunny, β when rainy'. It is easily checked that a mixed strategy is a symmetric equilibrium strategy of this normal form if and only if it induces (as in (6.1.2)) the behavior strategy from (9.5.2), hence, p is a symmetric equilibrium strategy if and only if

$$p(HH)+p(HD)=b_s(H) \quad \text{and} \quad p(HH)+p(DH)=b_r(H).$$

Consequently, there are no isolated symmetric equilibrium strategies and Corollary 9.2.5 shows that the normal form does not have an ESS.

Note that the nonexistence of ESS in the game of Fig. 9.5.4 is caused by the spurious duplication inherent in mixed strategies: there are many mixed strategies that induce the same behavior strategy. This problem does not arise in the game of Fig. 9.5.2 since in that case the ESS are pure strategies, however, as soon as there is randomization the problem does arise. Therefore, it is important to avoid this spurious duplication and to define ESS directly by means of behavior strategies. Before giving this formal definition, the general asymmetric game model is introduced. We follow Selten [1980].

Assume an individual can find itself in one of several *information situations*. Such an information situation provides a complete description of the state (role) of the individual itself, as well as a (possibly incomplete) description of the opponent's state. (In Fig. 9.5.1, there are 2 information situations, viz. owner and intruder; in Fig. 9.5.3, the situations are rain and shine.) The (finite) set of all information situations is denoted by U and, if $u \in U$, then C_u denotes the (finite) set of choices at u.

A *contest situation* is identified with a pair, uv, of information situations. If, in this situation, the u-individual chooses $c_u \in C_u$ and the v-individual chooses c_v, then the *payoffs* are

$$A_{uv}(c_u, c_v) \quad \text{and} \quad A'_{uv}(c_u, c_v), \tag{9.5.3}$$

to the u-individual and the v-individual, respectively, where A_{uv} and A'_{uv} are matrices of dimension $|C_u| \times |C_v|$. It is assumed that such matrices are given for every possible contest situation. If both players are in the same information situation, the game must be symmetric hence

$$A_{uu}(c_u, c'_u) = A'_{uu}(c'_u, c_u) \quad (\text{all } u, c_u, c'_u). \tag{9.5.4}$$

For an individual in information situation u, there is a probability $p(v|u)$ of having a contest with a v-individual. The choice taken at u will depend on these probabilities as well as on what possible opponents will do. Define a local strategy at u as a probability distribution on C_u and let a *behavior strategy* b be a map that assigns a local strategy b_u to every $u \in U$. Write B_u for the set of all local strategies at u and B for the set of all behavior strategies. The local payoff function A_u is defined by

$$A_u(b_u, b') := \sum_v p(v|u) A_{uv}(b_u, b'_v), \tag{9.5.5}$$

where $b_u \in B_u$, $b' \in B$ and

$$A_{uv}(b_u, b'_v) := \sum_{c \in C_u} \sum_{c' \in C_v} b_u(c) b'_v(c') A_{uv}(c, c'). \tag{9.5.6}$$

$A_u(b_u, b')$ is the expected payoff to an u-individual playing b_u whose opponents play b'.

Next, assume that the probability distributions $\{p(\cdot|u); u \in U\}$ are *consistent* in the sense of Harsanyi [1968], i.e. they can be derived from a common prior, hence, contest situations can be thought of as being generated randomly. Furthermore, introduce 2 artificial *players* and for each contest situation allocate the players randomly to the information situations of the contest. If we write $p(u, v)$ for the probability that contest situation uv arises with player 1 in information situation u, then

$$p(u, v) = p(v, u) \quad (\text{all } u, v \in U), \tag{9.5.7}$$

hence, p is a symmetric probability distribution on $U \times U$. By restricting U if necessary, we can assume that

$$p(u) := \sum_{v} p(u, v) > 0 \quad (\text{all } u \in U), \tag{9.5.8}$$

and the requirement of consistency implies that

$$p(v|u) = \frac{p(u, v)}{p(u)} \quad (\text{all } u, v \in U). \tag{9.5.9}$$

Now, if b and b' are the behavior strategies played by player 1 and player 2 respectively, then the ex-ante expected payoff to player 1 is equal to

$$A(b, b') := \sum_{u} p(u) A_u(b_u, b'). \tag{9.5.10}$$

In view of (9.5.4) and (9.5.7), the expected payoff to player 2 equals

$$A'(b, b') := \sum_{u} p(u) A_u(b'_u, b) \tag{9.5.11}$$

and we obviously have

$$A(b, b') = A'(b', b) \quad (\text{all } b, b'),$$

so that (9.5.10) – (9.5.11), when restricted to pure strategies, define a *symmetric normal form* game. However, we have already seen that it is more appropriate to define evolutionary stability by means of behavior strategies. The formal definition involves the analogues of the conditions (9.1.8) and (9.1.9) and is given below. Definition 9.5.1 summarizes the above construction.

Definition 9.5.1. An asymmetric contest is a quintuple $\Gamma = (U, C, p, A, A')$ where U is a finite nonempty set, $C = \{C_u;\ u \in U\}$ is a collection of nonempty sets, p is a symmetric probability distribution on $U \times U$ and (A, A') is a collection of bimatrices as in (9.5.3), satisfying (9.5.4), where 2 matrices are defined for every pair (u, v) with $p(u, v) > 0$. The (symmetric) normal form of Γ is the game defined by (9.5.10) – (9.5.11).

Definition 9.5.2. A behavior strategy b^* is an ESS of the asymmetric contest Γ if (b^*, b^*) is a symmetric equilibrium, i.e.

$$A(b^*, b^*) = \max_{b \in B} A(b, b^*), \tag{9.5.12}$$

which satisfies the following stability condition

$$\text{if } b \in B \setminus \{b^*\} \text{ and } A(b, b^*) = A(b^*, b^*), \text{ then } A(b^*, b) > A(b, b). \tag{9.5.13}$$

It is easily verified that, the behavior strategy from (9.5.2) is indeed an ESS of the game of Fig. 9.5.3 and that this is the unique ESS. Furthermore, in the game of Fig. 9.5.1, the bourgeois strategy *HD* and the paradoxial strategy *DH* are the only 2 strategies that are ESS.

In the next section it will be investigated why in the first example a normal form analysis was appropriate and why this was not the case in the second example. In that section it will also be motivated why evolutionary stability is defined by means of the global payoff function from (9.5.10), rather than by using the local payoff from (9.5.5).

9.6 Asymmetric Contests: Results

In this section it is first shown that the stability requirement in the definition of an ESS cannot be replaced by purely local conditions: a locally stable strategy need not be an ESS. Next, truly asymmetric contests are considered, i.e. contests in which role identification cannot occur. It is shown that, in such a contest, an ESS must be a strict equilibrium. This result considerably simplifies the analysis of many situations and it also shows that evolutionary stability is a very stringent requirement. In the second part of the section it is investigated whether this requirement is perhaps too restrictive.

Let $\Gamma = (U, C, p, A, A')$ be an asymmetric contest, let $b^* \in B$, $u \in U$ and $b_u \in B_u$. If b^* is an ESS of Γ, then it follows from (9.5.8) and (9.5.12) that

$$A_u(b_u, b^*) \leqq A_u(b^*_u, b^*) . \tag{9.6.1}$$

If we have equality in (9.6.1), then $b^*\backslash b_u$ is a best reply against b^*, hence, if $b_u \neq b^*_u$, then b^* should fare better against $b^*\backslash b_u$ than $b^*\backslash b_u$ itself

$$\text{if } b_u \neq b^*_u \text{ and } A_u(b_u, b^*) = A_u(b^*_u, b^*), \text{ then } A_u(b_u, b^*\backslash b_u) < A_u(b^*_u, b^*\backslash b_u) . \tag{9.6.2}$$

Following Selten [1983], let us call a behavior strategy b^* a *locally stable strategy* (shortly LSS) if it satisfies the conditions (9.6.1) and (9.6.2) for every $u \in U$ and $b_u \in B_u$. The concept of LSS is the relevant one when different information situations are under separate genetic control, the ESS concept allows for linkage between information sets.

Define the *local game at u induced by* b^* as that symmetric bimatrix game with pure strategy set C_u and with fitness matrix $\bar{A}_u$ defined by

$$\bar{A}_u(c, c') = A_u(c, b^*\backslash c') \quad (c, c' \in C_u) . \tag{9.6.3}$$

Clearly, b^* is a locally stable strategy if and only if b^*_u is an ESS of the local game at u induced by b^* for every $u \in U$. The local conditions (9.6.1) – (9.6.2) for an LSS are easier to check than the global conditions (9.5.12) – (9.5.13) for an ESS. Unfortunately, although the local conditions are necessary ones, they are not sufficient for evolutionary stability.

Theorem 9.6.1. *Every ESS is an LSS, but not conversely.*

Proof. Above we have already shown that an ESS is an LSS, hence, it suffices to construct an example of an LSS that is not an ESS. Consider the contest in which there are 2 information situations, $U = \{u, v\}$, in which the probability distribution

p is given by $p(u,u)=p(u,v)=p(v,u)=p(v,v)=1/4$ and in which the bimatrices of the contest situations uu, uv and vv are given by, respectively,

uu	L_u	R_u
L_u	0, 0	2, 0
R_u	0, 2	0, 0

uv	L_v	R_v
L_u	2, 2	0, 0
R_u	0, 0	2, 2

vv	L_v	R_v
L_v	0, 0	2, 0
R_v	0, 2	0, 0

Note that this indeed describes an asymmetric contest: the conditions (9.5.4) and (9.5.7) are satisfied. We claim that $b^*=R_uR_v$ is an LSS of this game that is not an ESS. Since the situation is symmetric with respect to u and v, it suffices to verify that (9.6.1) and (9.6.2) are satisfied for u. Now, the local game at u induced by b^* has fitness matrix

	L_u	R_u
L_u	0	1
R_u	1	1

and R_u is indeed an ESS of this local game. Hence, b^* is indeed an LSS. From the local game we see that L_u is an alternative best reply at u against b^*, hence, $b=L_uL_v$ is an alternative best reply against b^*

$$A(b, b^*)=A(b^*, b^*)=1 .$$

However, b fares better against b than b^* does

$$A(b, b)=1 \quad \text{and} \quad A(b^*, b)=0,$$

so that condition (9.5.13) is violated, b^* is not an ESS. □

Recall that, in the previous section, we found there was a big difference between the game of Fig. 9.5.1 and that of Fig. 9.5.3. In the first the ESS are pure and a normal form analysis is appropriate, while in the second the ESS involves randomization so that the normal form produces spurious duplication. The difference is caused by the fact that in the former role identification cannot occur, whereas in the latter this is inevitable. Let us call a contest *truly asymmetric* if differences can be determined unambiguously, i.e.

$$p(u, u)=0 \quad \text{for all } u\in U . \tag{9.6.4}$$

In a truly asymmetric contest, an u-individual never meets an u-individual, and in this case (9.5.5) defines an $|U|$-person game, in which there is 1 player for every information situation of the contest. This game will be called the *agent normal form* of the contest (cf. Sect. 6.4).

If (b^*, b^*) is a Nash equilibrium of a truly asymmetric contest and if b_u is an alternative best reply at u against b^*, then a mutant playing $b^* \backslash b_u$ is not selected against since it never meets itself. Consequently, an LSS of such a game must be strict and, hence, an ESS. This is formally proved in the next theorem, which also demonstrates that, in truly asymmetric contests, there is no discrepancy between the normal form and the agent normal form.

Theorem 9.6.2 (Selten [1980]). *For a truly asymmetric contest Γ, the following assertions are equivalent*

(i) *b^* is an ESS of Γ,*
(ii) *b^* is an LSS of Γ,*
(iii) *b^* is a strict equilibrium of the agent normal form of Γ,*
(iv) *(b^*, b^*) is a strict equilibrium of the normal form of Γ, and*
(v) *b^* is an ESS of the normal form of Γ.*

Proof. (i)$\rightarrow$(ii): This follows from Theorem 9.6.1.
(ii)$\rightarrow$(iii): If Γ is truly asymmetric, then the local fitness matrix at u has constant rows for every $u \in U$, (for every $b^* \in B$ and $c \in C_u$, the payoff $A_u(c, b^* \backslash c')$ is independent of c'). It is easily checked that in any such game an ESS must be a strict equilibrium. Hence, if b^* is an LSS, then b_u^* is the unique best reply against b^* at every u, therefore, b^* is a strict equilibrium of the agent normal form.
(iii)$\rightarrow$(iv): This implication follows from (9.5.8). Since all information situations are possible, b is a best reply against b^* if and only if b prescribes a local best reply at every information set (cf. Theorem 6.2.1).
(iv)$\rightarrow$(v): This follows from Lemma 9.2.2.
(v)$\rightarrow$(i): The Defs. 9.1.1 and 9.5.1 imply that, if s is an ESS of the normal form and b is the behavior strategy induced by s, then b is an ESS of Γ, hence, this implication holds for every asymmetric contest. □

In the proof above we have seen that in general (iv)$\rightarrow$(v)$\rightarrow$(i)$\rightarrow$(ii), and we already know that the reverse implications need not hold for games that are not truly asymmetric. The power of Theorem 9.6.2 lies in the implication (iii)$\rightarrow$(i): in truly asymetric contest, to find the set of ESS, one just has to check all pure strategies and this is an easy task which can be performed without actually computing the agent normal form or the normal form. The theorem shows that, in situations in which asymmetries can be determined unambigously, the ESS concept is very restrictive. It will now be shown that as a consequence it is easy to construct examples without ESS.

Let us consider the Battle of the Sexes as described in Dawkins [1976]. This game deals with the male/female conflict on how to divide the effort of rearing offspring. Specifically, the question is considered of whether the female, by insisting on a long courtship period, can prevent the male from leaving after

Male \ Female	Coy	Fast
Faith.	$V-C-c$; $V-C-c$	$V-C$; $V-C$
Phil.	0 ; 0	V ; $V-2C$

Fig. 9.6.1. Battle of the Sexes

having copulated. The male has an incentive to leave since, by doing so, he avoids the costs of rearing and increases the chance of finding a new mate.

For simplicity, assume that both males and females can adopt only 2 strategies. Females can be coy or fast, males can be faithful or philanderers. Coy females insist on a long courtship period, fast ones copulate immediately. Faithful males are willing to go through a long courtship period and they will help in raising the young. Philanderers go off immediately if a female does not copulate right away and, after copulation, they do not stay, but go off searching for fresh females. Let V be the value (to both partners) of having offspring, let $2C$ be the cost of rearing (which is to be shared among the partners) and let c be the cost of prolonged courtship. Then, the agent normal form of the contest is given by the bimatrix in Fig. 9.6.1. If $V>2C$, it pays the female to raise the offspring on her own and the unique ESS requires females to be fast and males to be philanderers. A more interesting situation arises if

$$0<c<C<C+c<V<2C.$$

In this case, there is no strict equilibrium and, consequently, there is no ESS. The game admits a unique equilibrium in completely mixed strategies, however, and this is regular, hence, it is stable in many ways (cf. the surveys). Therefore, it is appropriate to consider the question of whether this equilibrium failing to be an ESS is perhaps caused by Def. 9.6.2 being too restrictive. To answer this question, we will first investigate whether this equilibrium corresponds to a stable state of a dynamical system as in (9.4.3).

It will be convenient to consider the entire class of non-degenerate 2×2 bimatrix games with a unique, completely mixed equilibrium. Let the payoff matrices be A and A' and let p and q be mixed strategies of player 1 and player 2, respectively. Then the analogue of (9.4.3) is the system

$$\dot{p}_i=p_i[e_i^T Aq-p^T Aq] \quad (i=1, 2), \tag{9.6.5a}$$

$$\dot{q}_i=q_i[p^T A'e_i-p^T A'q] \quad (i=1, 2). \tag{9.6.5b}$$

These equations have been studied in Schuster and Sigmund [1981]. Note that (9.6.5a) is invariant under transformation as in (9.4.6) and that (9.6.5b) is invariant under similar transformations applied to the rows of A'. Therefore, we can assume that the bimatrix game (A, A') is as in Fig. 9.6.2: up to a permutation

a / 0	0 / b
0 / c	d / 0

Fig. 9.6.2. The class of 2×2 bimatrix games with a unique completely mixed equilibrium $(a, b, c, d>0)$

of the strategies, every game in the class under consideration is equivalent to exactly one game from Fig. 9.6.2.

Let us write $x=p_1$ and $y=q_1$ for the probabilities that the players choose their first strategies. The unique equilibrium of the game of Fig. 9.6.2. is given by

$$x^*=\frac{c}{b+c},\quad y^*=\frac{d}{a+d} \tag{9.6.6}$$

and the dynamic (9.6.5) reduces to

$$\dot{x}=(a+d)x(1-x)(y-y^*), \tag{9.6.7a}$$

$$\dot{y}=-(b+c)y(1-y)(x-x^*). \tag{9.6.7b}$$

Clearly, the Nash equilibrium (x^*, y^*) is a fixed point of this dynamic. We claim that it is a stable fixed point that is not asymptotically stable. Namely, consider the function Z

$$Z(x, y):=x^c(1-x)^b y^d(1-y)^a \quad (x, y\in[0, 1]), \tag{9.6.8}$$

which has its unique maximum at (x^*, y^*) (cf. the map Z in (9.4.8)). Then it is easily checked that

$$\dot{Z}=\frac{\partial Z}{\partial x}\dot{x}+\frac{\partial Z}{\partial y}\dot{y}=0,$$

so that (9.6.7) is a conservative system: The orbits are the level curves of Z and hence, are closed. Finally note that (9.6.7a) is equivalent to

$$\frac{d}{dt}\log\frac{x}{1-x}=(a+d)(y-y^*),$$

so that, if $\{x(t), y(t)\}_{t=0}^{T}$ is a closed orbit with length T in the interior of the strategy space

$$0=\log\frac{x}{1-x}\bigg|_0^T=(a+d)\int_0^T(y(t)-y^*)\,dt,$$

hence

$$y^*=\frac{1}{T}\int_0^T y(t)\,dt.$$

Similarly,

$$x^* = \frac{1}{T} \int_0^T x(t)\,dt,$$

so that the equilibrium (x^*, y^*) is the time average of every orbit (cf. Theorem 9.4.10). We have derived

Theorem 9.6.3 (Schuster and Sigmund [1981]). *Let Γ be a 2×2 bimatrix game with a unique and completely mixed Nash equilibrium. This equilibrium is a stable fixed point of the system (9.6.5) that is not asymptotically stable. The orbits of (9.6.5) are closed and have the Nash equilibrium as their time average.*

This theorem allows us to conclude that nonconverging oscillations will occur in the battle of the sexes game of Fig. 9.6.1 if the dynamic (9.6.5) is appropriate. This supports the conclusion that the mixed equilibrium is not qualified for an ESS. However, it is not completely clear that (9.6.5) is the appropriate dynamic. Namely, the argument from the beginning of Sect. 9.4 suggests that the right hand sides of (9.6.5) should be divided by p^TAq and $p^TA'q$, respectively, and in this asymmetric situation (if $p^TAq \neq p^TA'q$) this does not produce an equivalent system. This is most easily seen by taking $a=b=c=d=1$ in Fig. 9.6.2. In this case, the function Z from (9.6.8) is a Lyapunov function for the modified dynamic since $\dot{Z}>0$ as the reader easily checks. Hence, the Nash equilibrium $(\frac{1}{2}, \frac{1}{2})$ is an asymptotically stable fixed point of the modified dynamic. In Hofbauer [1985] it is shown that, for this modified dynamic, the mixed equilibrium of the game of Fig. 9.6.2 is always asymptotically stable.

Theorem 9.6.2 crucially depends on the game being truly asymmetric. In the concluding example of this section, we show that if there is only a very small probability that role identification is possible, an ESS can be in mixed strategies. This shows that Theorem 9.6.2 is not robust with respect to perturbations in the model. However, whether such perturbations are biologically meaningful must be determined on a case by case basis. For example, if the asymmetry concerns sex (as in Fig 9.6.1), it probably is not sensible to assume that the female can mistake itself for a male, but if the asymmetry concerns size or ownership, such errors in role identification can very well occur. For further elaborations on errors in role assessment, the reader is referred to Parker and Rubenstein [1981] and Hammerstein and Parker [1982].

Assume 2 animals are contesting for a resource; one animal owns the resource, the other is an intruder. The resource has the same value for the intruder as for the owner. Furthermore, the owner and the intruder have the same actions available. Hence, there only exists an asymmetry in roles, not in payoffs or strategies. Assume that the payoff matrices are given by Fig. 9.6.3.

In the game of Fig. 9.6.3, the unique equilibrium is completely mixed, hence, if the asymmetry can be determined unambiguously, there is no ESS. More precisely, if $\Gamma = (U, C, p, A, A')$ is the contest defined by $U = \{u, v\}$ (u=owner, v=intruder), $C_u = C_v = \{1, 2, 3\}$

$$p(u, u) = p(v, v) = 0, \quad p(u, v) = p(v, u) = \tfrac{1}{2}$$

and (A_{uv}, A'_{uv}) is the bimatrix game from Fig. 9.6.3, then Γ does not have an ESS. However, now assume that there is a small probability ε that the intruder mistakes

0, 0	1, 0	0, 1
0, 1	0, 0	1, 0
1, 0	0, 1	0, 0

Fig. 9.6.3. A game without an ESS, but with a neutrally stable strategy that becomes an ESS when errors of role identification occur

itself for the owner and vice versa. Then the situation is better represented by the contest $\Gamma(\varepsilon)$ which differs from Γ only in the probability distribution p which is now given by

$$p(u, u) = p(v, v) = \varepsilon, \quad p(u, v) = p(v, u) = \tfrac{1}{2}(1-2\varepsilon)\,.$$

(In the contest situations uu and vv the payoffs are again given by Fig. 9.6.3.) Note that the equilibrium strategy $(1/3, 1/3, 1/3)$ of the game of Fig. 9.6.3 is an ESS of the matrix A_{uv}. From this it easily follows that the strategy that requires both the owner and the intruder to play $(1/3, 1/3, 1/3)$ is an LSS of the contest $\Gamma(\varepsilon)$ for every $\varepsilon > 0$. With a little more effort one can show that this strategy is in fact an ESS of $\Gamma(\varepsilon)$ for every $\varepsilon > 0$, hence, the ESS correspondence is discontinuous at $\varepsilon = 0$. (Note that it is essential that both the owner and the intruder make mistakes; if only one of them does, there is no ESS.)

9.7 Contests in Extensive Form: Definitions

In the final two sections of this chapter, contests that involve repeated interaction are analysed. In this section it is first shown that such a contest is not completely specified by an extensive form, but that one also has to know which symmetry is the appropriate one. Next, it is argued that, in dynamic contests, Def. 9.5.2 is too restrictive and a more liberal ESS concept based on perturbed games (i.e. incorporating small mistake probabilities) is introduced.

The section is based on Selten [1983].

Once again, consider the game of Fig. 9.5.1, but now take the special case with $V = v = 10$ and $C = 20$. The symmetrization of this game can be represented by means of an extensive form in which there is an initial chance move determining which player becomes the owner followed by a Hawk-Dove subgame. This extensive form game is given in Fig. 9.7.1, in which 'l' should be interpreted as 'player 1 is owner'.

However, if, in Fig. 9.7.1, one interprets the chance choice l as sunshine and r as rain, the game represents the contest from Fig. 9.5.3 (again with $V = v = 10$ and

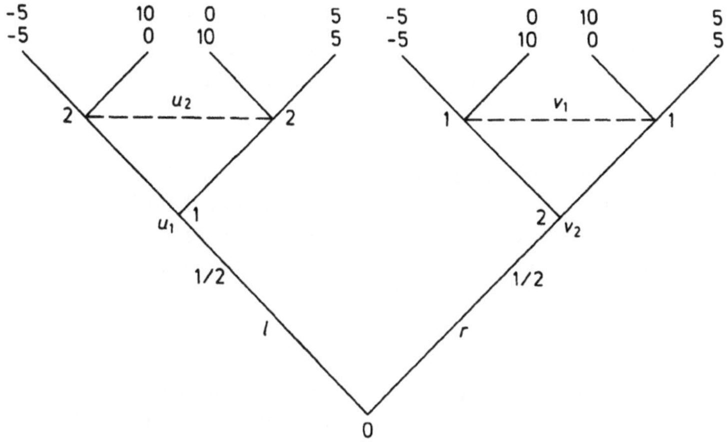

Fig. 9.7.1. An extensive form game with 2 symmetries that represents both the contest from Fig. 9.5.1 and the contest of Fig. 9.5.3

$C=20$). Hence, the tree in itself does not uniquely specify a contest, rather one has to know how to interpret the choices in the tree. In other words, one has to know which choices of player 1 correspond to which choices of player 2, hence, which symmetry of the game is the appropriate one. The game of Fig. 9.7.1 admits 2 symmetries: a horizontal one yielding the interpretation of Fig. 9.5.1 (choosing L at u_1 corresponds to choosing L at v_2) and a vertical one yielding the interpretation from Fig. 9.5.3. The obvious conclusion is that a dynamic contest is not merely represented by an extensive form game, but by such a game together with a symmetry.

Recall that a *2-person extensive form game* is a septet $\Gamma = (K, P, U, C, p, a, a')$, where K, P, U, C and p are as in Sect. 6.1 and where a and a' are the payoffs to player 1 and player 2, respectively. Throughout, attention will be restricted to games with perfect recall (Assumption 6.1.1), hence, it is assumed that evolution can adjust behavior to any past experience. Following Selten [1983], a *symmetry* of Γ is defined as a mapping $(\cdot)^T$ from choices to choices with the following properties (C_i denotes the set of choices of player i in Γ).

if $c \in C_0$, then $c^T \in C_0$ and $p(c) = p(c^T)$, (9.7.1a)

if $c \in C_i$, then $c^T \in C_j$ $(i \neq j \in \{1, 2\})$, (9.7.1b)

$(c^T)^T = c$ for all c, (9.7.1c)

for every information set u there exists an information set u^T such that every choice at u is mapped onto a choice at u^T, (9.7.1d)

for every endpoint z there exists an endpoint z^T such that if z is reached by the sequence $c_1 c_2 \ldots c_k$, then z^T is reached by (a permutation of) $c_1^T c_2^T \ldots c_k^T$, and (9.7.1e)

$a(z) = a'(z^T)$ for every endpoint z. (9.7.1f)

A *symmetric extensive form game* is a pair (Γ, T) where Γ is an extensive form game and where T is a symmetry of Γ. If b is a behavior strategy of player 1 in (Γ, T) then the *symmetric image* of b is the behavior strategy b^T of player 2 defined by

$$b_u^T(c) := b_{u^T}(c^T) \quad (u \in U_2, c \in C_u). \tag{9.7.2}$$

If b_1, b_2 are behavior strategies of player 1, then the probability that the endpoint z is reached when (b_1, b_2^T) is played is equal to the probability that z^T is reached when (b_2, b_1^T) is played (use (9.7.1a) and (9.7.1e)). Therefore, in view of (9.7.1f), the expected payoff to player 1 when (b_1, b_2^T) is played is equal to player 2's expected payoff when (b_2, b_1^T) is played.

$$A(b_1, b_2^T) = A'(b_2, b_1^T). \tag{9.7.3}$$

Equation (9.7.3), when restricted to pure strategies, defines the *symmetric normal form* of (Γ, T). However, we have already seen that evolutionary stability cannot be sensibly defined by means of this normal form. (Note that the extensive form model encompasses all models discussed earlier in this chapter.) It will now be shown that also a direct analogue to Def. 9.5.1 produces unsatisfactory results. Specifically, define a *direct ESS* of (Γ, T) as a behavior strategy $\bar{b}$ of player 1 that satisfies

$$A(\bar{b}, \bar{b}^T) = \max_{b \in B} A(b, \bar{b}^T), \tag{9.7.4}$$

and

$$\text{if } b \in B \setminus \{\bar{b}\} \text{ and } A(b, \bar{b}^T) = A(\bar{b}, \bar{b}^T), \text{ then } A(b, b^T) < A(\bar{b}, b^T). \tag{9.7.5}$$

(B denotes the set of all behavior strategies of player 1). A simple example will demonstrate that in many games all intuitively acceptable solutions will fail to satisfy the conditions of a direct ESS.

Consider the game of Fig. 9.7.2 (taken from Selten [1983]), which describes an aspect of the male-female conflict on raising offspring and of which the

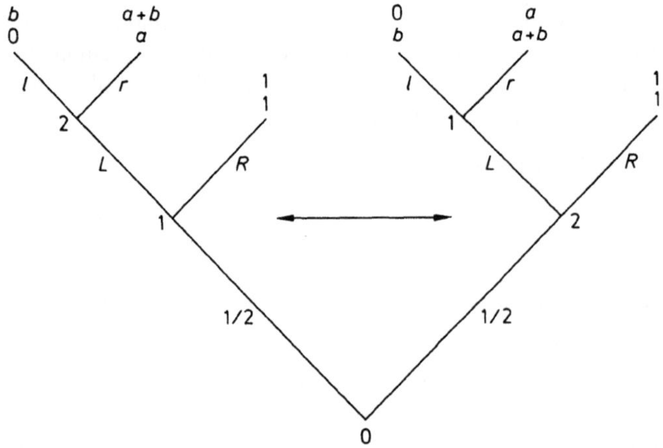

Fig. 9.7.2. A male desertion game showing that the concept of direct ESS is too restrictive. (The symmetry is indicated by the *double arrow*)

interpretation is as follows. The initial random move determines which sex is called player 1 (if chance chooses left, player 1 is the male). The male has a choice between helping to raise the offspring ($=R$) and leaving ($=L$). If the male leaves, the female has to choose between rearing the young on her own ($=r$) or to leave as well ($=l$). For simplicity, it is assumed that it is impossible for the female to find a second mate and that she will help raising in case the male does the same. It is also assumed that the value of offspring is 1, that the probability that the female on her own successfully raises the young is a and that the male's expected value of finding a new mate is b (with $0<a, b<1$).

First suppose $a+b>1$. In this case, there is a unique subgame perfect equilibrium, viz. Lr (the male leaves and the female raises the offspring on her own) and it is easily checked that this is a direct ESS. There is a second (imperfect) equilibrium (viz. Rl, in which the female threatens to leave in case the male does so), but this is not a direct ESS: since the female's information set is not reached, r is an alternative best reply against R and Rr violates condition (9.7.5).

Next, consider the case in which $a+b<1$. There again is a unique subgame perfect equilibrium, viz. Rr, but this is not a direct ESS. Since the female is not called upon to move, Rl is an alternative best reply that violates (9.7.5). One should, however, admit that the female never having to make a decision is only due to the overexactness of our model. Actually, one should also take into account all kinds of rare events such as the male having a fatal accident, or him being eaten by a predator, and these cause that the female has to make a decision, if only with a small probability, in case the male intends to choose R. These rare events may exert sufficient selective pressure to force the female to choose r: if she is actually reached, r is better than l and females choosing l are selected against. Hence, Rr should be considered evolutionary stable and the concept of direct ESS is too restrictive.

The above discussion suggests a natural and meaningful way to define a coarser ESS concept in extensive form games: Look at limit ESS, i.e. at limit points of direct ESS of perturbed games. A perturbed game can be viewed as a model that allows for rare events without forcing the analyst to model these in detail, hence, this approach is both convenient and sensible.

Recall that perfect equilibria have also been defined by means of perturbed games and that in that context it has been important that all mistake probabilities be positive. In the current situation, however, we want to relax the concept of direct ESS and, therefore, in Def. 9.7.1, the mistake probabilities are allowed to be zero (this is also justified by Theorem 9.8.3 and Corollary 9.8.6). Hence, we come to the following definitions.

Definition 9.7.1. Let (Γ, T) be a symmetric extensive form game. A perturbance of (Γ, T) is a mapping η from the set of choices into the reals such that

$$\eta_c \geqq 0, \quad \eta_c = \eta_{cT} \text{ (all } c\text{)} \quad \text{and} \quad \sum_{c \in C_u} \eta_c < 1 \quad \text{(all } u\text{)}. \tag{9.7.6}$$

The *perturbed game* (Γ, T, η) is the game with the same structure as (Γ, T) but in which a behavior strategy b is admissible only if $b_u(c) \geqq \eta_c$ for all u and c. The set of all such admissible strategies of player 1 is denoted by $B(\eta)$ and $B_u(\eta)$ denotes the set of local strategies admissible at u.

Definition 9.7.2. Let (Γ, T) be a symmetric extensive form game and let (Γ, T, η) be an associated perturbed game. A strategy $\bar{b} \in B(\eta)$ is a *direct ESS* of (Γ, T, η) if it satisfies

$$A(\bar{b}, \bar{b}^T) = \max_{b \in B(\eta)} A(b, \bar{b}^T), \tag{9.7.7}$$

and

$$\text{if } b \in B(\eta) \setminus \{\bar{b}\} \text{ and } A(b, \bar{b}^T) = A(\bar{b}, \bar{b}^T), \text{ then } A(b, b^T) < A(\bar{b}, b^T). \tag{9.7.8}$$

A strategy $\bar{b} \in B$ is a *limit ESS* of (Γ, T) if there exists a sequence $\{(\eta^k, b^k)\}_{k \in \mathbb{N}}$ such that b^k is a direct ESS of (Γ, T, η^k), with $\eta^k \to 0$ and $b^k \to \bar{b} (k \to \infty)$.

Note that the special case of $\eta = 0$ is not excluded in Def. 9.7.1, hence, every game can be viewed as a perturbed game. In particular, every direct ESS of (Γ, T) is a limit ESS of (Γ, T). In the next section, some more interesting properties will be derived.[10] This section is concluded with a couple of definitions. It will be convenient to make the following innocuous assumption.

Assumption 9.7.3. At every information set in Γ there are at least 2 choices.

This assumption implies that a symmetry T of Γ induces a mapping on the set of subgames of Γ, i.e. if Γ_x is a subgame, then there exists a point x^T in the tree such that $(\Gamma_x)^T = \Gamma_{x^T}$ (Selten [1983], Theorem 1). The subgame Γ_x is said to be a *symmetric subgame* of (Γ, T) if $x = x^T$, otherwise it is said to be asymmetric. Hence, Γ_x is symmetric if it is mapped onto itself. In this case, T induces a symmetry T_x on Γ_x, hence, (Γ_x, T_x) is again a symmetric game. If Γ_x is symmetric and b is a behavior strategy of player 1, then the (b, x)*-truncation* of (Γ, T) is the game $(\Gamma_{-x}(b), T_{-x})$ that results from (Γ, T) by replacing the subgame Γ_x by an endpoint with payoffs equal to the conditional payoffs of (b, b^T) after x (T_{-x} is just the restriction of T to $\Gamma_{-x}(b)$). If Γ_x is asymmetric, then the (b, x)-truncation is defined similarly, but now both the subgame Γ_x and the subgame Γ_{x^T} are replaced by their respective conditional payoffs. Hence, the (b, x)-truncation is always a symmetric game, irrespective of whether Γ_x is symmetric. If b' is a behavior strategy of player 1, then we can decompose b' as $b' = (b'_{-x}, b'_x)$ where b'_{-x} (resp. b'_x) is that part of b' relevant to $\Gamma_{-x}(b)$ (resp. Γ_x). Note that, by splitting a perturbance η as $\eta = (\eta_{-x}, \eta_x)$ one can also speak of subgames and truncations of perturbed games. Finally, Γ_x is said to be an *essential subgame* if it contains at least one information set of player 1 or player 2.

Let b_i be a behavior strategy of player i in the game (Γ, T). We write $\mathbb{P}(x|b_1, b_2)$ for the probability that the decision point x is reached when (b_1, b_2) is played. $\mathbb{P}(u|b_1, b_2)$ denotes the probability of the information set u being reached. b_1 is said to block the information set u if $\mathbb{P}(u|b_1, b_2) = 0$ for every b_2. The strategy b_1 is *pervasive* if it does not block any information set of player 1 or player 2. Pervasiveness of strategies of player 2 is defined similarly.

9.8 Contests in Extensive Form: Results

In this section, it is first shown that a direct ESS is pervasive, i.e. every information set is reached when a direct ESS and its symmetric image are played. This result allows us to derive necessary local conditions, both for direct ESS as for limit ESS. Loosely speaking, the local conditions amount to saying that an ESS must satisfy the backwards induction principle from dynamic programming. These conditions were first derived in Selten [1983], in which it was also claimed that they are sufficient to guarantee evolutionary stability for the special class of simultaneity games. Unfortunately, the latter is not correct as the concluding example shows.

Before proving that every direct ESS is pervasive, some preliminary results are derived. Assume a symmetric game (Γ, T) is given.

Lemma 9.8.1. *$b \in B$ is pervasive if and only if $\mathbb{P}(u|b, b^T) > 0$ for every information set u of player* 1.

Proof. By symmetry, b is pervasive if and only if b^T is pervasive. Assume b is pervasive and let $u \in U_1$. Then b chooses all choices leading to u with positive probability. Since b^T does not block u, this implies that $\mathbb{P}(u|b, b^T) > 0$.

If $\mathbb{P}(u|b, b^T) > 0$ for every $u \in U_1$, then both b and b^T do not block any information set of player 1. By symmetry, b does not block any information set of player 2. Hence, b is pervasive. □

Lemma 9.8.2. *If $\mathbb{P}(u|b, b^T) > 0$, then $\mathbb{P}(u|b, (b\backslash c)^T) > 0$ for every $c \in C_u$.*

Proof. Assume that $\mathbb{P}(u|b, b^T) > 0$, but $\mathbb{P}(u|b, (b')^T) = 0$ where $b' = b\backslash c$ and $c \in C_u$. Then $\mathbb{P}(u|b', (b')^T) = \mathbb{P}(u^T|b', (b')^T) = 0$. Let $x \in u$ and $\mathbb{P}(x|b, b^T) > 0$. Since the realization probability of x depends on what is chosen at u^T, the point x must come after some decision point, say y, in u^T. Then $\mathbb{P}(y|b, b^T) > 0$ so that the realization probability of y depends on what is chosen at u, therefore, y comes after some point, say z, in u. We have $z \neq x$, hence, there exists a path that intersects u at least twice, but this is impossible. Therefore, $\mathbb{P}(u|b, (b\backslash c)^T) > 0$ for every $c \in C_u$. □

Theorem 9.8.3. *If b is a direct ESS of the perturbed game (Γ, T, η), then b is pervasive.*

Proof. Assume that $\mathbb{P}(u|b, b^T) = 0$ for some information set u of player 1. Then any admissible local strategy at u is a best reply against b^T. Hence, if $b'_u \in B_u(\eta) \backslash \{b_u\}$ and $b' = b\backslash b'_u$, then b' is a best reply. The previous lemma shows that u and u^T cannot be reached when $(b', (b')^T)$ is played. Therefore $A(b', (b')^T) = A(b, (b')^T)$ and b' violates condition (9.7.8). (Note that we used Assumption 9.7.3.) Hence, if $\mathbb{P}(u|b, b^T) = 0$ for some $u \in U_1$, then b cannot be an ESS. This proves the theorem in view of Lemma 9.8.1. □

Theorem 9.8.3 shows that the concept of direct ESS is very restrictive: Any information set should be reached when a direct ESS is played. Consequently, for a large class of (unperturbed) games there will be no direct ESS. Note, however, that a limit ESS need not be pervasive: an information set might be reached by mistake, hence, the concept of limit ESS is much more liberal a concept. Below we will show that the pervasiveness of direct ESS entails an important property of limit ESS, viz. every limit ESS is a sequential equilibrium.

Let (Γ, T) be a symmetric game and let $b \in B$ be a pervasive strategy. In view of Lemma 9.8.1 and Lemma 9.8.2, the local payoff $A_u(b\backslash c, (b\backslash c')^T)$ is defined for every $u \in U_1$ and $c, c' \in C_u$ (cf. Eq. (6.2.3)). The *local game at u induced by b* is the symmetric bimatrix game $\Gamma_{ub} = \langle C_u, C_u, A_{ub}, A'_{ub} \rangle$ of which the payoff matrix of player 1 is given by

$$A_{ub}(c, c') := A_u(b\backslash c, (b\backslash c')^T) . \tag{9.8.1}$$

If η is a perturbance as in (9.7.6), then (Γ_{ub}, η_u) is the perturbed game with payoffs A_{ub} but in which the players are restricted to choose strategies in $B_u(\eta)$.

Next, let (Γ, T, η) be a perturbed game and assume that $\bar{b} \in B(\eta)$ is a direct ESS of (Γ, T, η). Then $\bar{b}$ is pervasive and it follows from Theorem 6.2.1 that condition (9.7.7) is equivalent to

$$A_{u\bar{b}}(\bar{b}_u, \bar{b}_u) = \max_{b_u \in B_u(\eta)} A_{u\bar{b}}(b_u, \bar{b}_u) \quad (\text{all } u \in U_1) , \tag{9.8.2}$$

hence, the equilibrium condition (9.7.7) in the definition of a direct ESS can be replaced by a set of local conditions. Since local conditions are easier to check than global ones, it would be convenient if also the stability conditions (9.7.8) could be replaced by local ones. Now, if (9.7.8) is satisfied, then we must obviously have

$$\begin{aligned} &\text{if } b_u \in B_u(\eta) \backslash \{\bar{b}_u\} \text{ and } A_{u\bar{b}}(b_u, \bar{b}_u) = A_{u\bar{b}}(\bar{b}_u, \bar{b}_u), \\ &\text{then } A_{u\bar{b}}(b_u, b_u) < A_{u\bar{b}}(\bar{b}_u, b_u) . \end{aligned} \tag{9.8.3}$$

Let us call $\bar{b}$ a direct locally stable strategy (or *direct LSS*) of (Γ, T, η) if $b \in B(\eta)$ is a pervasive strategy that induces an ESS in every local game (Γ_{ub}, η_u), hence, that satisfies the conditions (9.8.2) and (9.8.3). The above discussion can be summarized by saying that every direct ESS is a direct LSS. Since the current model encompasses that of Sect. 9.6, the converse of this statement is not correct (cf. Theorem 9.6.1): although (9.8.2) is equivalent to (9.7.7), condition (9.8.3) is weaker than (9.7.8). The concluding example of the chapter will demonstrate that this also holds for games with a very special structure (viz. for the so called simultaneity games). Hence, we have

Theorem 9.8.4. *Every direct ESS is a direct LSS, but not conversely.*

Next, some consequences of Theorem 9.8.4 will be derived. Call an information set u *image confronted* if there is a path in the tree that intersects both u and u^T, otherwise call u *image detached.* If u is image detached, then player 1's local payoff at u is independent of what player 2 chooses at u^T, hence, the matrix A_{ub} from (9.8.1) has constant rows. Therefore, if u is image detached and b_u is an ESS of (Γ_{ub}, η_u), then b_u must be the unique best reply against b_u in (Γ_{ub}, η_u).

Consequently, we have the following result that generalizes a part of Theorem 9.6.1.

Corollary 9.8.5. *If b is a direct ESS of (Γ, T, η) and u is an image detached information set, then (b_u, b_u) is a strict equilibrium of (Γ_{ub}, η_u).*

Assume b is a limit ESS of (Γ, T) and, for $\eta \geqq 0$, let $b(\eta)$ be a direct ESS of (Γ, T, η) such that $b(\eta) \to b$ as $\eta \to 0$. Let $\mu(\eta)$ be the system of beliefs induced by $(b(\eta), b(\eta)^T)$ and let $\mu(\eta) \to \mu$ as $\eta \to 0$. Then (b, μ) is a consistent assessment and by taking the limit in (9.8.2) we see that b_u is a local best reply against (b, μ) at every information set u, hence, (b, b^T) is a sequential equilibrium of Γ. Furthermore, if u is image detached, then it follows from Corollary 9.8.5 that b_u must be a pure strategy. Finally, by taking the limit in (9.8.3), it follows that b_u is a neutrally stable strategy of Γ_{ub} if u is an image confronted information set for which this local game is well defined, i.e. which is not blocked by b^T. Hence, we have shown

Corollary 9.8.6. *If b is a limit ESS of (Γ, T), then*

(i) *(b, b^T) is a sequential equilibrium of Γ,*

(ii) *b_u is a pure strategy if u is image detached, and*

(iii) *b_u is a neutrally stable strategy of Γ_{ub} if u is an image confronted information set that is not blocked by b^T.*

In part (iii) of the above corollary, neutrally stable cannot be replaced by evolutionarily stable, as is illustrated by the symmetric bimatrix game with the fitness matrix from Fig. 9.8.1. As long as R is mistakenly chosen with a positive probability, L is the unique best reply, hence, $(1-\varepsilon)L+\varepsilon R$ is a direct ESS of the perturbed game (Γ, η) with $\eta = (0, 0, \varepsilon)$ if $\varepsilon > 0$. Consequently, L is a limit ESS of Γ. However, L is not an ESS of Γ, since a mutant choosing M is not selected against if R does not occur in the population.

Corollary 9.8.6 implies that a limit ESS induces an equilibrium in every subgame. It will now be shown that much more can be said. Let (Γ, T, η) be a perturbed game, let Γ_x be an essential subgame of Γ and let b be a direct ESS of $(\Gamma,$

	L	M	R
L	1	1	1
M	1	1	0
R	0	0	0

Fig. 9.8.1. L is a neutrally stable strategy, that is a limit ESS, but not a direct ESS

T, η). Since b is pervasive and Γ_x contains at least one personal information set, it follows that $\mathbb{P}(x|b, b^T) > 0$, hence Γ_x can be reached. Assume Γ_x is a symmetric subgame. Since the conditions (9.7.7) and (9.7.8) must be satified for every strategy $b \backslash b'_x$ that differs from b only in the subgame Γ_x it follows that b_x must be a direct ESS of the subgame (Γ_x, T_x, η_x). Furthermore, since (9.7.7) and (9.7.8) must be satisfied for every strategy $b \backslash b'_{-x}$ that differs from b only in the truncated game $\Gamma_{-x}(b)$, we also have that b_{-x} is a direct ESS of $(\Gamma_{-x}(b), T_{-x}, \eta_{-x})$. Next, assume that Γ_x is asymmetric. Then every information set in Γ_x is image detached and it follows from Corollary 9.8.5 that a unique best reply is prescribed at every information set in Γ_x. Therefore, (b_x, b^T_{xT}) is a strict equilibrium of the game (Γ_x, η_x) (cf. the proof of (iii)$\rightarrow$(iv) in Theorem 9.6.2). Similarly as above, it can again be shown that b_{-x} is a direct ESS in the (b, x)-truncation of (Γ, T, η). hence, we have shown

Theorem 9.8.7. *If b is a direct ESS of (Γ, T, η) and Γ_x is an essential subgame of Γ, then*

(i) $\mathbb{P}(x|b, b^T) > 0$.
(ii) *if Γ_x is symmetric, then b_x is a direct ESS of (Γ_x, T_x, η_x),*
(iii) *if Γ_x is asymmetric, then (b_x, b^T_{xT}) is a strict equilibrium of (Γ_x, η_x), and*
(iv) *b_{-x} is a direct ESS of $(\Gamma_{-x}(b), T_{-x}, \eta_{-x})$.*

Now, assume that b is a limit ESS of (Γ, T) and that Γ_x is an essential subgame of Γ. Obviously, it need not be true that Γ_x is reached when b is played. From property (ii) in the above theorem, one can conclude that b_x is a limit ESS of (Γ_x, T_x) when Γ_x is symmetric and from property (iii) it follows that b_x is pure in case Γ_x is asymmetric. One is tempted to conclude from property (iv) that b_{-x} must be a limit ESS of $(\Gamma_{-x}(b), T_{-x})$, but this need not be true. Namely, consider the game of Fig. 9.7.2 with $a+b=1$ and $0<a, b<1$. Then Rr is the unique limit ESS, but if one truncates the last information set, one obtains the game

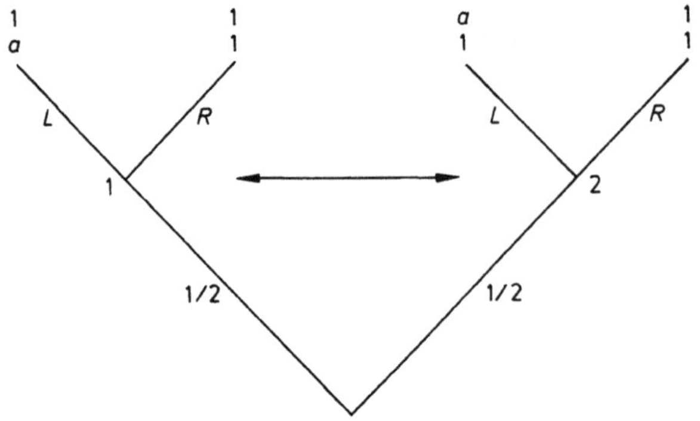

and since now choosing L is equivalent to choosing R, a limit ESS does not exist. However, in this truncated game, R trivially satisfies the conditions of a neutrally stable strategy and it is not difficult to see that a similar property will hold in general. Hence, we have the following corollary to Theorem 9.8.7.

Corollary 9.8.8. *If b is a limit ESS of (Γ, T)* and *Γ_x is an essential subgame of Γ, then*

(i) *if Γ_x is symmetric, then b_x is a limit ESS of (Γ_x, T_x),*
(ii) *if Γ_x is asymmetric, then (b_x, b_{xT}^T) is a pure equilibrium of Γ_x,*
(iii) *b_{-x} is a neutrally stable strategy of $(\Gamma_{-x}(b), T_{-x})$.*

Theorem 9.8.7 shows that a direct ESS must satisfy the backwards induction principle from dynamic programming and Corollary 9.8.8 indicates that in nondegenerate cases (such as $a+b\neq 1$ in the game of Fig. 9.7.2) the same will be true for limit ESS. Since these local conditions are quite strong, the conjecture suggests itself that they are also sufficient for evolutionary stability. Hence, one might conjecture that for a broad class of games, also the converses to Theorems 9.8.4, 9.8.7 and Corollaries 9.8.6, 9.8.8 are correct, or, equivalently, that any strategy found by dynamic programming using these necessary conditions is indeed evolutionarily stable. In Selten [1983] one finds several theorems (specifically, the theorems 8, 9, 11 and 12) claiming that this conjecture is correct for the class of *simultaneity games*, i.e. for games consisting of several rounds such that at the beginning of each round everything that happened in the past is common knowledge. (Note that this excludes the game discussed in Theorem 9.6.1). In the concluding example of this chapter, we show that this claim is not justified: even for simultaneity games, the local conditions are not sufficient to guarantee evolutionary stability.

Example 9.8.9. The converses to Theorems 9.8.4, 9.8.7 and Corollaries 9.8.6, 9.8.8 are not correct: the local conditions are not sufficient to guarantee evolutionary stability.

Consider the symmetric extensive form game from Fig. 9.8.2, in which the symmetry is indicated by the double arrows. This is a 2-stage game and since the

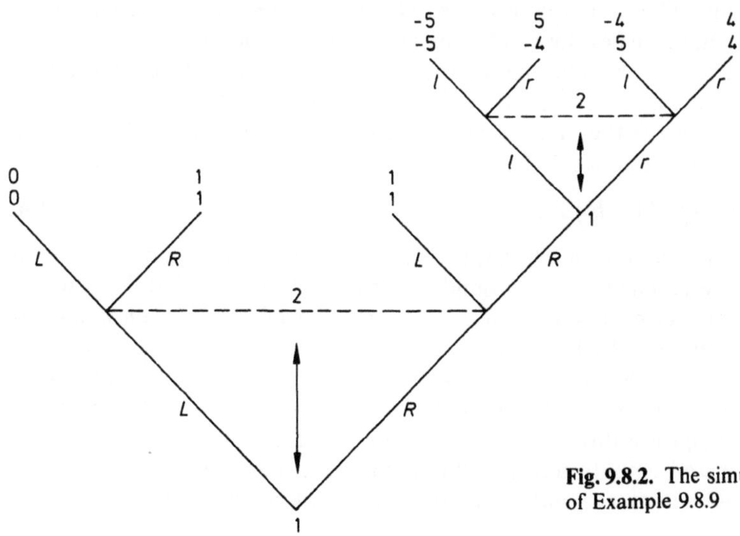

Fig. 9.8.2. The simultaneity game of Example 9.8.9

subgame of the second stage is reached only if both players choose R in the first stage, it is a simultaneity game.

The game of Fig. 9.8.2 admits one essential, symmetric subgame and the fitness matrix associated with this subgame is given by

	l	r
l	-5	5
r	-4	4

Hence, in the subgame there is a unique ESS, viz. $\frac{1}{2}l+\frac{1}{2}r$ and this results in the payoff 0. Taking the truncation with respect to the ESS of this subgame, we obtain the game with fitness matrix

	L	R
L	0	1
R	1	0

and this again has a unique ESS, viz. $\frac{1}{2}L+\frac{1}{2}R$. Consequently, there is exactly one strategy that satisfies the conditions (i) – (iv) in Theorem 9.8.7 with $\eta=0$, namely the strategy b that requires to randomize $(\frac{1}{2}, \frac{1}{2})$ both at stage 1 and at stage 2. We claim that b is not a direct ESS of (Γ, T), so that the converse to Theorem 9.8.7 is false. To verify the claim, consider the strategy b' that prescribes to choose R at stage 1 and r at stage 2. This is a best reply against b and we have

$$A(b', b') = 4 > A(b, b') = 2.75 \tag{9.8.4}$$

so that b' violates condition (9.7.8). (Note that b fares better against b' when the second round is reached, but this round is reached only with probability $\frac{1}{2}$ when b is played.) Furthermore, note that b is a direct LSS of (Γ, T), hence, also the converse to Theorem 9.8.4 is false.

From Theorem 9.8.7 we can conclude that the game (Γ, T) does not posses a direct ESS, so let us investigate limit ESS. Let (Γ, T, η) be a perturbed game and assume that $b(\eta)$ is a direct ESS of this game. Since the subgame has a unique, completely mixed equilibrium, it follows from Theorem 9.8.7 that $b(\eta)$ must prescribe $\frac{1}{2}l+\frac{1}{2}r$ for every η sufficiently small. By invoking the theorem once more, we see that it must also prescribe $\frac{1}{2}L+\frac{1}{2}R$ in this case. Hence, $b(\eta)=b$ for all η sufficiently small, but since the inequality in (9.8.4) is strict, b cannot be a direct ESS of (Γ, T, η). Consequently, (Γ, T) does not possess a limit ESS.

It is clear that b satisfies the conditions listed in Corollaries 9.8.6 and 9.8.8. In fact, b satisfies much stronger conditions: b prescribes a regular ESS at every local game. (Equivalently, b prescribes a regular ESS in the subgame and a regular ESS in the truncated game.) Obviously, the converses to the Corollaries 9.8.6 and 9.8.8 are false. □

The above example is fully generic and it clearly demonstrates that there is not much hope of obtaining a pure local characterization of direct ESS or limit ESS. However, local conditions that are sufficient for a global ESS can be established. For example, if b is a strategy that prescribes a unique best reply against (b, b^T) at every information set, then b is a direct ESS, but clearly this sufficient condition is much more restrictive than the necessary conditions derived in Theorems 9.8.4, 9.8.7 and Corollary 9.8.5. In Selten [1988b] a sufficient local condition has been derived that requires somewhat less than strictness at every information set, and by invoking this condition one can show that the main conclusions from Selten [1983] are indeed valid. Hence, the importance of Selten's 1983 paper should not be underestimated. The fact that an ESS necessarily satisfies the principle of dynamic programming simplifies the analysis considerably since it usually leaves one with just a small set of ESS candidates. Our example only shows that one might have to resort to a global analysis to verify which of these candidates are indeed ESS.[11]

Notes

1. A recent survey of the material discussed in this chapter is Hammerstein and Selten [1991]. A survey that covers some topics discussed in the first part of this chapter to some greater dept is Hines [1987].
2. For reformulation of the ESS conditions for the case of finite population, see Riley [1979a] and Schaffer [1988]. Schaffer shows that in finite populations the ESS leads to 'over-competitive' behavior: An ESS-strategist is not trying to maximize his payoff, he is trying to 'beat the average' (also cf. Shubik [1954]).
3. For a discussion on the role of sex, see Eshel [1989] and the references cited therein.
4. A more general model is the so-called 'playing the field model' introduced in Maynard Smith [1982, pp. 23–27]. In this model each individual's payoff is determined by the individual's own strategy and the frequencies of the other individuals' strategies, and fitness need not be linear in the latter frequencies. By means of this model one can incorporate 'n-player' – or 'n-population' – interactions. See Crawford [1990a] for the relationship between Nash equilibria and ESS in large – and finite – population 'playing the field models'. Crawford shows that if the fitness of a strategy depends discontinuously on

the population's strategy, then there may be strict Nash equilibria that are not ESS. Crawford [1990b] discusses an interesting game (due to Van Huyck, Battalio and Beil) where such a discontinuity exists.

5. Theorem 9.3.4 was not contained in Bomze [1986].
6. For an excellent survey on the mathematical theory of evolution and the dynamical systems involved, see Hofbauer and Sigmund [1988]. For a complete overview of the discrepancies between evolutionary stability and dynamic stability in a class of 3×3 games, see Weissing [1989].
7. Recent papers that discuss material that is related to the topic of this section are Friedman and Rosenthal [1986], Friedman [1987], Samuelson [1988], Nachbar [1990] and Boylan [1990]. Friedman/Rosenthal consider a dynamic equation, different from (9.4.3) of which the steady states need not be Nash equilibria. Friedman does not insist on the rate of increase of a strategy being proportional to how much this strategy is better than the average, hence, he studies a class of dynamical processes that include (9.4.3) as a special case. He finds that the Corollaries 9.4.2 and 9.4.5 remain true for a large class of dynamics. Samuelson also studies a class of dynamics that is more general than (9.4.3) and he establishes a generalization of Theorem 9.4.6. Nachbar works in the same spirit as Friedman and Samuelson. He also gives an interesting example of a dominance solvable game of which the solution (i. e. the strategy pair that remains after all weakly dominated strategies have been iteratively eliminated) is an unstable fixed point of the replicator equation. Finally, Boylan defines the concept of 'evolutionary equilibria' (as roughly those stationary points of processes more general than (9.4.3) that are stable against small mutation rates). He shows that every regular equilibrium is an evolutionary equilibrium and that every evolutionary equilibrium is perfect.
8. This question is also addressed in Bomze and Van Damme [1990] and Thomas [1985].
9. Schlag [1990] defines a generalized concept of eESS for games in which equivalent strategies exist (The 'e' stands for 'equivalent'), Schlag's basic idea is to look at the ESS of the 'reduced game' in which a player can pick an equivalence class of strategies. He shows that with this generalized definition the problems due to spurious duplication disappear.
10. In Samuelson [1989b] it is shown that, in a truly asymmetric contest, a strategy b^* is a limit ESS if and only if (i) (b^*, b^*) is a pure strategy perfect equilibrium and (ii) there is no strategy $b \neq b^*$ that is payoff equivalent to b^* (cf. Theorem 9.6.2). See Schlag [1990] for a discussion that focuses on equivalent strategies.
11. For examples of ESS analysis in extensive form games, see Gardner and Morris [1989a, b].

10 Strategic Stability and Applications

In this chapter, we study the concept of strategic stability that has been introduced in Kohlberg and Mertens [1986]. The first part of the chapter motivates this concept and studies its general properties. In the second half, the concept is applied to specific games.

Section 10.1 argues that a solution should be independent of modelling details and that, therefore, knowledge of the reduced normal form (or the reduced agent normal form, in case agents are not coordinated) should be sufficient to determine the solution. In Sect. 10.2 it is shown that, if one wants a normal form solution concept that produces self-enforcing outcomes in any extensive form game having this normal form, one is forced to a set-valued solution concept. Section 10.3 then introduces the Kohlberg/Mertens stability concept (which, loosely speaking, is the set-valued analogue of strict perfectness) and it is shown that this concept has several desirable properties. The first three sections are based on Kohlberg and Mertens [1986].

Sections 10.4 – 10.6 are devoted to signalling games. Section 10.4 introduces this class of games as well as the notation that is used. In Sect.10.5 it is investigated how strategic stability is related to various concepts that are based on 'intuitive' arguments for eliminating 'unreasonable' equilibria. In particular, it is shown that every stable equilibrium passes the test of Cho and Kreps [1987] and that it is (universally) divine in the sense of Banks and Sobel [1987]. This section also discusses the perfect sequential equilibrium concept of Grossman and Perry [1986a]. Section 10.6 (which is based on Kreps [1984]) illustrates these refined equilibrium concepts in a specific example, viz. Spence's job market signalling model (Spence [1973, 1974]).

Sections 10.7 and 10.8 illustrate how the concepts and intuitive arguments from Sect. 10.5 may be applied in dynamic games. No general theory is developed, only some examples are discussed. In Sect. 10.7, which is based on Kreps and Wilson [1982b], it is shown that the chain store paradox of Selten [1978] may be resolved if one allows for slight incomplete information. Section 10.8 demonstrates that, in repeated games with complete information, requiring stability may produce interesting and unexpected results, from which one may conjecture that the Folk Theorem (Theorem 8.7.7) does not remain valid when one requires stability.

10.1 Equivalence of Games

In this section, the concept of a reduced normal form is introduced and it is argued that, if each player can fully coordinate his agents, then two games with the same reduced normal form should be considered equivalent. Clearly, one wants equivalent games to have the same solutions and it is shown that requiring this kind of invariance together with sequential rationality leads to the concept of strict perfectness (i.e. stability against all perturbations in strategies). The discussion is based on Kohlberg and Mertens [1986].

In Chap. 1 we described the aim of Game Theory as finding a solution for every game, hence, the aim is to characterize 'rational behavior'; to prescribe a set of rules that tell each player how to behave in every situation that may arise. If the game is noncooperative (i.e. if binding commitments are impossible), the solution should be self-enforcing, i.e. if the solution is recommended to the players, nobody should have an incentive to deviate from it. If, furthermore, no communication is possible, this implies that a solution has to be a Nash equilibrium. In general, however, not every Nash equilibrium is self-enforcing since a Nash strategy need not be consistently optimal: It may prescribe non-optimal behavior at contingencies which arise when some player deviates from his recommendation. This observation was the starting point of our analysis and it has led to a variety of solution concepts (such as (subgame) perfect equilibria, sequential equilibria and proper equilibria) which all impose additional conditions that may be necessary for strategic stability, but neither of which actually guarantees that the equilibrium is self-enforcing.

In addition to the requirements that a solution exist for every game and that it be self-enforcing, we also want 'equivalent' games to have the same solutions. The trouble with this latter requirement is that we don't have a good theory about which games are actually equivalent and, in fact, this equivalence may depend on how a game is interpreted.[1] Namely, if each player can fully coordinate his agents and, hence, has the possibility of planning his strategy in advance, then it seems that two games with the same normal form should be considered equivalent. On the other hand, if a player just is a mathematical abstraction for a collection of independent agents (types) who have the same payoffs, but whose actions are not directly coordinated (as is usually the case in games of incomplete information), then this normal form does not seem relevant and two games seem to be equivalent if they have the same agent normal form. The game of Fig. 10.1.1 shows that it may make a crucial difference which interpretation is appropriate (also see Harsanyi [1968, Sects. 6, 7, and 11] and Aumann and Maschler [1972]).

We claim that, in the game of Fig. 10.1.1, only $(L_1 l, L_2)$ is self-enforcing when the agents 1_a and 1_b are coordinated by a central player 1, but that $(R_1 r, R_2)$ is self-enforcing too in case there is no communication between these agents.[2] Namely, in the first situation, player 1 will never choose $L_1 r$ since this is dominated by R_1, so that player 2, whenever he is reached, should conclude that $L_1 l$ has been chosen, hence, he should respond with L_2. (Note that $(L_1 l, L_2)$ is the strategy pair that remains after successive elimination of dominated strategies in the normal form). In the second situation, however, this dominance argument no longer applies and, when $(R_1 r, R_2)$ is recommended, no agent will have an incentive to deviate since he cannot be sure that the other agents will deviate as well. (This verbal argument is sustained by the concept of strategic stability from Sect. 10.3: the outcome (2, 2) is stable in the agent normal form, but not in the normal form.)

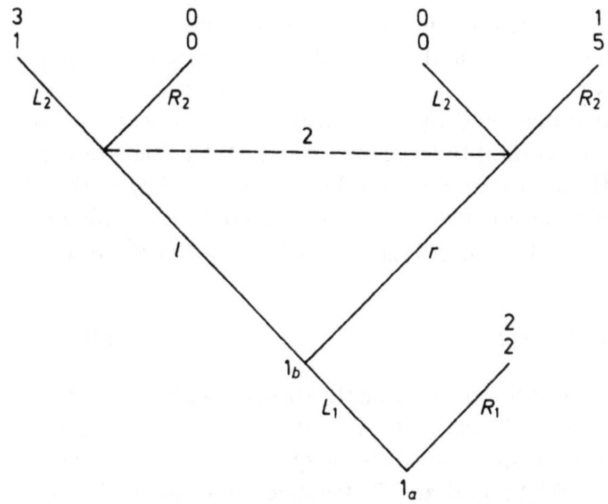

	L_2	R_2
$L_1 l$	3, 1	0, 0
$L_1 r$	0, 0	1, 5
$R_1 l$	2, 2	2, 2
$R_1 r$	2, 2	2, 2

Fig. 10.1.1. An equilibrium may be self-enforcing in the agent normal form, without having this property in the normal form

In the literature, it has been argued that the normal form should be used with care since this representation seems to presume immediate commitment on the part of the players, i.e. each player can commit himself at the beginning of the game, before any move has occurred. Specifically, in the normal form a player makes his decision before receiving any information whereas, in the extensive form, the decision is made only afterwards, and as Aumann and Maschler write, although this does not change the courses of action that a player has at his disposal, it may change his outlook. This point is especially relevant, for games of incomplete information since in that case the 'types' only come into existence after the beginning of the game (see Harsanyi [1986, Sects. 6, 7, and 11], Aumann and Maschler [1972], also cf. the distinction between ex-ante efficiency and interim efficiency in Holström and Myerson [1983]).

In the papers mentioned above, it has been clearly demonstrated that the issue of commitment is important and it has been convincingly argued that the normal form representation has certain drawbacks. However, it is this author's opinion that it would be premature to conclude that strategic stability cannot be studied by means of the normal form. Namely, note that the decisive element in the discussion of the game of Fig. 10.1.1 was not whether player 1 could commit himself or not, but whether the agents of player 1 were coordinated or not. In fact, one can argue that, as long as there is a central player coordinating his agents, the normal form contains all essential information to judge whether or not an equilibrium is self-enforcing. Namely, if a strategy combination is not self-enforcing in the extensive form, then there must exist an agent who has an incentive to deviate. Clearly, the player to whom this agent belongs can foresee this possibility in advance and, since the interests of the agent coincide with those of the player, the player has an incentive to deviate in the normal form, hence, the strategy combination is not self-enforcing in this normal form.[3]

In this chapter, we will investigate the consequences of the above point of view, hence, it will be assumed that there are no independent decision makers except the players and two games with the same normal form will be considered equivalent. Now, the normal form usually contains a lot of redundancy since many strategies just appear as duplications of others, and by eliminating such duplications one does not change the strategic situation. (In the game of Fig. 10.1.1, the strategies $R_1 l$ and $R_1 r$ are duplicates). Hence, two games should be considered *equivalent* if the same normal form results after all such duplications have been eliminated. Formally, 2 pure strategies φ_i and φ_i' of player i in the game Γ with payoffs $R=(R_1,\ldots,R_n)$ are *duplicates* if

$$R(\varphi\backslash\varphi_i)=R(\varphi\backslash\varphi_i') \quad \text{for all } \varphi\in\Phi \tag{10.1.1}$$

and the normal form game that results when all duplications have been removed will be called the *semi-reduced normal* form of Γ. A strategy in this semi-reduced normal form corresponds to a set of strategies of the original game that forms an equivalence class of relation (10.1.1) and any 2 strategies that are realization equivalent (cf. (6.1.3)) belong to the same equivalence class. In other words: a strategy of the semi-reduced normal form does not completely specify a strategy in the extensive form; it only prescribes definite choices at certain information sets, however, irrespective of how one extends this to a complete strategy, one always gets the same probability distribution on payoffs.

The following theorem completely describes the partition of the set of games into equivalence classes with the same semi-reduced normal form. By investigating whether the transformations described are indeed innocuous, the reader can judge the reasonableness of the equivalence relation defined above.[4]

Theorem 10.1.1 (Thompson [1952], Kohlberg and Mertens [1986]). *Let Γ, Γ' be extensive form games (not necessarily having perfect recall). Γ' has the same semi-reduced normal form as Γ if and only if Γ' results from Γ by successive application of transformations from the following list (or their inverses):*

(i) *Inflation: Split an information set of a player into two parts which are such that the player can deduce anyhow in which part he is on the basis of his strategy,*

(ii) *Addition of a Superfluous Decision Point: Let u be an information set and let z be an endpoint that does not come after u. If y is a node on the path to z, then add a decision point $x\in u$ immediately before y, such that after x, any choice $c\in C_u$ leads to (a copy of) y and such that no opponent gets to hear which choice was taken at x.*

(iii) *Agent Splitting: Split the choice set C_u at u into C and its complement $\bar{C}$ and split the agent at u into three agents, the first one choosing C or $\bar{C}$, the second one choosing a choice in C and the third taking a choice from $\bar{C}$.*

(iv) *Interchanging Simultaneous Moves: if the decisions at the information sets u and v have to be made simultaneously, then, in the game tree one is forced to draw one before the other, but the two representations should be considered equivalent.*

(v) *Truncation: If a chance move leads only to endpoints, replace this move by an endpoint with the corresponding expected payoffs.*

Note that uniting agents (the inverse of (iii)) is innocuous only if the decisions of these agents are coordinated as the game of Fig. 10.1.1 has shown. If an agent is independent, then he should be treated as a separate decision maker (player), hence, the proper way to proceed is to first identify the actual players and then apply the transformations from Theorem 10.1.1. Consequently, one can always find an equivalent normal form: If all agents are coordinated (which we assume in the first three sections of this chapter), this is the standard (semi-reduced) normal form, if they are all independent (as in the second half of this chapter), it is the (reduced) agent normal form.[5]

At this point it should also be noted that a concept of strategic stability cannot be simultaneously invariant with respect to agent splitting and satisfy subgame consistency, the latter requirement being that the solution of a subgame should depend only on the structure of the subgame itself (Selten [1973]). Namely, invariance with respect to agent splitting implies that (l, L_2) is the solution of the subgame in the game of Fig. 10.1.1, however, if agent 1_a would have the payoffs of player 2, then the same dominance argument as used before would lead to (r, R_2) as the solution of the subgame, although the subgame as such has remained unchanged. (For an elaboration, see Harsanyi and Selten [1988] or Pearce [1984].)[6] Hence, it seems that a subgame cannot be treated as a separate game since the way in which the subgame was reached can signal how a player intends to play in this subgame.

Next, we will investigate the consequences of requiring a solution (i) to exist, (ii) to be self-enforcing and (iii) to depend only on the semi-reduced normal form. Now, a necessary condition for an equilibrium to be self-enforcing is that it is sequential and from Theorem 6.5.2 we see that every proper equilibrium of the semi-reduced normal form induces a sequential equilibrium in every extensive form game having that semi-reduced normal form. (More precisely: the equivalence class of strategies of Γ corresponding to a proper equilibrium of the semi-reduced normal form of Γ contains a sequential equilibrium of Γ.) Unfortunately, not every sequential equilibrium is 'reasonable' and, consequently, not every normal form proper equilibrium is self-enforcing. For example, in the game of Fig. 10.1.1, the equilibrium $(R_1 r, R_2)$ is proper in the normal form, but it is not self-enforcing. This drawback of the properness concept is related to the fact that the set of proper equilibria is not invariant with respect to all inessential transformations of a game. Namely, a player always has the possibility of randomizing pure strategies, so that adding such a mixture as an additional pure strategy does not change the strategic situation and, hence, it should not affect the solution. However, the set of proper equilibria is affected by this transformation as is illustrated by the games of Fig. 10.1.2*a*, *b*. The game of *a* is the semi-reduced normal form of Fig. 10.1.1 and the game of *b* results from *a* by adding the mixture $M = \lambda R_1 + (1-\lambda) L_1 l$. We have that (R_1, R_2) is a proper equilibrium of *a*, but it is not proper in *b* when $\lambda \geqq 1/2$.

Let us call the game that results from the normal form of a game Γ after all pure strategies have been deleted that are just mixtures of others, the *reduced normal form*, to be denoted by $\mathscr{RN}(\Gamma)$. Formally, a pure strategy φ_i of player i is deleted when there exists a mixed strategy s_i with $\varphi_i \notin C(s_i)$ such that

$$R(\varphi \backslash \varphi_i) = R(\varphi \backslash s_i) \quad \text{for all } \varphi \in \Phi. \tag{10.1.2}$$

Equation (10.1.2) defines an equivalence relation on strategy combinations and this induces an equivalence relation on extensive form games. The following corollary, which is an easy consequence of Theorem 10.1.1 describes the equivalence classes of this latter relation.

Corollary 10.1.2 (Kohlberg and Mertens [1986]). *Two games Γ and Γ' have the same reduced normal form if and only if one results from the other by stepwise application of the five transformations from* Theorem 10.1.1 *together with*

(vi) *Addition of a Superfluous mixture: Add a choice c' to the choice set C_u, such that immediately after c' chance chooses between choices originally available at C_u and such that, for any $c \in C_u$, no opponent gets informed about whether chance or the player personally was responsable for choosing c.*

The preceding argument forces us to consider games as equivalent when they have the same reduced normal form, hence, we want a (reduced) normal form solution concept that induces a sequential equilibrium in every game having this reduced normal form. Since proper equilibria are not invariant with respect to adding mixtures, this concept does not satisfy the requirement: The game Γ of Fig. 10.1.3 has the game of Fig. 10.1.2*a* as its reduced normal form (the semi-reduced normal form is given in Fig. 10.1.2*b*, hence, (R_1, R_2) is a proper equilibrium of $\mathscr{RN}(\Gamma)$. However, when $\lambda > 1/2$, the unique subgame perfect equilibrium of Γ is $(L_1 l, L_2)$ (since M dominates R_1) and this is not induced by (R_1, R_2).

In the game under consideration in this section, the unique 'reasonable' equilibrium $(L_1 l, L_2)$, also is the unique strictly perfect equilibrium. We will now present a heuristic argument showing that the requirements of invariance (the

	L_2	R_2
$L_1 l$	3, 1	0, 0
$L_1 r$	0, 0	1, 5
R_1	2, 2	2, 2

a

	L_2	R_2
$L_1 l$	3, 1	0, 0
$L_1 r$	0, 0	1, 5
R_1	2, 2	2, 2
$L_1 m$	$3-\lambda$, $1+\lambda$	2λ, 2λ

b

Fig. 10.1.2. The set of proper equilibria is not invariant with respect to adding mixed strategies as pure ones

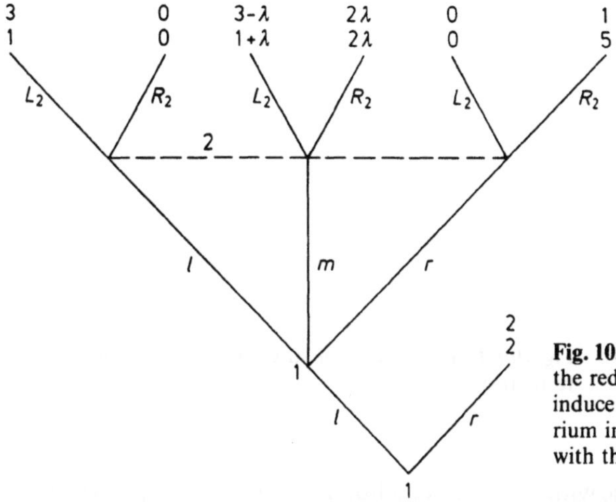

Fig. 10.1.3. A proper equilibrium of the reduced normal form need not induce a subgame perfect equilibrium in an extensive form game with this reduced normal form

solution should be defined on the reduced normal form) and sequential rationality (a sequential equilibrium should be induced in any game having this reduced normal form) naturally lead to the idea of strict perfectness.

First of all, one easily verifies that a strictly perfect equilibrium of a reduced normal form induces a strictly perfect equilibrium (hence, under a mild regularity condition, a proper equilibrium (cf. Theorem 2.3.8)) in every normal form game having this reduced normal form, so that (Theorem 6.5.2) a sequential equilibrium is induced in every extensive form game having this reduced normal form.

Next, we claim that (again under mild regularity conditions) one has the following converse: if a quasi-strict equilibrium s of the reduced normal form induces a perfect equilibrium in every game with this reduced normal form, then s must be strictly perfect. (The following argument is taken from Kohlberg and Mertens [1986], Appendix *D*.) Consider a perturbation η of the reduced normal form Γ, construct the completely mixed strategy combination $\sigma := s * \eta$ as in the proof of Theorem 2.4.3; for each player i, add σ_i as an additional pure strategy and, finally, split each player into 3 agents as in Theorem 10.1.1 (iii) with $C = C(s_i)$. The resulting game must have an ε-perfect equilibrium $s(\varepsilon)$ close to s and, if ε and η are small, the third agent of player i has σ_i as its unique best response against $s(\varepsilon)$ since s is quasi-strict. Consequently, σ_i is chosen with probability close to 1, hence $s_i^k(\varepsilon)/\eta_i^k \to 1$ as $\varepsilon \to 0$ for all $k \notin C(s_i)$. Now, $s(\varepsilon)$ induces a completely mixed strategy combination $\bar{s}(\varepsilon)$ in the reduced normal form Γ and this is an equilibrium of any perturbed game $(\Gamma, \eta(\varepsilon))$, with $\eta_i^k(\varepsilon) = \bar{s}_i^k(\varepsilon)$ for all $k \notin C(s_i)$. Hence, for any small η there exists η' close to η and an equilibrium s' of (Γ, η') close to s. Under some regularity conditions, this implies that s is strictly perfect.

Although the concept of strictly perfect equilibria satisfies invariance and sequential rationality, it is not satisfactory since it is unduly restrictive: strictly perfect equilibria need not exist and not every self-enforcing equilibrium has to be strictly perfect. This is illustrated by the game of Fig. 10.1.4.

	L_2	R_2
L_1	a, a	0, 0
R_1	0, 0	a, a
A	2, 2	2, 2

Fig. 10.1.4. The concept of strict perfectness is unduly restrictive

First, consider the case with $a=1$. The equilibria are the strategy pairs in which player 1 chooses A, but neither of these is strictly perfect since player 2's best response depends on how player 1 makes his mistakes. Nevertheless, all equilibria are self-enforcing: Player 1 definitely has no incentive to deviate and player 2, knowing this, is happy to do anything.

Next, suppose $a=3$. In this case, (L_1, L_2) and (R_1, R_2) are strictly perfect, but no equilibrium with payoff $(2, 2)$ satisfies this criterion. Namely, a perturbed game with $\eta_1(L_1) > \eta_1(R_1)$ (resp. $\eta_1(L_1) < \eta_1(R_1)$) only has equilibria close to (L_1, L_2), (R_1, R_2) and $(A, (1/3, 2/3))$ (resp. and $(A, (2/3, 1/3))$). Nevertheless, if no communication between the players is possible, $(2, 2)$ is a self-enforcing outcome. For example, in the extensive form game with structure as in Fig. 10.1.1, but with this reduced normal form, player 2 can deduce that player 1 has deviated from the recommendation A, but he can never know which response player 1 elicits, hence, he will randomize and player 1 will loose by deviating from A. Consequently, when it is recommended that 1 should choose A and that 2 should randomize (assigning probability at most 2/3 to either choice), no player has an incentive to deviate.

10.2 Requirements for Strategic Stability

In the previous section, it has been shown that the requirements of invariance and sequential rationality more or less naturally lead to the concept of strictly perfect equilibria, but that this concept is unduly restrictive. In this section, it is shown that the trouble is caused by the single-valuedness of the strict perfectness concept: If one wants to satisfy minimal requirements of strategic stability, one is forced to accept set-valued solution concepts. The discussion is based on Kohlberg and Mertens [1986].

Two basic requirements that one might want a solution concept to satisfy are *existence* (every game should have a solution) and *invariance* (2 games with the same reduced normal form should have the same solutions). Furthermore, a solution should be self-enforcing and, for this to be the case, it definitely should be a sequential equilibrium. Hence we want

Sequential Rationality: Every solution should be a sequential equilibrium.[7] (10.2.1)

Unfortunately, this requirement is incompatible with existence and invariance as the game of Fig. 10.2.1 illustrates. G is a reduced normal form game in which all strategy pairs in which player 1 chooses A are equilibria. $G(\lambda)$ is a normal form game with G as its reduced normal form (M is the mixture $\lambda A + (1-\lambda) L_1$). By splitting player 1 in $G(\lambda)$ into 2 agents, the first one choosing between A and $\{L_1, M, R_1\}$ and the second one choosing an element from $\{L_1, M, R_1\}$, and by deleting the superfluous decision point of player 2 after A, one obtains an extensive form game $\Gamma(\lambda)$ with the structure from Fig. 10.1.3. When $\lambda \in [0, 1/2)$, the subgame of $\Gamma(\lambda)$ has a unique equilibrium and this has player 2 choosing L_2 with probability $q(\lambda) = (1-2\lambda)(2-\lambda)^{-1}$, hence, (10.2.1) together with existence imply that $\Gamma(\lambda)$ has a unique solution and that this solution depends on λ ($q(\lambda)$ varies between 0 and 1/2). However, this violates invariance since the latter requires the solution to be independent of the irrelevant parameter λ.

The above example shows that it is even impossible to satisfy simultaneously existence, invariance and subgame perfectness hence, the exact definition of consistency is irrelevant for establishing this result. Furthermore, it shows that, by requiring invariance, one is forced to accept a set-valued solution concept: In G all strategies of player 2 are equally good as long as player 1 takes the (unique) rational choice A, however, different strategies can be rationalized by considering various extensive form games having G as their reduced normal form. (We only

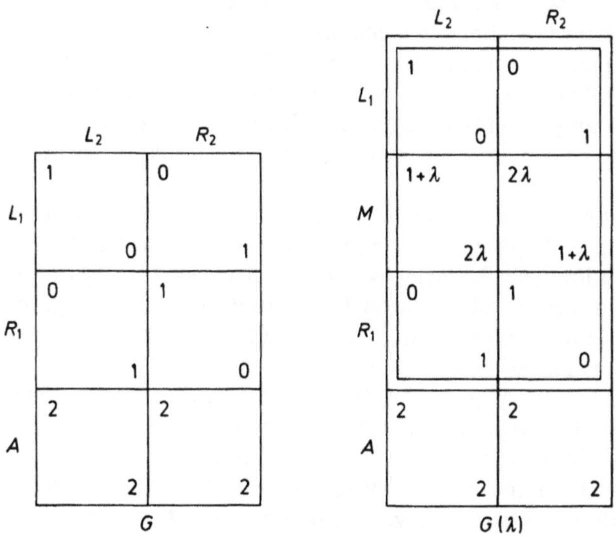

Fig. 10.2.1. It is impossible to satisfy existence, invariance and (10.2.1) simultaneously. (The subgame in $\Gamma(\lambda)$ is indicated by the double framework)

rationalized mixed strategies in which player 2 chooses L_2 with probability at most 1/2, by adding mixtures of A and R_1 one can rationalize the other mixed strategies.) This can be paraphrased by saying that different strategies of player 2 can be rationalized by different thought processes of player 1 which finally lead this player to the conclusion that he should play A. Since only actual choices matter and not the processes that produce them, it seems that we are forced to consider all equilibria in this example as being equivalent, hence, the set of all equilibria should be considered as the solution. Obviously, even the set of all equilibria does not satisfy (10.2.1) in its present form: If one wants existence and invariance, one can only ask for inclusion in (10.2.1).

We will now present a second argument, based on iterated dominance, also leading to a set-valued solution concept. Clearly, one wants a game theoretic solution to be consistent with standard Bayesian decision theory. Now this theory postulates that a player will never use a (weakly) dominated strategy, hence, we also want to impose this requirement. As another justification, note that, if a solution would prescribe a dominated strategy for a player, then, even if this player would only have the slightest doubt about the others obeying their recommendations, he would have an incentive to deviate. Taking this argument one step further one can argue that, if it is common knowledge that a player will never use a dominated strategy, then eliminating such strategies should have no effect on the solution. Hence, we want to require

Admissibility: Every solution should be undominated.[8] (10.2.2)

Elimination of Dominated Strategies: A solution should remain a solution after a dominated strategy has been deleted.[9] (10.2.3)

However, the latter requirement cannot be satisfied by a single-valued solution concept. For example, in the game G of Fig. 10.2.1 by eliminating L_1 first and then R_2, one retains (A, L_2). However, if one first eliminates R_1, one retains (A, R_2). Hence, as stated, (10.2.3) implies that the solution can only be (A, L_2) and that it can only be (A, R_2). Obviously, invariance with respect to elimination of dominated strategies can only be satisfied by a set-valued solution concept and even in this case one has to be content with inclusion ('remain' should be replaced by 'contain' in (10.2.3)).

Note that the admissibility requirement (10.2.2) implies that the solution correspondence cannot depend upper-semi-continuously on (the payoffs of) the game, nor lower-semi-continuously. Namely, by slightly perturbing the payoffs in Fig. 2.1.1 one can obtain a game in which player 1 chooses his second strategy in the unique equilibrium. However, the unique undominated equilibrium of the unperturbed game has this player choosing his first action.

Requirement (10.2.3) is very powerful for generic games with perfect information (Gale [1953]): In this case, it leads to the unique subgame perfect equilibrium outcome. Also in the game of Fig. 10.1.1, this condition eliminates all equilibria except $(L_1 l, L_2)$. In general, however, (10.2.3) is a relatively weak requirement and inessential changes in the game can easily lead to the criterion being ineffective. For example, if, in the game of Fig. 10.1.1, one replaces the endpoint $(0, 0)$ associated with $(L_1 r, L_2)$ by the zero-sum subgame with payoffs

4, -4	-4, 4
-4, 4	4, -4

then the resulting game is fully equivalent, but (10.2.3) is ineffective since the reduced normal form does not contain any dominated strategies.

A natural extension of the reasoning leading to (10.2.3) is the idea that, whether an equilibrium is self-enforcing or not, should not depend on the strategies that are not best replies against this equilibrium. Hence, one might wish to require

Elimination of Non-best Replies: A solution should remain a solution when a strategy that is not a best reply against the solution is deleted. (10.2.4)

As stated, condition (10.2.4) is incompatible with admissibility. Namely, the extensive form game of Fig. 10.2.2 has a unique subgame perfect equilibrium viz. $(L_1 l, R_2)$ and we definitely want to allow this as a solution. However, $R_1 l$ is not a best reply against R_2 and when $R_1 l$ has been deleted, $(L_1 l, R_2)$ is no longer undominated.

The Nash equilibria of the game of Fig. 10.2.2 are the strategy pairs in which player 1 chooses L_1 and player 2 chooses L_2 with probability at most 1/2 and this set satisfies an appropriate version of (10.2.4) with inclusion: $R_1 r$ is the unique pure strategy that is not a best reply against any element in this set and when this strategy is eliminated, the Nash equilibrium (L_1, R_2) remains as the unique

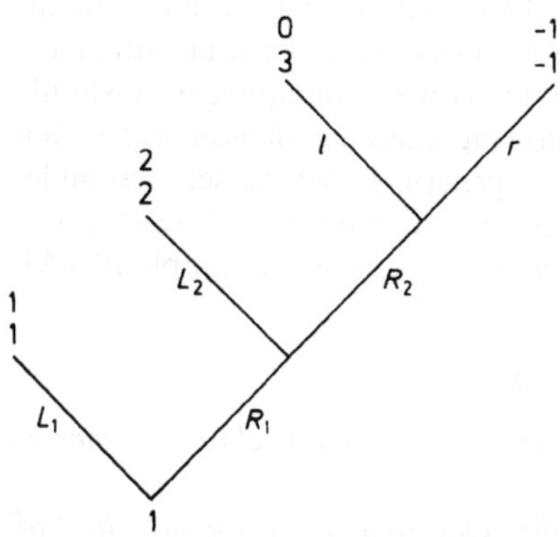

	L_2	R_2
L_1	1, 1	1, 1
$R_1 l$	2, 2	0, 3
$R_1 r$	2, 2	-1, -1

Fig. 10.2.2. For single-valued solution concepts, (10.2.2) and (10.2.4) are incompatible

undominated strategy pair. We see that a requirement as (10.2.4) again leads to a set-valued solution concept, but the reader may feel somewhat uncomfortable about this in the present context. After all, the game of Fig. 10.2.1 is of perfect information and it has a unique subgame perfect equilibrium, so what is the rationale for player 2 choosing L_2 with positive probability when he has to make a decision. Two arguments can be offered. First, if player 2 has to move, then player 1 has proved that he is irrational, so what guarantees that he will make a rational choice at his second move?[10] Secondly, and more fundamentally, player 2 choosing a randomized strategy is a consequence of invariance. Namely, by adding a mixture of L_1 and R_1r to the reduced form from Fig. 10.2.2 and by drawing a tree as in Fig. 10.1.3 one can construct an extensive form game with the same reduced normal form, but in which the unique subgame perfect equilibrium requires player 2 to choose L_2 with an arbitrary preprescribed probability between 0 and 1/2.

Above, 3 arguments for a set-valued solution concept have been presented and it has been demonstrated that only weak versions of (10.2.1), (10.2.3) and (10.2.4), just requiring for inclusion, are compatible with the basic requirements of existence and invariance. However, if one just asks for inclusion, the requirements become very weak: The set of all undominated equilibria of the reduced normal form satisfies them. Hence, in order to get a meaningful concept, one has to require some kind of minimality, loosely speaking, one wants a set of solutions to contain only equilibria that are in some sense 'equivalent'. In the examples of this section, it is intuitively clear that the equilibria under consideration are indeed equivalent: they can be continuously transformed into each other by changing the value of some irrelevant parameter (hence, they always form a connected set) and they always lead to the same payoff vector (in the extensive forms, 'equivalent' equilibria even lead to the same outcome). These observations might suggest to consider equilibria of as equivalent when they form a connected set with the same payoff. This equivalence concept is too narrow, however, since in nongeneric games equilibria may be equivalent without having identical payoffs. For example, if in the game G of Fig. 10.2.1, one adds a dummy player 3 who gets 0 when player 2 chooses L_2 and 1 otherwise, then all equilibria still should be considered equivalent (the strategic situation has not changed), but they don't have identical payoffs. Furthermore, one can construct more complicated examples of nongeneric games in which equivalent equilibria even differ along the equilibrium path. Hence, having the same payoff or leading to the same outcome is not necessary for equivalence. Consequently, the concept of equivalence should solely be based on connectedness. The following theorem demonstrates that equivalence can be defined in a way which, for generic games, closely resembles our intuition. (Actually, the concept of 'connected component' refines our intuition since 2 components might have the same outcome, see Example 10.3.4).

Theorem 10.2.1 (Kohlberg and Mertens [1986]).

(i) *The set of Nash equilibria of a game consists of finitely many closed, connected components,*

(ii) *Almost all games (within any set with the same structure) are such that all equilibria within a connected component lead to the same payoffs.*

Proof. The first statement follows from a general proposition concerning the structure of the set of solutions of a finite system of algebraic inequalities (Van der Waerden [1973], Satz 1, p.123). The second statement follows from the first together with Theorem 6.2.2: Since payoffs depend continuously on strategies and since generic games have only finitely many equilibrium outcomes, the payoffs must be constant on a connected set of equilibria of a generic game. □

Theorem 10.2.1 suggests to consider equilibria as equivalent when they are in the same connected component. Note that this equivalence actually is a reduced normal form concept: E is a connected component of $E(\Gamma)$ (i.e. E is a maximal connected set of equilibria of Γ) if and only if $\mathcal{RN}(E)$ is a connected component of $\mathcal{RN}(\Gamma)$, where $\mathcal{RN}(E)$ is the set of equilibria that E induces in the reduced normal form. However, this equivalence concept is too wide since, generally, a component will contain both dominated and undominated equilibria and (10.2.2) forces us to distinguish among these. (If one deletes L_1 from the game of Fig. 10.2.1, then the set of all equilibria is the unique component but only (A, R_2) is undominated). Hence, it is not completely clear which equilibria are actually equivalent, but our intuitive ideas imply that equivalence classes are connected. Consequently, a solution should be connected as well.

We conclude this section by reformulating the previous requirements for strategic stability (self-enforcingness) in terms of a set-valued solution concept, incorporating the fact that we can only ask for inclusion in (10.2.1), (10.2.3) and (10.2.4).

Requirements for Strategic Stability (Kohlberg and Mertens [1986]):

(i) *Existence*: Every game has a solution.
(ii) *Invariance*: Two games with the same reduced normal form have the same solutions.
(iii) *Sequential Rationality*: Every solution contains a sequential equilibrium.
(iv) *Admissibility*: Every element of a solution is undominated.
(v) *Elimination of Dominated Strategies*: A solution contains a solution of a game obtained by eliminating a dominated strategy.
(vi) *Elimination of Non-best Replies*: A solution contains a solution of a game obtained by eliminating a strategy that is not a best reply against any element of the solution.[11]
(vii) *Connectedness*: A solution is connected. (10.2.5)

10.3 Stable Equilibria

In the previous section, it has been shown that several minimal requirements of strategic stability can only be satisfied by a set-valued solution concept. In Sect. 10.1 it was argued that an equilibrium as a singleton is strategically stable if and only if it is strictly perfect. Combining these observations naturally leads to the following definition.

Definition 10.3.1. (Kohlberg and Mertens [1986]). A set E of equilibria of a game Γ is *stable* if it is a minimal set with the following property:

E is a closed set of equilibria and for every $\varepsilon>0$ there exists some $\hat{\eta}>0$ such that every perturbed game $(\mathcal{RN}(\Gamma), \eta)$ associated with the reduced normal form of Γ with $0<\eta<\hat{\eta}$ has an equilibrium within ε-distance of E. (10.3.1)

In this section, various properties of this concept of stability will be investigated.[12] We follow Kohlberg and Mertens [1986]. Let us first investigate what stability yields in the examples discussed thus far.[13] In the game of Figs. 10.1.1, 10.1.2 and 10.1.3 only $\{(L_1 l, L_2)\}$ is stable. In the game of Fig. 10.1.4 with $a<2$, the set of all equilibria satisfies (10.3.1), but this is not a minimal set with this property: The set of extreme points $\{(A, L_2), (A, R_2)\}$ also satisfies (10.3.1) and this set is minimal. Hence, this is the unique stable set.[14] Note, therefore, that requiring minimality precludes connectedness. Below we will see that minimality is imposed to guarantee admissibility. If $2<a<4$, then Fig. 10.1.4 admits 3 stable sets, viz $\{(L_1, L_2)\}$, $\{(R_1, R_2)\}$ and the extreme points of the set of equilibria with payoff (2, 2). If $a>4$, then the game admits 3 regular equilibria and these are stable. If G is the reduced normal form game of Fig. 10.2.1, then any perturbed game associated with G has an equilibrium close to (A, L_2) or close to (A, R_2), and one can construct perturbed games with unique equilibria close to either one. Hence, the unique stable set of G is $E=\{(A, L_2)\}\cup\{(A, R_2)\}$ and this set is not connected. Moreover, as we have seen in Sect. 10.2, there exists an extensive form game with reduced normal form G that has $(A, (1/2, 1/2))$ as its unique subgame perfect equilibrium, and this is not contained in E, hence, also sequential rationality is violated in this case. Finally, in the game of Fig. 10.2.2, the extreme points of the set of equilibria with payoff (1,1) constitute the unique stable set.

From these examples one sees that the concept of stable equilibria from Def. 10.3.1 does not satisfy connectedness, nor sequential rationality. However, in all examples, any stable set was completely contained within one component and stability always yielded the correct (intuitive) outcomes. In Example 10.3.4 we will see that these properties do not hold in general. First, however, we show that stability satisfies the remaining properties from (10.2.5).

By definition, the stability concept satisfies invariance. Furthermore, if Γ is a normal form game and $PE(\Gamma)$ is the set of perfect equilibria of Γ, then $PE(\Gamma)$ is a closed set as is easily verified. By definition, if $s\in E(\Gamma)\backslash PE(\Gamma)$, then the equilibria of any perturbed game (Γ, η) remain bounded away from s. Consequently, when E satisfies (10.3.1), then also $E\cap PE(\Gamma)$ satisfies this condition. Hence, when E is minimal, $E=E\cap PE(\Gamma)$ and we see that every element of a stable set is perfect, hence, undominated. We have shown

Corollary 10.3.1. *Every stable set consists of normal form perfect equilibria, hence, every element of a stable set is undominated.*

Next, let us consider existence of stable sets. Theorem 10.2.1 shows that the equilibrium set of a game Γ consists of finitely many closed connected components and it can be shown that at least one of these components (say C) is essential, i.e. that any game Γ' with payoffs close to Γ has an equilibrium close to C (see Jiang Jia-He [1963], Kohlberg and Mertens [1986]). This implies (cf. the proof of Theorem 2.4.3) that every perturbed game (Γ, η) with sufficiently small η has an equilibrium close to C, hence C satisfies (10.3.1). C need to be minimal, (it may contain dominated strategies; for example, in the game of Fig. 2.1.1, the set of all equilibria is the essential component), but this does not present a real problem: A standard application of Zorn's lemma shows that C must contain a stable set. Namely, consider the collection $\mathscr{C}$ of closed subsets of C satisfying (10.3.1) ordered by set inclusion. By compactness, the intersection of any ordered chain in $\mathscr{C}$ is nonempty and, hence, this again belongs to $\mathscr{C}$, so that Zorn's Lemma guarantees the existence of a minimal element in $\mathscr{C}$. We have derived

Corollary 10.3.2. *For every game there exists a stable set that is contained within a connected component of the set of Nash equilibria.*

Next, we show that the concept of stable equilibria satisfies the requirements 10.2.5(v), (vi) with respect to elimination of dominated strategies, resp. elimination of non-best replies.

Theorem 10.3.3. (a) *A stable set contains a stable set of any game obtained by eliminating a dominated strategy.*

(b) *A stable set contains a stable set of any game obtained by eliminating a strategy that is not a best reply against any element in this set.*

Proof. Let E be a stable set of the game Γ. Assume k is a strategy of player i that either is dominated or that is not a best reply against E, let Γ' be the game in which k is deleted and let $\varepsilon > 0$. For any perturbation η of Γ' and $\delta > 0$, consider the perturbation $\eta(\delta)$ of Γ that coincides with η for all strategies different from k and that has $\eta_i^k(\delta) = \delta$. For η and δ sufficiently small, let $s(\eta, \delta)$ be an equilibrium of $(\Gamma, \eta(\delta))$ within ε-distance from E. When ε is chosen sufficiently small, k is not a best reply against $s(\eta, \delta)$, so that $s_i^k(\eta, \delta) = \delta$. Hence $s(\eta) := \lim_{\delta \downarrow 0} s(\eta, \delta)$ is an equilibrium of (Γ', η) within ε-distance of $E' := \{s \in E;\ s_i^k = 0\}$. Consequently, E' satisfies (10.3.1). Zorn's lemma guarantees that E' contains a stable set. □

It is argued in Harsanyi [1976] and Harsanyi and Selten [1988] that, besides dominated strategies, also *inferior strategies* should be eliminated, where a strategy s is inferior if there exists some other strategy s' such that s' is a best reply when s is, but not conversely. Any weakly dominated strategy is inferior, but the converse is not true. Also note that any pure strategy that is a mixture of other pure strategies is inferior, so that some inferior strategies are already eliminated by considering the reduced normal form. Therefore, the reader might be inclined to think that a stable set contains a stable set of any game obtained by elimination of an inferior strategy. The game of Fig. 10.3.1, however, shows that this need not be the case

The set of equilibria of Fig. 10.3.1 consists of 2 connected components, viz. $\{(L_1, L_2)\}$ and $\{(R_1, qL_2 + (1-q)R_2);\ q \leq 1/2\} \cup \{(pL_1 + pM_1 + (1-2p)R_1,$

	L_2	R_2
L_1	2, 1	0, 0
M_1	1, 0	1, 1
R_1	0, 2	2, 2

Fig. 10.3.1. A stable set need not be 'invariant' with respect to elimination of inferior strategies

$1/2L_2+1/2R_2$; $p \leqq 1/2\}$. The latter component contains a completely mixed, hence, strictly perfect equilibrium, and this, taken as a singleton, is stable. However, M_1 is an inferior strategy (it is optimal only against $1/2L_2+1/2R_2$) and, after M_1 has been deleted, only (L_1, L_2) remains as a stable equilibrium.

We have shown that stability satisfies all properties from (10.2.5) except 'sequential rationality' and 'connectedness'. In fact, Corollary 10.3.2 shows that one can also satisfy a weak version of 'connectedness'. One could be satisfied with this if it would be true that any component of a reduced normal form game G that contains a stable set induces a sequential equilibrium *outcome* in any game Γ with $\mathcal{RN}(\Gamma)=G$, but this is an open problem. The following example (attributed to Faruk Gul in Kohlberg and Mertens [1986]) shows that there may exist stable sets that overlap with different components and that do not induce sequential equilibrium outcomes. Hence, there are good reasons to restrict oneself to stable sets that lie entirely within one component.[15]

Example 10.3.4. A stable set may contain points from different components and it need not induce a subgame perfect equilibrium outcome.

Consider the extensive form game Γ in which player 3 moves first. He can choose between terminating the game immediately (which yields each player a payoff 2) or playing the following simultaneous move subgame

L_3:

	L_2	R_2
L_1	3, 3, 0	1, 1, 5
R_1	1, 1, 5	0, 0, 1

R_3:

	L_2	R_2
L_1	0, 0, 1	1, 1, 5
R_1	1, 1, 5	3, 3, 0

It is easily verified that the subgame has a unique Nash equilibrium, viz. the strategy combination in which all players choose $(1/2, 1/2)$ and that this yields player 3 a payoff 11/4. Consequently, Γ has a unique subgame perfect equilibrium and in this player 3 chooses to play the subgame.

This subgame perfect equilibrium is one component of the set of Nash equilibria. There are two other components and in each of these player 3 chooses T, i.e. he terminates the game immediately. Namely, if we write p_i for the probability that player i chooses $L_i (i=1, 2)$, then it is optimal for player 3 to choose T if and only if

$$4(p_1+p_2)-9p_1p_2 \leqq 1 \quad \text{and} \quad 5(p_1+p_2)-9p_1p_2 \leqq 2$$

and the set of solutions to these inequalities indeed consists of 2 connected components. We claim that the set $\{(T, L_1, L_2)\} \cup \{(T, R_1, R_2)\}$, which contains one point from each of these components is stable. To verify this claim, note that when player 3 is expected to choose T, the payoffs to the player 1 and 2 depend only on the mistake probabilities for L_3 and R_3 (to be denoted by ε and δ), hence, the players 1 and 2 face the following game

	L_2	R_2
L_1	3ε, 3ε	$\varepsilon+\delta$, $\varepsilon+\delta$
R_1	$\varepsilon+\delta$, $\varepsilon+\delta$	3δ, 3δ

No matter what the (positive) values of ε and δ are, this game always has (L_1, L_2) or (R_1, R_2) as a strict equilibrium and irrespective of which of these is played, the unique best reply for player 3 is to choose T.

The above example shows that 'sequential rationality' can only be satisfied by a concept that also satisfies some form of 'connectedness'. Besides the arguments given in the previous section for why 'connectedness' is important one can also argue that this requirement follows from continuity. Namely, it seems reasonable to require that, off the equilibrium path, players should adjust their actions continuously as their beliefs vary and, whenever a stable set is disconnected, this is impossible. In the game of Example 10.3.4, if the players 1 and 2 want to force player 3 to choose T, then no matter what their beliefs are (i.e. for any ε and δ), they should either choose (L_1, L_2) or (R_1, R_2) and, furthermore, they have to jump from one value to the other for certain values of ε and δ. If they would adjust their actions continuously, then there always exist values for ε and δ such that it is optimal for player 3 to play the subgame hence, in this case the outcome in which 3 terminates the game is no longer stable. (For an interesting application of this idea of continuity, see Gul, Sonnenschein and Wilson [1986].)

The above discussion leads to the conclusion that if one wants to satisfy all properties from (10.2.5), one should require an additional regularity condition in (10.3.1). An immediate suggestion is to replace 'closed' by 'closed and connected'. This modified concept satisfies existence, invariance, connectedness and the properties of Theorem 10.3.3 (the proofs given above can be repeated verbatimyl), however, it does not satisfy sequential rationality.[15] J. F. Mertens has suggested another kind of regularity condition and he has shown that his modified concept satisfies all requirements from (10.2.5). (Mertens [1988, 1989]). Intuitively speaking, this new concept formalizes the idea of 'strategies varying smoothly as beliefs vary', hence, the distinction between this concept (of say stable* equilibria) and the concept of stable equilibria corresponds to the distinction between strictly proper equilibria and strictly perfect ones (cf. Defs. 2.2.7 and 2.3.7).[16,17]

In the remainder of this chapter, games of incomplete information will be studied and the reader may recall from Sect. 10.1 that for such games, the agent normal form is a more adequate description of the situation than the normal form is, hence, it would seem that one should investigate stability in this agent normal form. Actually, the distinction between these 2 forms is important only when there is an actual coordination problem, i.e. when 2 agents of the same player move in succession. If the coordination problem does not arise, then the agent normal form and the normal form have the same stable sets. Hence, we state without proof.[18]

Theorem 10.3.5. *If Γ is an extensive form game such that along every path each player moves at most once, then the normal form of Γ has the same stable sets as the agent normal form of Γ.*

The signalling games of Sects. 10.4 – 10.6 satisfy the condition of Theorem 10.3.5 since 2 types (agents) of the same player cannot be simultaneously active, hence, we can work with whatever normalization in most convenient. In the remainder of the chapter we will also abuse terminology a little and call a set of equilibria stable if it lies within one component and satisfies (10.3.1). Hence, minimality will not be required. Furthermore, in generic games an outcome will be called stable if there exists a stable component that induces this outcome.

10.4 Signalling Games: Introduction

In the next four sections, signalling games are studied.[19] This section provides an introduction and some examples. In the next one it will be investigated how various 'intuitive arguments' that have been used to eliminate equilibria are related to the stability concept of Kohlberg and Mertens. In Sects. 10.6 and 10.7 the ideas will be illustrated by means of two examples, viz. Spence's job market signalling model and Selten's chain store game.

Many economic models concerning asymmetric information contain the following situation as an essential ingredient: 'There are 2 parties, one informed about the state of nature and one uninformed; the informed party takes an action, which the uninformed party can observe; the latter then draws certain inferences and takes an action in response; the payoffs to both parties depend on the state of nature and the actions taken'. Examples abound in the theory of the labor market (e.g. Spence [1973, 1974, 1976], Akerlof [1970]), the insurance market (e.g. Rothschild and Stiglitz [1976], Wilson [1977]), the credit market (e.g. Ross [1977]), industrial organization (e.g. Kreps and Wilson [1982b], Milgrom and Roberts [1982, 1986]), accounting (Demski and Sappington [1987]), finance (Harris and Raviv [1982]), the economics of law (Cho [1986], Schweizer [1989]) and bargaining with incomplete information (e.g. Cramton [1983, 1984], Fudenberg and Tirole [1983], Rubinstein [1985ab] and Grossmann and Perry [1986]), etc. etc.[20]

A *signalling game* is a stylized model of this situation. Formally, it is described by the following rules

(i) Nature draws the type t of player 1 from a finite set T according to a probability distribution π with $\pi(t)>0$ for all t.
(ii) Player 1, knowing t, selects a message m from a finite set M.
(iii) Player 2, knowing only m, chooses a response, a, from a finite set A.
(iv) The payoffs are $u(t, m, a)$ and $v(t, m, a)$ to the players 1 and 2, respectively.

It is assumed above that all types t have the same message space, M, and that the same set of actions, A, can be chosen in response to each message, but this is for notational convenience only. The theory can equally well be developed in the more general case and, in fact, in some examples, we will allow the response set to depend on the message sent. It is also assumed that the rules of the game are common knowledge, hence, both players know the sextuple (T, M, A, π, u, v). The only asymmetry concerns the fact that player 1 knows his type but player 2 does not.

Below, we will work with the agent normal form of the signalling game.[21] As far as player 1 is concerned, this is consistent with our discussion in Sect. 10.1. The use of the agent normal form is also justified by Theorem 10.3.5. The agent that is responsible for player 2's choice after the message m will be consistently referred to as agent m and, similarly, type t is the agent responsible for the choice of player 1 after chance has chosen t.

Next, we introduce the notation that will be used for signalling games. Let $\Gamma=(T, M, A, \pi, u, v)$ be such a game. For any finite set F, denote by $P(F)$ the set of all probability distributions on this set. Elements of $P(T)$, $P(M)$ and $P(A)$ will be denoted by τ, μ and α, respectively. In particular, note that beliefs of agent m are denoted by τ (and not by μ); a system of beliefs of player 2 will be denoted by $\underline{\tau}$, where $\underline{\tau}=(\tau_m)_m$. A (behavior) strategy of player 1 is an array $p=(p_t)_t$ where $p_t \in P(M)$ for all t and a (behavior) strategy of player 2 is an array $q=(q_m)_m$ with $q_m \in P(A)$ for all $m \in M$. We write

$$u(t, \mu, q) := \sum_{m,a} \mu(m) q_m(a) u(t, m, a) \tag{10.4.1}$$

for the expected payoff to type t when he sends the mixed message μ and player 2 uses the strategy q. Similarly

$$v(\tau, m, \alpha) := \sum_{t,a} \tau(t) \alpha(a) v(t, m, a) \tag{10.4.2}$$

is the expected payoff for agent m when he has beliefs τ and responds with the mixed action α. We define

$$BR_t(q) := \underset{\mu}{\operatorname{argmax}}\, u(t, \mu, q) \tag{10.4.3}$$

as the set of best replies of type t against q and

$$BR_m(\tau) := \underset{\alpha}{\operatorname{argmax}}\, v(\tau, m, \alpha) \tag{10.4.4}$$

as the set of best responses of agent m against the beliefs τ. If p is a strategy of player 1, write

$$p(m) := \sum_t \pi(t) p_t(m)$$

for the probability that message m is chosen. By Bayes' rule, the beliefs of agent m induced by p are given by

$$\tau_m^p(t) := \frac{\pi(t) p_t(m)}{p(m)} \quad \text{if } p(m) > 0. \tag{10.4.5}$$

A strategy pair (p, q) is a *Nash equilibrium* if each strategy is a best reply to the other, i.e.

$$p_t \in BR_t(q) \quad \text{for all } t \in T, \tag{10.4.6}$$

and

$$q_m \in BR_m(\tau_m^p) \quad \text{for all } m \in M \text{ with } p(m) > 0. \tag{10.4.7}$$

Note that the Nash concept does not impose any restrictions on the actions of agent m when message m is not chosen. The concept of sequential equilibria does impose restrictions for such agents. Formally, a triple $(p, q, \underline{\tau})$ is a *sequential equilibrium* when the following 3 conditions are satisfied

$$p_t \in BR_t(q) \quad \text{for all } t \in T, \tag{10.4.8}$$

$$q_m \in BR_m(\tau_m) \quad \text{for all } m \in M, \tag{10.4.9}$$

and

$$\tau_m = \tau_m^p \quad \text{for all } m \text{ with } p(m) > 0. \tag{10.4.10}$$

Frequently, we will also say that (p, q) is a sequential equilibrium whenever there exists some $\underline{\tau}$ such that the triple $(p, q, \underline{\tau})$ satisfies (10.4.8) – (10.4.10).

Throughout, emphasis will be on the (generic) class of games for which, at every equilibrium, the conditions (10.4.6) – (10.4.7) are independent and which, consequently, have only finitely many equilibrium outcomes. To be more specific about this nondegeneracy condition, note that one can rewrite (10.4.6) – (10.4.7) as a system of equations $F(p, q) = 0$ (cf. Sect. 2.5). For almost all games, the derivative $DF(p, q)$ has maximal rank at every equilibrium, and whenever this is the case, the game has finitely many Nash outcomes (cf. Theorem 6.2.2, Kreps and Wilson [1982, Appendix 3]). Note that if Γ is such a generic game and (p, q) is an equilibrium with $p(m) > 0$ for all m, then $\{(p, q)\}$ is a stable component. Hence, in generic games, only unsent messages pose a test for stability. All refined equilibrium concepts that we will encounter in the next section indeed focus attention on such unsent messages.

Recall that the set of Nash equilibria consists of finitely many components, of which at least 1 is stable, and that in generic games all equilibria within a component yield the same outcome. Hence, if E is a component of a generic signalling game, then $E_t = \{p_t\}$ is a singleton for each t, $E_m = \{q_m\}$ is a singleton for each agent m with $p(m) > 0$ and E_m is the convex polyhedron $\{\alpha \in P(A); u(t, m, \alpha) \leqq u_E(t)$ for all $t\}$ for any unreached agent m, where $u_E(t) = u(t, p_t, q)$ is the equilibrium payoff of type t.

In Sect. 10.3 we have seen that, if E is a stable component, then the set of perfect equilibria in E is stable as well (Recall our convention from the end of Sect. 10.3 that we don't require minimality). Now, every perfect equilibrium is undominated, hence, it prescribes on undominated action for each agent m. Furthermore, it is easily seen that α is dominated (i.e. $v(t, m, \alpha) < v(t, m, \alpha')$ for all t, some α') if and only if α is not a best response against some posterior (i.e. $\alpha \notin BR_m(P(T))$). Hence, for signalling games, every perfect equilibrium of the normal form is a sequential equilibrium in the extensive form. We have shown

Theorem 10.4.1. *If Γ is a signalling game, then*

(i) *If E is stable, then the set of sequential equilibria in E is stable as well.*
(ii) *Every stable set contains a sequential equilibrium.*

The following example will demonstrate our conventions concerning the graphical display of signalling games. It will also illustrate that not every Nash equilibrium need be sequential and that not every sequential equilibrium is reasonable. (The game is nongeneric, but this is irrelevant for our present purposes.)

In the game of Fig. 10.4.1, there are 2 types of player 1, viz. t_1 and t_2 (we did not specify the prior distribution π since this is irrelevant for the present discussion), and they can send 2 messages, m_1 and m_2. If message m_1 is taken, the game terminates with player 1 receiving 1 and player 2 getting 2 (irrespective of the type of player 1). If message m_2 is sent, then player 2 has to choose a_1 or a_2 and the payoffs depend on type, message and response as in the matrix at the right of Fig. 10.4.1. For example, $u(t_1, m_2, a_1) = 0$, $v(t_1, m_2, a_1) = x$, $u(t_2, m_2, a_1) = -1$, $v(t_2, m_2, a_1) = 0$.

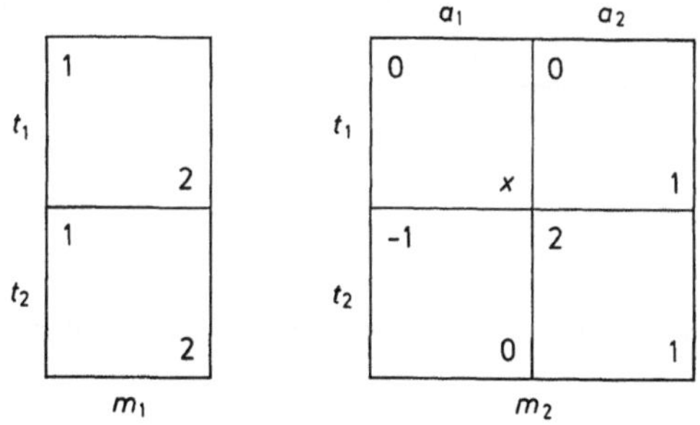

Fig. 10.4.1. A signalling game

First, consider the case $x=0$. Then there exists a Nash equilibrium in which both types of player 1 send m_1. Namely, this message is optimal when agent m_2 chooses a_1. However, this equilibrium is not sequential: No matter what beliefs agent m_2 has, it is always suboptimal to choose a_1. Consequently, he will choose a_2 and the unique sequential equilibrium involves type t_i choosing m_i $(i=1, 2)$ and agent m_2 responding with a_2.

Next, let $x=2$. Then there exists a sequential equilibrium in which both types sent m_1. Namely, in this case, a_1 is optimal for agent m_2 when he believes that type t_1 is more likely to send m_2 than t_2 is. (More precisely, (p, q, τ) with $p_{t_i}=m_1$ $(i=1, 2)$, $q_{m_2}=a_1$ and $\tau_{m_2}(t_1) \geqq 1/2$ is a sequential equilibrium.) However, such beliefs are nonsensical: message m_2 is dominated by m_1 for t_1 but not for t_2, so that only t_2 can have an incentive to deviate. However, when agent m_2 believes that most likely he is facing t_2, then he should choose a_2 and this upsets the equilibrium since type t_2 will deviate. Consequently, only the equilibrium in which type t_i sends m_i $(i=1, 2)$ and in which agent m_2 responds with a_2 is 'reasonable'.

The above example has shown that the concept of sequential equilibria is too weak since it does not restrict in any way the beliefs τ_m associated with messages m that are not sent in equilibrium. The following example illustrates this phenomenon once more and it, furthermore, demonstrates that the dominance argument used above is not powerful enough in general.

Example 10.4.2. Kreps' Beer-Quiche Example (Cho and Kreps [1987]).

Player 1 can be either strong or weak. Player 2, who initially assesses a probability $\pi(s)=.9$ that player 1 is strong, must decide whether to challenge player 1 in a duel or not. Before making this decision, player 2 gets to see whether player 1 has beer or quiche for breakfast and he knows that the strong type prefers beer whereas the weak type prefers quiche (each type gets a utility of $+1$ from having its preferred breakfast). It is also known that player 1, regardless of his type, does not like to duel (not having to duel yields 2 additional units of utility) and that player 2 profits from challenging if and only if he faces the weak type (player 2 gets $+1$ (resp. -1) if he challenges the weak (resp. strong) type).

This situation can be represented by the signalling game shown in Fig. 10.4.2. Common sense tells us that player 2 should challenge if and only if quiche is taken

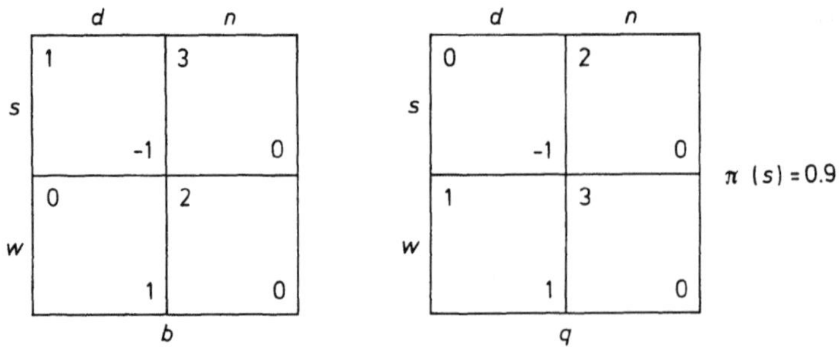

Fig. 10.4.2. Kreps' Beer-Quiche Example (s=strong, w=weak, d=duel, n=not duel, b=beer, q=quiche)

for breakfast. When player 2 behaves in this way, both types of player 1 will take beer (the weak type is better off by mimicking the strong type than by revealing that he is weak) and this indeed is a sequential equilibrium (since $\pi(s) \geqq 1/2$). However, there is a second sequential equilibrium in which both types have quiche for breakfast and in which player 2 challenges if and only if beer is taken for breakfast. This equilibrium is supported by the beliefs that, if player 1 has beer, then he is more likely to be weak. (More precisely: $p_s = p_w = q$, $q_b = d$, $q_q = n$, $\tau_b(s) \leqq 1/2$, $\tau_q = \pi$ is a sequential equilibrium.) However, these beliefs are counterintuitive: Player 2 begins by believing that 1 most likely is strong, but if 1 has beer for breakfast, this is taken as a sign that 1 is weak even though it is the strong type that likes beer.

Cho/Kreps give the following intuitive argument for rejecting the second type of equilibrium. Suppose that both types of player 1 are expected to choose quiche, that 2 is supposed to respond to beer by challenging and that 1 indeed is strong. Then 1 could make the following speech.

> I will have beer for breakfast and you ought to conclude from this that I am strong. For, if you so conclude, then you will not challenge me and I, being strong, will be better off. Furthermore, you should conclude that I am strong, for if I were weak, I would have no incentive whatsoever to take beer: If I were weak, then, no matter what conclusion you would draw from me taking beer, I would be worse off than at the original equilibrium.

Furthermore, Cho/Kreps argue that even if player 1 is unable to make such a speech, the logic behind it should lead player 2 to the same conclusion. Consequently, the second equilibrium does not seem viable and we are left only with the first, intuitive, equilibrium.

Note that this intuitive argument actually is a dominance argument, although it is different from the one used in the game of Fig. 10.4.1. It is not the case that taking beer is dominated by eating quiche for the weak type, but, *given that after quiche one will not be challenged,* drinking beer is indeed dominated. Equilibrium dominance is a more powerful concept than ordinary dominance but it also is less intuitive. Namely, the given equilibrium plays a crucial role in the criterion, yet, when the argument works, it rejects this equilibrium. Consider the following counterspeech[22] of player 2 in response to the above speech of player 1:

> Your argument is based on the presumption that the weak type is guaranteed a payoff of 3 if he takes quiche, i.e. that I will not challenge after quiche. However, when I accept your argument I know that quiche is a sure sign of weakness to which I, therefore, should respond by dueling. Consequently, quiche yields the weak type only 1 and also the weak type is better off by drinking beer (given that this will not be met with a fight). Hence, your speech cannot convince me that you are strong.

If the payoffs are as in Fig. 10.4.2 then the strong player 1 might counter by saying that, even if both types have beer, it is still optimal for 2 not to duel so that he will still be better off, however, this is not very convincing. Namely, consider the modification of the game in which player 2 gets 10 if 1 is weak, drinks beer and there is a duel ($v(w, b, d) = 10$, all other payoffs unchanged). Then 1's original speech remains unchanged, but 2 might continue his above speech with.

> If I accept your argument, I come to the conclusion that both types will have beer for breakfast. Therefore, I will respond to beer by challenging (which is optimal given my beliefs) and this leaves both types worse off than at the original equilibrium. Consequently, nobody should actually have deviated and I cannot assign any 'rational meaning' to a breakfast of beer. All in all, I am quite confused by your speech and I feel it is safest (for me and, therefore, also for you) to stick to the original equilibrium.

The reader should judge himself which of these intuitive arguments are indeed acceptable. In the next section we formalize the equilibrium dominance argument (as well as several relates ones) and it is hoped that this will make the judgement easier. It will also be shown that all such intuitive criteria and dominance arguments are less demanding than stability.

10.5 Signalling Games: Dominance, Intuitive Arguments and Stability

The weakness of the sequential equilibrium concept is that it virtually does not restrict beliefs at information sets off the equilibrium path. The concept allows all beliefs that can be explained by means of 'small mistakes', it does not require a player to ask himself whether perhaps an opponent is trying to signal something by taking an unexpected move, hence, a player does not have to scrutinize the possibility of someone intentionally deviating from the equilibrium. Many sequential equilibria are supported by beliefs that do not survive this kind of scrutinity, hence, they are not viable since the treathened behavior off the equilibrium path is optimal only against beliefs that are incredible. In this section, some procedures are described that have been proposed in the literature to reduce the set of 'plausible' beliefs off the equilibrium path and, hence, to eliminate 'unintuitive' equilibria. For expositional clarity, attention is restricted to signalling games, but the ideas can be fruitfully applied in general extensive form games (see Cho [1986, 1987]). The relationships between these refined equilibrium concepts and Kohlberg/Mertens stability are pointed out and the section concludes with a characterization of stability in generic signalling games. The discussion is based on Cho and Kreps [1987], Banks and Sobel [1987] and Grossman and Perry [1986a]. Because of space restrictions, detailed proofs will not be provided. For these, the reader is urged to consult the original papers.[23]

Let a signalling game $\Gamma = (T, M, A, \pi, u, v)$ be given. If the beliefs of agent m belong to the set $\mathscr{B}_m$, he will choose an action in

$$BR_m(\mathscr{B}_m) := \bigcup_{\tau \in \mathscr{B}_m} BR_m(\tau)\,. \tag{10.5.1}$$

The concept of sequential equilibria allows all beliefs in $P(T)$ for an unreached agent m. The procedures defined below have in common that they restrict the set of beliefs $\mathscr{B}_m$ that this agent can sensibly have. First, we describe dominance procedures. These are based on the idea that agent m should assign probability zero to the event that he is facing a type that cannot gain by sending message m. Formally, let $\mathscr{A}=(\mathscr{A}_m)_{m\in M}$ be a collection of nonempty, closed sets with $\mathscr{A}_m \subset P(A)$ and assume that, for one reason or another, it is known that each agent m will take an action in $\mathscr{A}_m$. If

$$\max_{\alpha\in\mathscr{A}_m} u(t,m,\alpha) < \min_{\alpha\in\mathscr{A}_{m'}} u(t,m',\alpha) \quad \text{for some } m', \tag{10.5.2}$$

then type t definitely will not choose m. We say that *m is $\mathscr{A}$-dominated for t* if (10.5.2) is satisfied and we write $T_m(\mathscr{A})$ for the set of types for which m is not $\mathscr{A}$-dominated. Whenever $T_m(\mathscr{A})$ is nonempty, the beliefs of agent m should be concentrated on this set. Hence, his beliefs should belong to

$$\mathscr{B}_m(\mathscr{A}) := \{\tau\in P(T);\ \tau(t)=0 \text{ if } t\notin T_m(\mathscr{A})\}, \tag{10.5.3}$$

and he should choose an action in $BR_m(\mathscr{B}_m(\mathscr{A}))$. Repeated application of this argument leads to the following iterative procedure:

Iterative Elimination of $\mathscr{A}$-dominated Actions:

$$\mathscr{A}_m^0 := \mathscr{A}_m,\quad \mathscr{B}_m^0 := P(T), \tag{10.5.4a}$$

$$\mathscr{A}_m^{k+1} := \begin{cases} BR_m(\mathscr{B}_m^k)\cap\mathscr{A}_m^k & \text{if } \mathscr{B}_m^k \neq \varnothing \\ \mathscr{A}_m^k & \text{otherwise,} \end{cases} \tag{10.5.4b}$$

$$\mathscr{B}_m^{k+1} := \begin{cases} \mathscr{B}_m(\mathscr{A}^k)\cap\mathscr{B}_m^k & \text{if } \mathscr{A}_{m'}^k \neq \varnothing \text{ for all } m'\in M \\ \mathscr{B}_m^k & \text{otherwise,} \end{cases} \tag{10.5.4c}$$

$$\mathscr{A}_m^\infty := \bigcap_k \mathscr{A}_m^k,\quad \mathscr{B}_m^\infty := \bigcap_k \mathscr{B}_m^k. \tag{10.5.4d}$$

When it is common knowledge that each agent m will take an action in $\mathscr{A}_m$, then $\mathscr{B}_m^\infty$ represents the set of beliefs that agent m can sensibly have and $\mathscr{A}_m^\infty$ is the set of actions that this agent can reasonably take. Consequently, only equilibria $(p,q,\underline{\tau})$ with $q_m\in\mathscr{A}_m^\infty$ and $\tau_m\in\mathscr{B}_m^\infty$ for all m can make sense (at least, when both limit sets are nonempty), and such equilibria will be called $\mathscr{A}$-admissible. Formally, however, it is more convenient (and if one accepts the argumentation in Sect. 10.2, also more natural) to define $\mathscr{A}$-admissibility of equilibrium components (i.e. equilibrium outcomes).

Definition 10.5.1. Let E be an equilibrium component of a signalling game, let E_m be the associated equilibrium actions of agent m and let $(\mathscr{A}_m)_{m\in M}$ be a collection of closed sets with $E_m\subset\mathscr{A}_m$ for all m. The component E is *$\mathscr{A}$-admissible* if $E_m\cap\mathscr{A}_m^\infty \neq \varnothing$ for all $m\in M$.

Let E be a stable component and let $\mathscr{A}$ be as in the above definition. Then in the first iteration round agent m only eliminates (i.e. puts zero weight on) types for which sending m is not a best reply against any element in E_m. Furthermore, he only eliminates own dominated actions ($\alpha \notin BR_m(P(T))$ if and only if α is strictly dominated by some α', since this is a 2-person game). Theorem 10.3.3 therefore implies that E induces a stable component E^1 in this reduced game. In particular, $E^1_m \subset \mathscr{A}^1_m$ for all m and by repeating this argument we obtain

Theorem 10.5.2. *If E is a stable component, then E is $\mathscr{A}$-admissible for any $\mathscr{A}$ with $E_m \subset \mathscr{A}_m$ for all $m \in M$.*

Several comments are in order concerning the above elimination procedure:

(i) The procedure assumes that all beliefs in $\mathscr{B}_m(\mathscr{A})$ do make sense when it is common knowledge that each agent m takes an action from $\mathscr{A}_m$. The concept of divine equilibria (see below) is based on the idea that only a subset of $\mathscr{B}_m(\mathscr{A})$ is viable in this case.

(ii) A slightly less restrictive procedure is obtained by requiring $\mathscr{A}^{k+1}_m$ to consist of those actions in $\mathscr{A}^k_m$ that are optimal against $\mathscr{B}^k_m$ subject to the restriction that the response should be in $\mathscr{A}^k_m$. This amounts to saying that, if a type has been eliminated in round k, then it is impossible that this type upsets the equilibrium in round k' with $k' > k$. The concepts of CK-admissibility and (universal) divinity would not be changed by this less restrictive requirement, but KM-admissibility is changed (see Fig. 10.5.2).

(iii) One can show that if an equilibrium component E is $\mathscr{A}$-admissible, then it is also $\bar{\mathscr{A}}$-admissible for any $\bar{\mathscr{A}}$ with $\mathscr{A}_m \subset \bar{\mathscr{A}}_m$ for all m. However, for this monotonicity to hold it is essential that E is a whole component and not just a set of equilibria. For example in the game of Fig. 10.4.1 with $x=2$, the equilibrium in which both types send m_1 is $\mathscr{A}$-admissible if $\mathscr{A}_{m2} = \{a_1\}$, but it is not $\bar{\mathscr{A}}$-admissible when $\bar{\mathscr{A}}_{m2} = \{a_1, a_2\}$.

(iv) The procedure is a generalization of ordinary elimination of (strictly) dominated actions. Namely, the action α is strongly dominated for agent m if and only if $\alpha \notin BR_m(P(T))$, from which it follows that elimination of strictly dominated actions corresponds to procedure (10.5.4) initialized by setting

$$\mathscr{A}^r_m := P(A) \quad (m \in M)\,. \tag{10.5.5}$$

If $\mathscr{A}^r$ is as in (10.5.5), then E is $\mathscr{A}^r$-admissible if and only if E is *rationalizable* in the sense of Bernheim [1984] and Pearce [1984] (also cf. the concept of dominance solvability from Moulin [1979]). In Sect. 10.4 we have already seen that not every equilibrium is rationalizable and that not every rationalizable outcome is reasonable. One could try to remedy the latter drawback by eliminating also weakly dominated actions, but this will have no effect in generic signalling games. Also the suggestion of McLennan [1985] the agent m should assign positive probability only to those types that have m as a best response in some sequential equilibrium, in general is not powerful enough to eliminate most of the unreasonable equilibria. Formally, McLennan argues that type t will not send message m if m is $\mathscr{A}^j$-dominated for t where

$$\mathscr{A}^j_m := \{\alpha \in P(A); \alpha = q_m \text{ for some sequential equilibrium } (p, q, \underline{\tau})\} \tag{10.5.6}$$

and he calls an equilibrium outcome *justifiable* if it is $\mathscr{A}^j$-admissible. In the game of Fig. 10.4.2, the outcome in which both types have quiche is justifiable since the weak type has beer for breakfast in some sequential equilibrium so that beer is not $\mathscr{A}^j$-dominated for this type.

It should not be surprising that the concepts of rationalizability and justifiability are only moderately successful. The procedures invoked in these concepts should be viewed as preliminary steps in the analysis of a game, eliminating only those equilibria that are patently implausible. If one wants to examine whether a specific equilibrium outcome is plausible, however, one should analyze the game by taking this outcome as a starting point and one should investigate whether the presence of this outcome induces a more refined dominance test. One such test has been proposed in Cho and Kreps [1987]. Let Γ be a generic signalling game and suppose that the component E has been suggested as the solution of the game, let $u_E(t)$ be the equilibrium payoff to type t and assume that message m is not sent in equilibrium. If the agents m' with $p(m')>0$ will not deviate from E, then type t is guaranteed $u_E(t)$ if he conforms, hence, this type has no incentive to choose m in case

$$\max_{\alpha\in P(A)} u(t, m, \alpha) < u_E(t) .$$

Cho and Kreps argue that agent m should assign probability zero to any such type and that he should choose his response accordingly. Now, note that this inequality is satisfied if and only if m is $\mathscr{A}^E$-dominated for t, where $\mathscr{A}^E$ is defined by

$$\mathscr{A}^E_{m} := \begin{cases} E_m & \text{if } p(m)>0, \\ P(A) & \text{otherwise} \end{cases} , \qquad (10.5.7)$$

($p(m)$ is the probability that m is chosen in any equilibrium of the component E) so that repeated application of the argument leads to the conclusion that E is plausible only if E is $\mathscr{A}^E$-admissible. For convenience, let us call a component *CK-admissible* if it satisfies this requirement. Clearly, any CK-admissible outcome is justifiable, hence, rationalizable. CK-admissibility is a more stringent requirement, however, since in the game of Fig. 10.4.2, it is satisfied only by the outcome in which both types have beer for breakfast. CK-admissibility corresponds to the intuitive argument discussed at the end of Sect. 10.4 (in fact, it is slightly more restrictive than the intuitive criterion). We will further illustrate its power in Sect. 10.6. However, in this author's view, not every CK-admissible equilibrium is plausible.

As an illustration, consider the game of Fig. 10.5.1. One sequential equilibrium outcome has both types choosing m_1 and agent m_2 responding with a_1. This outcome is CK-admissible, since m_2 is not dominated for any type. However, note that to support the outcome, player 2 should believe that type t_1 is more likely to send m_2 than t_2 is and this seems unreasonable since it is type t_2 that has the 'greater' incentive to deviate: No matter what action agent m_2 takes in response, if type t_1 is better off than he would have been had he chosen m_1, then type t_2 is better off as well. Hence, it seems that player 2 should not revise downward the probability that he is facing t_2 after hearing the message m_2, hence, $\tau_{m_2}(t_2)>1/2$ and m_2 should choose a_2, thereby upsetting the equilibrium.

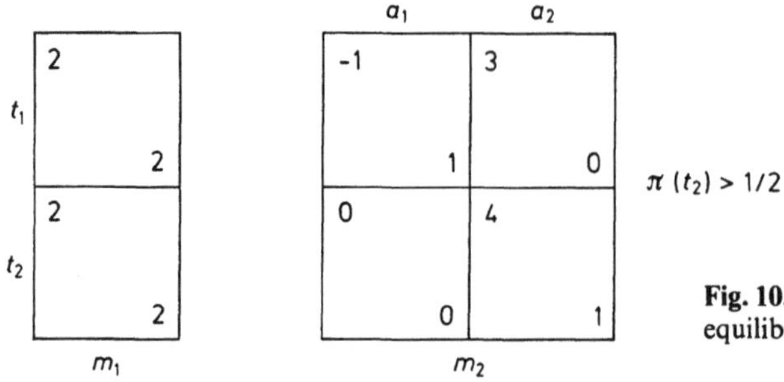

Fig. 10.5.1. Not every CK-admissible equilibrium is plausible

The reader might think that a properness kind of argument can eliminate the pooling equilibrium with outcome (2, 2) in the game of Fig. 10.5.1: Since t_1 has more to loose by choosing m_2 than t_2 has, it should be considered much less likely that t_1 chooses m_2. However, note that this argument involves a comparison of utility between the two types of player 1 and this need not make any sense. Furthermore, (and this is closely related to the previous point), as we argued in Sect. 10.1, this game should be analysed by means of the agent normal form and the equilibrium under consideration is a proper equilibrium of this agent normal form. Finally, it should be noted that the equilibrium is even proper in the normal form in case $\pi(t_2)$ is sufficiently large. Namely, although type t_2 has less to loose, his losses are weighted more heavily, so that in terms of ex-ante payoffs, player 1 might be better off in case only t_1 chooses m_2 as in case only t_2 chooses m_2.

Note that the argument that we invoked to eliminate the pooling equilibrium outcome with payoff (2, 2) did not use any intertypal comparison of utility: If we apply separate positive affine transformations to the utility functions of the 2 types, then it is still true that t_2 gains from deviating whenever t_1 does so (but not vice versa). This monotonicity argument underlies the concept of divine equilibria which has been proposed in Banks and Sobel [1987] and that will be discussed now.

Again suppose that a component E has been proposed as the solution of a generic signalling game, let $u_E(t)$ be the equilibrium payoff to type t and let m be an unsent action. When it is common knowledge that agent m will choose α in case he is reached then type t will deviate from E and choose m instead with a probability $\lambda(t, \alpha)$ given by

$$\lambda(t,\alpha)\begin{cases} =0 & \text{if } u(t,m,\alpha) < u_E(t)\,, \\ =1 & \text{if } u(t,m,\alpha) > u_E(t)\,, \quad \text{and} \\ \in [0,1] & \text{if } u(t,m,\alpha) = u_E(t)\,. \text{ and} \end{cases} \tag{10.5.8}$$

Let $\mathscr{B}_m(\pi, \alpha)$ be the set of posteriors obtained from π by updating with λ as in (10.5.8), hence

$$\mathscr{B}_m(\pi,\alpha) := \{\tau \in P(T);\ \tau(\cdot) = c\lambda(\cdot,\alpha)\pi(\cdot) \text{ for some } c>0 \text{ and } \lambda \text{ as in (10.5.8)}\}\,, \tag{10.5.9}$$

and, for any nonempty set $\mathscr{A}_m$ of mixed actions, define

$$\mathscr{B}_m(\pi, \mathscr{A}_m) := \text{co}\left[\bigcup_{\alpha\in\mathscr{A}_m} \mathscr{B}_m(\pi,\alpha)\right], \tag{10.5.10}$$

where 'co' denotes the 'convex hull'. Banks and Sobel argue that, if E is the proposed solution and if it is common knowledge that agent m chooses an action in $\mathscr{A}_m$, then agent m should have beliefs in $\mathscr{B}_m(\pi, \mathscr{A}_m)$ whenever this set is nonempty. The reader may wonder whether all such beliefs are plausible, specifically, why convexification is allowed in (10.5.10). For the moment, it suffices to remark that this is needed to guarantee existence, but we will return to this issue later (Figs. 10.5.3 and 10.5.4). If one accepts that only beliefs in $\mathscr{B}_m(\pi, \mathscr{A}_m)$ are sensible, it follows that agent m should choose an action in $BR_m(\mathscr{B}_m(\pi, \mathscr{A}_m)) \cap \mathscr{A}_m$ and the procedure can be repeated taking this set as the starting point. Hence, we obtain an iterative procedure as in (10.5.4), initialized by taking $\mathscr{A}^E$ as in (10.5.7), but with (10.5.4c) replaced by

$$\mathscr{B}_m^{k+1}(\pi) := \begin{cases} \mathscr{B}_m(\pi, \mathscr{A}_m^k) \cap \mathscr{B}_m^k(\pi) & \text{if } \mathscr{A}_m^k \neq \varnothing, \\ \mathscr{B}_m^k(\pi) & \text{otherwise.} \end{cases} \tag{10.5.11}$$

Let $\mathscr{A}_m^\infty(\pi)$ (resp. $\mathscr{B}_m^\infty(\pi)$) be the limit set of actions (resp. beliefs) of agent m associated with the procedure (10.5.7), (10.5.4a, b, d), (10.5.11). Banks and Sobel call the component (outcome) E *divine* if $E_m \cap \mathscr{A}_m^\infty(\pi) \neq \varnothing$ for every message m, and they show that every stable outcome is divine and that every divine outcome is CK-admissible (Banks and Sobel [1985]). In the game of Fig. 10.5.1 (with $\pi(t_2) > 1/2$) only the equilibrium in which both types send m_2 is divine so that divinity is strictly more restrictive than CK-admissibility.

Again consider the game of Fig. 10.5.1 but now with $\pi(t_2) < 1/2$. In this case, the equilibrium outcome in which both types send m_1 is divine. Namely, let $\bar{\alpha}$ be the mixed action of agent m_2 in which a_1 is chosen with probability 1/4. If agent m_2 plays $\bar{\alpha}$, then t_1 is indifferent between m_1 and m_2 whereas t_2 prefers m_2, hence

$$\mathscr{B}_{m_2}(\pi, \bar{\alpha}) = \{\tau \in P(T); \tau(t_2) \geqq \pi(t_2)\}.$$

Since $\mathscr{B}_{m_2}(\pi, \alpha) \subset \mathscr{B}_{m_2}(\pi, \bar{\alpha})$ for all α, we obtain $\mathscr{B}_{m_2}^1(\pi) = \mathscr{B}_{m_2}(\pi, \bar{\alpha})$ as the set of admissible beliefs of agent m_2 after the first step of the procedure for verifying divinity. Since any action of agent m_2 is a best response against a belief in this set $\mathscr{A}_{m_2}^2(\pi) = P(A)$, so that the sets of admissible beliefs and admissible actions remain constant after the first step. This implies that the outcome is indeed divine. Hence, this outcome is divine for some priors, but not for all. Using the terminology of Banks and Sobel we say that this outcome is not universally divine. Formally the latter concept is defined as follows. For $\alpha \in P(A)$, let $\mathscr{B}_m(\alpha)$ be the set of beliefs which agent m can 'reasonably' associate with α irrespective of which (nondegenerate) prior he has

$$\mathscr{B}_m(\alpha) := \bigcap_{\pi \in \operatorname{int} P(T)} \mathscr{B}_m(\pi, \alpha). \tag{10.5.9a}$$

(where 'int' denotes 'interior') and, for $\varnothing \neq \mathscr{A}_m \subset P(A)$, define

$$\mathscr{B}_m(\mathscr{A}_m) := \operatorname{co}\left[\bigcup_{\alpha \in \mathscr{A}_m} \mathscr{B}_m(\alpha)\right]. \tag{10.5.10a}$$

Then, the outcome E is said to be *universally divine* if $E_m \cap \mathscr{A}_m^\infty \neq \varnothing$ for any message m, where $\mathscr{A}_m^\infty$ is the limit set of actions of agent m associated with the iterative procedure (10.5.4), initialized as in (10.5.7), but with (10.5.4c) replaced by

$$\mathscr{B}_m^{k+1} := \begin{cases} \mathscr{B}_m(\mathscr{A}_m^k) \cap \mathscr{B}_m^k & \text{if } \mathscr{A}_m^k \neq \varnothing, \\ \mathscr{B}_m^k & \text{otherwise.} \end{cases} \tag{10.5.11a}$$

It is possible to give a more convenient characterization of the set $\mathscr{B}_m(\mathscr{A}_m)$. Let $\mathscr{D}(t, \mathscr{A}_m)$ be the set of actions in $\mathscr{A}_m$ that could possibly cause t to choose m

$$\mathscr{D}(t, \mathscr{A}_m) := \{\alpha \in \mathscr{A}_m;\ u(t, m, \alpha) \geqq u_E(t)\} \tag{10.5.12}$$

and let $\mathscr{D}^0(t, \mathscr{A}_m)$ be the subset for which the inequality is strict. It is easily seen that, if for any action that possibly causes t to defect, some other type deviates for sure, then $\tau(t) = 0$ for all $\tau \in \mathscr{B}_m(\mathscr{A}_m)$. In generic games, the converse will also be true so that in this case

$$\mathscr{B}_m(\mathscr{A}_m) = \left\{\tau;\ \tau(t) = 0 \text{ if } \mathscr{D}(t, \mathscr{A}_m) \subset \bigcup_{t' \neq t} \mathscr{D}^0(t', \mathscr{A}_m)\right\}. \tag{10.5.13}$$

This characterization also makes it clear that in every iteration step for checking universal divinity of the component E, agent m only eliminates types for which sending m is not a best response against any element of E. Consequently, every stable outcome is universally divine.

Universal divinity is quite a stringent requirement and the reader might conjecture that it is equivalent to requiring invariance with respect to elimination of "never weak best responses", but this conjecture is false. The latter may be more restrictive due to the fact that universal divinity (initially) considers too many actions for unreached agents as reasonable. Recall that the procedure for verifying universal divinity is initialized by setting $\mathscr{A}_m^0 = P(A)$ for every unreached agent m, whereas requiring the component E to be invariant with respect to elimination of never weak best responses amounts to demanding that E be E-admissible (where $E = (E_m)_m$ with E_m the associated equilibrium actions of agent m). Let us call a component *KM-admissible* if it satisfies this latter requirement. Note the crucial difference that KM-admissibility restricts unreached agents to equilibrium actions right from the beginning.

KM-admissibility implies universal divinity and in the game of Fig. 10.5.1 the requirements are equivalent, although the arguments are different (they are more or less dual). To support the (2, 2) outcome, agent m_2 should choose a_2 with probability at most 1/2 and, given that m_2 chooses such an equilibrium action, m_2 is strictly dominated for t_1 but not for t_2. Consequently, m_2 should believe that he is facing t_2, hence, he should choose a_2, which upsets the equilibrium: The outcome (2, 2) is not KM-admissible.

In general, however, KM-admissibility is a more stringent requirement. This is illustrated by the game of Fig. 10.5.2, which is taken from Cho and Kreps [1987] (also see Banks and Sobel [1987]).

We claim that the equilibrium in which both types choose m_1 is universally divine. Let $\mathscr{A}_2^k, \mathscr{B}_2^k$ $(k \in \mathbb{N})$ be the sets of actions (beliefs) of player 2 generated by the procedure for checking universal divinity. Then (10.5.13) shows that $\mathscr{B}_2^1 = P(T)$ since for each type t there exists an action α_t of m_2 such that only t

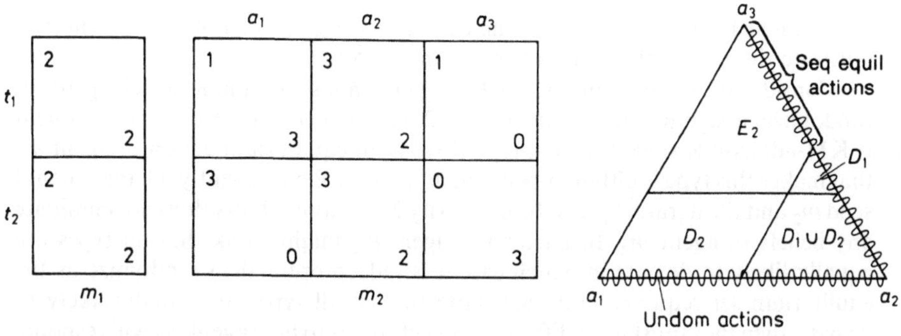

Fig. 10.5.2. The pooling outcome (2, 2) is universally divine, but not KM-admissible. The figure *on the right* is the space of mixed actions of agent m_2, divided into equilibrium actions (E_2) and actions in D_i that induce t_i to deviate from m_1

gains by deviating when m_2 chooses $\alpha_{t'}$. Furthermore, $\mathscr{A}_2^1$, the set of undominated actions of m_2, consists of all convex combinations of a_1 and a_2 together with all mixtures of a_2 and a_3. Now it is easily seen that $\mathscr{D}(t, \mathscr{A}_2^1) \not\subset \mathscr{D}(t', \mathscr{A}_2^1)$ for $t \neq t'$ so that $\mathscr{B}_2^2 = P(T)$ in view of (10.5.13). But this implies that $\mathscr{B}_2^\infty = P(T)$ and $\mathscr{A}_2^\infty = \mathscr{A}_2^1$. In particular, there exist equilibrium actions of m_2 in $\mathscr{A}_2^\infty$ and the outcome is universally divine.

Next, let $\bar{\mathscr{A}}_2^k$, $\bar{\mathscr{B}}_2^k$ $(k \in \mathbb{N})$ be the sets generated by procedure (10.5.4) for checking whether the component E is KM-admissible. Then $\bar{\mathscr{A}}_2^0 = E_2 = \{(p_1, p_2, p_3);\ p_2 \leqq 1/2,\ p_3 \geqq 1/3\}$. The constraint that deviating to m_2 should not be profitable is binding for both types so that $\bar{\mathscr{B}}_2^1 = P(T)$. Furthermore, $\bar{\mathscr{A}}_2^1$ are the sequential equilibrium actions in $\bar{\mathscr{A}}_2^0$, in this case $\bar{\mathscr{A}}_2^1 = \{(0, p, 1-p);\ p \leqq \frac{1}{2}\}$. Given that m_2 chooses an action in $\bar{\mathscr{A}}_2^1$, sending m_2 is dominated for t_2 but not for t_1, hence, $\bar{\mathscr{B}}_2^2 = \{t_1\}$. However, given that he is facing t_1, agent m_2 should choose a_1 and this does not belong to $\bar{\mathscr{A}}_2^1$. Hence, $\bar{\mathscr{A}}_2^2 = \bar{\mathscr{A}}_2^\infty = \varnothing$ and the outcome is not KM-admissible.

It is this author's view that, if one accepts Kreps' intuitive argument (hence CK-admissibility), then one should also accept KM-admissibility. The weaker concepts discussed before are all based on half-way procedures that assume that agents which expect to be reached will not deviate from their equilibrium recommendations whereas agents who expect not to be reached will completely ignore these recommendations. To discuss (im-)plausibility of an equilibrium outcome one should search for contradictions within the recommendations and within these recommendations alone, anything else is irrelevant. Specifically, for an outcome E to be self-enforcing it is necessary that *no* agent m has an incentive to deviate. For self-enforcing outcomes, therefore, only types for which m is not E-dominated can possibly have an incentive to send m and, consequently the beliefs of agent m should belong to $\mathscr{B}_m(E)$. Repeated application of the argument leads to the conclusion that KM-admissibility is necessary for self-enforcingness.

The next example may demonstrate that not all KM-admissible outcomes need be self-enforcing. Basically, this is because of the fact that KM-admissibility allows different types to have different conjectures about which equilibrium action an unreached agent might take. The example, the game of Fig. 10.5.3, will also illustrate the role of the convexification in (10.5.10).

Consider the component in which all types choose m_1 and in which agent m_2 randomizes assigning probability at most 2/3 to each pure action. This component is KM-admissible since for each type there is an equilibrium action of agent m_2 that makes this type indifferent between m_1 and m_2. Consequently, all types could send m_2 and the iterative procedure to verify KM-admissibility does not eliminate any beliefs of agent m_2. In particular, agent m_2 might think that all types are equally likely to deviate in which case it is indeed optimal to randomize as the equilibrium set requires. However, note that, if all types are equally likely to deviate, then they must have different conjectures as to how agent m_2 will respond. If the response of agent m_2 would be common knowledge, then 2 types simultaneously choosing m_2 would be impossible since for each response there is only one type who can possible gain. One might argue that in this case only beliefs that assign probability 1 to one of the types make sense (this amounts to dropping the convexification in (10.5.10)) and that, therefore, agent m_2 should choose a pure action. Obviously, the only pure action that makes sense is a_1 and this induces t_1 to deviate. Hence, the component does not seem viable. Note that the component in question is not stable in the sense of Kohlberg and Mertens: A perturbed game (of the agent normal form) in which t_1 involuntarily chooses m_2 with a much higher probability than t_2 and t_3, does not have an equilibrium close to the component. Namely, if it would, then player 2 would choose a completely mixed action and so he should be indifferent between his 3 pure actions, which in turn implies that both t_2 and t_3 should voluntarily choose m_2, but this is impossible as we have seen above (to induce both t_2 and t_3 to choose m_2 voluntarily, agent m_2 should choose both a_2 and a_3 with a probability of at least 2/3).

The pooling equilibrium outcome of the game of Fig. 10.5.3 in which both types send m_1 is also eliminated by the concept of *perfect sequential equilibria* (P.S.E) that has been introduced in Grossman and Perry [1986a] and that is closely related to the concept of *neologism proof equilibria* of Farrell [1984]. We will now describe the basic ideas underlying the P.S.E. concept.[24]

	m_1	m_2: a_1	m_2: a_2	m_2: a_3
t_1	2, 2	3, 3	0, 0	0, 0
t_2	2, 2	0, 0	3, 0	0, 3
t_3	2, 2	0, 0	0, 3	3, 0

Fig. 10.5.3. Not all KM-admissible outcomes are stable

Consider an equilibrium e of a signalling game Γ and assume m is a message that is not sent in equilibrium. For a mixed action α, define the set of posteriors at m, derived from π, consistent with e and α similarly as in (10.5.9), but now with $u_e(t)$ instead of $u_E(t)$ in (10.5.8). A belief τ of agent m is said to be *consistent* with e and prior π if there exists some $\alpha \in BR_m(\tau)$ with $\tau \in \mathscr{B}_m(\pi, \alpha)$. Hence, if m has the consistent belief τ then his best response α could lead to a confirmation of τ. (Note that this notion of consistency is much stronger than the one discussed for sequential equilibria in Sect. 6.3). Denote the set of beliefs that are consistent with e at m by $\mathscr{B}_m(\pi, e)$. A sequential equilibrium $e = (p, q, \underline{\tau})$ is said to be a *perfect sequential equilibrium* (P.S.E.) if the beliefs are consistent whenever possible, hence

$$\tau_m \in \mathscr{B}_m(\pi, e) \quad \text{whenever } \mathscr{B}_m(\pi, e) \neq \varnothing . \tag{10.5.14}$$

It will be convenient to also introduce a slightly stronger concept, based on a similar idea. Call an action α of agent m *consistent* with e and π if there exists some $\tau \in \mathscr{B}_m(\pi, \alpha)$ such that $\alpha \in BR_m(\tau)$. Hence, if α is consistent, then player 1 could deviate in such a way as to generate τ and, thereby justify playing α. Denote the set of actions that are consistent with e and π with $\mathscr{A}_m(\pi, e)$ and define $e = (p, q, \underline{\tau})$ to be a P.S.E.* if the equilibrium actions are consistent whenever possible, i.e.

$$q_m \in \mathscr{A}_m(\pi, e) \quad \text{whenever } \mathscr{A}_m(\pi, e) \neq \varnothing . \tag{10.5.15}$$

It is clear that every outcome induced by a P.S.E.* is also a P.S.E.-outcome but the converse does not hold. Namely, consider the equilibrium in Fig. 10.5.2 in which both types send m_1 and let $\pi(t_1) < 1/3$. Then it is easily verified that for m_2 only the posterior τ with $\tau(t_1) = 1/3$ is consistent with e and π (the associated α is the mixture of a_2 and a_3 that puts weight 2/3 on a_2 so that t_2 is indifferent between m_1 and m_2). Now the belief $\tau = (1/3, 2/3)$ can support the equilibrium: a_3 is a best response against τ and this leads both types to choose t_1. Hence, e is a P.S.E. when $\pi(t_1) < 1/3$. However, the action a_3 itself is not consistent with e and π: If $\pi(t_1) < 1/3$, then only the response $\alpha = (0, 2/3, 1/3)$ is consistent, but this does not support the equilibrium (type t_1 is sure to deviate). Consequently, the outcome cannot be supported by a P.S.E.*

The concepts of P.S.E. and P.S.E.* can be motivated by intuitive arguments, similarly as was done for CK-admissibility in Sect. 10.4. To motivate the stronger concept, consider an equilibrium $e = (p, q)$ that is not a P.S.E.* and for which there exists an unsent message m such that $\mathscr{A}_m(\pi, e) = \{\alpha_m\}$ with $\alpha_m \neq q_m$ and for which there exists exactly one type, t_m, that profits by sending m when α_m is taken in response rather than q_m. Then this type might upset the equilibrium by making the following speech:

> I am sending message m which you might not have expected me to send. When considering which action to take in response, bear in mind that I can mimic your chain of reasoning so that you can equally well assume that I can predict which response α you will take. Therefore, your beliefs should be in $\mathscr{B}_m(\pi, \alpha)$ (since all my types predict the same response). Furthermore, rationality on your part requires α to be a best response against your beliefs, hence, α should be consistent. Now, there is a unique consistent action, viz. α_m, hence I predict

you choose α_m. Furthermore, you can conclude that I am actually type t_m, but this just reinforces the conclusion that you should play α_m.

The above speech sounds convincing and the logic behind it could persuade agent m even if communication is not possible. Note, however, that also in this case player 2 could make several counterspeeches (cf. Sect. 10.4). Furthermore, the argument assumes that there exists a unique consistent action and it is not clear to this author that the argument is equally convincing in case $|\mathscr{A}_m(\pi, e)| > 1$. It could be that different types prefer different consistent actions and that a mixture of consistent actions, resulting from the confusion of the agent about which of these he is expected to take, might make all types worse off, so that all of them might rather prefer to stick to the equilibrium. An example is provided by the game of Fig. 10.5.4 (also cf. the discussion of the game of Fig. 10.1.4).

In the game of Fig. 10.5.4, the outcome in which both types send m_1 (and in which agent m_2 chooses each pure action with a probability of at most 2/3) is stable: If type t_i sends m_2 involuntary with a higher probability than does t_j, then by choosing a_j with probability 2/3, agent m_2 can induce t_j to voluntarily send m_2 such that as a result both types send it with the same probability, in which case the equilibrium action is indeed optimal.

However, no equilibrium yielding this outcome is a P.S.E. Namely, for each such equilibrium e and for each nondegenerate prior π we have that only beliefs that assign probability 1 to either type are consistent, $\mathscr{B}_2(\pi, e) = \{(1,0), (0,1)\}$, and none of these supports the outcome.

It is this author's view that the pooling equilibrium with outcome (2, 2) should be considered viable in the pure noncooperative context in which no communication is possible: In this case, agent m_2 can never know which type has deviated, all beliefs are reasonable and it is perfectly sensible to choose both actions with positive probability.

The above discussion might suggest to the reader that the concept of P.S.E. should be weakened, for example, by allowing convexification in (10.5.14) – (10.5.15) or (even weaker) by requiring $\tau_m \in \mathscr{B}_m(\pi, e)$ (resp. $q_m \in \mathscr{A}_m(\pi, e)$) only when this set is a singleton. The following example shows that such a weakening is also desirable because of the drawback that a P.S.E. need not exist. The example also shows that allowing convexification in (10.5.14) is not sufficient to guarantee existence. It is unknown whether existence is guaranteed when (10.5.14) is required only when $\mathscr{B}_m(\pi,e)$ is a singleton. (Recently, Myerson

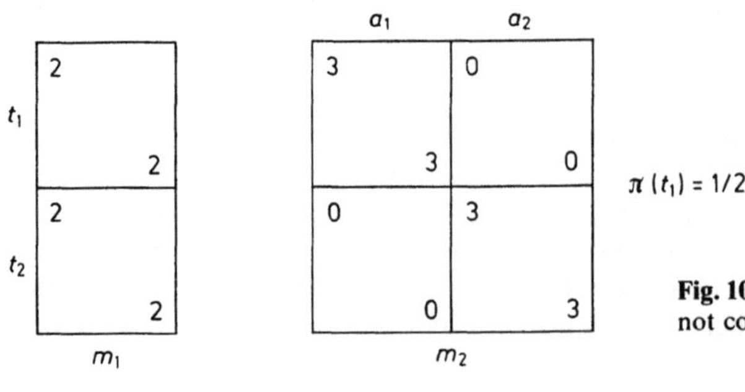

Fig. 10.5.4. A stable component need not contain a P.S.E

	a_1	a_2	a_3
t_1	2+, 3	1+, 0	0+, 2
t_2	1, 0	0, 3	2, 2

t_1

	a_1	a_2	a_3
t_1	2, 3	1, 0	0, 2
t_2	1+, 0	0+, 3	2+, 2

t_2

Fig. 10.5.5. A perfect sequential equilibrium need not exist (+ denotes $+\varepsilon$, where $0<\varepsilon<1$)

[1989] has proved existence for a related concept in games where communication is allowed.)

Consider the following guessing game, taken from Farrell [1984]. Player 1 can be of 2 types, both possibilities being equally likely ex-ante, and player 2 has to guess the type. Player 2 gets paid 3 if he guesses correctly, 0 if the guess is wrong and 2 if he admits that he does not know. Type t_1 receives 2 if the guess is right, 1 if it is wrong and 0 if player 2 admits that he does not know. (Hence, t_1 wants player 2 to know his type). Type t_2 receives 2 if player 2 admits that he does not know, 1 if the guess is wrong and 0 if it is correct (hence, t_2 wants to mimic t_1). Before the guess is made, player 1 can try to influence player 2 by telling his type. Of course, player 1 is not obliged to tell the truth, but if he does, then he gets a bonus ε. The payoffs are tabulated in Fig. 10.5.5.

In this game, type t_1 cannot succeed in screening himself from t_2, consequently the best that he can do is to tell the truth. Hence, the unique sequential equilibrium has both types saying that they are t_1 and player 2 admitting that he cannot tell which type he is facing. More formally, it can be shown that in equilibrium both types cannot randomize simultaneously and, once this has been established, it follows easily that any equilibrium must be a pure pooling equilibrium, hence, both types should announce t_1. However, this unique equilibrium outcome is not a P.S.E. Namely, it can be verified that exactly 2 beliefs are consistent with the equilibrium, viz. $\tau(t_1)=1$ or $\tau(t_1)=2/3$. However, if player 2 chooses a best response against any of these beliefs, then at least one type of player 1 will deviate to announce that he is t_2. (If player 2 believes that he is facing t_1 then he should choose a_1 and this indeed leads t_1 to deviate; If $\tau(t_1)=2/3$, then player 2 should randomize among a_1 and a_3, but, for any such randomization, at least one type gains from deviating). Note that even allowing convexification in (10.5.14) would not lead to existence: If player 2 chooses a best response against τ with $\tau(t_1)\geqq 2/3$, then still player 1 has an incentive to deviate.

The games of the Figs. 10.5.4 and 10.5.5 have shown that a stable component need not contain a perfect sequential equilibrium. We also have already seen that a P.S.E. need not be in a stable component (in Fig. 10.5.2 the (2, 2)-outcome is a P.S.E. for $\pi(t_1)<1/3$). The next example shows that a P.S.E.* need not be in a stable component. However, note that the game from Fig. 10.5.6 is nongeneric. It is

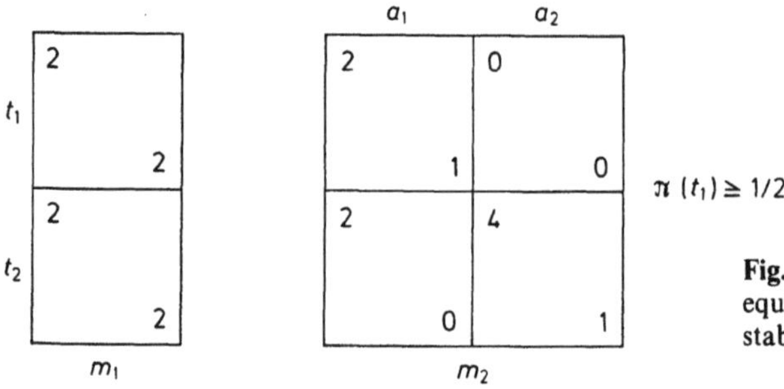

Fig. 10.5.6. A perfect sequential equilibrium outcome need not be stable

unknown to this author whether in generic signalling games, every P.S.E.* outcome is stable.

By eliminating weakly dominated actions one can see that the unique stable equilibrium of the game of Fig. 10.5.6 has type t_i sending message m_i ($i=1, 2$) and agent m_2 responding with action a_2. However, the pooling equilibrium in which both types send m_1 and in which agent m_2 chooses a_1 is also a P.S.E.*, since the action a_1 that supports this equilibrium is consistent with the equilibrium.

Although a P.S.E.* need not be stable, every P.S.E.* outcome is KM-admissible as the following theorem shows. Furthermore, every P.S.E. belongs to a CK-admissible component.

Theorem 10.5.3.
(i) *Every P.S.E.* belongs to a KM-admissible component.*
(ii) *Every P.S.E. belongs to a CK-admissible component.*

Proof. We will only demonstrate (i). The proof of (ii) is along the same lines and has been given in Grossman and Perry [1986a]. Note that Fig. 10.5.2 has shown that a P.S.E. need not be KM-admissible.
Let E be an equilibrium component and let $e \in E$. We will show that e cannot be a P.S.E.* if E is not KM-admissible. For a message m, denote by T_m the set of types for which m is not E-dominated. There exists an unsent message m such that for any belief concentrated on T_m and for any best response against such a belief, there exists a type in T_m that is better off by sending m than by sticking to E. This implies that for any $e \in E$ the response e_m is not consistent with e, hence, it suffices to show that there exists a consistent action at m. To that end, consider the following auxiliary game: Player 1's type is drawn from T_m (according to the conditional of π); there are 2 possible messages, viz. e which guarantees the equilibrium payoffs and m in which case player 2 has to respond with an action from E_m. This auxiliary game admits a sequential equilibrium e^* (since E_m is a convex polyhedron), and if e^* is played, then agent m is reached with positive probability (since E is not KM-admissible). This guarantees that the response e^*_m is consistent with the original equilibrium. □

To conclude this section, we provide a characterization of stability in generic signalling games. As an introduction, consider once more the games of Figs. 10.5.3 and 10.5.4. In the former, the pooling equilibrium with outcome (2, 2) is not stable since it is destabilized by the belief that it is type t_1 who sent m_2. In the game of Fig. 10.5.4 it might seem that this belief also destabilizes the equilibrium. However, this is not the case since by playing appropriately, agent m_2 can induce type t_2 to voluntarily send m_2. Therefore, the initial (problematic) belief that it is t_1 who deviates can be transformed into believing that both types are equally likely to deviate and the latter is not problematic since the equilibrium action is optimal against it.

The characterization of stability given below is a formalization of this insight: For an equilibrium outcome in a generic signalling game to be stable it is necessary and sufficient that each unreached agent can transform any belief into a belief against which an equilibrium action is optimal.

Formally, let Γ be a generic signalling game, let E be a connected component of the set of Nash equilibria and let E_m be the associated equilibrium actions of agent m. For $\alpha \in E_m$, write $\mathscr{B}^i_m(\alpha)$ for the beliefs concentrated on the set of types for which sending m yields exactly the equilibrium payoff when agent m responds with α, hence

$$\mathscr{B}^i_m(\alpha) := \{\tau \in P(T);\ \text{if } \tau(t) > 0, \text{ then } u(t, m, \alpha) = u_E(t)\}. \qquad (10.5.16)$$

Furthermore, let $\mathscr{B}^*_m(\alpha)$ be the set of beliefs that justify playing α

$$\mathscr{B}^*_m(\alpha) :=$$
$$\{\tau^* \in P(T); \alpha \in BR_m(\lambda\tau^* + (1-\lambda)\tau) \text{ for some } \lambda \in (0,1] \text{ and } \tau \in \mathscr{B}^i_m(\alpha)\}. \qquad (10.5.17)$$

If agent m has initial beliefs τ^*, then by playing α he can induce some types to send m voluntarily and he can adjust his beliefs to $\lambda\tau^* + (1-\lambda)\tau$ for any $\lambda \in (0, 1]$ and $\tau \in \mathscr{B}^i_m(\alpha)$. Consequently, as long as the initial beliefs are in $\mathscr{B}^*_m(\alpha)$, agent m has no incentive to deviate from α. No unreached agent will have an incentive to deviate from the equilibrium if any initial belief can be covered by some equilibrium action, i.e. if

$$\bigcup_{\alpha \in E_m} \mathscr{B}^*_m(\alpha) = P(T) \quad \text{for any } m \text{ with } p(m) = 0. \qquad (10.5.18)$$

For the component E to be stable, it is definitely necessary that (10.5.18) is satisfied. In generic signalling games, only unsent messages can pose problems and it can be shown that in this case, (10.5.18) is sufficient for stability as well. The proof of sufficiency is far from trivial, however, and for this the reader should consult the Banks/Sobel paper.

Theorem 10.5.2 (Cho and Kreps [1987], Banks and Sobel [1987]). *In a generic signalling game, a component E is stable if and only if condition* (10.5.18) *is satisfied.*

Since E_m is a convex polyhedron and since $\mathscr{B}^*_m(\alpha) \subset \mathscr{B}^*_m(\alpha_1) \cap \mathscr{B}^*_m(\alpha_2)$ when α is a convex combination of α_1 and α_2, it follows that to verify stability, it suffices to check the extreme points of E_m. To illustrate the theorem, consider the pooling equilibrium outcome with payoff (2, 2) in the game of Fig. 10.5.4. The set E_2 has 2 extreme points, viz. $\alpha_1 = (1/3, 2/3)$ and $\alpha_2 = (2/3, 1/3)$. We have $\mathscr{B}^i_2(\alpha_i) = \{t_i\}$

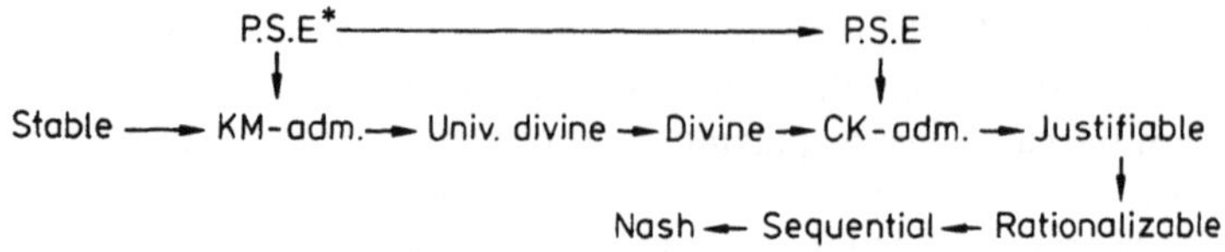

Fig. 10.5.7. Overview of the concepts discussed in this section. (All inclusions are strict)

and it is easy to check that $B_2^*(\alpha_i) = \{\tau^*; \tau^*(t_i) \geqq 1/2\}$ so that (10.5.18) is satisfied, hence, the outcome is stable. In the game of Fig. 10.5.3, the pooling outcome with payoff (2, 2) is not stable since the belief that assigns probability 1 to t_1 is not covered by any equilibrium action of agent m_2.

Figure 10.5.7 provides an overview of the relationships between the concepts discussed in this section.

10.6 Spence's Job Market Signalling Model

In this section we will consider a simple version of the job market signalling model of Spence [1973, 1974] to illustrate some of the concepts that were introduced in the previous section. Spence considers the question of whether education can serve as a filter to screen high ability workers from low ability ones. His model admits a large number of equilibria but (with 2 types) only one of these is CK-admissible and in this equilibrium workers of different ability are indeed separated. The discussion is based on Cho and Kreps [1987].

Consider the game $\Gamma(\pi)$ described by the following rules:

(i) Chance determines whether the worker has marginal productivity 1 (this occurs with probability $1-\pi$) or marginal productivity 2, the outcome being told only to the worker,
(ii) The worker selects an education level $y\,(y \geqq 0)$,
(iii) Simultaneously, 2 employers offer nonnegative wages $w_1(y)$ and $w_2(y)$ respectively, based on the observed education level y,
(iv) The worker selects an employer.

It is assumed that the cost of acquiring the education levely y is y/t for a worker with marginal productivity (= type) t. Hence, cost are negatively correlated with ability. It is also assumed that players are risk neutral and attention will be restricted to subgame perfect equilibria in which the worker always chooses the employer offering the highest wage, randomizing equally in case both employers offer the same wage. Hence, if the worker is of type t and chooses education level y then, if wages w_1 and w_2 are offered, the payoffs are

$$\max(w_1, w_2) - y/t \quad \text{and} \quad \delta_i(w_1, w_2)(t - w_i)\,, \tag{10.6.1}$$

to the worker and employer i, respectively. ($\delta_i(w_1, w_2)$ denotes the probability that the worker chooses employer i when w_1 and w_2 are offered). Note that education is not directly productive in this model. The model differs from that in previous section since now the action spaces are infinite and there are 2 players that react to the signal, but these are irrelevant changes we can still use the concepts from Sect. 10.5.[25]

We first consider Nash equilibria of $\Gamma(\pi)$. It will be clear that both employers have to offer the same wage for any choice of education that is made in equilibrium. Without loss of generality, it is assumed that $w_1(y) = w_2(y)$ for all y and we write $w = w_1$. The pure Nash equilibria are of 2 sorts: *screening equilibria*, in which different types choose different education levels, and *pooling equilibria*, in which both types choose the same education level. If y_t is the choice of type t in a screening equilibrium, then we must have

$$w(y_t) = t \quad \text{and} \quad w(y_t) - y_t/t = \max_y \{w(y) - y/t\} \quad (t = 1, 2)\,.$$

Namely, in a screening equilibrium, the worker reveals his type and the Bertrand competition forces the employers to offer the worker his marginal productivity. Furthermore, type t should indeed by willing to choose y_t. The above condition can be satisfied if and only if it can be satisfied by setting $w(y) = 0$ for $y \notin \{y_1, y_2\}$: The optimal threat of the employers is to pay nothing in case the worker does not choose a prespecified education level. Therefore, (y_1, y_2) can be obtained as the outcome of a screening equilibrium if and only if

$$1 - y_1 \geqq 2 - y_2, \quad 2 - y_2/2 \geqq 1 - y_1/2 \quad \text{and} \quad 1 - y_1 \geqq 0\,,$$

(the fourth condition, that type 2 prefers y_2 to 0 is redundant), and this set is equivalent to

$$0 \leqq y_1 \leqq 1 \quad \text{and} \quad 1 + y_1 \leqq y_2 \leqq 2 + y_1\,. \tag{10.6.2}$$

Note that in such an equilibrium, there can be useless investment in education. Specifically, these screening equilibria can be ordered according to the Pareto criterion and we will see that the Pareto optimal element of this set $(y_1 = 0, y_2 = 1)$ is the unique CK-admissible equilibrium of the game. Similarly it is seen that $\bar{y}$ is the education choice in a pure pooling equilibrium if and only if $w(\bar{y}) = 1 + \pi$ (hence, the employers pay the expected marginal productivity) and no type has an incentive to deviate given that all other education levels garner wages zero. Hence, $\bar{y}$ is a pooling equilibrium outcome if and only if

$$0 \leqq \bar{y} \leqq 1 + \pi\,. \tag{10.6.3}$$

Next, let us consider *mixed equilibria*, i.e. at least one type of worker randomizes between 2 different education levels. (Note that there is no need for the employers to randomize.) Since education is costly, a worker will randomize among y and y' with $y < y'$ only if $w(y) < w(y')$. Consequently, if type 1 randomizes between y and y' in equilibrium, then also type 2 must choose y' with positive probability. Since type 1 must be indifferent between y and y', type 2 strictly prefers y' so that $w(y) = 1$. Similarly, if type 2 randomizes among y' and y'' with $y' < y''$, then $w(y'') = 2$, so that in equilibrium each type randomizes among at most 2 different levels. Let us denote by $\langle p_1, y_1, p_2, y_2, \bar{y}\rangle$ the outcome in which each type t chooses y_t with probability p_t and $\bar{y}$ with probability $1 - p_t$. Then $\langle p_1,$

$y_1, p_2, y_2, \bar{y}\rangle$ is a Nash equilibrium outcome if and only if

$$w(y_t)=t\ ,\ w(\bar{y})=(1-\pi)(1-p_1)+2\pi(1-p_2)\ , \tag{10.6.4a}$$

$$w(y_t)-y_t/t \geqq w(\bar{y})-\bar{y}/t \quad \text{if } p_t>0\ , \tag{10.6.4b}$$

$$w(y_t)-y_t/t \leqq w(\bar{y})-\bar{y}/t \quad \text{if } p_t<1\ , \tag{10.6.4c}$$

$$0 \leqq y_1 \leqq 1 \quad \text{and} \quad 0 \leqq \bar{y} \leqq w(\bar{y})\ . \tag{10.6.4d}$$

Clearly, the conditions (10.6.4) include the conditions (10.6.2) and (10.6.3) as special cases. The above makes it clear that the game admist a large number of Nash equilibria. It will now be investigated to what extend the concepts from the previous section reduce the number of equilibria. We start by applying the logic of the sequential equilibrium concept.

According to this concept, threathening with a wage of zero is not credible: For each education level y (and not just for the one that is actually chosen in equilibrium), the employers should construct a probability $\pi(y)$ that the worker is of type 2 given that y is observed and they should offer the corresponding wage $w(y)=1+\pi(y)$. (Sequential equilibrium requires both employers to construct the same probability $\pi(y)$.) Consequently, in a sequential equilibrium, the optimal threat of the employers is to believe that unexpected education levels are only chosen by low ability workers and to set $w(y)=1$ for such y. Hence, $\langle p_1, y_1, p_2, y_2, \bar{y}\rangle$ is a sequential equilibrium if and only if (10.6.4abc) are satisfied together with

$$y_1=0 \quad \text{and} \quad 1+\bar{y} \leqq w(\bar{y})\ . \tag{10.6.5}$$

Consequently, in a sequential equilibrium, the low ability worker does not invest in education when he is screened out, but the high ability worker might still invest in useless education (it is possible that $y_2>1$) because of the fear that the employers might believe that he is of low ability in case he does not do so. Hence, the game admits many sequential equilibria.

Equation (10.6.5) shows that player 1 has a payoff of at least 1 in any sequential equilibrium and this implies that type 1 will never choose an education level $y>1$. Since the high ability type chooses $y>1$ in some sequential equilibria, the logic of McLennan's justifiability concept requires that the employers offer a wage $w(y)=2$ for any such education level. Hence, justifiability implies that type 2 does not invest in more education as is necessary to screen himself from type 1 and, consequently type 2 is guaranteed a payoff of at least $1\frac{1}{2}$ in any justifiable equilibrium. This implies that only the Pareto optimal screening equilibrium is justifiable if $\pi<1/2$ since, in this case, every other equilibrium yields the high ability type a payoff strictly less than $1\frac{1}{2}$ (cf. Fig. 10.6.1a).

If $\pi \geqq 1/2$, then the concept of justifiable equilibria does not lead to a unique solution. Besides the Pareto optimal separating equilibrium also pooling equilibria $\bar{y}$ with $\bar{y} \leqq 2p-1$ and some mixed equilibria in which the low ability type is not screened out are justifiable. (Note that mixed equilibria in which type 1 randomizes yield type 2 a utility less than $1\frac{1}{2}$ so that these are not justifiable, cf. Fig. 10.6.1b.) These pooling and mixed equilibria can be eliminated by invoking CK-admissibility. Consider a pooling equilibrium $\bar{y}$ with $\bar{y} \leqq 2\pi-1$. Then type 1's

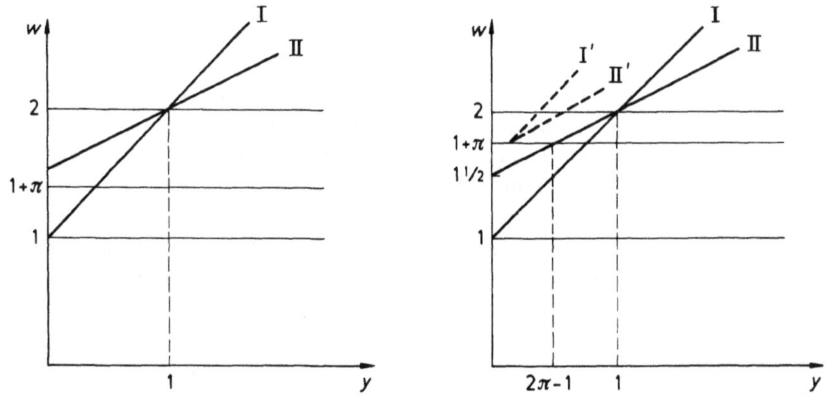

Fig. 10.6.1. If $\pi<\frac{1}{2}$, then only the Pareto optimal screening equilibrium is justifiable, but if $\pi>\frac{1}{2}$, then pooling equilibria $\bar{y}$ with $\bar{y}\leqq 2\pi-1$ (as well as some mixed equilibria) are also justifiable. I (II) is the indifference curve of type 1 (2) through (0, 1) (resp. (1, 2))

equilibrium payoff is $1+\pi-\bar{y}$, so that, according to the intuitive criterion of Cho and Kreps, type 1 will never choose an ecucation level y with $y>1-\pi+\bar{y}$. (This results in a lower payoff even if it garners the maximal wage of 2). On the other hand, type 2 would profit from choosing y with $y<2(1-\pi)+\bar{y}$ if this choice would result in the maximal wage (cf. the indifference curves I′ and II′ in Fig. 10.6.1b). Consequently, CK-admissibility requires employers to view education choices y with

$$1-\pi+\bar{y}<y<2(1-\pi)+\bar{y}$$

as evidence that the worker is of high ability, hence, they should respond to these by offering $w(y)=2$. This upsets the equilibrium, however, since in this case type 2 is sure to deviate: No pooling equilibrium is CK-admissible. Exactly the same argument establishes that none of the remaining mixed equilibria is CK-admisible, so that we have proved:

Theorem 10.6.1. *Spence's job market signalling model $\Gamma(\pi)$ admits exactly one equilibrium that is CK-admissible, viz. the Pareto optimal screening equilibrium.*[26]

The Pareto optimal screening equilibrium is stable in the sense of Kohlberg and Mertens (at least in discretizations of the model that are sufficiently fine). Namely, for $0<y<1$, the binding equilibrium constraint is that type 1 should not gain from deviating to y, so that (a continuous version of) KM-admissibility requires employers to believe that only type 1 workers choose such education levels. Hence, KM-admissibility requires $w(y)=1$ if $y\in(0,1)$ and, similarly $w(y)=2$ if $y>1$, which reinforces the equilibrium.

However, the Pareto optimal screening equilibrium does not satisfy the Grossman/Perry conditions of a P.S.E. if $\pi>\frac{1}{2}$. Namely, consider an education level y with $0<y<2\pi-1$. To support the equilibrium such y should garner wages at most $1+y$, however, no such wage is consistent with the equilibrium. (If $w(y)<1+y$, then no type has an incentive to deviate, if $w(y)=1+y$ only type 1 has such an incentive, but then the wage should actually equal 1.) Nevertheless, there exists a wage that is consistent with equilibrium in the sense of Grossman and Perry, viz. $w(y)=1+\pi$. This wage is optimal against the belief that both types have deviated from the equilibrium and, when this wage is offered, these beliefs are indeed confirmed. (Note that there is a second wage that is consistent with equilibrium, viz. $w(y)=1\frac{1}{2}+y/2$; this makes type 2 indifferent between deviating or not.) Since the wages that are consistent with the equilibrium do not support it, we see that the separating equilibrium is no P.S.E. in case $\pi>1/2$. In fact, the results of the previous section imply that there is no P.S.E. in this case and that, if $\pi<1/2$, the Pareto efficient separating equilibrium is the unique P.S.E.

Corollary 10.6.2. *If $\pi<1/2$, then the Pareto efficient separating equilibrium is the unique P.S.E. of Spence's job market signalling model. If $\pi>1/2$, then no P.S.E. exists.*

The above can be generalized to the case in which there is any finite number of types of workers and in which there is a negative correlation between ability and cost of education. However, in this case requiring CK-admissibility (or divinity) does not guarantee uniqueness of the equilibrium outcome. Universal divinity (or KM-admissibility) again leads to a unique outcome: the least able do not invest in education and types with higher ability just acquire enough to screen themselves from the less able ones (also see Riley [1975, 1979b]).[27]

The insights obtained above can also be fruitfully applied in related models with asymmetric information such as insurance markets with adverse selection (see Rothschild and Stiglitz [1976] and Wilson [1977, 1980]). In fact, Corollary 10.6.2 is closely related to the Rothschild/Stiglitz assertion that no 'equilibrium' exists in their model.[28]

It should be pointed out that the complete information case cannot be viewed as the limit of the incomplete information case: If $\pi=1$, then the high ability type does not acquire education ($y_2=0$), but he chooses $y_2=1$ even if the employers only assess a small probability that the worker has marginal productivity 1 ($y_1=1$ if $\pi<1$). This kind of discontinuity[29] will also be encountered in Sect. 10.7. Finally, it should be remarked that the results depend crucially on the sequencing of the moves: If the employers move both first (by offering wage schedules) and last (by accepting or rejecting worker applications), then only the Pareto optimal pooling equilibrium is stable. For details concerning this issue the reader is referred to Hellwig [1986, 1987].[30]

10.7 The Chain Store Paradox

The chain store game that has been introduced in Selten [1978] is a simple extensive form game that yields an inconsistency between game theoretical reasoning and plausible human behavior since well-informed players can be expected to disobey game theoretical recommendations. It has been argued in Rosenthal [1981], Kreps and Wilson [1982b] and Milgrom and Roberts [1982b]

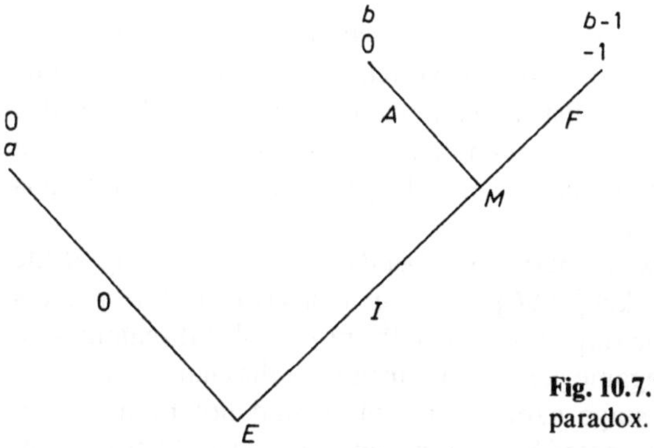

Fig. 10.7.1. The stage game of the chain store paradox. It is assumed that $a>1$ and $0<b<1$

that the paradox is caused by the fact that all parameters in Selten's game are common knowledge, a condition that is only rarely satisfied in real life. Kreps/Wilson and Milgrom/Roberts have also shown that, by including a small amount of incomplete information, the game theoretical solution is much more in agreement with observed behavior. In this section, we discuss this chain store game since it gives an opportunity to illustrate how the concepts from Sect. 10.5 can be used to eliminate implausible beliefs in dynamic games. The discussion is based on Kreps and Wilson [1982b]

Consider the following game between a (potential) entrant and an incumbent monopolist. The entrant moves first, choosing either to enter the market ($=I$) or to stay out ($=O$). If the entrant stays out, the incumbent is not called upon to move; Following entry, the monopolist has to decide whether to acquiesce ($=A$) or to Fight ($=F$). Payoffs are assumed to be as in Fig. 10.7.1 (with the entrant's payoffs on top).

The game of Fig. 10.7.1 has a unique subgame perfect equilibrium, viz. (I, A), the entrant enters and the monopolist acquiesces. There is a second Nash equilibrium in which the monopolist threatens to fight entry and in which the entrant stays out, but this involves an incredible threat of the monopolist.

Next, consider the situation in which the game from Fig. 10.7.1 is repeated a finite number of times: There is 1 monopolist, the owner of a chain store, who is challenged by N different entrants; The entrants move in succession (with entrant N first and entrant 1 last), each one being fully informed about all outcomes at previous stages of the game. This is version 1 of the *chain store game* of Selten [1978]. Common sense suggests that in this situation the monopolist should try to acquire a reputation of being tough: He should fight entry in early stages of the game in order to deter entry later. By behaving in this way, the monopolist incidentally will have a negative payoff, but, since (presumably) most entrants will learn their lesson and stay out, the overall result will be quite attractive. This behavior is indeed part of a Nash equilibrium of the game (recall that threatening to fight is an equilibrium strategy of the stage game), but this is not subgame perfect. Namely, the analysis of the stage game implies that entrant 1 (who is the last one to move) will always enter and that the monopolist will acquiesce in the last stage. Hence, the monopolist cannot build a reputation in the second to last stage and, since this is common knowledge, entrant 2 will enter and the monopolist will also acquiesce in this stage. Continuing in this way, one sees that the game unravels to the beginning: The unique subgame perfect equilibrium recommends all entrants to enter and the monopolist to acquiesce in every stage. Hence, this simple model cannot explain the real life phenomenon of reputation seeking and we face a paradox, the *chain store paradox*, since the game theoretical solution is not in agreement with our intuition.

Let us briefly investigate how the paradox depends on the sequencing of the moves in the stage game (see Trockel [1986] for more on this issue). If the players would move simultaneously in the stage game, then the monopolist threatening to fight entry would be part of a subgame perfect equilibrium of the chain store game (since (O, F) is Nash in the stage game), but this would not be a perfect equilibrium, since F is weakly dominated by A. If the monopolist would move first, thereby committing himself to either A or F, then the result would be completely different. In this case, the monopolist commits himself to F and the entrant chooses $O(I)$ after $F(A)$ in the unique subgame perfect equilibrium of the stage game, from which it follows that all entrants will decide to stay out in the repeated game. The sequencing as in Fig. 10.7.1 seems to be most natural, however, and, therefore, the paradox is serious.

Let us return to the stage game as in Fig. 10.7.1. The reason that we did not get reputation effects in the repeated game is that different stages are essentially independent: There is nothing that entrant n can learn from the monopolist's reaction to entrant n' since he is already fully informed from the beginning. It is more realistic, however, to assume that at the outset players are not yet completely informed so that some learning might take place. Specifically, let us assume that entrants are somewhat uncertain about the monopolist's profit function and that they consider it possible that the monopolist actually is better off by fighting entry than by acquiescing. Let us call the monopolist strong or weak depending on whether or not he is better off by fighting in the stage game. Clearly, an entrant will decide to stay out if he assesses a sufficiently high probability that the monopolist is strong. However, he might decide to stay out even if this probability is small, because he fears that also the weak monopolist will fight entry. Namely, a strong monopolist will always fight (it is better in the short run and (provided entrants draw the proper conclusions) it prevents more entry in the long run), so that a monopolist immediately reveals that he is weak if he does not fight. Once a monopolist has revealed that he is weak, he will be challenged in all coming rounds and this gives him a bad payoff. It is this consideration that gives the weak monopolist an incentive to mimic the strong one and to fight entry: Although fighting is costly in the short run, it leads entrants of later stages to revise upward the probability that the monopolist is strong and so they might decide not to enter in which case the initial losses will be recouped. Hence, we see that, with incomplete information, the reputation effect might come to life.[31]

To make the above argument precise, assume that the monopolist can be weak (in which case the payoffs are as in Fig. 10.7.1 in every stage) or strong (in which case the stage game is as in Fig. 10.7.2. Note that the games differ only with respect to the monopolist's payoffs). The monopolist knows which type he is, but the entrants do not, and at the outset they assess probability δ ($\delta > 0$, δ small) that the monopolist is strong.

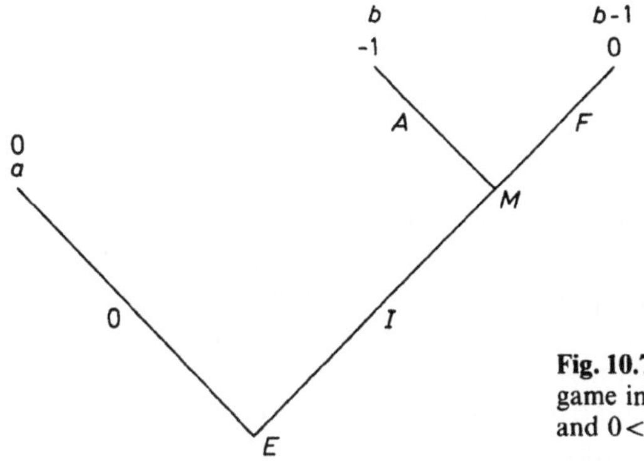

Fig. 10.7.2. The stage game of the chain store game in case the monopolist is strong ($a>1$ and $0<b<1$)

The above describes a dynamic game with incomplete information. (This is version 1 of the *chain store game with incomplete information* of Kreps and Wilson [1982b].) It is assumed that all entrants are perfectly informed about the outcomes in previous stages of this game and that the monopolist has perfect recall. Consequently, each information set, h_n, of entrant n contains 2 nodes, which correspond to the uncertainty about the monopolist's type. When it is his turn to move and his information is h_n, entrant n assesses the probability, $p(h_n)$, that the monopolist is strong and he enters with a probability $e(h_n)$ that is optimal given this assessment. When challenged, the weak and the strong monopolist fight with a probability equal to $f^w(h_n)$ and $f^s(h_n)$, respectively and we write $f(h_n)$ for the probability that entry is fought $(f(h_n) = (1-p(h_n)f^w(h_n) + p(h_n)f^s(h_n))$.

A 4-tuple of functions (p, e, f^w, f^s), each one defined on the set of all histories, is a *sequential equilibrium* of this game provided that:

(i) each entrant n chooses e to optimize against p, for each h_n,
(ii) both types of the monopolist choose f to optimize against e, for each h_n and
(iii) the probability assessment p is consistent with the monopolist's strategy.

We will not formalize (i) and (ii), but it is somewhat illustrative to write out (iii). Let $f = (f^w, f^s)$ be the monopolist's strategy. To satisfy (iii), the initial assessement must be consistent with our data, hence $p(h_N) = \delta$ where h_N denotes the empty history. Furthermore, when there is no entry, nothing can be learned so that we must have $p(h_n, O) = p(h_n)$ for each h_n (O denotes the event that entrant n stays out). Finally, as long as the evolution of the game is consistent with the strategy f being played, Bayes' rule should be used for updating beliefs:

$$p(h_n, F) = \frac{p(h_n)f^s(h_n)}{f(h_n)} \qquad \text{if } f(h_n) > 0, \tag{10.7.1}$$

and

$$p(h_n, A) = \frac{p(h_n)(1-f^s(h_n))}{1-f(h_n)} \qquad \text{if } f(h_n) < 1, \tag{10.7.2}$$

where (h_n, F) denotes the history h_n followed by an entry that is fought. These conditions are not only necessary, but they are also sufficient[32] for p to be consistent with f, hence, if (10.7.1) – (10.7.2) are satisfied, $p(h_N) = \delta$ and $p(h_n, O) = p(h_n)$, then the beliefs described by p can be explained in terms of small deviations from f.

The chain store game with incomplete information does not have a unique sequential equilibrium, in fact, it admits many implausible equilibria. For example, consider the case with $N=2$ and $\delta=.9$ and assume that entrant 2 enters. Then the monopolist's reaction to this entry serves as a signal to entrant 1 and the game between the monopolist and entrant 1 is very much like Kreps' Beer/Quiche game from Example 10.4.1. In particular, the monopolist acquiescing to entry at stage 2 and entrant 1 entering the market if and only if the monopolist fights entry at stage 2 is part of a sequential equilibrium.[33] To exclude such implausible equilibria, one has to use one of the refined equilibrium concepts of Sect. 10.5. Below we will show that the game admits an (essentially) unique KM-admissible equilibrium and that the reputation effect comes to life in this equilibrium.[34]

Below we restrict ourselves to strategies that depend on h_n only through $p(h_n)$, hence, the probability that the monopolist is strong is used as a state variable. It is not always possible to construct equilibria with this property (see Kreps and Wilson [1982b]), but if $a>1$, this is possible and the restriction is without loss of generality. This also allows a simplification in notation. If $p(h_n)=p$ is the belief of entrant n, then we write $e_n(p)$ for the probability that this entrant enters and $f_n^w(p)$ (resp. $f_n^s(p)$) for the probability that a weak (strong) monopolist fights entry. $v_n^w(p)$ (resp. $v_n^s(p)$) denotes the total expected payoff to the weak (strong) monopolist in stages $n, n-1, \ldots, 1$. Furthermore, we write $p_{n-1}(p, F)$ for the beliefs of entrant $n-1$ after fought entry in stage n, and $v_{n-1}^w(p, F)$ for the associated expected payoff of the weak monopolist, i.e. $v_{n-1}^w(p, F):= v_{n-1}^w(p_{n-1}(p, F))$.

Theorem 10.7.1 (Kreps and Wilson [1982b]). *If $\delta \neq b^k$ for $k=1, \ldots, n$, then the n-stage chain store game with incomplete information has exactly one sequential equilibrium outcome that is KM-admissible. The equilibrium strategies are given by*

$$f_k^s(p)=1, \tag{10.7.3}$$

$$f_k^w(p)=\begin{cases} \dfrac{p(1-b^{k-1})}{b^{k-1}(1-p)} & \text{if } p \leqq b^{k-1}, \\ 1 & \text{if } p > b^{k-1}. \end{cases} \tag{10.7.4}$$

$$e_k(p)=\begin{cases} 1 & \text{if } p<b^k, \\ 1-1/a & \text{if } p=b^k, \text{ and} \\ 0 & \text{if } p>b^k. \end{cases} \tag{10.7.5}$$

The equilibrium beliefs satisfy

$$p_k(p, O)=p \tag{10.7.6}$$

$$p_k(p, F)\begin{cases} \in [0, b^k] & \text{if } p=0, \\ =b^k & \text{if } 0<p<b^k, \\ =p & \text{if } p \geqq b^k, \end{cases} \tag{10.7.7}$$

$$p_k(p, A)\begin{cases} =0 & \text{if } p<b^k, \text{ and} \\ \in [0, b^k] & \text{if } p \geqq b^k. \end{cases} \tag{10.7.8}$$

Finally, the expected equilibrium payoffs to both types of the monopolist are equal to

$$v_k^w(p)=\begin{cases}0 & \text{if } p<b^k,\\ 1 & \text{if } p=b^k,\\ a+v_{k-1}^w(p) & \text{if } p>b^k,\end{cases} \tag{10.7.9}$$

$$v_k^s(p)=\begin{cases}v_{k-1}^s(b^{k-1}) & \text{if } p<b^k,\\ 1+v_{k-1}^s(b^{k-1}) & \text{if } p=b^k, \text{ and}\\ a+v_{k-1}^s(p) & \text{if } p>b^k.\end{cases} \tag{10.7.10}$$

Proof. It is straightforward to verify that Eqs. (10.7.3) – (10.7.8) describe a sequential equilibrium and that the associated payoffs are given by (10.7.9) – (10.7.10). Also note that the strategy f from (10.7.3) – (10.7.4) determines p completely (via (10.7.1) – (10.7.2)) except in two instances: (i) $p=0$ and the monopolist fights entry and (ii) $p \geqq b^{k-1}$ and the monopolist acquiesces. In both situations, we set beliefs such that entrant k enters (or randomizes) in order to ensure that the weak monopolist does not have an incentive to deviate. Hence, off the equilibrium path, the beliefs and equilibrium actions need not be unique.

Next, it will be shown that only the equilibrium outcome described in the theorem can be KM-admissible. The proof is by induction with respect to the number of stages of the game, where use is made of the fact that, if an outcome is KM-admissible, then it must induce a KM-admissible outcome in every (generalized) subgame.[35]

The assertions are easily verified in case $n=1$. Note that, if $n=1$ and $\delta=b$, the entrant is indifferent between entering or not, so that there exist multiple equilibria and the monopolist's payoffs are undetermined. However, below we will see that if this case arises in a subgame after previous fought entry, then entrant 1 should stay out with a probability of exactly $1/a$.

Next, let $n>1$ and suppose that it has already been shown that the $(n-1)$-stage game has a unique KM-admissible outcome for every prior belief p with $p \neq \delta^k$ $(k=1,\ldots,n-1)$ and that Eqs. (10.7.9) – (10.7.10) hold for this outcome. Let (e, p, f) be a KM-admissible equilibrium of the n-stage game. It is easily verified that (10.7.3) – (10.7.10) must be satisfied for $k=n$ and $p=0$, so assume $p>0$. We will first show that the strong monopolist fights with probability 1 and the weak monopolist fights with positive probability, hence

$$f_n^w(p)>0 \quad \text{and} \quad f_n^s(p)=1 \quad \text{if } p>0. \tag{10.7.11}$$

Assume $f_n^w(p)=0$ and $f_n^s(p)>0$. Then consistency implies $p_{n-1}(p,F)=1$, hence, $v_{n-1}^s(p,F)=(n-1)a$. If the strong monopolist acquiesces his payoff is strictly less so that we must have $f_n^s(p)=1$, hence, $p_{n-1}(p,A)=0$ and $v_{n-1}^w(p,A)=0$. But this implies that the weak monopolist is better off by fighting in round n (since $-1+v_{n-1}^w(p,F)=-1+v_{n-1}^w(1)>0$), so this cannot be an equilibrium. Next, assume $f_n^w(p)=f_n^s(p)=0$, so that $p_{n-1}(p, F)$ is not determined by the monopolist's strategy. Then we must have

$$-1+v_{n-1}^s(p,A) \geqq v_{n-1}^s(p,F), \tag{10.7.12}$$

since otherwise the strong monopolist would prefer to fight. The analogous condition for the weak monopolist is

$$v^w_{n-1}(p, A) \geqq -1 + v^w_{n-1}(p, F) . \tag{10.7.13}$$

Now, (10.7.9) – (10.7.10) show that (i) both $v^s_{n-1}(\cdot)$ and $v^w_{n-1}(\cdot)$ are (weakly) increasing and (ii) that $v^s_{n-1} - v^w_{n-1}$ is (weakly) decreasing, from which it follows that (10.7.13) is satisfied whenever (10.7.12) is, but not conversely. Consequently, (10.7.12) is a binding constraint, but (10.7.13) is redundant and, therefore, the KM-admissibility concept requires entrant $n-1$ to believe $p_{n-1}(p, F) = 1$. However, if $p_{n-1}(p, F) = 1$, then $v^s_{n-1}(p, F) = (n-1)a$ and it is impossible to satisfy (10.7.12).

Thus far, we have shown that in a KM-admissible equilibrium it is impossible that both types of the monopolist acquiesce and that it is impossible that only the strong monopolist fights with positive probability. Consequently, in a KM-admissible equilibrium, $f^w_n(p) > 0$ if $p > 0$. Therefore, the reverse inequality from (10.7.13) must be satisfied and (10.7.12) cannot hold. Hence, in a KM-admissible equilibrium, the strong monopolist must fight with probability 1. This establishes (10.7.11).

Next, assume $f^w_n(p) < 1$. Then (10.7.13) must be satisfied with equality. Since $p_{n-1}(p, A) = 0$ by consistency, we have $v^w_{n-1}(p, A) = 0$, hence, $v^w_{n-1}(p, F) = 1$. Therefore, $p_{n-1}(p, F) = b^{n-1}$ in view of (10.7.9), so that, again by consistency

$$p_{n-1}(p, F) = \frac{p}{p + (1-p) f^w_n(p)} = b^{n-1} .$$

This shows that $f^w_n(p)$ is given by (10.7.4) in case $f^w_n(p) < 1$. By combining this with (10.7.11), one sees that (10.7.4) is generally valid. Given that (10.7.3) – (10.7.4) are valid, it follows that entry of entrant n will be fought with probability $f_n(p) = \min(1, p/b^{n-1})$. Entrant n should enter if $f_n(p) < b$ (hence, $p < b^n$) and this is indeed what (10.7.5) requires him to do. (Note once more that entrant n may randomize arbitrarily if $\delta = b^n$, but, after fought entry in the previous stage entrant n should stay out with probability exactly $1/a$ in order to make the weak monopolist indifferent between fighting and acquiescing at stage $n+1$.)

We have shown that in any KM-admissible equilibrium, the strategies are as in (10.7.3) – (10.7.5). This completes the proof, since at least one KM-admissible outcome exists. □

It should be noted that in Kreps and Wilson [1982b] actually a slightly different result is established. In that paper, it is shown that there exists a unique sequential equilibrium path satisfying the following monotonicity condition

$$p_n(p, F) \geqq p_n(p, A) \quad \text{for all } p, n , \tag{10.7.14}$$

which states that fighting should be interpreted as positive evidence that the monopolist is strong. The Kreps/Wilson result can be obtained by mimicking the above proof. In fact, the proof is a little bit simpler since in order to establish the essential equation (10.7.11) one only has to use that v^w_{n-1} and v^s_{n-1} are (weakly) increasing. The reason why we proved the result using the KM-admissibility concept is that we didn't want to consider another ad hoc concept as that of Eq. (10.7.14).

In this model, there again is a discontinuity in the outcome as the uncertainty vanishes (if $\delta > 0$, then $f_n^w(\delta) = 1$ if n is large enough, but $f_n^w(0) = 0$ for all n; also cf. the model of the previous section). It has been shown in Harrington [1985] that the outcome is again qualitatively different when the entrant is uncertain about his payoffs. Benoit [1984] discusses a model in which the results are exactly opposite: In the complete information case, there is never entry, but if there is a little bit of incomplete information, then entry might occur in almost all periods. Hence, the game theoretic solution is very sensitive with respect to informational 'details' and modelling assumptions.

10.8 Repeated Games

In the previous sections we have illustrated the strength of stability in games of incomplete information. In this section, it will be shown that signalling can also play an important role in games with complete information and that, therefore, stability has much to say about equilibria in such games as well. Specifically, we will demonstrate that, in repeated games, deviations at early stages can signal unambiguously how a player intends to play later.[36] The subgame perfectness concept allows players to ignore such signals (they could have been sent by mistake), but stability requires that they be given their proper meaning. In Chapt. 8 we completely ignored signalling and, therefore, it should not come as a surprise that many equilibria used in the proof of the perfect folk theorem (Theorem 8.7.7) fail to be stable. Our aim is to demonstrate this phenomenon by means of some simple examples. Most of these examples are symmetric, but this is for notational convenience only, the arguments extend to nonsymmetric games.[37]

We will use the notation and terminology of Chap. 8, however, this time it will be more convenient to assume that players can only observe the outcomes of the random moves. Attention will be restricted to 2-fold repetitions. Two important lessons that can be drawn from Chap. 8 are:

if s is a Nash equilibrium of Γ, then (s, s) is a subgame perfect equilibrium path of $\Gamma(2)$, and (10.8.1)

if (s, t) is a subgame perfect equilibrium path of $\Gamma(2)$, then t is a Nash equilibrium of Γ, but s need not be an equilibrium; it is sufficient that every profitable deviation at stage can be countered by switching to a less attractive equilibrium at stage 2. (10.8.2)

From (10.8.2) it follows that, to verify subgame perfectness, one has to worry only deviations that are profitable in the one-shot game. Our first example shows that this is no longer true if one wants to verify stability: In this case, also unprofitable deviations have to be considered, since these can serve a signalling purpose. As a consequence, also (10.8.1) is no longer correct, when 'perfect' is replaced by 'stable'.

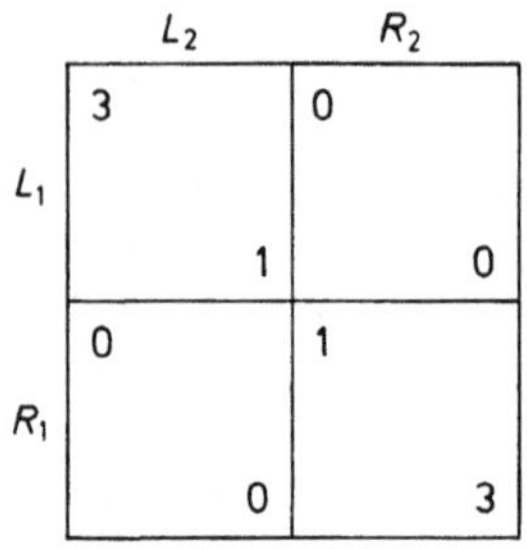

Fig. 10.8.1. Playing (L_1, L_2) twice is not a stable path in $\Gamma(2)$

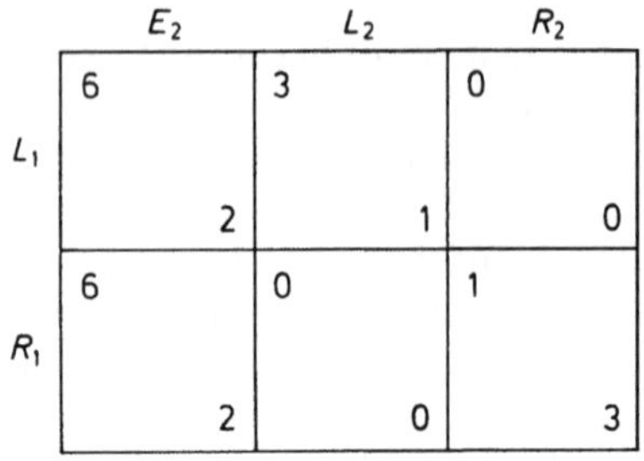

Fig. 10.8.2. The relevant game matrix in the subgame (L_1, R_2) given the players agreed on the path (L, L)

The unanimity game Γ of Fig. 10.8.1 has 3 equilibria, viz. $L=(L_1, L_2)$, $R=(R_1, R_2)$ and the mixed equilibrium $M=(M_1, M_2)=(3/4L_1+1/4R_1, 1/4L_2+3/4R_2)$. All equilibria are regular, hence, stable. However, we claim that the path (L, L) is not stable in the 2-fold repetition of this game. (To prove this claim, we will first give an intuitive argument, next we will formally show that the path is not CK-admissible.)

Suppose the players have agreed on the path (L, L), but that player 2 deviates to R_2 at stage 1. Player 1 knows that player 2 was guaranteed of a payoff 2 if he had not deviated and he also knows that an equilibrium of Γ has to be played at stage 2. If he assumes that player 2 deviates in order to obtain a higher payoff, then the unevitable conclusion must be that he should play R_1 at stage 2: Since both L and M yield player 2 a payoff less than 2, deviating is profitable only if R is played at stage 2. Hence, player 2's deviation signals that he will play R_2 and, therefore, player 1 should respond with R_1: The path is not stable.

A more formal argument can be provided by using the auxiliary game matrix from Fig. 10.8.2. This game matrix is the relevant one to analyse the subgame (L_1, R_2) at stage 2, given that the players agreed on the path (L, L). The bimatrix displays the payoffs, conditioned on the fact that the subgame (L_1, R_2) has been reached (i.e. that player 2 deviated from (L, L) at stage 1), together with the equilibrium strategy E_2 of player 2 (i.e. not deviating from (L, L)). The normal form of $\Gamma(2)$ does contain the game of Fig. 10.8.2 as an essential 'submatrix' and it is easily seen that for (L, L) to be a stable path, it is necessary that the payoff $(6, 2)$ is stable in Fig. 10.8.2. However, after elimination of dominated strategies in the latter game, only (R_1, R_2) remains, which shows that the payoff $(6, 2)$ is not stable.

It is easy to see that, if Γ is the game of Fig. 10.8.1 and (s^1, s^2) is a subgame perfect equilibrium path of $\Gamma(2)$, then both s^1 and s^2 have to be Nash equilibria[38] of Γ. Hence, $\Gamma(2)$ admits 9 subgame perfect equilibrium paths. We have already shown that (L, L) is not stable and obviously the same is true for (R, R). Also the path in which L (or R) is played at $t=1$ and M (the mixed equilibrium) at $t=2$ is unstable. The argument is exactly as above, the only difference is that in Fig. 10.8.2 one has to replace the payoff $(6, 2)$ by $(15/4, 7/4)$. However, the path in which the players alternate between L and R is stable. In this case, if a player deviates in

	L_2	R_2	A_2
L_1	3, 1	0, 0	0, 0
R_1	0, 0	1, 3	0, 0
A_1	0, 0	0, 0	1, a

Fig. 10.8.3. Playing (L_1, L_2) twice is stable in $\Gamma(2)$ if and only if $2<a\leqq 6$

round 1, then his total payoff is sure to be less than the payoff he was guaranteed in the equilibrium, so that one cannot draw any 'rational' inferences from a deviation. Furthermore, any path in which the players start with the mixed equilibrium is stable, since in this case one cannot detect any deviations (recall our assumption that only outcomes of random moves can be observed). Hence, we obtain the following classification of the equilibrium paths:

	Equilibrium Paths of $\Gamma(2)$
Stable	(M, L), (M, R), (M, M), (L, R), (R, L)
Unstable	(L, L), (L, M), (R, R), (R, M)

It should be noted that the equilibria in which M is played in the first round are not persistent. Both pure equilibrium paths are persistent, however. (It can be shown that any game possesses an outcome that is both stable and persistent.)

As a second illustration of stability, consider the game of Fig. 10.8.3, which is an extension of the game discussed above.

Consider the path (L, L) in the 2-fold repetition of the game of Fig. 10.8.3. To verify stability of this path, one has to investigate whether player 2 can profit by deviating. (Clearly, player 1 has no incentive whatsoever to deviate.) Hence, one has to consider the bimatrix of payoffs in the subgame (L_1, R_2) appended with a constant column of payoffs $(6, 2)$, representing the situation that player 2 sticks to his equilibrium strategy E_2. Clearly, in this auxiliary game, L_2 is strictly dominated by E_2. Furthermore, given that player 2 will not play L_2 in the subgame (L_1, R_2), it follows that L_1 is dominated by a mixture of R_1 and A_1. Hence, by eliminating dominated strategies, we see that the path (L, L) is stable only if the payoff $(6, 2)$ is stable in the game of Fig. 10.8.4.

If $a\leqq 2$ then further elimination of dominated strategies reduces the game of Fig. 10.8.4 to (R_1, R_2), in which case the payoff $(6, 2)$ is not stable. Furthermore, if $a>6$, then $(6,2)$ is not even an equilibrium payoff in the game of Fig. 10.8.4.

	E_2	R_2	A_2
R_1	6, 2	1, 3	0, 0
A_1	6, 2	0, 0	1, a

Fig. 10.8.4. The relevant game matrix in the subgame (L_1, R_2) given the players agreed on (L, L) in the 2-fold repetition of the game of Fig. 10.8.3

	L_2	R_2
L_1	3, 1	0, 0
R_1	a, a	1, 3

a

	E_2	L_2	R_2
L_1	6, 2	3, 1	0, 0
R_1	6, 2	a, a	1, 3

b

Fig. 10.8.5. If $a < 2\frac{1}{2}$, then playing L twice is not stable in the 2-fold iteration of the game on the left

Hence, if $a \leqq 2$ or $a > 6$ then the path (L, L) is not stable in the 2-fold repetition of the game of Fig. 10.8.3. If $2 < a \leqq 6$, however, then this path is stable: If player 2 deviates in round 1, then player 1 can never know whether player 2 intends to play R_2 or A_2 in round 2. He could think that both possibilities are equally likely, in which case it is justified to play the mixed equilibrium strategy of the one-shot game that assigns probability $a/(3+a)$ to R_1 and $3/(3+a)$ to A_1, and this results in an actual loss for player 2 when he deviates.

In the above examples, stability could be verified by eliminating dominated strategies (or, equivalently, by invoking Kreps' intuitive argument). In the next example, there is an equilibrium path that remains after elimination of dominated strategies, but that is not stable since it is not KM-admissible.

If $a < 3$, then the game of Fig. 10.8.5a has 3 equilibria, viz. L, R and a mixed one which yields each player the payoff $3/(4-a)$. Consider the path (L, L) in the 2-fold repetition. Stability depends only on whether player 2 can profitably deviate and this in turn depends on whether or not the payoff $(6, 2)$ is stable in the game of Fig. 10.8.5b. Elimination of dominated strategies shows that the path is not stable if $a \leqq 2$. Next, assume $2 < a < 3$. Write p for the probability with which player 1 chooses L_1 in the subgame (L_1, R_2). A deviation of player 2 is unprofitable if and only if p is sufficiently high. Specifically, Fig. 10.8.5b shows that

$$(R_2, L_2) \text{ is unprofitable iff } p \geqq (a-2)/(a-1)\,, \tag{10.8.3}$$

and

$$(R_2, R_2) \text{ is unprofitable iff } p \geqq 1/3\,. \tag{10.8.4}$$

	L_2	R_2	A_2
L_1	4, 4	0, a	−1, −1
R_1	a, 0	1, 1	−1, −1
A_1	−1, −1	−1, −1	0, 0

Fig. 10.8.6. If $4<a<5$, then the cooperative outcome $(4, 4)$ cannot be obtained in a stable path of $\Gamma(2)$

Note that

$$(a-2)/(a-1)\geqq 1/3 \quad \text{iff} \quad a\geqq 5/2,$$

so that (10.8.3) is a redundant constraint for the set of equilibria supporting the path (L, L) if $a<5/2$. Consequently, if $a<5/2$, then KM-admissibility requires player 1 to believe that player 2 will choose R_2 in the subgame (L_1, R_2). Hence, in this subgame, player 1 should choose R_1 (his best response against R_2) and since this does not support the path (L, L), this path is not stable. If $a>5/2$, however, then KM-admissibility requires player 1 to believe that player 2 will play L_2 in the subgame (L_1, R_2) (the binding equilibrium constraint is (10.8.3)) and, in this case, the best response L_1 supports the path (L, L). In fact, it is not difficult to show that (L, L) is stable if and only if $5/2\leqq a<3$, i.e. if and only if the mixed equilibrium of the game of Fig. 10.8.5a yields a payoff of at least 2. Hence, the path is stable whenever there is more than one 'rational inference' that player 1 can draw after a deviation of player 2.

The above examples might have given the impression that to verify stability, one only has to worry about unprofitable one-shot deviations. In the final example, a modified prisoners' dilemma game, we want to show that this impression is misleading.

The game of Fig. 10.8.6 has 3 equilibria, viz. R, A and a mixed one, M, that yields each player a payoff $-1/3$. Consider the path (L, R) in the 2-fold repetition of this game. If $4<a<5$ then this is a subgame perfect equilibrium path, since the threat to switch to A deters deviations. However, according to stability this threat is not credible. Namely, if player i deviates to R_i, then R is the unique equilibrium continuation that renders this deviation profitable so that stability requires player $j(j\neq i)$ to interpret the deviation as a signal that R should be played in round 2. Hence, if $4<a<5$, then the path (L, R) is not stable: If player j responds 'rationally', player i gains from deviating.

If $5<a<16/3$, then (L, R) is still a subgame perfect equilibrium path (since threatening with M deters deviations) and in this case the path is stable. Namely, a deviation of player i in round 1 is profitable if either R or A is played in round 2 and player j's uncertainty about which of these is called for justifies playing the mixed equilibrium strategy M_j which in turn renders the deviation unprofitable.[39]

Notes

1. For an in-depth discussion on equivalence of games the reader is referred to Mertens [1987]. Mertens argues that, as far as the self-enforcing aspect is concerned, two games are equivalent if for each player and every prior that this player might have on his opponents' strategies, his set of best replies is the same. Mertens shows that the requirement of ordinality (equivalent games have the same set of (self-enforcing) solutions) implies that the solutions can depend only on the reduced normal form.
2. This point of view is rejected in Harsanyi and Selten [1988], see the examples in the postscript of that book.
3. This point is also made in the introduction of Mertens [1987].
4. By means of transformation (i) a game with perfect recall can be transformed into a game that does not have this property. In Elmes and Reny [1989] the question is addressed whether games of perfect recall with the same (semi-)-reduced normal form can be transformed into each other via a sequence of games that all have perfect recall.
5. One might argue that in signalling games the types of the informed player should indeed be considered as separate players, but that the agents of the receiver should be considered as being centrally coordinated. This intermediate game form yields the same stable outcomes (Theorem 10.3.5 and note 15), working with different game forms yields different 'intuitive criteria', however. We return to this issue in note 22.
6. Also see Mertens [1989].
7. Recall that this requirement (or the weaker one obtained by replacing 'sequential' with 'subgame perfect') has been criticized by some workers in the field. See notes 2 and 3 of Chapter 1.
8. Additional arguments in favor of the admissibility requirement are presented in Mertens [1987] and Hammond [1987]. Also see Blume, Brandenburger and Dekel [1991 a, b].
9. Some papers that address the interplay between the concepts of common knowledge and admissibility and that study the implications of the assumption that it is common knowledge that the players choose only admissible strategies are Börgers [1990], Brandenburger [1990], Dekel and Fudenberg [1987], Harstad [1990], Samuelson [1989a] and Stahl [1990b]. In Van Damme [1990] it is shown that the Harsanyi and Selten [1988] solution of a game does not necessarily satisfy requirement (10.2.3), not even in the case where the elimination process leads to a unique outcome that is independent of the elimination order.
10. See any of the papers mentioned in note 3 to Chap. 1.
11. This property was not formulated as a requirement in Kohlberg and Mertens [1986]. They call this property 'Forward Induction' and note that stable sets satisfy this property (Theorem 10.3.3). We prefer not to use the name 'Forward Induction' for this property since, in the economics literature, 'Forward Induction' is associated with a whole collection of concepts (not all

implied by stability) that all incorporate the idea that the way in which a subgame will be played will depend on the way in which the subgame is reached. See Van Damme [1989] and the references therein, as well as Okunu-Fujiwara and Postlewaite [1987], and the papers mentioned in note 33.

12. In Kohlberg and Mertens [1986] also other (related) notions of stability are introduced, but these were discarded since they don't satisfy the admissibility requirement. Specifically, hyperstable sets are defined similarly to Definition 10.3.1 but then with perturbed payoffs rather than perturbed strategies. The third concept of full stability requires that when each player's strategy set is slightly perturbed to an arbitrary convex compact polyhedron in the interior of the original set, then the resulting game has an equilibrium close to the given set. A hyperstable set contains a fully stable set, and the latter in turn contains a stable set. A modified stability concept was introduced in Mertens [1988].
13. In Wilson [1990b] an algorithm is described to compute a 'weakly' stable set that lies within a connected component. Define a singleton perturbation as one that imposes a positive lower bound on the probability of only one pure strategy. Wilson calls an equilibrium set weakly stable if each nearby game with a singleton perturbation has an equilibrium close to the set.
14. In Jansen, Jurg and Borm [1990] it is shown that, for bimatrix games, every stable set is finite.
15. Of course, it is not guaranteed that by restricting oneself to stable sets that lie entirely within one component one prevents the difficulties posed by Example 10.3.4. Hillas [1989] presents a slightly modified version of this game for which sequential rationality indeed is not obtained by requiring connectedness.
16. In Mertens [1988] stable sets are defined as "the sets (and also the Hausdorff limits of those sets) of limits at the true game of a semi-algebraic subset of the equilibrium correspondence for the perturbed games, with the properties of (a) the projection map to perturbed game is (homologically) nontrivial and (b) the subset is locally connected at the true game". For the special (simple) case of integer homology, this definition amounts to the requirement that the projection from the equilibrium correspondence to the set of perturbed games be of non-zero degree. Note that no minimality is required, Mertens shows that minimality conflicts with the invariance requirement. Mertens shows that this concept satisfies all properties from (10.2.5) as well as (i) player splitting (stable sets do not change when splitting a player into two agents, provided that these agents never have to act after one another) and (ii) decomposition (if disjoint player sets play different games in different rooms one can as well consider the compound game as the two separate games). Note that player splitting implies Theorem 10.3.5. In Mertens [1990] it is shown that stable sets satisfy a property that is even stronger than decomposition. Specifically, stable sets satisfy the 'small worlds' axiom: If the best reply correspondence to a subset of the players is unaffected by the actions of the outsiders, then the stable sets of the game between insiders are exactly the projections of the stable sets of the larger game. It is also shown in Mertens [1990] that every persistent retract (Kalai and Samet [1984]) contains a stable set.

17. Hillas [1989] defines a stable set of a game G as a minimal closed set of Nash equilibria of G satisfying the following property: For any game G' with the same reduced normal form as G there exists, for every $\varepsilon > 0$, some $\delta > 0$ such that any upper-hemi-continuous compact convex valued correspondence that is pointwise within δ-Hausdorff distance of the best reply correspondence of G' has a fixed point within ε of S. Hillas proves that this stability concept satisfies all properties from (10.2.5), but he does not know whether his concept satisfies Theorem 10.3.5.
18. However, see note 15.
19. Throughout the remainder of the chapter it will be assumed that signals are costly to send. From the seminal papers Crawford and Sobel [1982], Farrell [1984, 1990] and Grossman [1981], an extensive literature has sprung up that investigates what happens when players can engage in 'cheap talk', i. e. when the informed party can costlessly say things to influence the beliefs of the uninformed players. In 'cheap talk' games, stability is not very useful to refine the set of equilibria, but many ideas similar to the 'intuitive criteria' from Section 10.5 have been proposed that do indeed refine the equilibrium set. Recent papers in this area that provide a state of the art are Farrell and Gibbons [1989a, b], Forges [1988, 1990], Matthews et al. [1990], Myerson [1989], Rabin [1990] and Seidmann [1990].
20. Since the publication of the first edition many additional papers dealing with signalling have been written. These have been many applications in finance, and in this area mostly Grossman/Perry's PSE concept has been used. Examples, dealing with takeovers, are Fishman [1988, 1989] and Shleifer and Vishny [1986]. Cadsby et al. [1990] present results of an experiment conducted with the signalling game from Myers and Majluf [1984]. Many models have been relatively standard one-period, one-dimensional signalling games satisfying a monotonicity condition and the so-called 'single crossing' property for which stability related refinements select a unique outcome (see Cho and Sobel [1990]). Hence let me just mention some models in which (in a subset of the parameter space), one of these assumptions is violated. Güth and Van Damme [1990] study a signalling model of disarmament, Rosenthal [1990] argues that suicide attempts can sometimes be interpreted as signals to manipulate the environment, Reinganum and Wilde [1986] consider a model of settlement and litigation and, in political science, Banks [1990a, b] considers a signalling model of setting the voting agenda. Gertner et al. [1987] study a model in which the informed part sends the same signal to two uninformed players in the case in which the sender would like to send two different signals. Lutz [1989], Quinzi and Rochet [1983], Ramey [1988] and Wilson [1985] consider multidimensional signalling models (also see Engers [1987]). For a survey on signalling games, also see Kreps and Sobel [1991].
21. The intuitive criteria from Section 10.5, however, are not defined by means of the agent normal form, see also note 22.
22. Cho and Kreps attribute this type of critique to Joe Stiglitz. Okuno-Fujiwara and Postlewaite [1987] respond to the critique by demanding that the speech be a proposal of an alternative sequential equilibrium.

23. The reader is also urged to consult Sobel et al. [1990] for a better and more formal treatment of the material discussed in this section. Specifically, in this section, our use of the terms 'rationalizable' and 'justifiable' is somewhat imprecise, and we differ from Cho/Kreps and Banks/Sobel since we deal with all unsent messages simultaneously, rather than with each unsent message at a time. As Sobel et al. point out, sometimes more outcomes may be eliminated if one analyzes the game on a signal by signal basis. Sobel et al. also point out that for rationalizability, it makes a difference whether the game is analyzed as a 2-person 'incomplete' information game or as a $(|T|+1)$-person 'imperfect' information game, since in the first case all types of the uninformed party necessarily have the same beliefs about the responses of player 2. (Note that Sobel et al. do not introduce agents for the receiver, they always treat him as one player, hence, they do not work with the agent normal form.) Still, the idea underlying all of the 'intuitive' concepts is quite simple and it should come out clearly from the discussion in this Section. The unifying procedure is that, given an equilibrium, one first modifies the original game by replacing the equilibrium actions with an action that yields the equilibrium payoffs for sure. In the game that results in this way (or in its normal form, or in the agent normal form or in some intermediate game form) one then iteratively eliminates dominated strategies. If by this process the original equilibrium vanishes, then one rejects this equilibrium.
24. The PSE concept has been defined for a class of games of which the signalling games form a strict subset. In general, the PSE concept imposes the so-called 'support restriction', i. e. it is required that once an uninformed player believes that the type t of the informed player belongs to a subset K of T he continues to believe that $t \in K$ no matter what happens in the remainder of the game. Madrigal et al. [1987] have constructed an example in which no sequential equilibrium satisfies the support restriction, while Noldeke and Van Damme [1990b] argue that the support restriction has nothing compelling to it. Some papers in which the PSE concept is applied are Bagwell and Staiger [1988], Diamond [1989], Fishman [1988, 1989], Gertner et al. [1989], Grossman and Perry [1986b], Lutz [1989], Shleifer and Vishny [1986], Stoughton [1988] and Vincent [1990a].
25. The 'irrelevant changes' refers to the fact that the definitions of the intuitive criteria can be easily extended to games in which the signal space is a continuum. Of course, stability has been defined only for finite games and it may be misleading to identify the stable outcome of a continuum game with the outcome surviving application of a certain intuitive criterion. Specifically, the order in which limits are taken may matter. Consider, for example, the discretized version $\Gamma^d(\pi)$ of $\Gamma(\pi)$ in which wage offers have to be in integer multiples of a smallest money unit g and in which education choices have to be integer multiples of some $h>0$. If the ex ante expected productivity is at least $2-g$ (hence $\pi \geqq 1-g$) and if $h>2g$ then $\Gamma^d(\pi)$ has pooling at an education choice of 0 as a stable equilibrium outcome. The type 2 worker does not want to deviate since the maximal wage increase g does not offset the additional education cost of $h/2$.)

26. Actually only the equilibrium outcome is unique. CK-admissibility does not determine uniquely the wage the firms should offer at education choices that do not occur in equilibrium.
27. In Cho and Sobel [1990] it is investigated to what extent the results from this section can be generalized. In that paper first a monotonicity condition is identified under which (a milder version of) universal divinity is generically equivalent to strategic stability. Thereafter it is shown that additional assumptions (such as the 'single crossing property' and the signal and action spaces being compact intervals in $\mathbb{R}$) guarantee that the universally divine outcome is unique. The universally divine outcome is a separating one, except in the case when the signal space is not large enough to allow full separation. For example, if one modifies the game $\Gamma(\pi)$ of this section such that the maximal possible education level is Y with $Y<1$, then the universally divine outcome is pooling at Y.
28. The Rothschild/Stiglitz model is a 'screening' one, i.e. the firms move first by offering contracts. For the relation between signalling and screening, see Stiglitz and Weiss [1988] and Engers and Fernandez [1987].
29. However, recall note 25.
30. It is interesting to note that the Harsanyi/Selten [1988] solution of the Spence signalling game $\Gamma(\pi)$ coincides with the outcome proposed in Wilson [1977], i.e. if $\pi<1/2$, then the HS-solution coincides with the stable outcome identified in Theorem 10.6.1, if $\pi>1/2$, then the HS-solution is the equilibrium in which the types pool at $y=0$. See Güth and Van Damme [1989]. Note, therefore, that the HS-Solution is Pareto-efficient among the sequential equilibria in this game.
31. See Camerer and Weigelt [1988] for an experimental test of whether sequential equilibrium can predict players' behavior in a related reputation model.
32. See Fudenberg and Tirole [1989].
33. Note that this 'implausible' equilibrium would vanish if the strong monopolist did not have the possibility to acquiesce, i.e. if he were committed to fight. It can be shown that in this modified model the strategies from Theorem 10.7.1 constitute the unique sequential equilibrium. For generalizations of this result to models with a single long run player and multiple short run players and in which the long run player with a small probability is committed to use his Stackleberg action in every period, see Fudenberg and Kreps [1987] and Fudenberg and Levine [1989a, d].
34. Of course, the concept of KM-admissibility has been defined only for one-shot signalling games, hence, it cannot be directly applied to this dynamic incomplete information game. The chain store game can be viewed as a repeated signalling game and we will proceed in the spirit of KM-admissibility and use this concept in a backward inductive fashion. Hence, we will first compute the (unique) KM-admissible outcome for the last period signalling game (that starts with the reaction of the monopolist in the second to last period), then we substitute the resulting payoffs in the second to last period signalling game and again apply KM-admissibility, etc. For a justification of this procedure, albeit in a different context, see Cho [1990]. This procedure is also followed in Noldeke and Van Damme [1990a]. Some other recent papers

dealing with repeated signalling games are Admati and Perry [1987], Matthews and Fertig [1990] and Vincent [1990a, b].

35. This is not a 'fact', the statement is true since KM-admissibility in the multiperiod game is defined in this way, see the previous note.
36. Such signalling possibilities exist generally in games with a dynamic structure. We now mention some papers that have investigated the consequences of this type of signalling in simple games. Ben-Porath and Dekel [1989] and Van Damme [1989] allow players to publicly 'burn money' before playing a normal form game. The right to burn money may be very valuable: In the stable equilibrium, the player having this right may obtain his most preferred outcome without having to burn anything. Advertising has some aspects of money burning to it and applications of the above ideas to the role of advertising to coordinate decisions of firms and consumers can be found in Glazer and Weiss [1990] and Bagwell and Ramey [1990a]. Bagwell and Ramey [1990b] apply forward induction ideas to coordination problems in models of market entry, and Dekel [1989] shows that requiring stability may lead to inefficiency in a 2-period simultaneous bargaining model. In Ponssard [1990a, b, c] concepts of forward induction are developed that are not all implied by the Kohlberg/Mertens stability concept. Experimental tests of whether subjects learn to understand the forward induction logic have been reported in Abdalla et al. [1989] and Brandts and Holt [1989].
37. For some general results on stability in repeated games, the reader is referred to Osborne [1990].
38. This statement is wrong. There exists a (stable) equilibrium in which in the first period each player i chooses $\frac{1}{2} L_i + \frac{1}{2} R_i$, and in which in the second period players continue with M after (L_1, R_2) and after (L_2, R_1), with R after L and with L after R.
39. The reader may wonder to what extent requiring stability in repeated games is compatible with the ideas of renegotiation-proofness that were discussed in Section 8.8. The answer is that the two concepts are incompatible, see Van Damme [1988] for a simple example and Gale and Hellwig [1989] for an application to a model of sovereign debt.

References

Abakuks, A. [1980]. Conditions for evolutionarily stable strategies. J. Appl. Prob. *17*, 559–562.

Abdalla, A., R. Cooper, D. DeJong, R. Forsythe and T. Ross [1989]. Forward induction in coordination and battle of the sexes games: Some experimental results. University of Iowa WP 89–22.

Abreu, D. [1982]. Repeated games with discounting: A general theory and an application to oligopoly. Thesis, Princeton University.

Abreu, D. [1983]. Repeated games with discounting: A general theory. Mimeo, Harvard.

Abreu, D. [1986]. Extremal equilibria of oligopolistic supergames. Journal of Economic Theory *39*, 191–225.

Abreu, D. [1988]. On the theory of infinitely repeated games with discounting. Econometrica *56*, 383–396.

Abreu, D., P. Milgrom and D.G. Pearce [1990]. Information and timing in repeated partnerships. Mimeo, Stanford University.

Abreu, D. and D.G. Pearce [1984]. On the inconsistency of certain axioms on solution concepts for non-cooperative games. Journal of Economic Theory *34*, 169–174.

Abreu, D. and D.G. Pearce [1989]. A perspective on renegotiation in repeated games. Forthcoming in Selten [1991 b].

Abreu, D., Pearce, D. and E. Stachetti [1986]. Optimal cartel equilibria with imperfect monitoring. Journal of Economic Theory *39*, 251–269.

Admati, A. and M. Perry [1987]. Strategic delay in bargaining. Review of Economic Studies *54*, 345–364.

Akerlof, G. [1970]. The market for lemons. Quarterly Journal of Economics *84*, 488–500.

Asheim, G. [1990]. Extending renegotiation-proofness ot infinite horizon games. Mimeo, Norwegian School of Economics and Business Administration, forthcoming in Games and Economic Behavior.

Aumann, R.J. [1959]. Acceptable points in general cooperative *n*-person games, in R.D. Luce and A.W. Tucker (eds). Contributions to the theory of games IV, Ann. Math. Study 40, Princeton, N.J. pp. 287–324.

Aumann, R.J. [1960]. Acceptable points in games of perfect information. Pacific J. Mathematics *10* (1960), 381–417.

Aumann, R.J. [1964]. Mixed and behavior strategies in infinite extensive form games, in M. Dresher, L.S. Shapley and A.W. Tucker (eds.). Advances in game theory, Ann. Math. Study *52*, Princeton, N.J., pp. 627–650.

Aumann, R.J. [1974]. Subjectivity and correlation in randomized strategies. Journal of Mathematical Economics *1*, 67–96.

Aumann, R.J. [1976]. Agreeing to disagree. The Annals of Statistics *4*, 1236–1239.

Aumann, R.J. [1981]. Survey of repeated games, in Essays in Game Theory and Mathematical Economics in Honor of Oskar Morgenstern. Bibliographisches Institut Mannheim/Wien/Zürich, pp. 11–42.

Aumann, R.J. [1987 a]. Game Theory. In: The New Palgrave Dictionary of Economics, J. Eatwell, M. Milgate and P. Newman (eds.), 460–482.

Aumann, R.J. [1987 b]. What is game theory trying to accomplish? In: K. Arrow and S. Honkapohja (eds). Frontiers of Economics 28–100.

Aumann, R.J. [1987 c]. Correlated equilibrium as an expression of Bayesian rationality. Econometrica *55*, 1–19.

Aumann, R.J. [1989]. Nash equilibria are not self-enforcing. Mimeo, Hebrew University of Jerusalem.

Aumann, R.J. and S. Hart [1991]. Handbook of Game Theory. To appear with North-Holland (three volumes).
Aumann, R.J., Y. Katznelson, R. Radner, R.W. Rosenthal and B. Weiss [1983]. Approximate purification of mixed strategies. Mathematics of Operations Research *8*, 327–341.
Aumann, R.J. and M. Maschler [1972]. Some thoughts on the minimax principle. Management Science *18*, 54–63.
Aumann, R.J. and L.S. Shapley [1976]. Long term competition – a game theoretic analysis. Mimeo, Hebrew Univ.
Aumann, R.J. and S. Sorin [1989]. Cooperation and bounded recall. Games and Economic Behavior *1*, 5–39.
Bagwell, K. and G. Ramey [1990a]. Advertising and coordination. Northwestern University, CMSEMS DP 898.
Bagwell, K. and G. Ramey [1990b]. Capacity, entry and forward induction. University of California at San Diego DP 90–22.
Bagwell, K. and R. Staiger [1988]. The role of export subsidies when product quality is unknown. Journal of International Economics *27*, 69–89.
Balinski, M.L. [1961]. An algorithm for finding all vertices of a convex set. J. Soc. Ind. Appl. Math. *9*, 72–88.
Banks, J.S. [1990a]. Monopoly agenda control and asymmetric information. Quarterly Journal of Economics *105*, 445–464.
Banks, J.S. [1990b]. Two-sided uncertainty in the monopoly agenda setter model. University of Rochester WP 215.
Banks, J.S. and J. Sobel [1987]. Equilibrium selection in signalling games. Econometrica *55*, 647–663.
Baron, D. and E. Kalai [1990]. Dividing a cake by majority: The simplest equilibria. Mimeo, MEDS, Northwestern University.
Basu, K. [1988]. Strategic irrationality in extensive games. Mathematical Social Sciences, *15*, 247–260.
Basu, K. [1990]. On the non-existence of rationality definition for extensive games. International Journal of Game Theory *19*, 33–44.
Battigalli, P. [1990]. On rationalizability in extensive games. Mimeo, Università Commerciale L. Bocconi.
Beja, A. [1990]. Imperfect equilibrium. MIT WP 3126-90-EFA.
Bellman, R. [1957]. Dynamic programming. Princeton University Press, Princeton.
Bennett, E. and E. E. C. van Damme [1990]. Demand commitment bargaining – the case of apex games. Mimeo, CentER, Tilburg University, Forthcoming in Selten [1991c].
Benoit, J.-P. [1984]. Financially constrained entry in a game with incomplete information. Rand Journal of Economics *15*, 490–499.
Benoit, J.-P. and V. Krishna [1985]. Finitely repeated games. Econometrica *53*, 905–922.
Benoit, J.-P. and V. Krishna [1987]. Nash equilibria of finitely repeated games. International Journal of Game Theory *16*, 197–204.
Benoit, J.-P. and V. Krishna [1989]. Renegotiation in finitely repeated games. Mimeo, Harvard Business School.
Ben-Porath, E. and E. Dekel [1989]. Coordination and the potential for self sacrifice. Mimeo, Stanford University.
Bergin, J. and W.B. MacLeod [1989]. Efficiency and renegotiation in repeated games. Mimeo, Queen's University.
Bernheim, B.D. [1984]. Rationalizable strategic behavior, Econometrica *52*, 1007–1029.
Bernheim, B.D. [1986]. Axiomatic characterizations of rational choice in strategic environments. Scandinavian Journal of Economics *88*, 473–488.
Bernheim, B.D., B. Peleg and M.D. Whinston [1987]. Coalition-proof Nash equilibria I: Concepts. Journal of Economic Theory *42*, 1–12.
Bernheim, B.D. and D. Ray [1989]. Collective dynamic consistency in repeated games. Games and Economic Behavior *1*, 295–326.
Bernheim, B.D. and M.D. Whinston [1987]. Coalition-proof Nash equilibria II: Applications. Journal of Economic Theory *42*, 13–29.

Bicchieri, C. [1989]. Self-refuting theories of strategic interaction: A paradox of common knowledge. Erkenntnis *30*, 69–85.
Billingsley, P. [1968]. Convergence of probability measures. John Wiley and Sons, New York.
Binmore, K. [1985]. Bargaining and coalitions. Game Theoretic Models of Bargaining, ed. A. Roth, Cambridge University Press.
Binmore, K. [1987]. Modeling rational players, I. Economics and Philosophy *3*, 179–214.
Binmore, K. [1987 a]. Nash bargaining theory II, Chapter 4. Binmore and Dasgupta [1987].
Binmore, K. [1987 b]. Perfect equilibria in Bargaining models, Chapter 5. Binmore and Dasgupta [1987].
Binmore, K. [1987 c]. Nash bargaining and incomplete information, Chapter 8. Binmore and Dasgupta [1987].
Binmore, K. [1988]. Modeling rational players, II. Economics and Philosophy *4*, 9–55.
Binmore, K. [1990]. Foundations of game theory. Paper presented at the 6th World Congress in Barcelona.
Binmore, K. and P. Dasgupta [1987]. The Economics of Bargaining, Basil Blackwell, Oxford.
Binmore, K., M.J. Osborne and A. Rubinstein [1991]. Noncooperative Models of Bargaining. Forthcoming in Handbook of Game Theory, eds. R. Aumann and S. Hart, North Holland.
Binmore, K., A. Rubinstein and A. Wolinsky [1986]. The Nash bargaining solution in economic modelling. Rand Journal of Economics *17*, 176–188.
Binmore, K. and L. Samuelson [1990]. Evolutionary stability in repeated games played by finite automata. Paper presented at the 6th World Congress of the Econometric Society in Barcelona.
Bishop, D.T. and C. Cannings [1976]. Models of animal conflict. Adv. Applied Prob. *8*, 616–621.
Blume, A. [1987]. Renegotiation-proof theories in finite and infinite games. Mimeo, Department of Economics, University of California at San Diego.
Blume, L.E. [1986]. Lexicographic refinements of Nash equilibrium. Mimeo, Dept. of Economics, Univ. of Michigan.
Blume, L.E., A. Brandenburger and E. Dekel [1991 a]. Lexicographic probabilities and equilibrium refinements. Forthcoming in Econometrica.
Blume, L.E., A. Brandenburger and E. Dekel [1991 b]. Lexicographic probabilities and choice under uncertainty. Forthcoming in Econometrica.
Bohnenblust, H.F., S. Karlin and L.S. Shapley [1950]. Solutions of discrete two-person games, pp. 51–72 in H.W. Kuhn and A.W. Tucker (eds.), Contributions to the theory of games I (Annals of Mathematics Studies *24*), Princeton University Press, Princeton.
Bomze, I.M. [1986]. Non-cooperative 2-person games in biology: a classification. International Journal of Game Theory *15*, 31–59.
Bomze, I.M. and E.E.C. van Damme [1990]. A dynamical characterization of evolutionary stable states. DP 9045, CentER for Economic Research, Tilburg University.
Börgers, T. [1990]. Weak dominance and approximate common knowledge. Mimeo, Institut für Volkswirtschaft, University of Basel.
Borm, P. [1990]. On perfectness concepts for bimatrix games. Mimeo, Department of Economics, University of Nijmegen.
Borm, P., M. Jansen, J. Potters and S. Tijs [1988]. On the structure of the set of undominated equilibria in bimatrix games. Report 8819, Department of Mathematics, University of Nijmegen.
Boylan, R. [1990]. Evolutionary equilibria resistant to mutation. Mimeo, Washington University.
Brandenburger, A. [1990]. Lexicographic probabilities and iterated admissibility. Mimeo.
Brandenburger, A. and E. Dekel [1986 a]. On an axiomatic approach to refinements of Nash equilibrium: A characterization of perfect equilibrium. Mimeo, Dept. of Economics, Univ. California, Berkeley.
Brandenburger, A. and E. Dekel [1986 b]. Subjective utility, admissibility and sequential rationality. Mimeo, Dept. of Economics, Univ. California, Berkeley.
Brandts, J. and C. Holt [1989]. Forward induction: Experimental evidence from two-stage games with complete information. Mimeo, University of Virginia.
Bruyn, N.G. de [1958]. Asymptotic methods in analysis. North Holland, Amsterdam.
Burgess, J.W. [1976]. Social spiders. Scient. American, March, 100–106.

Cadsby, C.B. and M. Frank [1990]. Pooling, separating, and semiseparating equilibria in financial markets: Some experimental evidence. Review of Financial Studies *3*, 315–342.
Camerer, C. and K. Weigelt [1988]. Experimental tests of sequential equilibrium reputation model. Econometrica *56*, 1–36.
Canning, D. [1989]. Convergence to equilibrium in a sequence of games with learning. STICERD Discussion Paper.
Canning, D. [1990]. Social equilibrium. Mimeo, Pembroke College, Cambridge.
Carlsson, H. [1987]. A bargaining model where parties make errors. Mimeo, University of Lund, forthcoming in Econometrica.
Carlsson, H. and E. van Damme [1990]. Global games and equilibrium selection. CentER DP 9052.
Chatterjee, K., B. Dutta, D. Ray and K. Sengupta [1990]. A non-cooperative theory of cooperative bargaining. Mimeo, Penn State University.
Chatterjee, K. and L. Samuelson [1990]. Perfect equilibria in simultaneous-offers bargaining. International Journal of Game Theory *19*, 237–268.
Chin, H.H., T. Parthasarathy and T.E.S. Raghavan [1974]. Structure of equilibria in *n*-person non-cooperative games. International Journal of Game Theory *3*, 1–19.
Cho, I.K. [1986]. Refinement of sequential equilibrium: Theory and application, Ph. D. dissertation, Princeton University.
Cho, I.K. [1987]. A refinement of sequential equilibrium. Econometrica *55*, 1367–1390.
Cho, I.K. [1990]. Strategic stability in repeated signaling games. Mimeo, University of Chicago.
Cho, I.K. and D. Kreps [1987]. Signaling games and stable equilibria. Quarterly Journal of Economics *102*, 179–221.
Cho, I.K. and J. Sobel [1990]. Strategic stability and uniqueness in signaling games. Journal of Economic Theory *50*, 381–413.
Cooper, R., D. DeJong, R. Forsythe and T. Ross [1989]. Communication in the battle of the sexes game: Some experimental results. Rand Journal of Economics *20*, 568–587.
Cramton, P.C. [1984a]. Bargaining with incomplete information: An infinite horizon model with continuous uncertainty. Review of Economic Studies, *51*, 579–593.
Cramton, P.C. [1984b]. The role of time and information in bargaining. Research paper 729, Graduate School of Business, Stanford University.
Crawford, V.P. [1977]. A game of fair division. Review of Economic Studies *XLIV*, 235–247.
Crawford, V.P. [1979]. A procedure for generating Pareto-efficient egalitarian equivalent allocations. Econometrica *47*, 49–60.
Crawford, V.P. [1982]. A theory of disagreement in bargaining. Econometrica *50*, 607–637.
Crawford, V.P. [1990a]. Nash equilibrium and evolutionary stability in large – and finite – population playing the field models. Journal of Theoretical Biology *143*, 269–279.
Crawford, V.P. [1990b]. An evolutionary interpretation of Van Huyck, Battalio, and Beil's experimental results on coordination. Mimeo, University of California at San Diego, forthcoming in Games and Economic Behavior.
Crawford, V.P. and W.P. Heller [1979]. Fair division with indivisible commodities. Journal of Economic Theory *21*, 10–27.
Crawford, V.P. and J. Sobel [1982]. Strategic information transmission. Econometrica *50*, 1431–1451.
Dawkins, R. [1976]. The selfish gene. Oxford Univ. Press.
Debreu, G. [1970]. Economies with a finite set of equilibria. Econometrica, *38*, 387–392.
Dekel, E. [1988]. Simultaneous offers and the inefficiency of bargaining: A two-period example. Mimeo, University of California at Berkeley, forthcoming in Journal of Economic Theory.
Dekel, E. and D. Fudenberg [1987]. Rational behavior with payoff uncertainty. Mimeo, MIT, forthcoming in Journal of Economic Theory.
DeKoster, R., Peters, H.J.M., Tijs, S.H. and P. Wakker [1983]. Risk sensitivity, independence of irrelevant alternatives and continuity of bargaining solutions. Math. Soc. Sc. *4*, 295–300.
Demange, G. [1982]. Implementing efficient egalitarian equivalent allocations. Mimeo, Ecole Polytechnique, Laboratoire d'Econometrie, Paris.
Demski, J. and D. Sappington [1987]. Delegated Expertise. J. Accounting Research, 68–89
Diamond, D. [1989]. Reputation acquisition in debt markets. Journal of Political Economics *97*, 828–862.

Dieudonné, J. [1969]. Foundations of modern analysis. Academic Press, New York.
Dixit, A. and B. Nalebuff [1990]. Thinking strategically. Princeton University Press.
Dresher, M. [1961]. Games of strategy. Prentice-Hall, Englewood Cliffs, New Jersey.
Eilenberg, S. and N. Steenrod [1952]. Foundations of algebraic topology. Princeton Univ. Press, Princeton, New Jersey.
Elmes, S. and P. Reny [1989]. On the equivalence of extensive from games. Mimeo, Columbia University.
Engers, M. [1987]. Signaling with many signals. Econometrica *55*, 663–674.
Engers, M. and L. Fernandez [1987]. Market equilibrium with hidden knowledge and self-selection. Econometrica *55*, 425–440.
Eshel, I. [1989]. Game theory and population dynamics in complex genetical systems: The role of sex in short term and in long term evolution. DP 24, Research Group on Game Theory in the Behavioral Sciences, ZIF Bielefeld. Forthcoming in Selten [1991 a].
Evans, R. and E. Maskin [1989]. Efficient renegotiation-proof equilibria in repeated games. Games and Economic Behavior *1*, 361–369.
Farrell, J. [1983]. Credible repeated game equilibrium. Mimeo, MIT.
Farrell, J. [1984]. Credible neologisms in games of communication, Mimeo, MIT.
Farrell, J. [1987]. Cheap talk, coordination, and entry. Rand Journal of Economics *18*, 34–39.
Farrell, J. [1988]. Communication, coordination, and Nash equilibrium. Economic Letters *27*, 209–214.
Farrell, J. [1990]. Meaning and credibility in Cheap-talk games. Forthcoming in Mathematical Models in Economics, ed. M. Dempster, Oxford University Press.
Farrell, J. and R. Gibbons [1989 a]. Cheap talk can matter in bargaining. Journal of Economic Theory *48*, 221–237.
Farrell, J. and R. Gibbons [1989 b]. Cheap talk with two audiences. MIT Working paper 518.
Farrell, J. and E. Maskin [1989]. Renegotiation in repeated games. Games and Economic Behavior *1*, 327–360.
Farrell, J. and G. Saloner [1988]. Coordination through committees and markets. Rand Journal of Economics *19*, 235–252.
Fershtman, C., K. Judd and E. Kalai [1987]. Cooporation through delegation. Mimeo, MEDS, Northwestern University.
Fishman, M. [1988]. A theory of preemptive takeover bidding. Rand Journal of Economics *19*, 828–862.
Fishman, M. [1989]. Preemptive bidding and the role of the medium of exchange in acquisitions. Journal of Finance *44*, 41–57.
Forges, F. [1986]. An approach to communication equilibria. Econometrica *54*, 1375–1387.
Forges, F. [1988]. Universal mechanisms. CREQE WP 8810, forthcoming in Econometrica.
Forges, F. [1990]. Equilibrium with communication in a job market example. Quarterly Journal of Economics *105*, 375–398.
Fort, M.K. [1950]. Essential and non-essential fixed points. Am. J. Math. *72*, 315–322.
Friedman, D. [1987]. Evolutionary economic games. Mimeo, University of California at Santa Cruz. Forthcoming in Econometrica.
Friedman, J.W. [1971]. A noncooperative equilibrium for supergames. Review of Economic Studies *38*, 1–12.
Friedman, J.W. [1977]. Oligopoly and the theory of games. North Holland, Amsterdam.
Friedman, J.W. [1985]. Cooperative equilibria in finite horizon noncooperative supergames. Journal of Economic Theory *35*, 390–398.
Friedman, J.W. [1986]. Game Theory with Applications to Economics. Oxford Press.
Friedman, J.W. and R. Rosenthal [1986]. A positive approach to non-cooperative games. Journal of Economic Behavior and Organization *7*, 235–251.
Friedman, J.W. and L. Samuelson [1988]. Subgame perfect equilibrium with continuous reaction functions. Mimeo, Penn State University.
Fudenberg, D. and D. Kreps [1987]. Reputation and simultaneous opponents. Review of Economic Studies *54*, 541–568.
Fudenberg, D. and D. Kreps [1988]. Learning, experimentation and equilibrium in games. Mimeo, Stanford University.

Fudenberg, D., D. Kreps and D. Levine [1988]. On the robustness of equilibrium refinements. Journal of Economic Theory *44*, 354–380.

Fudenberg, D. and D. Levine [1983]. Subgame perfect equilibria of finite and infinite-horizon games. Journal of Economic Theory *31*, 251–268.

Fudenberg, D. and D. Levine [1986]. Limit games and limit equilibria. Journal of Economic Theory *38*, 261–279.

Fudenberg, D. and D. Levine [1989 a]. Reputation and equilibrium selection in games with a patient player. Econometrica *57*, 759–778.

Fudenberg, D. and D. Levine [1989 b]. Equilibrium payoffs with long-run and short-run players and imperfect public information. Mimeo, UCLA.

Fudenberg, D. and D. Levine [1989 c]. An approximate folk theorem with imperfect private information. Mimeo, UCLA.

Fudenberg, D. and D. Levine [1989 d]. Reputation, unobserved strategies, and active supermartingales. Mimeo, MIT and UCLA.

Fudenberg, D. and D. Levine [1990]. Steady state learning and self-confirming equilibrium. Mimeo, UCLA.

Fudenberg, D., D. Levine and E. Maskin [1989]. The folk theorem with imperfect public information. Mimeo, UCLA.

Fudenberg, D. and E. Maskin [1986]. The folk theorem in repeated games with discounting and with incomplete information. Econometrica, *54*, 533–554.

Fudenberg, D. and E. Maskin [1990]. Evolution and cooperation in noisy repeated games. Mimeo, Harvard University, forthcoming in American Economic Review.

Fudenberg, D. and J. Tirole [1983]. Sequential bargaining with incomplete information. Review of Economic Studies *50*, 221–247.

Fudenberg, D. and J. Tirole [1989 a]. Noncooperative game theory for industrial organization: An introduction and overview. In: R. Schmalensee and R. Willig (eds.) Handbook of Industrial Organization. North-Holland, Amsterdam.

Fudenberg, D. and J. Tirole [1989 b]. Perfect Bayesian equilibrium and sequential equilibrium. Mimeo, Harvard University, forthcoming in Journal of Economic Theory.

Fudenberg, D. and J. Tirole [1991]. Game theory. MIT Press (forthcoming)

Gale, D. [1953]. A theory of *n*-person games with perfect information. Proc. Nat. Acad. Sci. USA *39*, 396–501.

Gale, D. and M. Hellwig [1989]. Repudiation and renegotiation: The case of Sovereign Debt. International Economic Review *30*, 3–31

Gale, D. and S. Sherman [1950]. Solutions of finite two-person games, pp. 37–49 in H.W. Kuhn and A.W. Tucker (eds.). Contributions to the theory of games I (Annals of Mathematics Studies *24*), Princeton University Press, Princeton.

Garcia Jurado, I. and J. Prada Sánchez [1990]. A remark on Myerson's concept of proper equilibrium. International Journal of Game Theory *19*, 11–16.

Gardner, R. and M. Morris [1989 a]. The evolution of bluffing in animal contests: An ESS approach. Journal of Theoretical Biology *137*, 235–243.

Gardner, R. and M. Morris [1989 b]. Evolutionary stability in a class of extensive form games. DP, Research Group on Game Theory in the Behavioral Sciences, ZIF Bielefeld. Forthcoming in Selten [1991 a].

Gertner, R., R. Gibbons and D. Scharfstein [1987]. Simultaneous signaling to the capital and product markets. Mimeo, University of Chicago.

Gilboa, I. and A Matsui [1989]. Social stability and equilibrium. MEDS Discussion Paper, Northwestern University, Forthcoming in Econometrica.

Glazer, J. and A. Weiss [1990]. Pricing and coordination: Strategically stable bargaining. Games and Economic Behavior *2*, 118–128.

Greenberg, J. [1989 a]. Deriving strong and coalition-proof Nash equilibria from an abstract system. Journal of Economic Theory *49*, 195–202.

Greenberg, J. [1989 b]. An application of the theory of social situations to repeated games. Journal of Economic Theory *49*, 278–293.

Greenberg, J. [1990]. Social situations: A game-theoretic approach. Cambridge University Press.

Gretlein, R. [1983]. Dominance elimination procedures of finite alternative games. International Journal of Game Theory *12*, 107–113.

Grossman, S.J. [1981]. The informational role of warranties and the private disclosure about product quality. Journal of Law and Economics *24*, 461–483.

Grossman, S.J. and M. Perry [1986a]. Perfect sequential equilibrium. Journal of Economic Theory *39*, 97–119.

Grossman, S.J. and M. Perry [1986b]. Sequential bargaining under asymmetric information. Journal of Economic Theory *39*, 120–154.

Gul, F. [1989]. Rational strategic behavior and the notion of equilibrium. Mimeo, Stanford University.

Gul, F., H. Sonnenschein and R. Wilson [1986]. Foundations of dynamic monopoly and the Coase conjecture. Journal of Economic Theory *39*, 155–190.

Güth, W. and B. Kalkofen [1989]. Unique solutions for strategic games. Lecture Notes in Economic and Mathematical Systems *328*, Springer, Berlin.

Güth, W. and E.E.C. van Damme [1989]. Equilibrium selection in the Spence signaling model. CentER DP 8939, forthcoming in Selten [1991 b].

Güth, W. and E.E.C. van Damme [1990]. Gorby games – a game theoretic analysis of disarmament campaigns and the defense efficiency-hypothesis. CentER DP 9011, Tilburg University.

Güth, W. and E.E.C. van Damme [1986]. Auctions and distributional conflicts: A comparison of pricing rules. Social Choice and Welfare *3*, 177–198.

Güth, W., W. Leininger and G. Stephen [1988]. On supergames and folk theorems: A conceptual discussion. WP 19, Research Group Game Theory in the Behavioral Sciences, ZIF, Bielefeld. Forthcoming in Selten [1991 b].

Güth, W. and R. Tietz [1990]. Ultimatum bargaining behavior – a survey and comparison of experimental results. Journal of Economic Psychology *11*, 417–449.

Haigh, J. [1975]. Game theory and evolution. Adv. Applied Prob. *7*, 8–11.

Halmos, P.R. [1950]. Measure theory, Van Nostrand, Princeton, New Jersey.

Hammerstein, P. [1981]. The role of asymmetries in animal contests. Animal Behavior *29*, 193–205.

Hammerstein, P. and G.A. Parker [1982]. The asymmetric war of attrition. Journal of Theoretical Biology *96*, 647–682.

Hammerstein, P. and R. Selten [1991]. Game theory in biology. Aumann and Hart [1991].

Hammond, P. [1987]. Consequentialist foundations for expected utility. CORE DP 8716.

Harrington, J.E. [1986]. Limit pricing when the potential entrant is uncertain of its cost function, Econometrica *54*, 429–439.

Harris, C. [1984]. A note on the existence of optimal simple penal codes. Mimeo, Nuffield College.

Harris, C. [1985a]. Existence and characterization of perfect equilibrium in games of perfect information. Econometrica *53*, 613–628.

Harris, C. [1985b]. A characterization of the perfect equilibria of infinite horizon games. Journal of Economic Theory *37*, 99–125.

Harris, M. and A. Raviv [1983]. A sequential signalling model of convertible debt call policy. Technical Report 425. IMSSS, Stanford University.

Harsanyi, J.C. [1968]. Games with incomplete information played by "Bayesian" players, Parts I, II and III. Management Science *14*, 159–182, 320–332, 468–502.

Harsanyi, J.C. [1973a]. Games with randomly disturbed payoffs: a new rationale for mixed strategy equilibrium points. International Journal of Game Theory *2*, 1–23.

Harsanyi, J.C. [1973b]. Oddness of the number of equilibrium points: a new proof. International Journal of Game Theory *2*, 235–250.

Harsanyi, J.C. [1975]. The tracing procedure: a Bayesian approach to defining a solution for *n*-person noncooperative games. International Journal of Game Theory *4*, 61–94.

Harsanyi, J.C. [1976]. A solution concept for *n*-person noncooperative games. International Journal of Game Theory *5*, 211–225.

Harsanyi, J.C. [1977]. Rational behavior and bargaining equilibrium in games and social situations. Cambridge University Press.

Harsanyi, J.C. and R. Selten [1972]. A generalized Nash solution for two-person bargaining games with incomplete information. Management Science *18*, part II, 80–106.

Harstad, R.M. [1990]. Reasoning without common knowledge of reasoning. Mimeo, Virginia Commonwealth University.

Hellwig, M. [1986 a]. Some recent developments in the theory of competition in markets with adverse selection. Discussion paper A–82, Dept. of Economics, University of Bonn.
Hellwig, M. [1986 b]. A sequential approach to modelling competition in markets with adverse selection. Discussion Paper, Dept. of Economics, University of Bonn.
Hellwig, M. [1987]. Some recent developments in the theory of competition in markets with adverse selection. European Economic Review *33*, 319–325.
Hellwig, M. and W. Leininger [1987]. On the existence of subgame perfect equilibrium in finite action games of perfect information. Journal of Economic Theory *43*, 55–75.
Hellwig, M., W. Leininger, P. Reny and A. Robson [1991]. Subgame perfect equilibrium in continuous games of perfect information: An elementary approach to existence and approximation by discrete games. Forthcoming in Journal of Economic Theory.
Herrero, M. [1985]. A strategic bargaining approach to market institutions. Ph.-D. thesis, London School of Economics.
Heuer, G.A. and C.B. Millham [1976]. On Nash subsets and mobility chains in bimatrix games. Naval Res. Logist. Quart. *23*, 311–319.
Hillas, J. [1989]. On the definition of the strategic stability of equilibria. Mimeo, Ohio State University, forthcoming in Econometrica.
Hines, W.G.S. [1980]. Strategy stability in complex populations. J. Appl. Prob. *17*, 600–610.
Hines, W.G.S. [1987]. Evolutionary stable strategies: A review of basic theory. Theoretical Population Biology *31*, 195–272. Erratum *33*, 114.
Hirsch, M.W. and S. Smale [1974]. Differential equations, dynamical systems and linear algebra. Academic Press.
Hofbauer, J. [1985]. Game dynamics for asymmetric contests. Discussion Paper, Dept. of Math., University Wien.
Hofbauer, J. and K. Sigmund [1988]. The Theory of Evolution and Dynamical Systems. Cambridge University Press, Cambridge.
Holmström, B. and R.B. Myerson [1983]. Efficient and durable decision rules with incomplete information. Econometrica *51*, 1799–1819.
Jansen, M.J.M. [1981 a]. Maximal Nash subsets for bimatrix games. Naval. Res. Logist. Quart. *28*, 147–152.
Jansen, M.J.M. [1981 b]. Regularity and stability of equilibrium points of bimatrix games. Mathematics of Operations Research *6*, 530–550.
Jansen, M.J.M. [1987]. Regular equilibrium points in bimatrix games. Operations Research Spectrum *9*, 87–91.
Jansen, M.J.M., P. Jurg and P. Borm [1990]. On the finiteness of stable sets. Mimeo, University of Nijmegen.
Jansen, M.J.M. and Tijs [1981]. Arbitration games: A survey. DGOR Operations Research Proceedings 1980, Springer-Verlag, Berlin.
Jiang Jia-He [1963]. Essential equilibrium points of *n*-person non-cooperative games. Sci. Sinica *12*, 651–671.
Kakutani, S. [1941]. A generalisation of Brouwer's fixed point theorem. Duke Math. J. *8*, 457–459.
Kalai, E. [1977]. Nonsymmetric Nash solutions and replications of 2-person bargaining. International Journal of Game Theory *6*, 129–133.
Kalai, E. [1981]. Preplay negotiations and the prisoners dilemma. Math. Methods in the Social Sciences *1*, 375–379.
Kalai, E. [1985]. Solutions to the Bargaining Problem, pp. 77–107. In: Hurwicz, Schmeidler and Sonnenschein (eds). Social goals and social organization, essays in memory of Elishna Pazner. Cambridge University Press.
Kalai, E. and E. Lehrer [1990]. Rational learning leads to Nash equilibrium. Mimeo, Northwestern University.
Kalai, E. and R.W. Rosenthal [1978]. Arbitration of two-party disputes under ignorance. International Journal of Game Theory *7*, 65–72.
Kalai E. and D. Samet [1984]. Persistent equilibria in strategic games. Int. J.Game Theory *13*, 129–144.
Kalai, E. and D. Samet [1985]. Unanimity games and Pareto optimality. International Journal of Game Theory *14*, 41–50.

Kalai, E. and M. Smorodinsky [1975]. Other solutions to Nash's bargaining problem. Econometrica *43*, 513–518.

Kaneko, M. [1981]. A bilateral monopoly and the Nash cooperative solution. Journal of Economic Theory *24*, 311–327.

Kaneko, M. [1982]. Some remarks on the folk theorem in game theory. Mathematical Social Sciences *3*, 281–290; also Erratum in *5*, 1983, 233.

Kannai, Y. [1977]. Concavifiability and constructions of concave utility functions. Journal of Mathematical Economicsomics *4*, 1–56.

Kennan, J. and R. Wilson [1990]. Bargaining with private information. SITE Technical report No. 5, Stanford University.

Kihlstrom, R.E., A.E. Roth, and D. Schmeidler [1981]. Risk aversion and solutions to Nash's bargaining problem, pp. 65–71 in O. Moeschlin and D. Pallaschke (eds.), Game Theory and Mathematical Economics. North Holland Publishing Co., Amsterdam.

Kingman, J.F.C. and J.S. Taylor [1966]. Introduction to measure and probability. Cambridge University Press, London.

Kohlberg, E. [1989]. Refinement of Nash equilibrium: The main ideas. Mimeo, Harvard University.

Kohlberg, E. and J.-F. Mertens [1986]. On the strategic stability of equilibria. Econometrica *54*, 1003–1039.

Kohlberg, E. and P. Reny [1991]. Consistent assessments in sequential equilibria (provisional title).

Kojima, M., A. Okada and S. Shindoh [1985]. Strongly stable equilibrium points of *n*-person noncooperative games. Math. Operations Research *10*, 650–663.

Kolm, S.C. [1972]. Justice et Équité. Editions du Centre National de la Recherche Scientifique.

Kreps, D.M. [1984]. Signalling games and stable equilibria. Mimeo, Graduate School of Business, Stanford University.

Kreps, D.M. [1990 a]. Game Theory and Economic Modeling. Oxford University Press.

Kreps, D.M. [1990 b]. A Course in Microeconomic Theory. Princeton University Press.

Kreps, D.M., P. Milgrom, J. Roberts, and R. Wilson [1982]. Rational cooperation in the finitely-repeated Prisoners' Dilemma. Journal of Economic Theory *27*, 245–252.

Kreps, D.M. and G. Ramey [1987]. Structural consistency, consistency and sequential rationality. Econometrica *55*, 1331–1348.

Kreps, D.M. and J. Sobel [1991]. Signalling. In: R. Aumann and S. Hart (eds.) Handbook of Game Theory.

Kreps, D.M. and R. Wilson [1982 a]. Sequential equilibria. Econometrica *50*, 863–894.

Kreps, D.M. and R. Wilson [1982 b]. Reputation and imperfect information. Journal of Economic Theory *27*, 253–279.

Krohn, I., S. Moltzahn, J. Rosenmüller, P. Sudhölter and H.-M. Wallmeier [1989]. Implementing the modified LH-algorithm. IMW WP 179, University of Bielefeld.

Kuhn, H.W. [1953]. Extensive games and the problem of information. Annals of Mathematics Studies *28*, 193–216.

Kuhn, H.W. [1961]. An algorithm for equilibrium points in bimatrix games. Proc. Nat. Acad. Sci. U.S.A. *47*, 1656–1662.

Lefschetz, S. [1949]. Introduction to topology. Princeton University Press, Princeton.

Lehrer, E. [1985]. Two player repeated games with non-observable actions. Mimeo, Hebrew University.

Lehrer, E. [1990]. Nash equilibria of *n*-player repeated games with semi-standard information. International Journal of Game Theory *19*, 191–217.

Leininger, W. [1986]. Strategic equilibrium in games with perfect recall. Technical Memorandum, Bell Laboratories.

Leininger, W. [1988]. Escalation and cooperation in conflict situations – the dollar auctions revisited. Forthcoming in Journal of Conflict Resolution.

Leland, M.E. and D.M. Pyle [1977]. Informational asymmetries, financial structure and financial intermediation. Journal of Finance *32*, 371–387.

Lemke, C.E. and J.T. Howson [1964]. Equilibrium points of bimatrix games. J.Soc. Indust. Appl. Math. *12*, 413–423.

Levin, D. and R. Harstadt [1984]. Robustness and perfection in auctions with competitive bidding. Mimeo, University of Houston.

Luce, R.D. and H. Raiffa [1957]. Games and decisions: Introduction and critical survey. Wiley, New York.

Lutz, N. [1989]. Warranties as signals under consumer moral hazard. Rand Journal of Economics *20*, 239–255.

Madrigal, V., T. Tan and S. Werlang [1987]. Support restrictions and sequential equilibria. Journal of Economic Theory *43*, 329–334.

Mailath, G., L. Samuelson and J. Swinkels [1990]. Extensive form reasoning in normal form games. PennState Working paper.

Malawski, M. [1989]. Some learning processes in games. PhD Thesis, Bonn University.

Maskin, E. and J. Tirole [1989]. Markov equilibrium. Mimeo, Harvard University.

Masso, J. and R. Rosenthal [1989]. More on the Anti-Folk Theorem. Journal of Mathematical Economics *18*, 281–290.

Matsui, A. [1989]. Information leakage forces cooperation. Games and Economic Behavior *1*, 94–115.

Matthews, S. [1989]. Veto threats: Rhetoric in a bargaining game. Quarterly Journal of Economics *104*, 347–369.

Matthews, S. and D. Fertig [1989]. Advertising signals of product quality. CMSEMS DP 881, Northwestern University.

Matthews, S. and A. Postlewaite [1989]. Pre-play communication in two-person sealed-bid double auctions. Journal of Economic Theory *48*, 238–263.

Matthews, S., M. Okuno-Fujiwara and A. Postlewaite [1990]. Refining cheap-talk equilibria. Mimeo, Northwestern University.

Maynard Smith, J. [1982]. Evolution and the theory of games. Cambridge Univ. Press.

Maynard Smith, J. and G.A. Parker [1976]. The logic of asymmetric contests. Animal Behavior *24*, 159–175.

Maynard Smith, J. and G.R. Price [1973]. The logic of animal conflict. Nature, Lond., *246*, 15–18.

McAfee, R. and J. McMillan [1987]. Auctions and bidding. Journal of Economic Literature *25*, 699–738.

McLennan, A. [1989]. Consistent conditional systems in noncooperative game theory. International Journal of Game Theory *18*, 141–174.

McMillan, J. [1986]. Game Theory in International Economics. Harwood Academic Publishers, Chur, Switzerland.

McMillan, J. [1989]. Strategic decisions: Using game theory in management. Draft, University of California at San Diego.

Mertens, J.-F. [1986]. Localization of the degree of lower-dimensional sets. CORE DP 8605.

Mertens, J.-F. [1987]. Ordinality in non cooperative games. CORE DP 8728.

Mertens, J.-F. [1988]. Stable Equilibria – a reformulation. CORE DP 8838.

Mertens, J.-F. [1989]. Equilibrium and rationality: Context and history-dependence. Mimeo, CORE.

Mertens, J.-F. [1989]. Stable equilibria – a reformulation, Part I. Mathematics of Operations Research *14*, 575–625.

Mertens, J.-F. [1990]. The small worlds axiom for stable equilibria. CORE DP 9007.

Mertens, J.-F., S. Sorin and S. Zamir [1990]. Repeated games. Draft of a book.

Milgrom, P. [1987]. Auction theory. In: T. Bewley (ed.) Advances in economic theory. Fifth World Congress, Cambridge University Press, Cambridge.

Milgrom, P. and D.J. Roberts [1982a]. Limit pricing and entry under incomplete information: An equilibrium analysis. Econometrica *50*, 443–459.

Milgrom, P. and D.J. Roberts [1982b]. Predation, reputation and entry deterrence. Journal of Economic Theory *27*, 280–312.

Milgrom, P. and D.J. Roberts [1986]. Price and advertising signals of product quality. Journal of Political Economy *94*, 796–821.

Milgrom, P. and D.J. Roberts [1989a]. A theory of adaptive and sophisticated learning in repeated normal form games. Mimeo, Stanford University.

Milgrom, P. and D.J. Roberts [1989b]. Rationalizability, learning and equilibrium in games with strategic complementarities. Mimeo, Stanford University, forthcoming in Econometrica.
Milgrom, P. and R. Weber [1981]. Topologies on information and strategies in games with incomplete information, pp. 89–96 in O. Moeschlin and D. Pallaschke (eds.), Game Theory and Mathematical Economics, North Holland, Amsterdam.
Milgrom, P. and R. Weber [1985]. Distributional strategies for games with incomplete information. Mathematics of Operations Research *10*, 619–632.
Millham, C.B. [1974]. On Nash subsets of bimatrix games. Naval. Res. Logist. Quart. *21*, 307–317.
Mills, H. [1960]. Equilibrium points in finite games. J. Soc. Indust. Appl. Math. *8*, 397–402.
Milnor, J. [1965]. Topology from the differentiable viewpoint. The university press of Virginia, Charlottesville.
Mitzkewitz, M. [1988]. Equilibrium properties of experimentally based duopoly strategies. Working paper B 107, University of Bonn.
Monderer, D. and D. Samet [1989]. Approximating common knowledge with common beliefs. Games and Economic Behavior *1*, 170–190.
Moulin, H. [1979]. Dominance solvable voting schemes. Econometrica *47*, 1337–1351.
Moulin, H. [1982]. Bargaining and noncooperative implementation; A survey of recent results. Math. Social Sciences, *3*, 243–258.
Moulin, H. [1984]. Implementing the Kalai-Smorodinsky bargaining solution. Journal of Economic Theory *33*, 32–45.
Moulin, H. [1985]. Game Theory for the social sciences. New York University Press, 1985
Moulin, H. and J.P. Vial [1978]. Strategically zero-sum games: The class of games whose completely mixed equilibria cannot be improved upon. International Journal of Game Theory *7*, 201–221.
Muthoo, A. [1990]. Bargaining without commitment. Games and Economic Behavior *2*, 291–297.
Myers, S.C. and N.S. Majluf [1984]. Corporate financing and investment decisions when firms have information that investors do not have. Journal of Financial Economics *13*, 187–221.
Myerson, R.B. [1978]. Refinements of the Nash equilibrium concept. International Journal of Game Theory *7*, 73–80.
Myerson, R.B. [1983]. Mechanism design by an informed principal. Econometrica *51*, 1761–1797.
Myerson, R.B. [1986a]. Acceptable and predominant correlated equilibria. International Journal of Game Theory *15*, 133–154.
Myerson, R.B. [1986b]. Multistage games with communication. Econometrica *54*, 323–358.
Myerson, R.B. [1989]. Credible negotiation statements and coherent plans. Journal of Economic Theory *48*, 264–303.
Myerson, R.B. [1990]. Game theory: Analysis of conflict. Mimeo, Northwestern University.
Nachbar, J. [1990]. Evolutionary selection dynamics in games: Convergence and limit properties. Intenational Journal of Game Theory *19*, 59–89.
Nash, J.F. [1950]. Equilibrium points in *n*-person games. Proc. Nat. Acad. Sci. U.S.A. *36*, 48–49.
Nash, J.F. [1950]. The bargaining problem. Econometrica *18*, 155–162.
Nash, J.F. [1951]. Non-cooperative games. Ann. of Math. *54*, 286–295.
Nash, J.F. [1953]. Two-person cooperative games. Econometrica *21*, 128–140.
Neumann, J. von [1928]. Zur Theorie der Gesellschaftsspiele. Math. Ann. *100*, 295–320.
Neumann, J. von and O. Morgenstern [1947]. Theory of games and economic behavior. Princeton University Press, Princeton (first edition 1944).
Neyman, A. [1987]. Bounded common knowledge justifies cooperation in the finitely repeated Prisoners' Dilemma. Unpublished notes.
Noldeke, G. and E.E.C. van Damme [1990a]. Signaling in a dynamic labor market. Review of Economic Studies *57*, 1–24.
Noldeke, G. and E.E.C. van Damme [1990b]. Switching away from probability one beliefs. University of Bonn DP A-304.
Okada, A. [1981a]. On stability of perfect equilibrium points. International Journal of Game Theory, *10*, 67–73.
Okada, A. [1981b]. A note on the perfectness concept and information structures in games. International Journal of Game Theory *10*, 133–136.

Okada, A. [1982]. A note on quasi-strong equilibrium points of noncooperative games. Research report No. B–114, Department of System Science of the Tokyo institute of technology.

Okada, A. [1984]. Strictly perfect equilibrium points of bimatrix games. International Journal of Game Theory *13*, 145–154.

Okada, A. [1988]. Perfect equilibrium points and lexicographic domination. International Journal of Game Theory *17*, 225–239.

Okada, A. [1989]. Lexicographic domination in extensive games. Mimeo, Saitama University, Japan.

Okada, A. [1990]. A Noncooperative approach to the Nash bargaining problem. DP 15, ZIF, Bielefeld. Forthcoming in Selten [1991 c].

Okuno-Fujiwara, M. and A. Postlewaite [1987]. Forward induction and equilibrium refinement. University of Pennsylvania WP 8701.

O'Neill, B. [1986]. International escalation and the dollar auction. Journal of Conflict Resolution *30*, 33–50.

Osborne, M. [1990]. Signaling, forward induction, and stability in finitely repeated games. Journal of Economic Theory *50*, 22–36.

Osborne, M. and A. Rubinstein [1990]. Bargaining and markets. Academic Press, San Diego.

Owen, G. [1968]. Game theory. W.B. Saunders Company, Philadelphia.

Owen, G. [1971]. Optimal threat strategies of bimatrix games. International Journal of Game Theory *1*, 3–9.

Parker, G.A. and D.I. Rubenstein [1981]. Role assessment, reserve strategy and acquisition of information in asymmetric animal conflict. Animal Behavior, *29*, 221–240.

Parker, G.A. and P. Hammerstein [1985]. Game theory and animal behaviour, pp. 73–95 in Greenwood, Harvey and Sladkin (eds): Evolution: Essays in Honor of John Maynard Smith.

Pearce, D.G. [1982]. A problem with single-valued solution concepts. Mimeo. Inst. Math. Appl., Univ. of Minnesota, Minneapolis.

Pearce, D.G. [1984]. Rationalizable strategic behavior and the problem of perfection. Econometrica *52*, 1029–1051.

Pearce, D.G. [1987]. Renegotiation-proof equilibria: Collective rationally and intertemporal cooperation. Cowles Foundation DP 855.

Persson, T. and G. Tabelini [1990]. Macroeconomic policy. Credibility and Politics. Harwood Academic Publishers, Chur, Switzerland.

Peters, H. and E.E.C. van Damme [1987]. A characterization of the Nash bargaining solution not using IIA. Discussion Paper A 117, Dept. Economics, Univ. Bonn.

Ponssard, J.P. [1976]. On the subject of non optimal play in zero sum extensive form games: "the trap phenomenon". International Journal of Game Theory *5*, 105–115.

Ponssard, J.-P. [1990 a]. Self enforceable paths in games in extensive form: A behavioral approach based on interactivity. Theory and Decision *28*, 69–83.

Ponssard, J.-P. [1990 b]. Forward induction and sunk costs give average cost pricing. Forthcoming in Games and Economic Behavior.

Ponssard, J.-P. [1990 c]. A note on forward induction and escalation games with perfect information. Mimeo, Ecole Polytechnique, Paris.

Quinzii, M. and J.-C. Rochet [1985]. Multidimensional signalling. Journal of Mathematical Economics *14*, 261–284.

Rabin, M. [1990]. Communication between rational agents. Journal of Economic Theory *51*, 144–170.

Radner, R. [1978]. Can bounded rationality resolve the Prisoners' Dilemma? Mimeo, Bell Labs.

Radner, R. [1980]. Collusive behavior in noncooperative epsilon-equilibria of oligopolies with long but finite lives. Journal of Economic Theory *22*, 136–154.

Radner, R. and R.W. Rosenthal [1982]. Private information and pure strategy equilibria. Mathematics of Operations Research *7*, 401–409.

Ramey, G. [1985]. A note on structural consistency in sequential equilibrium. Mimeo, Stanford University.

Ramey, G. [1988]. Intuitive signaling equlibria with multiple signals and a continuum of types. Mimeo, University of California at San Diego.

Rasmusen, E. [1989]. Games and information: An introduction to game theory. Basil Blackwell, Oxford.

Ray, D. [1989]. Internally renegotiation-proof equilibrium sets: Limit behaviour with low discounting. Mimeo, Indian Statistical Institute.
Reinganum, J.F. and L.L. Wilde [1986]. Settlement, litigation, and the allocation of litigation costs. Rand Journal of Economics *17*, 557–566.
Reny, P. [1988a]. Backward induction, normal form perfection and explicable equilibria. Mimeo. University of Western Ontario.
Reny, P. [1988b]. Backward induction and common knowledge in games with perfect information. Mimeo, University of Western Ontario.
Riley, J. [1975]. Competitive signalling, Journal of Economic Theory, *10*, 174–186.
Riley, J. [1979a]. Evolutionary equilibrium strategies. Journal of Theoretical Biology *76*, 109–123.
Riley, J. [1979b]. Informational equilibrium. Econometrica *47*, 331–359.
Ritzberger, K. and K. Vogelsberger [1989]. The Nash field. Mimeo, Institute for Advanced Studies, Vienna.
Robson, A. [1986]. The existence of Nash equilibria in reaction functions for dynamic models of oligology. International Economic Review *27*, 539–544.
Robson, A. [1989]. Efficiency in evolutionary games: Darwin, Nash and the secret handshake. Mimeo, University of Western Ontario.
Rockefellar, R.T. [1979]. Convex analysis. Princeton Univ. Press, Princeton, N.J.
Rosen, J.B. [1965]. Existence and uniqueness of equilibrium points for concave *n*-person games. Econometrica *33*, 520–534.
Rosenmüller, J. [1971]. On a generalization of the Lemke-Howson algorithm to noncooperative *n*-person games. SIAM J. Appl. Math. *21*, 73–79.
Rosenmüller, J. [1981]. The theory of games and markets. North Holland, Amsterdam.
Rosenthal, R.W. [1976].An arbitration model for normal form games. Mathematics of Operations Research *1*, 82–88.
Rosenthal, R.W. [1981]. Games of perfect information, predatory pricing and the chain-store pradox. Journal of Economic Theory *25*, 92–100.
Rosenthal, R.W. [1990]. Suicide attempts and signalling games. Mimeo, Boston University.
Ross, S. [1977]. The determination of financial structure: The incentive-signalling approach. Bell Journal of Economics *8*, 23–40.
Roth, A.E. [1975]. Self-supporting equilibrium in the supergame, Mimeo.
Roth, A.E. [1979]. Axiomatic models of bargaining. Springer-Verlag, Berlin.
Roth, A.E. [1985]. A note on risk aversion in a perfect equilibrium model of bargaining. Econometrica *53*, 207–213.
Roth, A.E. [1988]. Laboratory experimentation in economics: A methodological overview. Economic Journal *98*, 974–1032.
Rothschild, M. and J. Stiglitz [1976]. Equilibrium in competitive insurance markets. Quarterly Journal of Economics *90*, 629–650.
Rubinstein, A. [1977]. Equilibrium in Supergames. Mimeo, The Hebrew University.
Rubinstein, A. [1978]. Equilibrium in supergames with the overtaking criterion. Journal of Economic Theory *21*, 1–9.
Rubinstein, A. [1980]. Strong perfect equilibrium in supergames. International Journal of Game Theory *9*, 1–12.
Rubinstein, A. [1982]. Perfect equilibrium in a bargaining model. Econometrica *50*, 97–109.
Rubinstein, A. [1985a]. A bargaining model with incomplete information about time preferences. Econometrica *53*, 1151–1173.
Rubinstein, A. [1985b]. The choice of conjectures in a bargaining game with incomplete information, in A.E. Roth (ed.), "Game theoretic models of bargaining", Cambridge University Press.
Rubinstein, A. [1986a]. Finite automata play the repeated Prisoners' Dilemma. Journal of Economic Theory *39*, 38–97.
Rubinstein, A. [1986b]. A sequential theory of bargaining, in T. Bewley (ed.): Advances in economic theory. Cambridge Univ. Press.
Rubinstein, A. [1988]. Comments on the interpretation of game theory. STICERD Discussion Paper. Forthcoming in Econometrica.
Rubinstein, A. [1989]. The electronic mail game: Strategic behavior under almost common knowledge. American Economic Review *79*, 385–391.

Rubinstein, A. and A. Wolinsky [1990]. Rationalizable conjectural equilibrium: Between Nash and rationalizability. Tel Aviv University WP 19–90.

Samuelson, L. [1988]. Evolutionary foundations of solution concepts for finite, two-player, normal-form games. In: M.Y. Vardi (ed.) Theoretical Aspects of Reasoning About Knowledge. Morgan Kaufmann, Los Altos.

Samuelson, L. [1989a]. Dominated strategies and common knowledge. Mimeo, Penn State University.

Samuelson, L. [1989b]. Evolutionary stability in asymmetric games. Working paper, PennState University.

Sard, A. [1942]. The measure of the critical values of differentiable maps. Bulletin of the American Mathematical Society *48*, 883–890.

Savage, L.J. [1954]. The foundations of statistics. John Wiley and Sons, New York.

Schaffer, M. [1988]. Evolutionary stable strategies for a finite population and a variable contest size. Journal of Theoretical Biology *132*, 469–478.

Schanuel, S., L. Simon and W. Zame [1991]. The algebraic geometry of the tracing procedure. Forthcoming in Selten [1991b].

Schelling, T.C. [1956]. An essay on bargaining. American Economic Review *46*, 281–306.

Schelling, T.C. [1960]. The strategy of conflict. Harvard University Press, Cambridge, Mass.

Schlag, K. [1990]. Evolutionary stability in games with equivalent strategies, mixed strategy types and asymmetries. Mimeo, MEDS, Northwestern University.

Schuster, P. and K. Sigmund [1981]. Coyness, philandering and stable strategies. Animal Behavior *29*, 186–192.

Schuster, P., Sigmund K., Hofbauer, J. and R. Wolff [1981]. Selfregulation of behavior in animal societies, parts I, II and III. Biol. Cybern. *40*, 1–8, 9–15, 17–25.

Schweizer, U. [1989]. Litigation and settlement under two-sided incomplete information. Review of Economic Studies *56*, 163–178.

Seidmann, D. [1990]. Effective cheap talk with conflicting interests. Journal of Economic Theory *50*, 445–458.

Selten, R. [1965]. Spieltheoretische Behandlung eines Oligopolmodells mit Nachfrageträgheit. Zeitschrift für die gesamte Staatswissenschaft *12*, 301–324.

Selten, R. [1973]. A simple model of imperfect competition, where 4 are few and 6 are many. International Journal of Game Theory *3*, 141–201.

Selten, R. [1975]. Reexamination of the perfectness concept for equilibrium points in extensive games. International Journal of Game Theory *4*, 25–55.

Selten, R. [1977]. A simple model of kidnapping, pp. 139–156 in R. Henn and O. Moeschlin (eds.), Mathematical Economics and Game Theory. Lecture notes in economics and mathematical system *141*, Springer-Verlag, Berlin.

Selten, R. [1978]. The chain store paradox. Theory and Decision *9*, 127–159.

Selten, R. [1980]. A note on evolutionarily stable strategies in asymmetric animal conflicts. J. Theor. Biol. *84*, 93–101.

Selten, R. [1983]. Evolutionary stability in extensive 2-person games. Math. Social Sciences *5*, 269–363.

Selten, R. [1988a]. Anticipatory learning in two-person games. Mimeo ZIF, Bielefeld, forthcoming in Selten [1991a].

Selten, R. [1988b]. Evolutionary stability in extensive 2-person games: correction and further development. Math. Social Sciences *16*, 223–266.

Selten, R. [1991a]. Game equilibrium models, Vol. I; evolution and game dynamics. Springer, Heidelberg.

Selten, R. [1991b]. Game equilibrium models, Vol. II; methods, morals and markets. Springer, Heidelberg.

Selten, R. [1991c]. Game equilibrium models, Vol. III; strategic bargaining. Springer, Heidelberg.

Selten, R., M. Mitzkewitz, G. Uhlich [1988]. Duopoly strategies programmed by experienced players. SFB 303 DP B 106, University of Bonn.

Shaked, A. and I. Sutton [1984]. Involuntary unemployment as a perfect equilibrium in a bargaining model. Econometrica *52*, 1351–1364.

Shapley, L.S. [1974]. A note on the Lemke-Howson algorithm. Math. Programming Study *1*, 175–189.
Shleifer, A. and R.W. Vishny [1986]. Greenmail, white knights, and shareholders' interest. Rand Journal of Economics *17*, 293–309.
Shubik, M. [1954]. Does the fittest necessarily survive? In: M. Shubik (ed.) Readings in Game Theory and Political Behavior. Doubleday Inc. Garden City, New York.
Shubik, M. [1981]. Perfect or robust noncooperative equilibrium: a search for the philosopher's stone?, pp. 153–179 in Essays in Game Theory and Mathematical Economics in Honor of Oskar Morgenstern. Bibliographisches Institut Mannheim/Wien/Zürich.
Shubik, M. [1982]. Game Theory in the social sciences, Concepts and solutions. M.I.T. Press. Cambridge.
Simon, L. [1987]. Local perfection. Journal of Economic Theory *43*, 134–156.
Simon, L. and M. Stinchcombe [1990]. Refinements in games with large strategies spaces. Paper presented at the 6th World Congress of the Econometric Society in Barcelona.
Smale, S. [1980]. The Prisoners' Dilemma and dynamical systems associated to non-cooperative games. Econometrica *48*, 1617–1635.
Smart, D.R. [1974]. Fixed point theorems. Cambridge University Press, London.
Sobel, J., L. Stole and I. Zapater [1990]. Fixed-equilibrium rationalizability in signaling games. University of California at San Diego DP 90–13, forthcoming in Journal of Economic Theory.
Sobel, J. and I. Takahashi [1983]. A multistage model of bargaining. Review of Economic Studies *50*, 411–426.
Sorin, S. [1986]. On repeated games with complete information. Mathematics of Operations Research *11*, 147–161.
Sorin, S. [1988]. Repeated games with complete information. CORE DP 8822, forthcoming in R. Aumann and S. Hart [1991].
Spence, M. [1973]. Job market signalling. Quarterly Journal of Economics *87*, 355–374.
Spence, M. [1974]. Competitive and optimal responses to signals: An analysis of efficiency and distribution. Journal of Economic Theory *7*, 296–332.
Spence, M. [1976]. Competition in salaries and signalling prerequisites for jobs. Quarterly Journal of Economics *90*, 51–74.
Stahl, D. [1989]. The graph of Prisoners' Dilemma supergame payoffs as a function of the discount factor. Mimeo, University of Texas at Austin.
Stahl, D. [1990a]. Entropy control costs and entropic equilibria. International Journal of Game Theory *19*, 129–138.
Stahl, D. [1990b]. Lexicographic rationality, common knowledge, and iterated admissibility. Mimeo, University of Texas.
Ståhl, I. [1972]. Bargaining Theory. Stockholm School of Economics, Stockholm.
Ståhl, I. [1977]. An *n*-person bargaining game in the extensive form, pp. 156–173 in R. Henn and O. Moeschlin (eds.), Mathematical Economics and Game Theory. Lecture notes in economics and mathematical systems *141*, Springer-Verlag, Berlin.
Stanford, W. [1986a]. Subgame perfect reaction function equilibria in discounted duopoly supergames are trivial. Journal of Economic Theory *39*, 226–232.
Stanford, W. [1986b]. On continuous reaction function equilibria in duopoly supergames with mean payoffs. Journal of Economic Theory *39*, 233–250.
Steinhaus, H. [1948]. The problem of fair division. Econometrica *16*, 101–104.
Stiglitz, J. and A. Weiss [1988]. Sorting out the difference between signaling and screening models. Forthcoming in: M. Bacharach (ed.) Oxford Essays in Mathematical Economics. Oxford University Press.
Stoughton, N. [1988]. The informational content of corporate merger and acquisition offers. Journal of Financial and Quantitative Analysis *23*, 175–197.
Sutton, J. [1985]. Noncooperative bargaining theory: An introduction. Review of Economic Studies LIII (5), No. 176, 709–725.
Taylor, P.D. and L.B. Jonker [1978]. Evolutionarily stable strategies and game dynamics. Math. Biosc. *40*, 145–156.
Thomas, B. [1985]. Evolutionary stable sets in mixed-strategist models. Theoretical Population Biology *28*, 332–341.

Thompson, F. [1952]. Equivalence of games in extensive form, Discussion Paper RM 759, the Rand Corporation.
Thomson, W. [1990]. Bargaining theory: Axiomatic approach. Academic Press, San Diego.
Thomson, W. [1991]. Cooperative models of bargaining. Forthcoming in Aumann and Hart [1991].
Tijs, S.H. and M.J.M. Jansen [1982]. On the existence of values for arbitration games. International Journal of Game Theory *11*, 87–104.
Tirole, J. [1988]. The theory of industrial organization. MIT Press, Cambridge, Massachusetts.
Trockel, W. [1986]. The chain-store paradox revisited. Theory and Decision *21*, 163–179.
Van Damme, E.E.C. [1980 a]. Some comments on Harsanyi's postulates for rational behavior in game situations. Methods of Operations Research *38*, 189–205.
Van Damme, E.E.C. [1980 b]. A characterization of perfect and proper equilibria in zero-sum games. Memorandum Cosor 80 – 18. Eindhoven University of Technology, Eindhoven.
Van Damme, E.E.C. [1981 a]. History-dependent equilibrium points in dynamic games, pp. 27–38 in O. Moeschlin and D. Pallaschke (eds.), Game Theory and Mathematical Economics, North-Holland, Amsterdam.
Van Damme, E.E.C. [1981 b]. Refinements of the Nash equilibrium concept for bimatrix games. Memorandum COSOR 81–08, Eindhoven University of Technology, Eindhoven.
Van Damme, E.E.C. [1981 c]. Regular equilibrium points of *n*-person games in normal form. Memorandum COSOR 81–12, Eindhoven University of Technology, Eindhoven.
Van Damme, E.E.C. [1982]. Refining the equilibrium concept for bimatrix games via control costs. Memorandum COSOR 82–02, Eindhoven University of Technology, Eindhoven.
Van Damme, E.E.C. [1983]. Refinements of the Nash equilibrium concept. Lecture Notes in Economics and Math. Systems, *219*, Springer-Verlag, Berlin.
Van Damme, E.E.C. [1984]. A relation between perfect equilibria in extensive form games and proper equilibria in normal form games. Int. J. Game Theory *13*, 1–13.
Van Damme, E.E.C. [1985]. Fair allocation of an indivisible commodity. Discussion paper, Dept. of Math., Delft University of Technology.
Van Damme, E.E.C. [1986]. The Nash bargaining solution is optimal. Journal of Economic Theory, *38*, 78–100.
Van Damme, E.E.C. [1987]. Equilibria in non-cooperative games. In: H.J.M. Peters and O.J. Vrieze (eds.) Surveys in Game Theory and Related Topics. CWI Tract 39, Amsterdam, 1–37.
Van Damme, E.E.C. [1988]. The impossibility of stable renegotiation. Economic Letters *26*, 321–324.
Van Damme, E.E.C. [1989 a]. Renegotiation-proof equilibria in repeated Prisoners' Dilemma. Journal of Economic Theory *47*, 206–217.
Van Damme, E.E.C. [1989 b]. Stable equilibria and forward induction. Journal of Economic Theory *48*, 476–496.
Van Damme, E.E.C. [1990]. On dominance solvable games and equilibrium selection theories. CentER DP 9046, Tilburg University.
Van Damme, E.E.C., R. Selten and E. Winter [1990]. Alternating bid bargaining with a smallest money unit. Games and Economic Behavior *2*, 188–201.
Van den Elzen, A. [1991]. Adjustment processes of find equilibria. PhD Thesis, Tilburg University.
Van den Elzen, A. and A. Talman [1991]. A procedure for finding Nash equilibria in bimatrix games. To appear in ZOR – Methods and Models in Operations Research.
Van der Waerden [1973]. Einführung in die Algebraische Geometrie. Springer-Verlag. Berlin, Heidelberg, New York.
Van Huyck, J., A. Gillette and R. Battalio [1988]. Credible assignments in non-cooperative Games. Texas A&M Working paper.
Vickrey, W. [1961]. Counterspeculation, auctions and competitive sealed tenders. Journal of Finance *16*, 8–37.
Vincent, D. [1990 a]. Dynamic Auctions. Review of Economic Studies *57*, 49–61.
Vincent, D. [1990 b]. Bilateral monopoly, non-durable goods and dynamic trading relationships. Mimeo, Northwestern University.
Vorob'ev, N.N. [1958]. Equilibrium points in bimatrix games. Theory Probability Appl. *3*, 297–309.

Weibull, J. [1990]. On self-enforcement in extensive-form games. Mimeo, Princeton University.

Weissing, F. [1989]. Evolutionary stability and dynamic stability in generalized "Rock-Scissors-Paper" games. DP, Research Group on Game Theory in the Behavioral Sciences, ZIF Bielefeld. Forthcoming in Selten [1991 a].

Wilson, C. [1977]. A model of insurance markets with incomplete information, Journal of Economic Theory *16*, 167–207.

Wilson, C. [1980]. The nature of equilibrium in markets with adverse selection. Bell Journal of Economics, 108–130.

Wilson, R. [1971]. Computing equilibria of n-person games. SIAM J. Appl. Math. *21*, 80–87.

Wilson, R. [1972]. Computing equilibria of two-person games from the extensive form. Management Science, *18*, 448–460.

Wilson, R. [1985]. Multi-dimensional signalling. Economic Letters *19*, 17–21.

Wilson, R. [1990 a]. Two surveys: Auctions and entry deterrence. SITE Technical Report No. 3, Stanford University. Forthcoming in Aumann and Hart [1991].

Wilson, R. [1990 b]. An algorithm for stable equilibria. Mimeo, Stanford Business School.

Winkels, H.M. [1979]. An algorithm to determine all equilibrium points of a bimatrix game, pp. 137–148 in O. Moeschlin and D. Pallaschke (eds.), Game Theory and related topics, North-Holland, Amsterdam.

Wu Wen-Tsün and Jiang Jia-He [1962]. Essential equilibrium points of n-person non-cooperative games. Sci. Sinica *11*, 1307–1322.

Yamamoto, Y. [1990]. A path-following procedure to find a proper equilibrium of finite games. Report No. 90636 – OR, University of Bonn.

Zeeman, E.C. [1980]. Population dynamics from game theory, pp. 471–497 in Z. Nitecki and C. Robinson (eds.) Global Theory of Dynamical Systems. Lecture Notes in Math. 819, Springer-Verlag.

Zeeman, E.C. [1981]. Dynamics of the evolution of animal conflict. J. Theor. Biol. *89*, 249–270.

Survey Diagrams

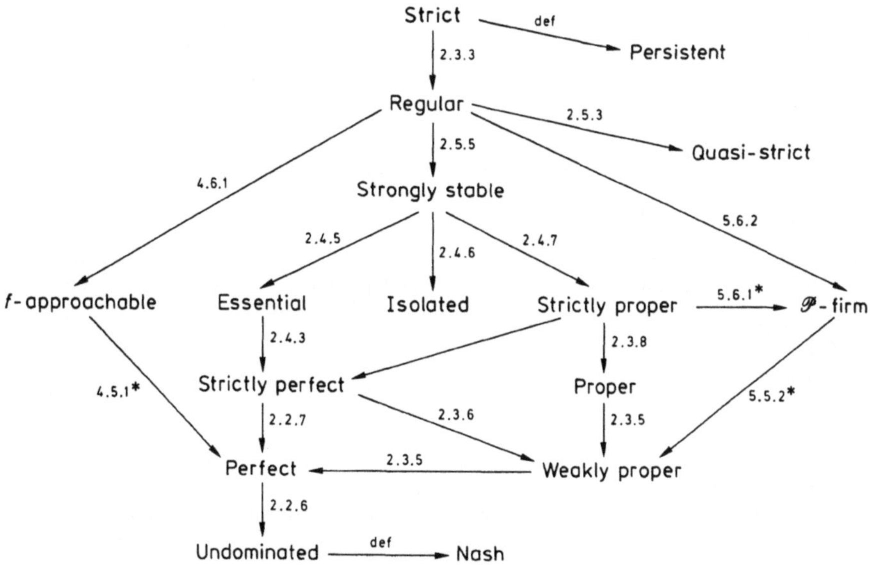

Diagram 1. An overview of the main relations between refinements of the Nash equilibrium concept for normal form games. Inclusions go in the direction of the arrows. Numbers represent the theorem in which the result is proved. A *star* denotes that the relation holds only if some additional condition is satisfied. All inclusions displayed are strict, except (possibly) the one which is not numbered. An equilibrium is regular if it is strongly stable and quasi-strict (Theorem 2.5.6)

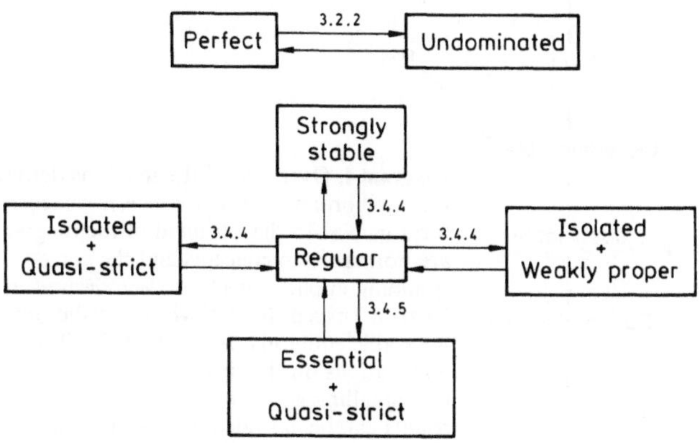

Diagram 2. Characterizations of regularity and perfectness in bimatrix games.

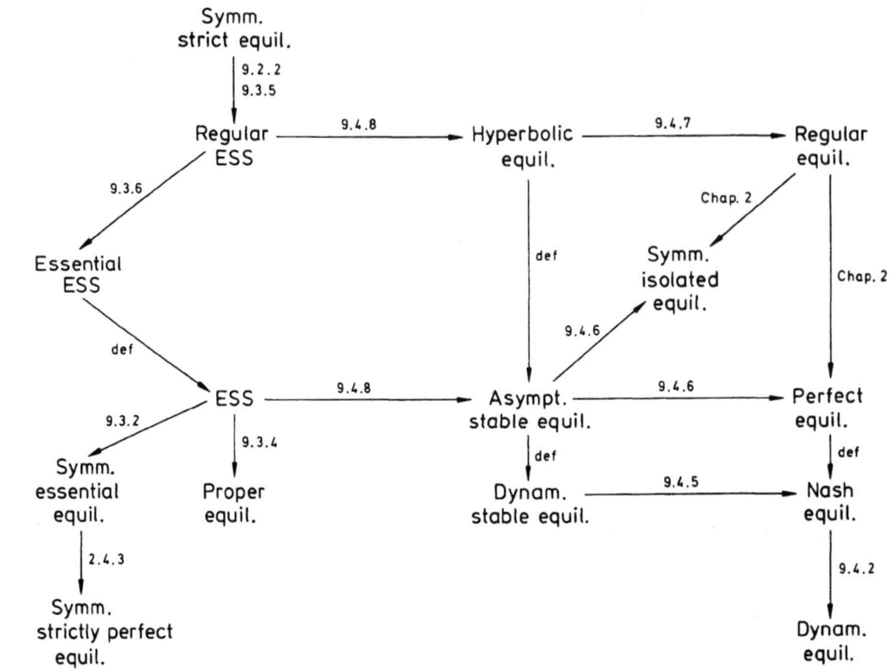

Diagram 3. Overview of the main relations derived in Chap. 9 for evolutionary stability in symmetric contests. All inclusions are strict (characterizations of regular ESS are given in Theorem 9.3.5; for truly asymmetric contests, every ESS is strict (Theorem 9.6.2))

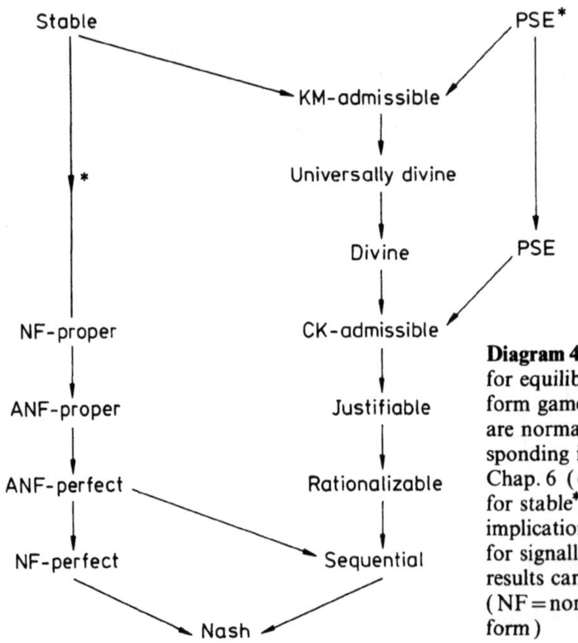

Diagram 4. Overview of the relations derived for equilibrium *outcomes* in *generic* extensive form games. In the left hand of the diagram are normal form concepts and the corresponding implications have been derived in Chap. 6 (except for (*) which is valid only for stable*equilibria, see Sect. 10.3). The implications on the right have been derived for signalling games in Sect. 10.5; these results can be extended to general games. (NF = normal form, ANF = agent normal form)

Subject Index

MIX
Papier aus verantwortungsvollen Quellen
Paper from responsible sources
FSC® C105338

If you have any concerns about our products,
you can contact us on
ProductSafety@springernature.com

In case Publisher is established outside the EU,
the EU authorized representative is:
Springer Nature Customer Service Center GmbH
Europaplatz 3, 69115 Heidelberg, Germany

Printed by Libri Plureos GmbH
in Hamburg, Germany